Inverse problems arise in practical situations like geophysical exploration, medical imaging, and nondestructive evaluation, where measurements made on the exterior of a body are used to determine properties of the inaccessible interior. In the last twenty years there have been substantial developments in the mathematical theory of inverse problems, and applications have expanded greatly. In this book, leading experts in the theory and applications of inverse problems offer extended surveys of such vital and rapidly expanding areas as microlocal analysis, reflection seismology, tomography, inverse scattering, and X-ray transforms.

Each article covers a particular topic or topics with an emphasis on accessibility and integration with the whole volume. Thus the collection can be at the same time stimulating to researchers and accessible to graduate students.

Mathematical Sciences Research Institute
Publications

47

Inside Out: Inverse Problems and Applications

Mathematical Sciences Research Institute Publications

Volumes 1–4 and 6–27 are published by Springer-Verlag

Inside Out:
Inverse Problems
and Applications

Edited by

Gunther Uhlmann
University of Washington

CAMBRIDGE UNIVERSITY PRESS
Cambridge, New York, Melbourne, Madrid, Cape Town,
Singapore, São Paulo, Delhi, Tokyo, Mexico City

Cambridge University Press
The Edinburgh Building, Cambridge CB2 8RU, UK

Published in the United States of America by Cambridge University Press, New York

www.cambridge.org
Information on this title: www.cambridge.org/9780521168748

First published 2003
First paperback edition 2011

A catalogue record for this publication is available from the British Library

ISBN 978-0-521-82469-9 Hardback
ISBN 978-0-521-16874-8 Paperback

Contents

Preface

Inverse problems arise in practical situations such as medical imaging, geophysical exploration, and nondestructive evaluation where measurements made on the exterior of a body are used to determine properties of the inaccessible interior. In this book leading experts in the theoretical and applied aspects of inverse problems have written extended surveys on some of the main topics of the inverse problems semester held at MSRI during the Fall of 2001. We describe here briefly the chapters of the book.

The chapter by Faridani is an introduction to *computed tomography* (CT), which is probably the inverse problem best known to the general public. In this imaging method the attenuation in intensity of an X-ray beam is measured, and the information from many X-rays from different sources is assembled and analyzed on a computer. Mathematically it is a problem of recovering a function from the set of its line integrals (or the set of its plane integrals). Radon found in the early part of the twentieth century a formula to recover a function from this information. The application to diagnostic radiology did not happen until the late 1960s with the aid of the increasing calculating power of the computer. In 1970 the first computer tomograph that could be used in clinical work was developed by G. N. Hounsfield. He and Allan M. Cormack, who independently proposed some of the algorithms, were jointly awarded the 1979 Nobel prize in medicine. In practice only integrals from finitely many lines can be measured, and the distribution of these lines is sometimes restricted. The focus of Faridani's chapter is on what features of a function f can be stably recovered from a given collection of line integrals of f. If a full reconstruction is not possible then one tries to detect the location of boundaries (jump discontinuities of f) from local tomographic data.

In a related topic, Finch writes about the *attenuated X-ray transform*. Mathematically the problem is again to recover a function from its line integrals, but this time with an exponential weight. This problem arises in Single Photon Emission Computed Tomography (SPECT), where one would like to find the distribution of a radiopharmaceutical f in a cross section of the body from measuring the radiation outside the body. During the past few years there have been substantial developments in the mathematical theory of the inverse problem of recovering f from its attenuated X-ray transform, assuming the attenuation is known. For instance, very simple reconstruction formulas have been obtained, thanks to the work of Boman, Bukhgeim, Natterer, Novikov, Strömberg, and others. Finch describes these developments, which are expected to have significant applications.

An important class of inverse problems are *inverse scattering* problems. The chapter by Colton is devoted to this topic; in particular to the developments deriving from the sampling method developed by Colton, Kirsch, Kress and others. In inverse scattering a wave field is generated far away from a target having unknown physical properties and propagates through the region containing the target. The scattered field is measured and from this one attempts to determine the properties of the scatterer. A classical example of this type of problem is the determination of an obstacle or an acoustic medium by measuring the response to time harmonic waves. The success of radar and sonar soon caused scientists to ask if more could be determined about a scattering object than simply its location. However, due to the lack of a mathematical theory of inverse problems, together with limited computational capabilities, further progress was not possible at the time. This situation was dramatically changed in the mid 1960s with Tikhonov's introduction of regularization methods for ill-posed problems and subsequently, starting in the 1980s, the mathematical basis for the inverse scattering problem together with numerical algorithms for its solution began to be developed. Colton describes the tremendous progress made on this subject in the last 20 years or so.

Optical tomography is a relatively new imaging technique with several potential applications in diagnostic imaging. The goal in this imaging method is to determine the optical absorption and diffusion coefficients in a highly absorbing body by making boundary measurements of near-infrared light transmitted through the body. The possibility of performing imaging with infra-red radiation opens up numerous new possibilities compared to traditional tomography, but requires handling much more complex mathematical problems. This difficulty is due to the fact that infra-red radiation does not travel along straight lines as, for example, X-rays. Rather, due to multiple scattering interactions, it travels along essentially random paths inside the interior of tissues and objects. As a consequence, the forward problem becomes a highly nonlinear function of the model parameters, and the inverse problem becomes much harder. In spite of this, there have been substantial developments in both the theory and applications of optical tomography. The chapter by Stefanov describes some recent work on the inverse problem for the linear Boltzmann equation, relevant to optical tomography. This equation is also known as the radiative transport equation and is used in neutron transport and other fields.

Carney and Schotland consider *near-field tomography*, which is, roughly speaking, the use of inverse scattering methods to reconstruct tomographic images in near-field optics. In particular, they study near-field scanning optical microscopy, total internal reflection microscopy, and photon scanning tunneling microscopy.

The presence of evanescent fields makes near-field tomography very ill-posed. Carney and Schotland show how to use the overdeterminacy of the problem to develop a singular value decomposition of the relevant scattering operators. The chapter by Ola, Päivärinta, and Somersalo deals with inverse boundary problems associated to time harmonic electromagnetic fields at fixed frequency. It is shown that the electromagnetic parameters like the electrical permittivity, electrical conductivity and magnetic permeability can be reconstructed if one measures the tangential component of the magnetic and electric fields at the boundary of the medium. This information is equivalent to the near-field at fixed energy. The reconstruction method relies on the construction of exponentially growing solutions for Maxwell's system (also called complex geometrical optics solutions) pioneered by Calderón and Faddeev and further developed by Sylvester and Uhlmann for the Schrödinger equation. They are related to the evanescent fields in the chapter of Carney and Schotland, and perhaps the connection should be explored further.

One of the fascinating aspects of inverse problems is the continuous interplay between pure and applied mathematics. This interplay has been particularly noticeable in the applications of microlocal analysis (MA) to inverse problems. MA, which is, roughly speaking, local analysis in phase space, was developed about 30 years ago by Hörmander, Maslov, Sato, and many others in order to understand the propagation of singularities of solutions of partial differential equations. The early roots of MA were in the theory of geometrical optics. Microlocal analysis has been used successfully in determining the singularities of medium parameters in several inverse problems ranging from X-ray tomography to reflection seismology and electrical impedance tomography. The chapter by Finch, Lan and Uhlmann considers several applications of the theory of paired Lagrangian distributions to inverse problems. A close study is made of the microlocal analysis in three dimensions of the *X-ray transform with sources on a curve*, a topic also mentioned in Faridani's chapter. Other applications include the inverse backscattering problem for conormal potentials, electrical impedance tomography, and the microlocal characterization of the range of generalized Radon transforms.

De Hoop gives an extensive review of applications of MA to *reflection seismology*. In this inverse method one attempts to estimate the index of refraction of waves in the earth from seismic data measured at the Earth's surface. The techniques used in reflection seismology are very relevant to imaging using ultrasound. Seismic imaging creates images of the Earth's upper crust using seismic waves generated by artificial sources and recorded into extensive arrays of sensors (geophones or hydrophones). The technology is based on a complex, and rapidly evolving, mathematical theory that employs advanced solutions to a wave equation as tools to solve approximately the general seismic inverse problem. The heterogeneity and anisotropy of the Earth's crust require advanced mathematics to generate wave-equation solutions suitable for seismic imaging. In his chapter, de Hoop describes several important developments using MA to generate these wave-solutions by manipulating the wavefields directly on their phase space. De Hoop also considers some recent applications of MA to global seismology.

The study of propagation of singularities of solutions of partial differential equations, a fundamental question in MA, connects classical and quantum mechanics. The next three chapters make explicit and use that connection in several contexts.

The chapter by Petkov and Stoyanov considers, as does Colton's, *inverse scattering by an obstacle*, but with a different emphasis. The authors, using MA, study the information obtained from the singularities of the scattering kernel, which is the Fourier transform in frequency of the scattering amplitude. This determines the sojourn or travel times of rays incoming in a given direction and outgoing in another. The authors also consider at length the inverse problem of recovering geometric information about the obstacle from the set of sojourn times, also called the scattering length spectrum.

Vasy gives an extensive survey of *many body quantum scattering*, which is the analysis of motion of several interacting particles. In particular the author studies in detail the singularities of the scattering operator and propagation of singularities in many-body scattering. Vasy makes the analogy between the studies of these singularities and propagation of singularities for the wave equation for manifolds with corners. Vasy uses the information so obtained to prove inverse scattering results for single cluster to single cluster scattering. He also considers some inverse results for the three-body problem at low energies for the case of two-cluster to two-cluster scattering, by using complex geometrical optics solutions similar to the two-body problem. Finally he describes some recent results on scattering on locally symmetric spaces which has several features analogous to the many-body problem.

Time reversal mirrors have attracted a lot of attention in recent years because of potential applications in medical imaging, underwater acoustics, and other areas. It has been seen experimentally that inhomogeneities, randomness, and ergodicity contribute to much better refocusing of waves at the source, which is the desired feature of the time reversal mirror. The chapter by Bardos gives a rigorous analysis of this phenomenon for the case of cavities for which the associated classical flow is ergodic. MA provides the tools for the study of the high-frequency asymptotics or propagation of singularities of solutions in this case.

I would like to express my deep gratitude to all of the MSRI staff for their invaluable assistance in organizing the inverse problems semester at MSRI in Fall 2001 and also for their efficient help in putting together this volume.

Gunther Uhlmann
Seattle, April 2003

Introduction to the Mathematics
of Computed Tomography

ADEL FARIDANI

ABSTRACT. Computed tomography (CT) entails the reconstruction of a
function f from line integrals of f. This mathematical problem is encountered in a growing number of diverse settings in medicine, science, and
technology. This introductory article is divided into three parts. The first
part is concerned with general theory and explores questions of uniqueness,
stability and inversion, as well as detection of singularities. The second
part is devoted to local tomography and is centered around a discussion of
recently developed methods for computing jumps of a function from local
tomographic data. The third part treats optimal sampling and has at its
core a detailed error analysis of the parallel-beam filtered backprojection
algorithm. Matlab source code for the filtered backprojection algorithm
and the Feldkamp–Davis–Kress algorithm is included in an appendix.

1. Introduction

Computed tomography (CT) entails the reconstruction of a function f from
line integrals of f. This mathematical problem is encountered in a growing
number of diverse settings in medicine, science, and technology, ranging from
the famous application in diagnostic radiology to research in quantum optics. As
a consequence, many aspects of CT have been extensively studied and are now
well understood, thus providing an interesting model case for the study of other
inverse problems. Other aspects, notably three-dimensional reconstructions, still
provide numerous open problems.

The purpose of this article is to give an introduction to the topic, treat some
aspects in more detail, and to point out references for further study. The reader
interested in a broader overview of the field, its relation to various branches of
pure and applied mathematics, and its development over the years may wish to
consult the monographs [6; 31; 32; 36; 62; 67; 78], the volumes [21; 22; 28; 33;
34; 76; 77], and review articles [42; 49; 56; 58; 66; 84; 89].

Work supported by NSF grant DMS-9803352 and NIH grant R01 RR 11800-4.

In practice only integrals over finitely many lines can be measured, and the distribution of these lines is sometimes restricted. The following presentation is centered around the question: *What features of f can be stably recovered from a given collection of line integrals of f?* For example, we may ask what resolution can be achieved with the available data. If a full reconstruction of f is not possible, we may try to detect the location of boundaries (jump discontinuities of f), or also the sizes of the jumps.

The exposition is divided into three parts. The first part is concerned with general theory. Its main themes are questions of uniqueness, stability and inversion for the x-ray transform, as well as detection of singularities. The second part is devoted to local tomography. The exposition is similar to [17] and is centered around a discussion of recently developed methods for computing jumps of a function from local tomographic data. The third part treats optimal sampling and has at its core a detailed error analysis of the parallel-beam filtered backprojection algorithm. The article conludes with three appendices containing basic results on wavelets, Matlab source code for the filtered backprojection algorithm and the Feldkamp–Davis–Kress algorithm, and some exercises.

2. The X-Ray and Radon Transforms

We begin by introducing some notation and background material. $\mathbb{R}^n$ consists of n-tuples of real numbers, usually designated by single letters, $x = (x_1, \ldots, x_n)$, $y = (y_1, \ldots, y_n)$, etc. The inner product and absolute value are defined by $\langle x, y \rangle = \sum_1^n x_i y_i$ and $|x| = \sqrt{\langle x, x \rangle}$. The unit sphere S^{n-1} consists of the points with absolute value 1. $C_0^\infty(\mathbb{R}^n)$ denotes the set of infinitely differentiable functions on $\mathbb{R}^n$ with compact support. A continuous linear functional on C_0^∞ is called a distribution. If X is a set, X° denotes its interior, $\overline{X}$ its closure, and X^c its complement. χ_X and χ_n denote the characteristic functions (indicator functions) of X and of the unit ball in $\mathbb{R}^n$, respectively (that is, $\chi_X(x) = 1$ if $x \in X$ and $\chi_X(x) = 0$ if $x \notin X$). $|X|$ denotes the n-dimensional Lebesgue measure of $X \subset \mathbb{R}^n$. However, when it is clear that X should be treated as a set of dimension $m < n$, $|X|$ is the m-dimensional area measure. Thus

$$|S^{k-1}| = 2\pi^{k/2}/\Gamma(k/2)$$

is the $(k-1)$-dimensional area of the $(k-1)$-dimensional sphere.

The convolution of two functions is given by

$$f * g(x) = \int_{\mathbb{R}^n} f(x-y)g(y) \, dy.$$

The Fourier transform is defined by

$$\hat{f}(\xi) = (2\pi)^{-n/2} \int_{\mathbb{R}^n} f(x)e^{-i\langle x, \xi \rangle} \, dx$$

for integrable functions f, and is extended to larger classes of functions or distributions by continuity or duality. For square-integrable functions f, g we have

$$f * g(x) = \int_{\mathbb{R}^n} \hat{f}(\xi)\hat{g}(\xi)e^{i\langle x,\xi\rangle}\, d\xi. \tag{2-1}$$

The integral transforms most relevant for tomography are the x-ray transform and the Radon transform.

DEFINITION 2.1. Let $\theta \in S^{n-1}$ and $\Theta^\perp$ the hyperplane through the origin orthogonal to θ. We parametrize a line $l(\theta, y)$ in $\mathbb{R}^n$ by specifying its direction $\theta \in S^{n-1}$ and the point y where the line intersects the hyperplane $\Theta^\perp$.

The x-ray transform of a function $f \in L_1(\mathbb{R}^n)$ is given by

$$Pf(\theta, y) = P_\theta f(y) = \int_{\mathbb{R}} f(y + t\theta)\, dt, \quad y \in \Theta^\perp.$$

The Radon transform of f is defined by

$$Rf(\theta, s) = R_\theta f(s) = \int_{\Theta^\perp} f(x + s\theta)\, dx, \quad s \in \mathbb{R}. \tag{2-2}$$

We see that $Pf(\theta, x)$ is the integral of f over the line $l(\theta, y)$ parallel to θ which passes through $y \in \Theta^\perp$, and that $Rf(\theta, s)$ is the integral of f over the hyperplane orthogonal to θ with signed distance s from the origin. In the following we will be mostly concerned with the x-ray transform. In two dimensions the two transforms coincide apart from the parameterization: We parametrize $\theta \in S^1$ by its polar angle φ and define a vector $\theta^\perp$ orthogonal to θ such that

$$\theta = (\cos \varphi, \sin \varphi), \quad \theta^\perp = (-\sin \varphi, \cos \varphi). \tag{2-3}$$

Then the points in the subspace $\Theta^\perp$ are given by $\Theta^\perp = \{s\theta^\perp : s \in \mathbb{R}\}$ and we have the relation $Pf(\theta, s\theta^\perp) = Rf(\theta^\perp, s)$. Also, when working in two dimensions, we will often use the simplified notation $Pf(\theta, s)$ or $P_\theta f(s)$ instead of $Pf(\theta, s\theta^\perp)$. Occasionally we will also replace θ by the polar angle φ according to (2–3) and write $Pf(\varphi, s)$.

We consider two examples. Let G be the Gaussian function $G(x) = e^{-\langle x,x\rangle/2}$. Then

$$PG(\theta, y) = e^{-\langle y,y\rangle/2} \int_{\mathbb{R}} e^{-\langle t\theta, t\theta\rangle/2}\, dt = (2\pi)^{1/2} e^{-\langle y,y\rangle/2}, \quad y \in \Theta^\perp.$$

For $\mathcal{X}_n$, the characteristic function of the unit ball in $\mathbb{R}^n$, we can use a geometrical argument. We obtain $P\mathcal{X}_n(\theta, y) = 0$ for $|y| > 1$ since then the line $l(\theta, y)$ does not intersect the unit ball. For $|y| \le 1$ observe that the intersection of the line $l(\theta, y)$ with the unit ball in $\mathbb{R}^n$ is a line segment of length $2\sqrt{1 - |y|^2}$ and that $P\mathcal{X}_n(\theta, y)$ is equal to this length.

The following relation between the Fourier transforms of $P_\theta f$ and f will prove to be useful:

THEOREM 2.2. *Under the hypotheses of Definition* 2.1,

$$\widehat{P_\theta f}(\eta) = (2\pi)^{1/2} \hat{f}(\eta), \quad \eta \in \Theta^\perp,$$

$$\widehat{R_\theta f}(\sigma) = (2\pi)^{(n-1)/2} \hat{f}(\sigma\theta), \quad \sigma \in \mathbb{R}$$

PROOF. This is a straightforward computation. We demonstrate it for the x-ray transform. Let $\eta \in \Theta^\perp$. Then

$$\widehat{Pf}(\theta, \eta) = (2\pi)^{(1-n)/2} \int_{\Theta^\perp} Pf(\theta, x) e^{-i\langle x, \eta\rangle} \, dx$$

$$= (2\pi)^{(1-n)/2} \int_{\Theta^\perp} \int_{\mathbb{R}} f(x + s\theta) \, ds \, e^{-i\langle x, \eta\rangle} dx$$

$$= (2\pi)^{(1-n)/2} \int_{\mathbb{R}^n} f(y) e^{-i\langle y, \eta\rangle} \, dy = \sqrt{2\pi}\, \hat{f}(\eta). \qquad \square$$

As we will see below, Theorem 2.2 can be used to explore questions of uniqueness, nonuniqueness, stability, and inversion.

Current medical scanners employ an x-ray source which moves around the patient. To describe this type of data collection, the parameterization of lines by $\theta \in S^{n-1}$ and $y \in \Theta^\perp$ is less convenient. It is more suitable to introduce the *divergent beam x-ray transform*

$$Df(a, \theta) = D_a f(\theta) = \int_0^\infty f(a + t\theta) \, dt, \quad \theta \in S^{n-1},$$

which gives the integral of f over the ray with direction θ emanating from the source point a.

The x-ray and Radon transforms are special cases of the general k-plane transform, which maps a function into its integrals over k-dimensional affine subspaces; see [42], for instance.

3. Uniqueness and Nonuniqueness

THEOREM 3.1 [89; 42]. *Let $f \in L_2(\mathbb{R}^n)$ have compact support, and suppose that $Pf(\theta, \cdot) \equiv 0$ for infinitely many θ. Then $f \equiv 0$.*

PROOF. The Fourier transform $\hat{f}$ is analytic and $\hat{f}(\eta) = \widehat{P_\theta f}(\eta) = 0$ on the hyperplanes $\langle \eta, \theta \rangle = 0$. Since no nontrivial entire function can vanish on an infinite set of hyperplanes through the origin, we must have $\hat{f} \equiv 0$. $\square$

As an application, consider the so-called limited angle problem. Let $Pf(\theta, \cdot)$ be given for infinitely many θ concentrated in a cone C. Then f is uniquely determined, even if C is very small. However, if $C \neq S^{n-1}$, the reconstruction is not stable. Indeed, the proof of the above theorem shows that reconstructing f is equivalent to analytic continuation of $\hat{f}$, and analytic continuation is known to be extremely unstable.

The uniqueness result requires an infinite number of directions, while in practice only a finite number can be measured. It was already recognized by the pioneers of CT that this entails the loss of uniqueness; see the example given in [3]. The next theorem shows that the nonuniqueness is quite extensive, i.e., given $Pf(\theta_j, \cdot)$ for finitely many directions θ_j, there are null functions which can be prescribed arbitrarily on a large portion of their domain.

THEOREM 3.2. ([89]) Let $\theta_1, \ldots, \theta_p \in S^{n-1}$, $K \subset \mathbb{R}^n$ compact, and $f \in C_0^\infty(K)$. Let $K_0 \subset U \subset K$ with U open and K_0 compact. Then there is $f_0 \in C_0^\infty(K)$, $f_0 = f$ on K_0, and $Pf_0(\theta_k, \cdot) \equiv 0$, $k = 1, \ldots, p$.

While this result makes it seem difficult to obtain reliable reconstructions in practice, it is not the end of the story. It turns out that the null functions for the x-ray transform are high-frequency functions [51; 52; 53; 60] , and that it is possible to suppress such functions in practical reconstructions.

THEOREM 3.3 [51; 52; 53]. Let $f_0 \in L_2(\mathbb{R}^2)$ with support contained in the unit disk. If $Pf_0(\theta_k, \cdot) \equiv 0$ for $k = 1, \ldots, p$, then

$$\hat{f}_0(\sigma\theta) = \sum_{m>p} i^m \sigma^{-1} J_{m+1}(\sigma) q_m(\theta),$$

where $\sigma \in \mathbb{R}$, $\theta \in S^1$, J_{m+1} the order $m+1$ Bessel function of the first kind and q_m a polynomial of degree m.

Since $J_l(t)$ is very small for $l > t$, it follows that if $P_\theta f$ vanishes for p distinct directions θ_j, then $\hat{f}(\xi)$ is almost entirely concentrated in the set $\{\xi \in \mathbb{R}^n : |\xi| > p\}$ [60]. This means that measuring $P_{\theta_j} f$ determines $\hat{f}(\xi)$ reliably for $|\xi| < p$. However, the reconstruction problem may still be severely unstable, e.g., when the directions are concentrated in a narrow range. In cases where sufficient stability is present, a low-pass filtered version of f may be recovered. A loose application of Shannon's sampling theorem yields that the reconstruction will resolve details of size $2\pi/p$ or greater.

REMARK 3.4. It follows that the influence of nonuniqueness may be avoided in practice under the following conditions:

(a) A-priori information that $|\hat{f}(\xi)|$ is small for $|\xi| > b$ is available.
(b) Data $P_{\theta_j} f$ for $p > b$ directions θ_j are measured.
(c) The reconstruction method used produces a function f_R with $|\hat{f}_R(\xi)|$ small for $|\xi| > b$.

Nonuniqueness theorems for the divergent beam x-ray transform have been proved in [48; 93]. A generalization to the general k-plane transform has been given in [42].

4. Inversion and Ill-Posedness

Calderón's operator Λ is defined in terms of Fourier transforms by

$$\widehat{\Lambda\varphi}(\xi) = |\xi|\hat{\varphi}(\xi), \quad \varphi \in C_0^\infty(\mathbb{R}^n).$$

It is extended by duality to the class of functions f for which $(1 + |x|)^{-1-n} f$ is integrable [14]. Note that

$$\Lambda^2 = -\Delta, \quad \Delta = \text{Laplacian}. \tag{4-1}$$

For $n \geq 2$, the inverse Λ^{-1} of Λ is given by convolution with the Riesz kernel R_1:

$$\Lambda^{-1} f = R_1 * f, \quad R_1(x) = (\pi |S^{n-2}|)^{-1} |x|^{1-n}.$$

In dimension $n = 1$ we have $\Lambda f = \mathcal{H} \partial f$, where ∂f denotes the derivative of f and $\mathcal{H}$ denotes the Hilbert transform

$$\mathcal{H} f(s) = \frac{1}{\pi} \int_{\mathbb{R}} \frac{f(t)}{s - t} \, dt \tag{4-2}$$

where the integral is understood as a principal value.

We can formally derive an inversion formula for Pf by combining Theorem 2.2 and the inverse Fourier transform. For simplicity we first consider dimension $n = 2$. Using the Fourier inversion formula, Theorem 2.2, the relation (2–3) and changing to polar coordinates we obtain

$$
\begin{aligned}
f(x) &= (2\pi)^{-1} \int_{\mathbb{R}^2} \hat{f}(\xi) e^{i\langle x, \xi \rangle} \, d\xi \\
&= (2\pi)^{-1} \int_0^{2\pi} \int_0^{\infty} \sigma \hat{f}(\sigma \theta^{\perp}) e^{i\langle x, \sigma \theta^{\perp} \rangle} \, d\sigma \, d\varphi \\
&= (4\pi)^{-1} \int_0^{2\pi} \int_{-\infty}^{\infty} |\sigma| \hat{f}(\sigma \theta^{\perp}) e^{i\langle x, \sigma \theta^{\perp} \rangle} \, d\sigma \, d\varphi \\
&= \tfrac{1}{2} (2\pi)^{-3/2} \int_0^{2\pi} \int_{-\infty}^{\infty} |\sigma| \widehat{P_\theta f}(\sigma) e^{i\sigma \langle x, \theta^{\perp} \rangle} \, d\sigma \, d\varphi \\
&= \tfrac{1}{2} (2\pi)^{-3/2} \int_0^{2\pi} \int_{\mathbb{R}} \widehat{\Lambda P_\theta f}(\sigma) e^{i\sigma \langle x, \theta^{\perp} \rangle} \, d\sigma \, d\varphi \\
&= (4\pi)^{-1} \int_0^{2\pi} \Lambda P_\theta f(\langle x, \theta^{\perp} \rangle) \, d\varphi \\
&= \frac{1}{4\pi^2} \int_0^{2\pi} \int_{\mathbb{R}} \frac{\partial P_\theta f(s)}{\langle x, \theta^{\perp} \rangle - s} \, ds \, d\varphi.
\end{aligned}
\tag{4-3}
$$

In the last step we made use of the relation $\Lambda g = \mathcal{H} \partial g$ mentioned above.

For general dimension n one uses the change of variables [89, Formula (9.2′)]

$$\int_{\mathbb{R}^n} g(\xi) \, d\xi = |S^{n-2}|^{-1} \int_{S^{n-1}} \int_{\Theta^{\perp}} |\eta| g(\eta) \, d\eta \, d\theta \tag{4-4}$$

and obtains

$$f(x) = (2\pi |S^{n-2}|)^{-1} \int_{S^{n-1}} \Lambda P_\theta f(E_{\Theta^{\perp}} x) \, d\theta \tag{4-5}$$

where $E_{\Theta^{\perp}} x$ denotes the orthogonal projection of x onto the subspace $\Theta^{\perp}$.

If we use the *backprojection* operator $P^\sharp$ defined by

$$P^\sharp g(x) = \int_{S^{n-1}} g(\theta, E_{\Theta^\perp} x)\, d\theta,$$

then (4–5) assumes the compact form

$$f(x) = \left(2\pi|S^{n-2}|\right)^{-1} P^\sharp \Lambda P f(x).$$

An inversion formula for the Radon transform can be derived in a similar way. For other inversion formulas see [62, § II.2].

From the last line of (4–3) we see that computation of $f(x)$ requires integrals over lines far from x, because the Hilbert transform kernel has unbounded support. Note that $P_\theta f(\langle x, \theta^\perp \rangle)$ is the integral over the line with direction θ which passes through x. Hence the inversion formula is not "local". A local inversion formula would utilize only values $P_\theta f(s)$ with s close to $\langle x, \theta^\perp \rangle$. We will discuss what can be done with local formulas in a later section.

Equation (4–3) gives us valuable information about the stability of the inversion. The factor $|\sigma|$ in the inverse Fourier integral will become arbitrarily large. This means that the inversion is unstable. In practice measurement and discretization errors will prevent accurate computation of $\widehat{P_\theta f}(\sigma)$ for large $|\sigma|$, and these errors are then amplified by multiplication with $|\sigma|$. In other words, due to the integration in P, Pf is smoother than f itself. The inversion has to reverse this smoothing and this makes it unstable. The extent of this instability will depend on the amount of smoothing inherent in P. This can be quantified using Sobolev norms. For functions f with compact support we define

$$\|f\|_{H_0^\alpha} = \left(\int_{\mathbb{R}^n} (1 + |\xi|^2)^\alpha \, |\hat{f}(\xi)|^2 \, d\xi \right)^{1/2},$$

$$\|Pf\|_\alpha = \left(\int_{S^{n-1}} d\theta \int_{\Theta^\perp} d\eta \, (1 + |\eta|^2)^\alpha \, |\widehat{P_\theta f}(\eta)|^2 \right)^{1/2}.$$

THEOREM 4.1 [62, p. 42]. *If $f \in C_0^\infty$ is supported in the unit ball, then there are constants $c(\alpha, n), C(\alpha, n)$ such that*

$$c(\alpha, n)\|f\|_{H_0^\alpha} \leq \|Pf\|_{\alpha + \frac{1}{2}} \leq C(\alpha, n)\|f\|_{H_0^\alpha}.$$

Hence the operator P smoothes by an order $\frac{1}{2}$ measured in a Sobolev scale. In order to see what the instability might mean in practice we assume that we have measured data g^ε such that $\|Pf - g^\varepsilon\|_{L_2} \leq \varepsilon$, and a-priori information about f of the form $\|f\|_{H^\beta} \leq \rho$. For $\beta > 0$ this excludes highly oscillatory functions, so this condition corresponds to condition (a) in Remark 3.4. Let f_1, f_2 be two candidate functions for reconstruction, i.e., f_1, f_2 both satisfy the a-priori condition and $\|Pf_i - g^\varepsilon\|_{L_2} \leq \varepsilon$. We are interested to know by how much f_1

and f_2 can differ. Since $\|P(f_1 - f_2)\|_{L_2} \leq 2\varepsilon$ and $\|f_1 - f_2\|_{H_0^\beta} \leq 2\rho$, we have the worst case error

$$\|f_1 - f_2\|_{L^2} \leq d(\varepsilon, \rho, \beta), \text{ with}$$

$$d(\varepsilon, \rho, \beta) = \sup \left\{ \|f\|_{L_2} : \|Pf\|_{L_2} \leq 2\varepsilon, \|f\|_{H_0^\beta} \leq 2\rho \right\}.$$

A natural choice for β is such that functions which are smooth except for jump discontinuities along smooth boundaries are in H_0^β. This leads to the condition $\beta < \frac{1}{2}$ [62, p. 92]. For the limiting case $\beta = \frac{1}{2}$ the worst case error satisfies

$$d\left(\varepsilon, \rho, \tfrac{1}{2}\right) \leq c(n)\sqrt{\varepsilon\rho}.$$

[62, p. 94]. This means that the reconstruction problem is moderately ill-posed. We expect a gain of $2k$ digits in data accuracy to yield k additional accurate digits in the reconstruction. In other words, the instability in the reconstruction causes a loss of half the number of accurate digits.

Another approach to quantify the degree of ill-posedness is provided by the singular value decomposition of P [60]. Here one looks at how fast the singular values converge to zero. Again, the assessment of moderate ill-posedness is confirmed.

In order to use the inversion formula in practice we have to stabilize it. This involves a well-known trade-off between stability and accuracy of the reconstruction. Here we give up the goal of recovering the function f itself, and aim instead at reconstructing an approximation $e * f$, where e is an approximate delta function. As the computation below shows, stabilization requires the Fourier transform $\hat{e}(\xi)$ to decay sufficiently fast for large $|\xi|$. The price to pay for the stabilization is limited resolution, so e must be chosen carefully, depending on the amount and accuracy of the available measurements. Note also that a proper choice of e helps to satisfy the condition (c) for avoiding the influence of nonuniqueness given in Remark 3.4.

As we will see later, it is sometimes advantageous to reconstruct $\Lambda^m f$ instead of f, with $m > -1$ an integer. The case $m = 0$ of course yields an approximation to the function f itself. Using the convolution theorem (2–1) for the Fourier transform we obtain, in a similar way as above,

$$
\begin{aligned}
e * \Lambda^m f(x) &= \int_{\mathbb{R}^n} \hat{e}(\xi)|\xi|^m \hat{f}(\xi) e^{i\langle x,\xi\rangle} \, d\xi \\
&= |S^{n-2}|^{-1} \int_{S^{n-1}} \int_{\Theta^\perp} |\eta|^{m+1} \hat{e}(\eta) \hat{f}(\eta) e^{i\langle x,\eta\rangle} \, d\eta \, d\theta \\
&= (2\pi)^{-1} |S^{n-2}|^{-1} \int_{S^{n-1}} \int_{\Theta^\perp} |\eta|^{m+1} \widehat{P_\theta e}(\eta) \widehat{P_\theta f}(\eta) e^{i\langle E_{\Theta^\perp} x, \eta\rangle} \, d\eta \, d\theta \\
&= \int_{S^{n-1}} (k * P_\theta f)(E_{\Theta^\perp} x) \, d\theta, \quad m \geq -1, \quad (4\text{–}6)
\end{aligned}
$$

with the convolution kernel

$$k(y) = (2\pi |S^{n-2}|)^{-1} \Lambda^{m+1} P_\theta e(y), \quad y \in \Theta^\perp. \tag{4-7}$$

If e is a radial function, $P_\theta e$ and the convolution kernel k are independent of θ.

A corresponding formula for the Radon transform can be derived by using polar coordinates in $\mathbb{R}^n$ instead of (4–4). For rigorous proofs and general conditions on e and f for which (4–6) is valid see [48], [90] and [59]. Of greatest interest are the case $m = 0$, which gives the formulas for reconstructing the function f itself, and the cases $m = \pm 1$. Letting $e \to \delta$ yields the exact inversion formula

$$\Lambda^m f(x) = (2\pi |S^{n-2}|)^{-1} \int_{S^{n-1}} \Lambda^{m+1} P_\theta f(E_{\Theta^\perp} x) \, d\theta.$$

A desirable property would be the possibility of local reconstruction, i.e., reconstruction at a point should require only lines passing through a small neighborhood of that point. Since the parameters θ and $y \in \Theta^\perp$ of a line passing through a point x must satisfy the equation $E_{\Theta^\perp} x = y$, reconstruction according to (4–6) will be local if the kernel k is supported in a small neighborhood of the origin. However, for m even and $\int_{\mathbb{R}^n} e(x) \, dx \neq 0$, $\hat{k}$ is not analytic, so k cannot have compact support. Hence ordinary tomography is global, not local. On the other hand, it follows from (4–7) and (4–1) that k has compact support if $m \geq -1$ is odd and e has compact support. This explains the interest in the cases $m = \pm 1$. Computing $\Lambda^{-1} f(x)$ consists of taking the average of all integrals over lines passing through x. This was done in early imaging techniques preceding CT. However, the result is a very blurry image of f which by itself is of limited usefulness; see the bottom left picture in Figure 1. Current local tomography, reviewed below, avoids this disadvantage by computing a linear combination of Λf and $\Lambda^{-1} f$.

If f is supported in the unit ball, and the source points a lie on a sphere A with center in the origin and radius $R > 1$, then the approximate inversion formula for the divergent beam x-ray transform reads as follows [90]:

$$e * \Lambda^m f(x) = R^{-1} \int_A \int_{S^{n-1}} D_a f(\theta) \, |\langle a, \theta \rangle| \, k(E_{\Theta^\perp}(x - a)) \, d\theta \, da, \tag{4-8}$$

with $m \geq -1$ and k as in (4–7).

We conclude this section with a few remarks on reconstruction algorithms. The *filtered backprojection algorithm* is the most popular reconstruction method. It is a computer implementation of the approximate reconstruction formulas (4–6) and (4–8) for parallel-beam and fan-beam sampling, respectively. We will discuss it in detail in a later section. For references on the filtered backprojection algorithm see, e.g., [49].

The *Fourier reconstruction algorithm* uses the Fast Fourier transform to compute

$$\widehat{P_{\theta_j} f}(\eta) = \sqrt{2\pi} \hat{f}(\eta), \quad \eta \in \Theta_j^\perp, \quad j = 0, \ldots, P-1.$$

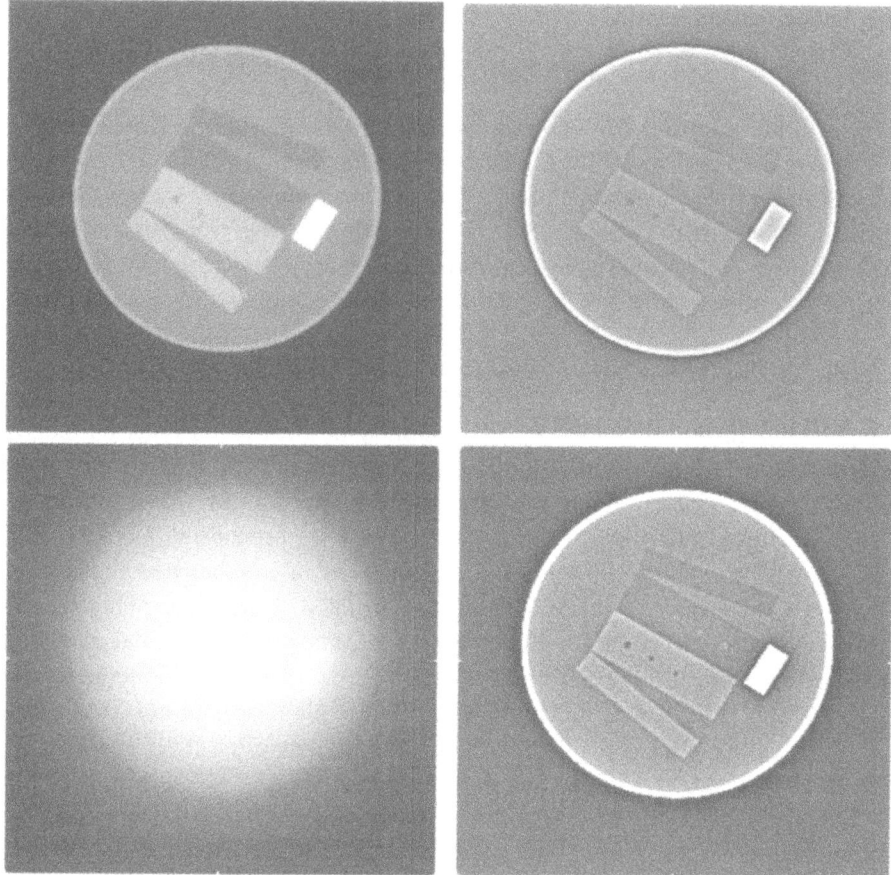

Figure 1. Top left: Global reconstruction of density $f(x)$ of calibration object. Top right: Reconstruction of Λf. Bottom left: Reconstruction of $\Lambda^{-1} f$. Bottom right: Reconstruction of $Lf = \Lambda f + \mu \Lambda^{-1} f$, with $\mu = 46$.

In 2D this gives values of $\hat{f}$ on a polar grid. These are now interpolated onto a rectangular grid and a 2D inverse FFT is used to obtain f. This is much faster than filtered backprojection, but the interpolation is problematic, i.e., prone to cause artifacts in the reconstructed image. For further discussion and references on methods to overcome these drawbacks see [66; 67].

Algebraic methods do not discretize an inversion formula or use the projection slice theorem, but start from an ansatz $f(x) = \sum_{i=1}^{N} c_i \psi_i(x)$ and then solve the linear system

$$\sum_{i=1}^{N} c_i P_{\theta_j} \psi_i(y_k) = g_{jk}, \quad j, k = 1, 2, \ldots$$

for the unknown coefficients c_i. Here $g_{jk} = P_{\theta_j} f(y_k)$ are the measured data. Often the basis functions are the characteristic functions of pixels or voxels, but this is of course not the only choice. Indeed, the advantage of such methods lies in their flexibility, e.g., in incorporating irregular sampling geometries or available a-priori information on f. The resulting linear systems are large and sparse and require special (usually iterative) algorithms for sufficiently fast solution. Stabilization can be achieved by limiting resolution or by stopping the iteration before convergence is achieved (see, e.g., Figure V.12 in [62]).

Numerous other reconstruction algorithms have been developed. For a survey see, e.g., [62, Chapter V] and [65; 66; 67].

5. Incomplete Data Problems and Detection of Singularities

Incomplete data problems arise when measurements of $P_\theta f(y)$ are unavailable for a certain range of arguments (θ, y). In dimension 2 the most common examples are the limited angle problem, the exterior problem, and the interior problem. Assume that f has compact support contained in the unit disk. In the limited angle problem, measurements $Pf(\varphi, s)$ are available only in an angular range $\varphi \in [\varphi_1, \varphi_2]$ with $|\varphi_1 - \varphi_2| < \pi$. Note that because of $Pf(\varphi, s) = Pf(\varphi + \pi, -s)$, an angular range of π is sufficient for complete data. It follows from Theorem 3.1 that f is uniquely determined by the limited angle data. The problem is lack of stability. We see from Theorem 2.2 that the data determine the Fourier transform $\hat{f}(\xi)$ in the cone $\{\xi = \sigma(\cos\varphi, \sin\varphi) : \varphi \in [\varphi_1 + \pi/2, \varphi_2 + \pi/2], \sigma \in \mathbb{R}\}$. Reconstructing f is therefore equivalent to accomplishing an analytic continuation of $\hat{f}$, and analytic continuation is severely ill-posed. A more detailed picture emerges from the singular value decomposition of the limited angle transform [54]. The severe ill-posedness is reflected in exponentially decaying singular values. However, the spectrum splits into two parts, one with singular values close to the singular values in the full-range case, and the other with singular values close to 0. The components of f corresponding to singular functions in the first part are therefore recoverable. The characterization of the unrecoverable singular functions in [54] allows to predict and recognize typical reconstruction artifacts.

In the exterior problem only line integrals $P_\theta f(s)$ with $|s| > a > 0$ are available. Uniqueness holds in the measured region but stability is missing. The singular value decomposition was given in [70; 73], and used to develop a reconstruction algorithm [74; 75].

Finally, the interior problem is characterized by measurements in the range $|s| \leq a < 1$. Uniqueness does not hold, not even inside the disk $|x| < a$ where for each point x all integrals over lines passing through a small neighborhood of x are measured. A singular value decomposition has been derived in [61]. Promising new methods for the interior problem also include the wavelet-based approaches of [2; 80] and pseudolocal tomography [41]. The wavelet localization method presented in [68] requires additional integrals over a small number of lines not intersecting the disk $|x| < a$.

None of these problems provides both uniqueness and stability. It is now natural to ask that if the function f itself cannot be recovered stably, what features of f can? One answer to this question is provided by the singular value decompositions, which tell us that components of f corresponding to singular functions with large singular values are recoverable. Another approach is to ask which *singularities* of f can be stably recovered. In many applications f can be considered to be approximately piecewise constant with jump discontinuities along the boundaries between different features. Identifying the singularities of f thus allows to determine the shape of such features. A general answer has been given in [75] based on the correspondence between the wavefront sets of f and Pf. In the special case of f being smooth except for jump discontinuities along a smooth curve Γ, a singularity at a point x is detected stably if and only if integrals over lines in a neighborhood of the tangent line to Γ at x are available. Applying this rule to the incomplete data problems mentioned above yields that in the limited angle and exterior problems not all jumps can be stably detected, since for any point x there are lines passing through x for which the data are not available. On the other hand, in case of the interior problem one can stably determine all singularities inside the disk $|x| < a$. It is thus interesting to note that the interior problem is the worst behaved of the three with respect to uniqueness, but is the best behaved with respect to detecting singularities.

For general reconstruction methods where the reconstruction preserves the stable singularities see [45]. Several methods have been suggested to detect singularities directly from the line integrals without first performing a reconstruction [39; 75; 79].

A problem of great practical interest which still poses many open problems is three-dimensional cone-beam reconstruction with sources on a curve. See, e.g., [97] for an inversion formula, [19] for a general stability result, [75] for conditions to detect singularities, and [8; 18; 23; 57; 64] for reconstruction algorithms and other developments. The approximate inversion formula (4–6) is very useful in two dimensions, but not so in three dimensions. It needs integrals over all lines, but in three dimensions the lines form a four parameter family, so (4–6) requires far more data than should be needed to determine a function of three variables. In practical 3D tomography an x-ray source moves on a curve, so only integrals over lines intersecting the curve are measured. This situation is modelled by the divergent beam x-ray transform $D_a f(\theta)$, where a runs through the curve Γ. The conditions on the source curve Γ for stable inversion are restrictive, so that in most practical situations one has an incomplete data problem. Based on the exposition in [75], we now state the precise definitions for the microlocal concepts mentioned above and apply them to this situation. The reader interested in a deeper treatment may wish to first read [75] and [27], and then proceed to articles such as [1; 24; 25; 26; 72].

The following concept of a wavefront set uses the fact that the Fourier transform of a C_0^∞ function decays rapidly. A local version of this fact can be obtained by first multiplying f with a C_0^∞ cut-off function Φ with small support, and seeing if the Fourier transform of the product Φf decays rapidly. The wavefront set gives even more specific, so-called microlocal information, inasmuch as it identifies the directions in which the Fourier transform of Φf does not decrease rapidly.

DEFINITION 5.1. Let f be a distribution and take $x_0, \xi_0 \in \mathbb{R}^n$ with $\xi_0 \neq 0$. Then (x_0, ξ_0) is in the wavefront set of f if and only if for each cut-off function Φ in C_0^∞ with $\Phi(x_0) \neq 0$, the Fourier transform of Φf does not decrease rapidly in any conic neighborhood of the ray $\{t\xi_0, t > 0\}$.

Loosely speaking, we say that a singularity of f can be stably detected from available x-ray data, if there exists a corresponding singularity of comparable strength in the data. The strength of a singularity can be quantified microlocally using Sobolev space concepts:

DEFINITION 5.2. A distribution f is in the Sobolev space H^s microlocally near (x_0, ξ_0) if and only if there is a cut-off function $\Phi \in C_0^\infty(\mathbb{R}^n)$ with $\Phi(x_0) \neq 0$ and function $u(\xi)$ homogeneous of degree zero and smooth on $\mathbb{R}^n \setminus \{0\}$ and with $u(\xi_0) \neq 0$ such that $u(\xi)\widehat{(\Phi f)}(\xi) \in L^2(\mathbb{R}^n, (1 + |\xi|^2)^s)$.

First, one localizes near x_0 by multiplying f by Φ, then one microlocalizes near ξ_0 by forming $u\widehat{\Phi f}$. and sees how rapidly $\widehat{\Phi f}$ decays at infinity.

For 3D tomography with sources on a curve we have the following result:

THEOREM 5.3 [75, Theorem 4.1], [1], [24]. *Let Γ be a smooth curve in $\mathbb{R}^3$ and f a distribution whose support is compact and disjoint from Γ. Then any wavefront set of f at (x_0, ξ_0) is stably detected from divergent beam x-ray data Df with sources on Γ if and only if the plane $\mathcal{P}$ through x_0 and orthogonal to ξ_0 intersects Γ transversally.*

If data are taken over an open set of rays with sources on Γ, then a ray in $\mathcal{P}$ from Γ to x_0 must be in the data set for stable detection to apply. In these cases f is in H^s microlocally near (x_0, ξ_0) if and only if the corresponding singularity of Df is in $H^{s+1/2}$.

We see that the corresponding singularities of Df are weaker by $\frac{1}{2}$ Sobolev order, but this is still strong enough to allow stable detection in practice.

Theorem 5.3 allows to analyze singularity detection in 3D tomography in the same way as described above in the two-dimensional case.

It is now interesting to ask if the available numerical algorithms can actually reconstruct all the stable singularities. The results for a general class of restricted x-ray transforms obtained in [24; 25; 26] show that microlocal analysis is also a powerful tool to answer such a question. For an introduction to these results see [27]. Explicit calculations analysing an algorithm for contour reconstruction proposed in [57] and some closely related methods have recently been given in [38; 47].

The algorithm of [57] aims to reconstruct the function

$$f_R = -\Delta D^* D f,$$ (5–1)

with

$$D^* g(x) = \int_\Gamma \|x - a\|^{-1} g\left(a, \frac{x - a}{\|x - a\|}\right) da.$$

The results in [24; 38; 47] show that the wavefront set of f_R consists of two parts. The first part contains those wavefronts (x, ξ) of f for which the plane through x and normal to ξ intersects Γ. The second part may introduce new singularities, namely on the line from a source point $a \in \Gamma$ to x, the location of the original singularity in f. This will happen if the plane through x and normal to ξ contains a and the tangent vector to Γ at a is orthogonal to ξ, i.e., the plane touches Γ but does not intersect Γ transversally. In addition, the acceleration vector of the curve at a should not be orthogonal to ξ. The Sobolev strength of these additional singularities is the same as the reconstructed part of the original wavefront set [25; 26; 38], and they appear as artifacts in numerical simulations [17; 35; 38].

An advantage of the formula (5–1) is that reconstruction of f_R is local, i.e., reconstruction at a point x requires only integrals over lines close to x. In [57] it is shown that f_R approximates Λf in certain cases. Another, and apparently the historically first method for 3D local tomography is an adaptation of the algorithm by Feldkamp et al. [18] developed by P.J. Thomas at the Mayo Clinic. While the details of this algorithm have not been published, it has been used in various papers, e.g., [94; 14].

6. Local Tomography

Often only part of an object needs to be imaged. In this case it would be preferable if only integrals over lines which intersect the region of interest (ROI) are needed. We know from the discussion of the interior problem above that we don't have uniqueness. However, it turns out that the null functions are nearly constant inside the ROI, and we know already that all singularities inside the ROI are stably determined. Several approaches have been developed in the literature. For example, the wavelet based method of [68] exploits the fact that the error contains mostly low frequencies, and that these can be recovered by supplementing the data with relatively few measurements outside the ROI. The method of [80] which will be discussed below, extrapolates the missing data and aims at reconstruction of f up to a constant error. Another method using extrapolation of the missing data is described in [62, § VI.4].

Lambda tomography, the main topic of this section, was introduced independently in [98] and [90]. It does not attempt to reconstruct the function f itself but instead produces the related function $Lf = \Lambda f + \mu \Lambda^{-1} f$. This has the advantage that the reconstruction is strictly local in the sense that computation of $Lf(x)$ requires only integrals over lines passing arbitrarily close to x. Local tomography has found applications in medical imaging [94], nondestructive testing [85; 99], and microtomography [15; 16; 83; 86]. Extensions to more general settings have been presented in [37; 45]. Other approaches include [2] and [41].

Intelligent use of Lambda tomography requires knowledge of what kind of useful information about f is retained in Lf. Let us consider an example. The upper left of Figure 1 shows an ordinary, global reconstruction of the density function f of a calibration object used by the Siemens company. The data come from an old generation Siemens hospital scanner. Units are such that the radius of the global reconstruction circle is one. The figure displays the reconstruction inside the rectangle $[-0.5, 0.5]^2$. The scanning geometry is a fan-beam geometry (7–10) with source radius $R = 2.868$, $p = 720$ source positions, and $2q = 512$ rays per source. The upper right of Figure 1 shows a reconstruction of Λf. Reconstructions of $\Lambda^{-1} f$ and $Lf = \Lambda f + 46\Lambda^{-1} f$ are shown in the lower left and lower right, respectively. The similarity between the images of f and Λf is at first glance surprising. We expect that a good local reconstruction method should detect the singularities of f, since these are stably determined by the data. Indeed, since Λ is an invertible elliptic pseudo-differential operator, f and Λf have precisely the same singular set. However, we see that Λf is cupped where f is constant, and that the singularities are amplified in Λf. The image of $\Lambda^{-1} f$ by itself seems less useful, but it provides a countercup for the cup in Λf. Thus, the image of Lf shows less cupping and looks even more similar to f than the image of Λf. For example, the image of Lf indicates that the density just inside the boundary of the object is larger than the density outside the object, while this can not be clearly seen from the image of Λf. To achieve this effect, a good selection of μ is necessary. A prescription for selecting μ can be found in [15].

A more detailed understanding of images of Λf or Lf is obtained from studying quantitative relations between Λf, $\Lambda^{-1} f$ and f [14; 15]. Some of the results for Λf are as follows. For corresponding results on Λ^{-1} see [14].

THEOREM 6.1. ([14]) *Let X and Y be measurable sets, $n \geq 2$, and let $(1 + |x|)^{-1-n} f$ be integrable.*

(a) *If $f_r(x) = f(x/r)$, then $\Lambda f_r(x) = r^{-1} \Lambda f(x/r)$.*

(b) *$\Lambda \chi_X(x) > 0$ on X°, and < 0 on $X^{c\circ}$; $\Lambda \chi_{X^c} = -\Lambda \chi_X$.*

(c) *$\Lambda \chi_X$ is subharmonic (Laplacian ≥ 0) on X°, and superharmonic on $X^{c\circ}$. This implies that $\Lambda \chi_X$ cannot have a local maximum in X°, nor a local minimum in $X^{c\circ}$.*

(d) *If x is outside the support of f, then*

$$\Lambda f(x) = \frac{1-n}{\pi |S^{n-2}|} \int_{\mathbb{R}^n} |x - y|^{-1-n} f(y) \, dy.$$

(e) *Near ∂X we have $|\Lambda \chi_X(y)| \sim 1/d(y, \partial X)$, where $d(x, \partial X)$ is the distance of x to ∂X.*

REMARK 6.2. The results for $\Lambda \chi_X$ are of practical interest, since in many applications the function f can be modeled as a linear combination of characteristic functions.

- As a consequence of (a), small features are amplified in images of Λf. This is beneficial for the detection of small, low contrast details. For example, in Figure 1 the small holes in the rectangular pieces are more clearly visible in the image of Λf than in the image of f.
- Part (b) indicates that the jumps of Λf at discontinuities of f have the same direction as those of f.
- Part (c) explains why there are no oscillations which could be mistaken for actual details in images of Λf.
- Part (d) shows that if f has compact support, then Λf cannot. This means that there are global effects in images of Λf in the sense that the value of $\Lambda f(x_0)$ depends on the values of f everywhere. However, Part d) implies that $\Lambda f(x)$ will decay at least as $O(|x|^{-1-n})$ for $|x| \to \infty$. More refined estimates are derived in [15].
- Part (e) shows that a finite jump in f causes an infinite jump in Λf. In a neighborhood of ∂X, Λf is not a function but a principal value distribution [14].

While Lf retains the signs of jumps in density, it does not give direct information about the size of these jumps. However, such information about density differences may be extracted in certain cases. In the following we will describe several methods. We assume that f is a linear combination of a smooth function and of characteristic functions of sets:

$$f = f_0 + \sum c_i \chi_{X_i}, \tag{6-1}$$

with $f_0 \in C_0^\infty$, $|\partial X_i| = 0$, $X_i = \overline{X_i^\circ}$, and $X_i^\circ \cap X_j^\circ = \varnothing$ if $i \neq j$.

We are interested in estimating $c_j - c_i$ when X_j, X_i have a common nontrivial boundary Γ,

$$\Gamma = \partial X_i \cap \partial X_j \cap W \neq \varnothing, \quad W = (X_i \cup X_j)^\circ.$$

We first discuss the method developed in [15]. It is based on Theorem 6.3 below. The theorem expresses the fact that for x sufficiently close to Γ, we have

$$c_j - c_i = \frac{\Lambda f(x)}{\Lambda \chi_{X_j}(x)} + O(d), \qquad |c_j - c_i| = \frac{|\nabla \Lambda f(x)|}{|\nabla \Lambda \chi_{X_j}(x)|} + O(d^2),$$

where d is the distance from x to Γ.

Recall that a set Y has curvature $\leq 1/r$ along a subset Y_0 of ∂Y if for each point $\bar{y} \in Y_0$ there are open balls $B \subset Y$ and $B' \subset Y^c$ of radius r with $\bar{y} \in \bar{B} \cap \bar{B}'$. The distance of a point x to a set Y is denoted by $d(x, Y)$.

THEOREM 6.3 [15]. *Let f be as in* (6–1). *Fix i, j, let $W = (X_i \cup X_j)^\circ$ and assume that*

$$\Gamma = \partial X_i \cap \partial X_j \cap W \neq \varnothing.$$

Let X_j have curvature $\leq 1/r$, $r > 0$, along a closed subset Γ_0 of Γ. Let $x \in W \setminus \Gamma$ be such that $d(x, \partial X_j) = d(x, \Gamma_0) = d$. Then

$$\left| \frac{\Lambda f(x)}{\Lambda \chi_{X_j}(x)} - (c_j - c_i) \right| \leq F_1(d/r) \left(\max |\Lambda f_0| + C_1 \frac{\max_{k \neq j} |c_k|}{d(x, \partial W)} \right) d, \qquad (6\text{--}2)$$

$$\left| \frac{|\nabla \Lambda f(x)|}{|\nabla \Lambda \chi_{X_j}(x)|} - |c_j - c_i| \right| \leq F_2(d/r) \left(\max |\nabla \Lambda f_0| + C_2 \frac{\max_{k \neq j} |c_k|}{d(x, \partial W)^2} \right) d^2. \quad (6\text{--}3)$$

The constants C_1 and C_2 and the functions F_1, F_2 can be given explicitly. For example, for $n = 2$, we have $C_1 = 2$ and $C_2 = 3$. Furthermore,

$$\lim_{t \to 0^+} F_1(t) = \lim_{t \to 0^+} F_2(t) = \pi.$$

The error terms on the right-hand sides of (6–2) and (6–3) indicate that in general the estimate (6–3) should be more accurate than (6–2) when d is small. The terms involving $d(x, \partial W)$ come from the influence of other boundaries than Γ.

Numerical implementation of (6–2) or (6–3) requires computation of reconstructions of Λf and $\Lambda \chi_{X_j}$ inside a region of interest R. In the following let $\bar{\Lambda} f$ and $\bar{\Lambda} \chi_{X_j}$ denote these reconstructions, rather than the functions Λf and $\Lambda \chi_{X_j}$ themselves. It is also assumed that f has the form (6–1) with sets X_i such that $X_i \subset R$ or $X_i \cap R = \varnothing$. This entails no loss of generality since any set X_i violating this condition can be replaced by the two sets $X_i \cap R$ and $X_i \cap R^c$. $\bar{\Lambda} \chi_{X_j}$ is computed using simulated x-ray data, after ∂X_j has been found from $\bar{\Lambda} f$. In principle, either (6–2) or (6–3) can be used, but as discussed above the method based on (6–3) is likely to be more accurate. This gives only $|c_j - c_i|$, but since the sign of $c_j - c_i$ is preserved in Λf, this is all that is needed.

The method consists of the following steps:

(i) Compute $\bar{\Lambda} f$ from local data inside a region of interest R.
(ii) Determine X_j by finding ∂X_j from $\bar{\Lambda} f$.
(iii) Compute $\bar{\Lambda} \chi_{X_j}$ inside the region of interest from simulated x-ray data, using the same sampling geometry as for the original data.
(iv) If $x \in \partial X_j$, take the ratio $|\nabla \bar{\Lambda} f(x)| / |\nabla \bar{\Lambda} \chi_{X_j}(x)|$ as an estimate for the magnitude of the density jump. It is advisable to use suitable averages of the gradients over points near the boundary of X_j instead of the gradient at a single point x. This reduces effects due to measurement noise.

Following [12], we demonstrate the method with x-ray data from a medical scanner. Additional applications of this method are reported in [15; 12; 83].

The top panel of Figure 2 shows again the global reconstruction of the calibration object. The region of interest R is indicated by the box. The picture in the lower left shows the local reconstruction $\bar{\Lambda} f$ inside the region of interest.

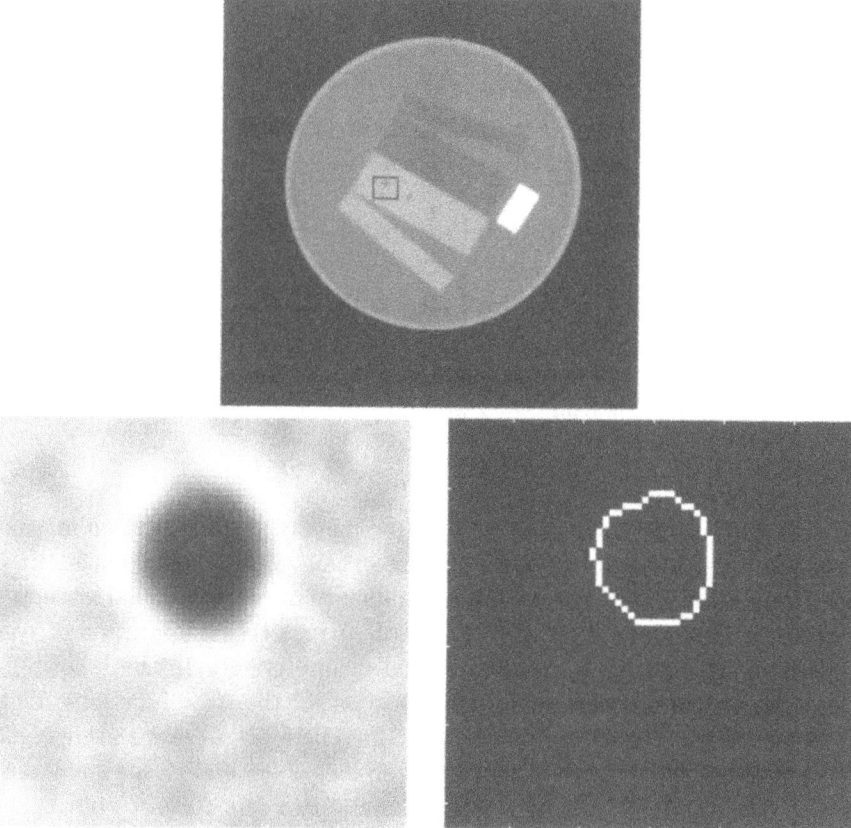

Figure 2. Top: Global reconstruction of density $f(x)$ of calibration object. Box indicates region of interest R. Bottom left: Local reconstruction $\bar{\Lambda}f$ inside region $R = [-0.14, -0.08] \times [0.008, 0.058]$ (contained in the small box in the top panel). Bottom right: Result of automatic edge detector applied to the image of $\bar{\Lambda}f$ shown in top right. Pixels where an edge is detected are white.

The goal is to determine the density difference between the small hole and its surroundings. Let X_j be the characteristic function of the hole.

Finding ∂X_j involves edge detection. This is currently done by the user of the method, who specifies the vertices of a polygon approximating ∂X_j. Matlab's image processing toolbox allows to do this selection conveniently. Our software gives the user the option to use either the reconstruction $\bar{\Lambda}f$ itself, an image of $|\nabla\bar{\Lambda}f|$, or the result of a standard automatic edge detector, for specifying the polygon. Which image is most convenient differs from application to application. In [15], where the method was applied to projection data from a human pelvis, the gradient image was most convenient. Here the result of the automatic edge detector (Matlab's edge command) applied to the reconstruction $\bar{\Lambda}f$ is satisfactory, as can be seen from the lower right of Figure 2.

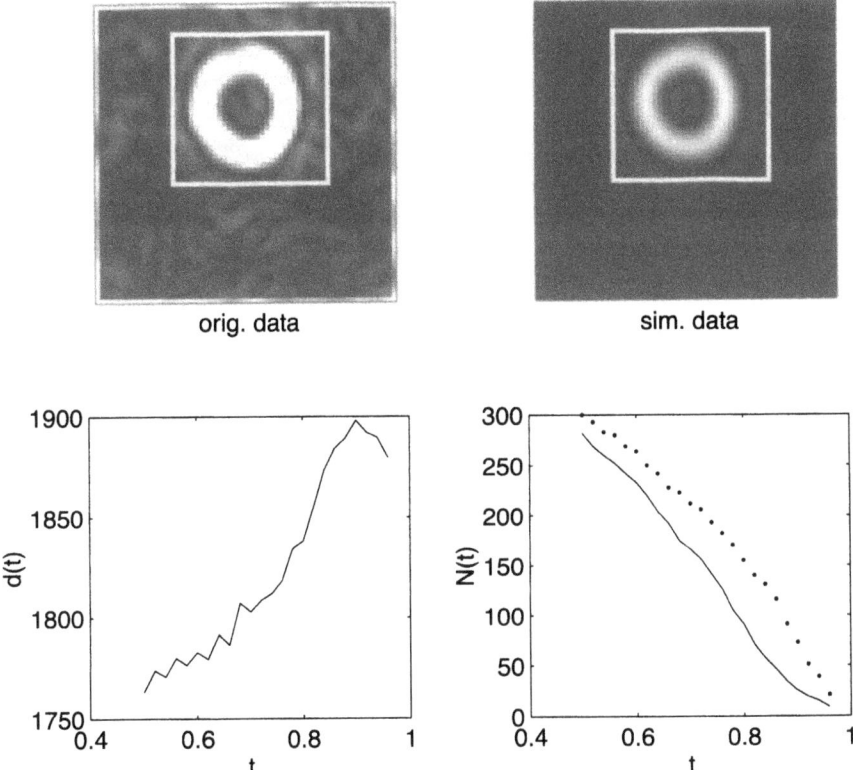

Figure 3. Top left: Image of $|\nabla\bar{\Lambda}f|$ inside R, f being the density function of the calibration object. Box indicates region R'. Top right: Image of $|\nabla\bar{\Lambda}\chi_{X_j}|$ inside R. Box indicates region R'. Bottom left: Graph of estimated density difference $d(t)$ for $0.5 \le d(t) \le 0.95$. Bottom right: Number $N(t)$ of points contributing to the averages of $|\nabla\bar{\Lambda}f|$ (solid line), and of $|\nabla\bar{\Lambda}\chi_{X_j}|$ (dotted line).

The top left of Figure 3 shows an image of $|\nabla\bar{\Lambda}f|$ inside the region of interest. The box indicates the subregion R' containing the part of the boundary which will be used to estimate the density jump. Here we average over the whole boundary of the small hole. Matlab's imcrop command allows convenient selection of R' by the user. The corresponding image for $|\nabla\bar{\Lambda}\chi_{X_j}|$ is shown in the top right part of the figure. This reconstruction was computed from simulated x-ray data using the same scanning geometry as in the reconstruction from real data. Having no specific information on the detectors, the effect of the positive detector width was modeled by averaging line integrals over the angular distance between two adjacent detectors.

The following averaging procedure was used to estimate the density difference. Let M be the maximum of $|\nabla\bar{\Lambda}f|$ in R'. Take the average of $|\nabla\bar{\Lambda}f(x)|$ over all points x in R' such that $|\nabla\bar{\Lambda}f(x)| > tM$ for some $t \in (0,1)$. The same averaging procedure is applied to $|\nabla\bar{\Lambda}\chi_{X_j}(x)|$, with M replaced by the maximum

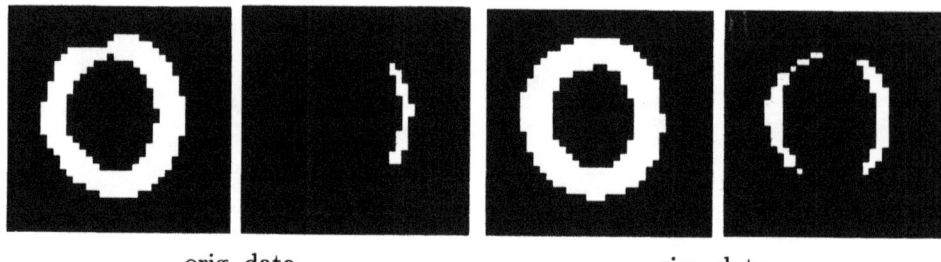

orig. data   sim. data

Figure 4. Two left panels: White points are those where $|\nabla \bar{\Lambda} f(x)|$ exceeds $t \max_{y \in R'} |\nabla \bar{\Lambda} f(y)|$ for $t = 0.6$ (leftmost) and $t = 0.9$ (middle left). Two right panels: White points indicate where $|\nabla \bar{\Lambda} \chi_{X_j}(x)| > t \max_{y \in R'} |\nabla \bar{\Lambda} \chi_{X_j}(y)|$ for $t = 0.6$ (middle right) and $t = 0.9$ (rightmost).

of $|\nabla \bar{\Lambda} \chi_{X_j}|$ in R'. The ratio of the two averages is the estimate $d(t)$ for the density difference. This estimate depends on the choice of t. If t is too close to 1, the average is taken over very few points, while a small t will include points too far from the boundary. So t should be chosen small enough to have sufficiently many points for averaging, but large enough so that only points close to the boundary contribute to the averages. The graph in the bottom left of Figure 3 displays the estimated density differences $d(t)$ for $0.5 \leq t \leq 0.96$. The bottom right shows the numbers $N(t)$ of points contributing to the averages of $|\nabla \bar{\Lambda} f|$ (solid line), and of $|\nabla \bar{\Lambda} \chi_{X_j}|$ (dotted line). If $t \geq 0.9$ very few points contribute to the average of $|\nabla \bar{\Lambda} f(x)|$. The corresponding estimates are therefore likely to be unreliable. On the other hand, for $t < 0.6$ points away from the boundary begin to contribute to the average. The binary images in Figure 4 show the location of the points considered for the averages in the case of $t = 0.6$, and $t = 0.9$, respectively. Hence reasonable estimates for the density difference are the values of $d(t)$ for $0.6 \leq t \leq 0.9$. These values lie between 1782 and 1898. The global reconstruction indicates that the true density difference is approximately 1854. Hence all of the acceptable estimates lie between 96 and 102 per cent of the true value. Since the x-ray data have been scaled by an unknown factor, the reconstructed values do not correspond to Hounsfield numbers.

When implementing the method described above a few parameters have to be chosen judiciously and a few comments on how to do this are in order. If the filtered backprojection algorithm is used the reconstruction $\bar{\Lambda} f$ will, apart from discretization errors, be equal to $e * \Lambda f$. The point-spread function e is assumed to be of the form $e = e_r$ defined in (7–5) and (7–8) below, usually with $\alpha = 11.4174$; compare the appendix of [14]. Choosing the point spread radius r entails the usual tradeoff between stability (larger r) and high resolution (smaller r) and will depend on the number of measured line integrals as well as on the accuracy of these measurements. In the example above $r = 0.0225$, which means that the minimum of the convolution kernel falls on the second detector; see [14, § 9].

The other important parameter is the spacing h of the grid of points where $|\nabla \bar{\Lambda} f| \simeq |\nabla(e * \Lambda f)|$ is computed. $|\nabla(e * \Lambda f)|$ varies rapidly near a boundary and h has to be sufficiently small so that the maximum of the gradient at the gridpoints is close to the overall maximum. The special case with f the characteristic function of a halfspace seems to give sufficient guidance for practical purposes. If f is the characteristic function χ_H of a halfspace H then both Λf and $e * \Lambda f$ can be computed as follows. For $x \notin \partial H$ one has ([14, Theorem 4.5])

$$\Lambda \chi_H(x) = (\pi \tilde{d}(x))^{-1},$$

where $\tilde{d}(x)$ is the signed distance of x from ∂H, i.e., $\tilde{d}(x) = d(x, \partial H)$ for $x \in H$, and $\tilde{d}(x) = -d(x, \partial H)$ for $x \notin H$. Computing $e * \Lambda \chi_H$ involves the Radon transform (2–2) of e. Since e is radial, $R_\theta e$ does not depend on θ. Therefore the subscript θ will be suppressed and $Re(s)$ viewed as a function of the one variable s. It now follows that

$$e * \Lambda \chi_H(x) = \mathcal{H} Re(\tilde{d}(x)), \tag{6-4}$$

where $\mathcal{H}$ denotes the Hilbert transform as defined in (4–2). Observing that for functions f of one variable $\Lambda f(t) = \frac{d}{dt} \mathcal{H} f(t)$ gives

$$|\nabla(e * \Lambda \chi_H(x))| = |\Lambda Re(\tilde{d}(x))|. \tag{6-5}$$

Inspection of the graph of ΛRe for e as in (7–8) and $\alpha = 11.4174$ now yields that the width of the interval where $|\Lambda Re(t)| > 0.98(\max_{s \in \mathbb{R}} |\Lambda Re(s)|)$ is approximately $r/20$. Hence a rule of thumb for choosing h would be to set $h = r/20$.

The method described above can be simplified by making a priori assumptions about the unknown boundary ∂X_j, so that the polygonal approximations and the reconstruction from simulated data are avoided. For example, X_j could be assumed to be a halfspace H. Replacing Λf and $\Lambda \chi_{X_j}$ in (6–2) and (6–3) by $e * \Lambda f$ and $e * \Lambda \chi_H$, and using (6–4) and (6–5) gives the approximate formulas

$$c_j - c_i \simeq \frac{e * \Lambda f(x)}{\mathcal{H} Re(\tilde{d}(x))}, \qquad |c_j - c_i| \simeq \frac{|\nabla(e * \Lambda f(x))|}{|\Lambda Re(\tilde{d}(x))|}. \tag{6-6}$$

These two formulas are the basis of two of the algorithms proposed in [40; 78] for dimension $n = 2$; see formulas (2.17) and (2.21) in [40]. The derivation in [40; 78] is different and employs an asymptotic expansion for Λf, where f is smooth except for jumps across smooth boundaries.

Another method to compute jumps of a function from essentially local data is *pseudolocal tomography* [41; 78]. We follow the presentation given in [4] which allows to understand the numerical implementation of this method in the framework of equations (6–6).

The starting point for pseudolocal tomography is the two-dimensional inversion formula from (4–3), which we repeat here:

$$f(x) = \frac{1}{4\pi} \int_{S^1} \mathcal{H} \partial P_\theta f(\langle x, \theta^\perp \rangle) \, d\theta = \frac{1}{4\pi^2} \int_0^{2\pi} \int_{\mathbb{R}} \frac{\frac{d}{ds} P_\theta f(s)}{\langle x, \theta^\perp \rangle - s} \, ds \, d\varphi.$$

Now truncate the Hilbert transform integral and define

$$f_d(x) = \frac{1}{4\pi^2} \int_0^{2\pi} \int_{\langle x,\theta^\perp\rangle-d}^{\langle x,\theta^\perp\rangle+d} \frac{\frac{d}{ds}P_\theta f(s)}{\langle x,\theta^\perp\rangle - s} \, ds \, d\varphi.$$

It was shown in [41] that $f - f_d$ is continuous, hence f_d has the same jumps as f. Recalling that $P_\theta f(\langle x,\theta^\perp\rangle)$ is the integral over the line in direction θ which passes through x, we see that computation of $f_d(x)$ requires only integrals over lines with distance at most d from x ("pseudo-local" reconstruction.)

In practice one has to use an approximate inversion formula and computes

$$f_{d,r}(x) = e_r * f_d(x) = \int_0^{2\pi} \int_{\mathbb{R}} \tilde{k}_{d,r}(\langle x,\theta^\perp\rangle - s) P_\theta f(s) \, ds \, d\varphi,$$

$$\tilde{k}_{d,r}(t) = \frac{1}{4\pi^2} \int_{t-d}^{t+d} \frac{\frac{d}{ds}P_\theta e_r(s)}{t - s} \, ds,$$

where e_r is a $radial$ function satisfying

$$e_r(x) = r^{-2}e_1(x/r), \quad e_1(x) = 0 \text{ for } |x| > 1, \quad \int_{\mathbb{R}^2} e_1 \, dx = 1.$$

Note that $\tilde{k}_{d,r}(t) = 0$ for $|t| > d+r$, i.e., computation of $f_{d,r}(x)$ requires integrals over lines with distance at most $d+r$ from x. Furthermore, $\lim_{d\to\infty} \tilde{k}_{d,r}(t) = (4\pi)^{-1}\mathcal{H}\partial P_\theta e_r(t)$. Hence (4–6) gives that $\lim_{d\to\infty} f_{d,r}(x) = e_r * f(x)$. Indeed, the convolution kernel $k_{d,r}$ can be obtained from the kernel k in (4–7) by letting $m = 0$ and truncating the Hilbert transform integral. The relation $f_{d,r} = e_r * f_d$ was shown in [41].

It turns out that for small d (i.e., local data), f_d is significantly different from zero only in a narrow region near a boundary [41, Figure 3], and that the convolution with the point spread function e_r alters these values so much that the jumps cannot just be simply read off the reconstructed image $f_{d,r}$. We need an algorithm to obtain information about the jumps of f. The methods developed by Katsevich and Ramm [41; 78] can be understood in the framework developed for Lambda tomography. According to equations (6–6) we have, for x close to Γ,

$$c_j - c_i \simeq \frac{E * \Lambda f(x)}{\mathcal{H}RE(\tilde{d}(x))}, \qquad |c_j - c_i| \simeq \frac{|\nabla E * \Lambda f(x)|}{|\Lambda RE(\tilde{d}(x))|}. \tag{6–7}$$

The task now is to find $E_{d,r}$ such that $E_{d,r} * \Lambda f = f_{d,r} = e_r * f_d$.

PROPOSITION 6.4 [78; 4]. *Define* $E_{d,r}$ *by*

$$P_\theta E_{d,r} = (P_\theta e_r) * M_d$$

with

$$M_d(s) = -\frac{1}{\pi} \ln(|s/d|) \chi_{[-d,d]}(s).$$

Then

$$f_{d,r}(x) = E_{d,r} * \Lambda f(x).$$

With this result, equations (6–7) give

$$c_j - c_i \simeq \frac{f_{d,r}(x)}{\mathcal{H}RE_{d,r}(\tilde{d}(x))}, \qquad |c_j - c_i| \simeq \frac{|\nabla f_{d,r}(x)|}{|\Lambda RE_{d,r}(\tilde{d}(x))|}, \qquad (6\text{–}8)$$

and we can apply the same algorithms for recovering the jumps as in Lambda tomography.

Some remarks are in order.

1. Because $E_{d,r}$ is radial, we have $\mathcal{H}RE_{d,r}(0) = 0$, so $f_{d,r}(x) \simeq 0$ for $x \in \Gamma$. This makes it difficult to use the first relation (6–8) in practice, since finding $\tilde{d}(x)$ is not easy. See the algorithm given [41] and further discussed in [4]; see also [17]. But since $|\nabla f_{d,r}|$ is maximal for $x \in \Gamma$, one can find the points $x \in \Gamma$ by looking for the local maxima of $|\nabla f_{d,r}|$ and then estimate the jump by

$$|c_j - c_i| \simeq \frac{|\nabla f_{d,r}(x)|}{|\Lambda RE_{d,r}(0)|}, \qquad x \in \Gamma.$$

This approach has essentially been used in [78] for pseudolocal tomography and in [40] for Lambda tomography.

2. The property that f_d has the same jumps as f is not used in the algorithm.

3. $E_{d,r}(x) = 0$ for $|x| > d + r$. Hence our derivation of the algorithm is only justified for $d + r$ sufficiently small. In practice the method seems to work also for much larger values of $d + r$.

Another method which can be used for region of interest tomography is the *wavelet-based multiresolution local tomography* of [80]. It illustrates the possible uses of wavelets to "localize" the x-ray transform, or, more precisely, to separate the features which are well determined by local data from those who are not. For readers unfamiliar with wavelets we have collected some basic facts in Appendix A.

Consider a (two-dimensional) multiresolution analysis of nested subspaces V_j, $j \in \mathbb{Z}$ of $L_2(\mathbb{R}^2)$. We assume a dilation matrix $M = 2I$ (see Definition A.2 below), where I is the identity matrix, and use the notation

$$f_{j,k}(x) = 2f(2^j x - k) \quad \text{for } j \in \mathbb{Z}, \ k \in \Gamma = \mathbb{Z}^2, \ x \in \mathbb{R}^2$$

(compare (A–1) in the Appendix). Let Φ be the scaling function and Ψ^μ, $\mu = 1, 2, 3$ the associated wavelets. Since the $\Phi_{j+1,k}$, $k \in \mathbb{Z}^2$ are a Riesz basis of the subspace V_{j+1}, a function $f \in V_{j+1}$ can be written as

$$f(x) = \sum_{k \in \mathbb{Z}^2} \tilde{A}_{j+1,k} \Phi_{j+1,k}(x).$$

The so-called *approximation coefficients* $\tilde{A}_{j,k}$ are given by

$$\tilde{A}_{j,k} = \langle f, \tilde{\Phi}_{j,k} \rangle$$

where $\langle , \rangle$ denotes the inner product in L_2 and $\tilde{\Phi}$ is the biorthogonal scaling function (Definition A.4). Alternatively we can use the relation $V_{j+1} = V_j + W_j$ and obtain the expansion

$$f(x) = \sum_{k \in \mathbb{Z}^2} \tilde{A}_{j,k} \Phi_{j,k}(x) + \sum_{\mu=1}^{3} \sum_{k \in \mathbb{Z}^2} \tilde{D}_{j,k}^{\mu} \Psi_{j,k}^{\mu}(x).$$

We can interpret the first sum as an approximation to f in $V_j \subset V_{j+1}$, i.e., at a lower resolution. The second sum supplies the missing detail information. Therefore the coefficients

$$\tilde{D}_{j,k}^{\mu} = \langle f, \tilde{\Psi}_{j,k}^{\mu} \rangle$$

are called *detail coefficients*. The Fast Wavelet Transform and its inverse (see Theorem A.6) allow efficient computation of the $\tilde{A}_{j,k}$ and $\tilde{D}_{j,k}^{\mu}$, $k \in \mathbb{Z}^2$ from the $\tilde{A}_{j+1,k}$, $k \in \mathbb{Z}^2$, and vice versa.

We now observe that the approximation and detail coefficients can be computed directly from the x-ray data. Let $f^{\vee}(x) = f(-x)$. Then

$$\tilde{A}_{j,k} = \langle f, \tilde{\Phi}_{j,k} \rangle = (f * (\overline{\tilde{\Phi}_{j,0}})^{\vee})(2^{-j}k),$$

and Similarly

$$\tilde{D}_{j,k}^{\mu} = \langle f, \tilde{\psi}_{j,k}^{\mu} \rangle = (f * (\overline{\tilde{\Psi}_{j,0}^{\mu}})^{\vee})(2^{-j}k).$$

Hence we can use the approximate inversion formula (4–6) with $e(x) = (\overline{\tilde{\phi}_{j,0}})^{\vee}(x)$ and reconstruction on the grid $x = 2^{-j}k$, $k \in \mathbb{Z}^2$, to obtain the approximation coefficients directly from the x-ray data. For the detail coefficients we let $e = (\overline{\tilde{\Psi}_{j,0}^{\mu}})^{\vee}$. Alternatively one could first compute the approximation coefficients $\tilde{A}_{j+1,k}$ by letting $e(x) = (\overline{\tilde{\phi}_{j+1,0}})^{\vee}(x)$ and choosing the finer grid $x = 2^{-j-1}k$, $k \in \mathbb{Z}^2$, and then use the Fast Wavelet Transform to obtain the approximation and detail coefficients at level j. Since the additional computational burden of applying the Fast Wavelet Transform is negligible compared to the effort required for the reconstruction from the x-ray data, this alternative method seems preferable, since only one point-spread function and corresponding convolution kernel need to be used. However, if not all coefficients on level j are needed, the first method will be more efficient.

The next question is how this approach allows to 'localize' the x-ray transform, i.e., to separate features which are determined by local data from those which are not. It was observed in [68] that the detail coefficients for sufficiently large j should be well determined by local data, if the wavelets Ψ^{μ} have vanishing moments. Let's see why.

DEFINITION 6.5. A function f of n variables has vanishing moments of order up to N, if

$$\int_{\mathbb{R}^n} x^{\alpha} f(x) \, dx = 0$$

for all multiindices $\alpha = (\alpha_1, \ldots, \alpha_n)$ with $|\alpha| = \sum \alpha_i \leq N$. Recall that the α_i are non-negative integers and that $x^\alpha = x_1^{\alpha_1} x_2^{\alpha_2} \ldots x_n^{\alpha_n}$.

The nonlocality in the approximate inversion formula comes from the convolution kernel k in (4–7) in case of $m = 0$. In two dimensions this is caused by the presence of the Hilbert transform in the formula $k = (4\pi)^{-1} \Lambda P_\theta e = (4\pi)^{-1} \mathcal{H} \partial P_\theta e$. The key observation now is that the Hilbert transform of a function with vanishing moments decays fast.

LEMMA 6.6. ([80, p. 1418]) Let $f(t) \in L_2(\mathbb{R})$ vanish for $|t| > A$ and have vanishing moments of order up to N. Then, for $|s| > A$,

$$|\mathcal{H}f(s)| \leq \frac{1}{\pi |s - A|^{N+2}} \int_{-A}^{A} |f(t) t^{N+1}| \, dt$$

The construction of wavelets with vanishing moments is well known, and it turns out that the functions $\partial P_\theta (\widetilde{\Psi}_{j,0}^\mu)^\vee$ inherit the vanishing moments from the $\widetilde{\Psi}^\mu$. Therefore the convolution kernels $k = (4\pi)^{-1} \mathcal{H} \partial P_\theta (\widetilde{\Psi}_{j,0}^\mu)^\vee$ will decay rapidly outside the support of $P_\theta (\widetilde{\Psi}_{j,0}^\mu)^\vee$.

So we see that the detail coefficients for large j, when $\widetilde{\Psi}_{j,0}^\mu$ has small support, are well determined by local data. This is intuitively plausible since these coefficients contain high-frequency information, and we know already from Lambda tomography that high-frequency information is well-determined. So the nonlocality shows its greatest impact in the approximation coefficients. Since the scaling function satisfies $\int \widetilde{\Phi}(x) \, dx = 1$, its zero order moment does not vanish. One could still choose $\widetilde{\Phi}$ so that the moments of order 1 through N vanish. It is shown in [80, p. 1419] that in such a case the resulting convolution kernel k satisfies

$$|k(s)| = O(s^{-2}) + O(s^{-N-3}).$$

It seems that this does not achieve much, since we cannot remove the leading $O(s^{-2})$ term. Nevertheless, the authors of [80] found that some scaling functions having vanishing moments lead to convolution kernels with sufficiently rapid decay for practical purposes. These scaling functions where found from one-dimensional scaling functions by the method of Definition A.8. In their reconstructions the authors of [80] also extrapolated the missing data by constant values, thus reducing cupping artifacts. While it is suggested in [80] to first compute the approximation and detail coefficients at level j and then use an inverse fast wavelet transform to obtain the approximation coefficients at level $j + 1$, numerical tests in [87] indicated that the simpler approach of directly computing the approximation coefficients at level $j+1$ yields equivalent results. We observe that this can be done without using wavelet theory, namely just by specifying the particular point spread function $e = (\widetilde{\Phi}_{j+1,0})^\vee$ in (4–6).

7. Sampling the 2D X-Ray Transform

We first consider the parallel-beam geometry in two dimensions. Our analysis of sampling and resolution will use techniques from Fourier analysis. These require both the domain of Pf as well as the sampling sets to have a group structure. In 2D we parameterize $\theta \in S^1$ by $\theta = (\cos\varphi, \sin\varphi)$, and let $\theta^{\perp} = (-\sin\varphi, \cos\varphi)$. We write $Pf(\varphi, s)$ for $Pf(\theta, s\theta^{\perp})$ and consider Pf to be a function on the group $\mathbb{T} \times \mathbb{R}$, where $\mathbb{T}$ denotes the circle group. We take the interval $[0, 2\pi)$ with addition modulo 2π as a model for $\mathbb{T}$.

Recall that for fixed φ, the values of $Pf(\varphi, s)$ for different s correspond to integrals over a collection of parallel lines. We first consider the case where Pf is measured at points

$$(\varphi_j, s_{jl}), \quad j = 0, \dots, P{-}1, \quad l \in \mathbb{Z}.$$

Since for each angle φ_j we measure integrals over a collection of parallel lines $l(\varphi_j, s_{jl})$, such an arrangement is called a parallel-beam geometry. We would like the set of points (φ_j, s_{jl}) to be a discrete subgroup of $\mathbb{T} \times \mathbb{R}$, and for practical reasons we require that more than one measurement is taken for each occurring angle φ_j. We call a sampling set which satisfies these requirements an *admissible sampling lattice* (ASL). There are several ways to parameterize such lattices [11; 13; 16]. Here we use the parametrization given in [16]. If $\mathbf{L}$ is an ASL, then there are $d > 0$ and integers N, P, such that $0 \le N < P$ and

$$\mathbf{L} = \mathbf{L}(d, N, P) =$$
$$\{(\varphi_j, s_{jl}) : \varphi_j = 2\pi j/P, \ s_{jl} = d(l + jN/P), j = 0, \dots, P{-}1; \ l \in \mathbb{Z}\}. \quad (7{-}1)$$

We see that P is the number of angles (views). For each view angle φ_j the values s_{jl}, $l \in \mathbb{Z}$, are equispaced with spacing d, hence d is the detector spacing. The parameter N characterizes an angle dependent shift of the detector array. We also see that there are P different lattices for given parameters d and P.

The most important lattices are the standard lattice

$$\mathbf{L}_S = \{(\varphi_j, s_l) : \varphi_j = 2\pi j/P, \ s_l = dl, \ j = 0, \dots, P{-}1, \ l \in \mathbb{Z}\},$$

which is obtained by letting $N = 0$, and the interlaced lattice

$$\mathbf{L}_I = \{(\varphi_j, s_{jl}) : \varphi_j = 2\pi j/P, \ s_{jl} = d(l + j/2), \ j = 0, \dots, P{-}1, \ l \in \mathbb{Z}\},$$

where P is even and $N = P/2$. We see that for the standard lattice the detector positions s_l do not change with the angle of view. For the interlaced lattice the detector array is shifted by one-half of a detector spacing when going from one angle of view to the next.

In practice one chooses $P = 2p$ for both lattices, and for the interlaced lattice one lets p be even. Then, because of the symmetry relation

$$Pf(\varphi, s) = Pf(\varphi + \pi, -s), \quad (7{-}2)$$

only the angles $\varphi_j \in [0, \pi)$ need to be measured. It can be shown that among all admissible lattices the standard and interlaced lattices are the only ones which fully exploit this symmetry [16].

We now describe the implementation of the filtered backprojection algorithm, which is based on discretizing the approximate inversion formula (4–6) with the trapezoidal rule. In two dimensions (4–6) reads

$$e * \Lambda^m f(x) = (\Lambda^m e) * f(x) = \int_0^{2\pi} \int_{\mathbb{R}} k(\langle x, \theta^{\perp} \rangle - s) P f(\varphi, s) \, ds \, d\varphi, \qquad (7\text{–}3)$$

$$k(s) = \frac{1}{4\pi} \Lambda^{m+1} P_{\theta} e(s).$$

We assume that we have sampled Pf on an admissible lattice. Discretizing (7–3) with the trapezoidal rule gives

$$e * f(x) \simeq \frac{2\pi}{P} \sum_{j=0}^{P-1} Q_j(\langle x, \theta_j^{\perp} \rangle), \qquad Q_j(t) = d \sum_l k(t - s_{jl}) P f(\varphi_j, s_{jl}),$$

with φ_j, s_{jl} as in (7–1), and $\theta_j^{\perp} = (-\sin \varphi_j, \cos \varphi_j)$. We assume that f is supported in the unit disk. Hence the sum in the discrete convolution is finite. The reconstruction is usually computed for values of x on a rectangular grid $x_{m_1 m_2} = (m_1/M_1, m_2/M_2)$, $|m_i| \le M_i$. Since computing the discrete convolution $Q_j(\langle x, \theta_j^{\perp} \rangle)$ for each occurring value of $\langle x, \theta_j^{\perp} \rangle$ would take too long, one first computes $Q_j(iH)$, $|i| \le 1/H$, and then obtains an approximation $I_H Q_j(\langle x, \theta_j^{\perp} \rangle)$ for $Q_j(\langle x, \theta_j^{\perp} \rangle)$ by linear interpolation with stepsize H. We assume that

$$H = d/(N'm) \quad \text{with } 0 < m, N' \in \mathbb{Z} \text{ and } N'N/P \in \mathbb{Z}.$$

This gives $H = d/m$ for the standard lattice ($N' = 1$) and $H = d/(2m)$ for the interlaced lattice ($N' = 2$). Then the effect of interpolating the convolution is the same as replacing the kernel k with the piecewise linear function $I_H k$ which interpolates k at the points Hl, $l \in \mathbb{Z}$; see, e.g., [11, p. 84]. Hence the algorithm computes the function

$$f_R(x) = \frac{2\pi}{P} \sum_{j=0}^{P-1} I_H Q_j(\langle x, \theta_j^{\perp} \rangle)$$

$$= \frac{2\pi d}{P} \sum_{j=0}^{P-1} \sum_{l \in \mathbb{Z}} I_H k \left(\langle x, \theta_j^{\perp} \rangle - s_{jl} \right) P f(\varphi_j, s_{jl}). \qquad (7\text{–}4)$$

If e is not radial, k and $I_H k$ will depend on θ, which is not explicitly reflected in our notation. Matlab source code implementing (7–4) for the standard lattice is provided for illustrative purposes (see Appendix B).

In practice one chooses a basic point spread function e_1 and then controls the resolution by using a scaled version

$$e(x) = e_r(x) = r^{-2} e_1(x/r). \qquad (7\text{–}5)$$

The corresponding kernels scale as

$$k_r(s) = r^{-2-m} k_1(s/r). \tag{7-6}$$

Examples for point spread functions and kernels are as follows. A popular choice for global tomography ($m = 0$) is the Shepp–Logan kernel [62, p. 111]:

$$k_1(s) = \frac{1}{2\pi^3} \frac{(\pi/2) - s \sin s}{(\pi/2)^2 - s^2}. \tag{7-7}$$

This kernel is bandlimited with bandwidth 1, i.e., the Fourier transform

$$\widehat{k_1}(\sigma) = \frac{1}{2}(2\pi)^{-3/2}|\sigma| \frac{\sin(\pi\sigma)}{\pi\sigma} \chi_{[-1,1]}(\sigma)$$

vanishes for $|\sigma| > 1$. It follows that the scaled kernel $k_r(s) = r^{-2} k_1(s/r)$ is bandlimited with bandwidth $b = 1/r$, and the same is true for the corresponding point-spread function e_r and hence for $e_r * f(x)$.

Since the kernel (7–7) does not have compact support, it is not useful for local tomography. There we start with a point spread function

$$e_1(x) = \begin{cases} C(1 - |x|^2)^{\alpha+1/2} & \text{for } |x| < 1, \\ 0 & \text{for } |x| \geq 1, \end{cases} \tag{7-8}$$
$$C = \Gamma(\alpha + 5/2)/(\pi\Gamma(\alpha + 3/2));$$

see [14, p. 482]. The corresponding kernel for computing Λf (i.e., $m = 1$) is

$$K_1(s) = \begin{cases} \dfrac{-1}{4\pi} \dfrac{d^2}{ds^2} P_\theta e_1(s) = \dfrac{2\Gamma(\alpha+5/2)}{\sqrt{\pi}\Gamma(\alpha+1)}(1-s^2)^{\alpha-1}(1-(2\alpha+1)s^2) & \text{for } |s| < 1, \\ 0 & \text{for } |s| \geq 1. \end{cases}$$

Here we used the fact that $\Lambda^2 = -d^2/ds^2$ in one dimension. Now the kernel is no longer bandlimited, but has compact support. The scaled kernels $K_r(s) = r^{-3} K_1(s/r)$ vanish for $|s| > r$. The kernel for global tomography generated by the point spread function (7–8) has a complicated analytic expression but a quickly convergent series expansion [88].

Discretization of (4–8) yields the filtered backprojection algorithm for the fan-beam sampling geometry. Recall that f is supported in the unit disk. Let $R > 1$, $a = R(\cos\alpha, \sin\alpha)$, $\theta = -(\cos(\alpha-\beta), \sin(\alpha-\beta))$, and $x - a = -|x-a| \times (\cos(\alpha-\gamma), \sin(\alpha-\gamma))$. Writing $Df(\alpha, \beta)$ for $D_a f(\theta)$, (4–8) becomes

$$e_r * \Lambda^m f(x) = R \int_0^{2\pi} \int_{-\pi/2}^{\pi/2} Df(\alpha, \beta) \cos(\beta) k_r(|x - a| \sin(\gamma-\beta)) \, d\beta \, d\alpha. \tag{7-9}$$

In order to evaluate the inner integral efficiently, a "homogeneous approximation" [46] is needed. It follows from (4–7) and (7–6) that

$$k_r(|x - a| \sin(\gamma-\beta)) = |x - a|^{-2-m} k_c(\sin(\gamma-\beta)), \quad c = r/|x - a|.$$

The approximation consists in replacing $c = r/|x - a|$ by a constant independent of x and a. This gives

$$e_r * \Lambda^m f(x) \simeq R \int_0^{2\pi} |x - a|^{-2-m} \int_{-\pi/2}^{\pi/2} Df(\alpha, \beta) \cos(\beta) k_c(\sin(\gamma - \beta)) \, d\beta \, d\alpha.$$

From here we can proceed as before by discretizing with the trapezoidal rule and inserting an interpolation step. The standard sampling lattice for the fan-beam geometry has the form

$$\mathbf{L}_{SF} = \{(\alpha_j, \beta_l) : \alpha_j = 2\pi j/p \text{ for } j = 0, \ldots, p - 1,$$
$$\beta_l = l \arcsin(1/R)/q \text{ for } l = -q, \ldots, q - 1\}. \quad (7\text{--}10)$$

The reconstruction of $\Lambda^{-1} f$ is not unstable, so convolution with e_r is not needed. One can directly discretize the formula

$$\Lambda^{-1} f(x) = (R/4\pi) \int_0^{2\pi} |x - a|^{-1} Df(\alpha, \gamma) \cos \gamma \, d\alpha,$$

which comes from letting $e_b \to \delta$ in (7–9).

In order to further analyze the parallel-beam algorithm we use Shannon sampling theory. We begin with some definitions. We define the Fourier transform of a function g with domain $\mathbb{T} \times \mathbb{R}$ by

$$\hat{g}(k, \sigma) = \frac{1}{2\pi} \int_0^{2\pi} \int_{\mathbb{R}} g(\varphi, s) e^{-i(k\varphi + \sigma s)} \, ds \, d\varphi, \quad (k, \sigma) \in \mathbb{Z} \times \mathbb{R}.$$

The corresponding inverse Fourier transform is given by

$$\tilde{G}(\varphi, s) = \frac{1}{2\pi} \int_{\mathbb{Z} \times \mathbb{R}} G(\zeta) e^{i\langle z, \zeta \rangle} \, d\zeta \quad (\text{where } z = (\varphi, s))$$
$$= \frac{1}{2\pi} \sum_{k \in \mathbb{Z}} \int_{\mathbb{R}} G(k, \sigma) e^{-i(k\varphi + \sigma s)} \, d\sigma. \quad (7\text{--}11)$$

The reciprocal lattice $\mathbf{L}^{\perp} \subset \mathbb{Z} \times \mathbb{R}$ is defined as

$$\mathbf{L}^{\perp}(d, N, P) = \begin{pmatrix} P & -N \\ 0 & 2\pi/d \end{pmatrix} \mathbb{Z}^2.$$

For $g \in C_0^\infty(\mathbb{T} \times \mathbb{R})$, $\mathbf{L} = \mathbf{L}(d, N, P)$ an ASL, $K \subset \mathbb{Z} \times \mathbb{R}$ compact define

$$Sg(z) = \frac{d}{P} \sum_{y \in \mathbf{L}} g(y) \tilde{\chi}_K(z - y), \quad z \in \mathbb{T} \times \mathbb{R},$$

where $\tilde{\chi}_K$ is the inverse Fourier transform of the characteristic function of K. We may view Sg as an approximation of g computed from sampled values of g on the lattice $\mathbf{L}$. The following classical sampling theorem gives an error estimate for this approximation:

THEOREM 7.1. *Let* $g \in C_0^\infty(\mathbb{T} \times \mathbb{R})$, $\mathbf{L}$ *an ASL and* K *be a compact subset of* $\mathbb{Z} \times \mathbb{R}$ *such that the translates* $K + \eta, \eta \in \mathbf{L}^\perp$ *are disjoint. Then*

$$|g(z) - Sg(z)| \leq \pi^{-1} \int_{(\mathbb{Z} \times \mathbb{R}) \setminus K} |\hat{g}(\zeta)| \, d\zeta.$$

This result is an adapted version of the multidimensional sampling theorem of Petersen and Middleton [69]. For a proof see, for example, [62, p. 62] or [11, Theorem 2.2].

If $\mathrm{supp}(\hat{g}) \subseteq K$, then $g = Sg$, i.e., g can be recovered exactly from its samples on the lattice $\mathbf{L}$.

In order for Sg to be close to g, the set K should be such that $\hat{g}$ is concentrated in K. The "sampling condition" that the translates $K + \eta$, $\eta \in \mathbf{L}^\perp$ be disjoint requires the reciprocal lattice $\mathbf{L}^\perp$ to be sufficiently sparse, and therefore the sampling lattice $\mathbf{L}$ to be sufficiently dense.

A suitable set K for sampling the 2D x-ray transform was given by Natterer [62] based on results by Lindgren and Rattey [81]:

THEOREM 7.2. [62, p. 71] *For* $b > 0$ *and* $0 < \vartheta < 1$ *let*

$$K_0(\vartheta, b) = \left\{ (k, \sigma) \in \mathbb{Z} \times \mathbb{R} : |\sigma| < b, \ |k| < \vartheta^{-1} \max(|\sigma|, (1 - \vartheta)b) \right\}. \quad (7\text{-}12)$$

Let $f \in C_0^\infty(\Omega)$. *Then*

$$\int_{(\mathbb{Z} \times \mathbb{R}) \setminus K_0} |\widehat{Pf}(\zeta)| \, d\zeta \leq \frac{8}{\pi^2 \vartheta} \int_{|\xi| > b} |\hat{f}(\xi)| \, d\xi + \|f\|_{L_1} \eta(\vartheta, b), \quad (7\text{-}13)$$

where $\eta(\vartheta, b)$ *decreases exponentially with* b, *satisfying an estimate*

$$0 \leq \eta(\vartheta, b) \leq C(\vartheta) e^{-\lambda(\vartheta) b}$$

with constants $C(\vartheta), \lambda(\vartheta) > 0$.

Usually the parameter ϑ is chosen close to 1. The parameter b plays the role of a cut-off frequency. If $|\hat{f}(\xi)|$ is sufficiently small for $|\xi| > b$ then the right-hand side of (7-13) will be small. In this case Theorem 7.1 imposes the condition that the sampling lattice $\mathbf{L}$ be such that the translated sets $K_0(\vartheta, b) + \eta$, $\eta \in \mathbf{L}^\perp$ are disjoint. In terms of the lattice parameters d, N and P these conditions are as follows:

For the standard lattice $N = 0$, and the reciprocal lattice $\mathbf{L}^\perp$ equals

$$\mathbf{L}^\perp = \{(Pk_1, 2\pi k_2/d) : k_1, k_2 \in \mathbb{Z}\}.$$

For reasons of efficiency as discussed above we let $P = 2p$ be even. The translated sets $K_0(\vartheta, b) + \eta$, $\eta \in \mathbf{L}^\perp$ are disjoint if and only if

$$d < \pi/b \quad \text{and} \quad p > b/\vartheta, \quad P = 2p. \quad (7\text{-}14)$$

For the interlaced lattice we again let $K = K_0(\vartheta, b)$ as in (7-12). For this lattice $P = 2p$ and $N = P/2 = p$. We always let p be even, so that because of the symmetry relation (7-2) only the angles $\varphi_j \in [0, \pi)$ need to be measured.

The reciprocal lattice is $\mathbf{L}^\perp = \{(p(2k_1-k_2), 2\pi k_2/d) : k_1, k_2 \in \mathbb{Z}\}$. It turns out [62; 11] that the sets $K_0(\vartheta, b) + \eta$, $\eta \in \mathbf{L}^\perp$ are disjoint if either the conditions (7–14) are satisfied, or if

$$\frac{\pi}{b} < d \le \frac{2\pi}{b} \quad \text{and} \quad p > \max\left(\frac{2\pi}{\vartheta d}, \frac{(2-\vartheta)b}{\vartheta}\right), \quad p \text{ even}, \quad P = 2p. \quad (7\text{–}15)$$

We see that the interlaced lattice allows for a maximal detector spacing of $d = 2\pi/b$ which is twice as large as the maximum allowed for standard lattice, with only a moderate increase in p. Sampling conditions for a general admissible sampling lattice have been given in [13]. We see that both (7–14) and (7–15) require p to be greater than the cut-off frequency b, which corresponds to condition (b) in Remark 3.4 on avoiding the effects of nonuniqueness.

It remains to investigate if the theoretically superior resolution of the interlaced lattice can be realized in practice. In principle there are two obvious approaches: One could first interpolate the missing data to a denser lattice and then use any reconstruction algorithm. This approach has been successfully tried in [10], but we will not discuss it here. The second approach would be to reconstruct directly from interlaced data. It turns out that the filtered backprojection algorithm is very suitable for this purpose [44; 11; 16]. We have the following error estimate, which extends the results of [44] and [11].

THEOREM 7.3 [16]. *Let e be radial and sufficiently smooth, let $f \in C_0^\infty(\mathbb{R}^2)$ be supported in the unit disk and let the sets $K_0(\vartheta, b) + \eta$, for $\eta \in \mathbf{L}^\perp$, be disjoint. Then*

$$f_R(x) = G_H * e * \Lambda^m f(x) + \sum_{i=1}^{4} E_i(x),$$

$$\widehat{G}_H(\xi) = (2\pi)^{-1} \operatorname{sinc}^2(H|\xi|/2)\chi_1(|\xi|/b),$$

$$|E_1(x)| \le c\|\hat{k}\|_\infty \int_{|\xi|>b} |\hat{f}(\xi)|\, d\xi,$$

$$|E_2(x)| \le c\|f\|_\infty \int_{|\sigma|>b} |\hat{k}(\sigma)|\, d\sigma,$$

$$|E_3(x)| \le (2\pi)^{3/2} \sup_\theta \int_{-b}^{b} (1 - \operatorname{sinc}^2(H\sigma/2)) \, |\hat{k}(\sigma)| \sum_{l\in\mathbb{Z}} \left|\hat{f}\left(\left(\sigma + \frac{2\pi l}{d}\right)\theta\right)\right| d\sigma,$$

$$|E_4(x)| \le c\|f\|_\infty \|\hat{k}\|_\infty \, \eta(\vartheta, b).$$

Here $\operatorname{sinc} x = (\sin x)/x$. The proof ([16]) is somewhat technical and will not be given here. However, it is worthwhile to note that apart from the interpolation step this is an estimate for the error of numerical integration by the trapezoidal rule. The estimate for this error is based on the Poisson summation formula for $\mathbb{T} \times \mathbb{R}$. This approach was first applied in the present context by Kruse [44].

We will discuss the origin and importance of the four error terms. The error E_1 is the so-called aliasing error stemming from the fact that f is not bandlimited, since it has compact support. If the cut-off frequency b is chosen sufficiently large, E_1 will be small. The sampling conditions then require that the number of data available is commensurate with b. The error E_2 is present when k is not bandlimited with bandwidth b. In global tomography, i.e., when $m = 0$, one can chose e and k to be b-bandlimited, so that E_2 vanishes. In local tomography one wishes k to have compact support, so k cannot be bandlimited.

The error E_3 is caused by the interpolation step and usually not a concern when using the standard lattice. This can be explained as follows ([11]): Consider the common parameter choice $d = H = \pi/b$. Since $\hat{f}(\xi)$ is assumed to be small for $|\xi| > b$, only the term with $l = 0$ in the sum will be significant, i.e., we have for $|\sigma| \le b$

$$\sum_{l \in \mathbb{Z}} \left| \hat{f}((\sigma + 2\pi l/d)\theta) \right| = \sum_{l \in \mathbb{Z}} \left| \hat{f}((\sigma + 2bl)\theta) \right| \simeq |\hat{f}(\sigma\theta)|.$$

Usually the density function f is non-negative so that $|\hat{f}(\sigma\theta)|$ has a sharply peaked maximum at $\sigma = 0$ and is very small for $|\sigma|$ close to b. In such a case the error E_2 will be small since the factor $1 - \text{sinc}^2(H\sigma/2)$ is small exactly where $|\hat{f}(\sigma\theta)|$ is large.

The error E_3 is of much greater concern when the interlaced lattice is used. Consider the choice of parameters $d = 2\pi/b$, $H = \pi/b$. Now the sum over l in the estimate for E_3 may have three significant terms for $|\sigma| < b$:

$$\sum_{l \in \mathbb{Z}} |\hat{f}((\sigma + 2\pi l/d)\theta)| = \sum_{l \in \mathbb{Z}} |\hat{f}((\sigma + bl)\theta)| \simeq |\hat{f}((\sigma - b)\theta)| + |\hat{f}(\sigma\theta)| + |\hat{f}((\sigma + b)\theta)|.$$

As discussed before the contribution of the term $\hat{f}(\sigma\theta)$ is largely cancelled by the factor $(1 - \text{sinc}^2(H\sigma/2))$. However, this is not the case for the other two terms. E.g., let σ be close to b. Then, assuming again that $\hat{f}$ is large near the origin, $|\hat{f}((\sigma - b)\theta)|$ will be large and is not attenuated by the factor $(1 - \text{sinc}^2(H\sigma/2))$ which will be close to 1. Therefore we expect considerable reconstruction errors for this choice of parameters. That this is indeed the case has been demonstrated for global tomography in [44; 11]. Hence when using the interlaced lattice one should choose $H \ll b$, so that $(1 - \text{sinc}^2(H\sigma/2))$ is small for $|\sigma| < b$. Typical choices in practice are $H = \pi/(16b)$ or smaller. Choosing $H < \pi/b$ has also a cosmetic side effect. If a b-bandlimited convolution kernel is used whose Fourier transform has a jump discontinuity at $|\sigma| = b$ (e.g., a scaled version of the Shepp–Logan kernel (7–7)), then ringing artifacts are caused by this discontinuity. In case of the standard lattice with the parameter choice $d = H = \pi/b$ these artifacts are practically removed by the additional smoothing from the interpolation. For smaller H this effect is lost. In this case the Fourier transform of k should taper off continuously to zero if smooth images are desired [11].

Another effect of the interpolation is the additional filtering with G_H. Since this alters only the higher frequencies, it is usually not a concern. In any case, the effect can be eliminated by choosing very small H.

The last error E_4 decreases exponentially with b, as indicated by the notation $\eta(\vartheta, b)$. Explicit estimates are as follows [16]. Let

$$\beta = \left(1 - \vartheta^2 |x|^2\right)^{3/2}.$$

For the standard lattice we have

$$|E_4(x)| \leq c\|f\|_\infty \|\hat{k}\|_\infty b \frac{e^{-\beta b/\vartheta}}{1 - e^{-\beta}}.$$

For the interlaced lattice we let $b' = (1 - \vartheta)b/\vartheta$, and obtain with β as above

$$|E_4(x)| \leq c\|f\|_\infty \|\hat{k}\|_\infty b\vartheta e^{-\beta b'} \left(\frac{1 + b'}{1 - e^{-\beta}} + \frac{e^{-\beta}}{(1 - e^{-\beta})^2}\right).$$

In both cases the error decays exponetially with b, but in case of $\vartheta|x|$ close to 1, when β is close to zero, the above estimates indicate that the error should be larger in case of the interlaced lattice due to the term involving $(1 - e^{-\beta})^{-2}$. This effect has been observed in [11]. It causes a thin ring artifact in the region $|x| \simeq 1$, i.e., at the boundary of the reconstruction region. It can be eliminated by choosing a smaller value for ϑ, i.e., by increasing the number of views p; compare equations (7–14), (7–15).

Numerical experiments for both global and local tomography, with simulated as well as real data [44; 11; 10; 12; 16] show that the higher efficiency of the interlaced lattice can at least be partly realized in practice. However, there are also some drawbacks. There is a somewhat reduced stability because of inaccurate convolutions. In case of the interlaced lattice the stepsize $d = 2\pi/b$ is not small enough to allow accurate computation of the convolutions. Because of the truly two-dimensional nature of the numerical integration, these errors cancel out during the discrete backprojection step. The sensitivity with regard to the interpolation stepsize H can be understood as coming from a disturbance of these cancellations by the additional interpolation. A second drawback is a requirement that the sampling condition with respect to the number of views P has to be strictly observed. The aliasing caused by violating this condition is usually quite moderate in case of the standard lattice but much more severe for the interlaced lattice. This can be easily seen from the pattern in which the translated sets $K_0 + \eta$ begin to overlap when the sampling condition for p is violated. Hence the interlaced lattice seems to be most useful when the detector spacing is the main factor limiting resolution.

Good reconstructions from the interlaced lattice can also be obtained by using the direct algebraic reconstruction algorithm [43], or by increasing the amount of data through interpolation according to the sampling theorem [10]. Results for the fan beam geometry can be found in [63; 64; 67]. As we have seen, the interpolation step can introduce significant errors in certain cases. It has also been shown [64] that the interpolation can be avoided by chosing the points x where the reconstruction is computed on a polar grid rather than on a rectangular grid, and interchanging the order of the two summations. This algorithm should work well for the interlaced lattice [100] and is particularly beneficial in case of the fan-beam sampling geometry [64], since the method also avoids the homogeneous approximation, whose influence on the reconstruction is difficult to estimate.

Appendix A. Basic Facts about Wavelets

We give a brief introduction to multidimensional biorthogonal wavelets. The discussion follows [87] and is based on the presentation in [96] for multidimensional orthonormal wavelets. For more details on wavelets and filter banks, see [7] or [95], for example.

DEFINITION A.1. A *lattice* Γ is a discrete subgroup of $\mathbb{R}^n$ given by integral linear combinations of a vector space basis $\{v_1, \ldots, v_n\}$ of $\mathbb{R}^n$.

DEFINITION A.2. Let Γ be a lattice and M be an $n \times n$ matrix such that $M\Gamma \subset \Gamma$ and that all eigenvalues λ of M satisfy $|\lambda| > 1$. M is called the *dilation matrix*. Let $m = |\det M|$. A *multiresolution analysis* with scaling function ϕ, $\int \phi(t)\, dt = 1$, is a sequence of subspaces V_j of $L_2(\mathbb{R}^n)$, $j \in \mathbb{Z}$, satisfying the following conditions:

1. $V_j \subset V_{j+1}$, $\bigcap V_j = \{0\}$ and $\overline{\bigcup V_j} = L^2(\mathbb{R}^n)$.
2. $f(t) \in V_j$ if and only if $f(Mt) \in V_{j+1}$.
3. $f(t) \in V_0$ if and only if $f(t - k) \in V_0$ for $k \in \Gamma$.
4. $\{\phi(t - k), k \in \Gamma\}$ is a Riesz basis of V_0.

CONVENTION. For notational convenience we set

$$f_{j,k}(t) := m^{j/2} f(M^j t - k), \quad \text{for } k \in \Gamma, \ j \in \mathbb{Z}. \tag{A-1}$$

It follows from the definition of a multiresolution analysis that $\{\phi_{j,k}(t), k \in \Gamma\}$ is a Riesz basis of V_j.

DEFINITION A.3. Consider a multiresolution analysis with lattice Γ and dilation matrix M. For $j \in \mathbb{Z}$, let W_0 be such that V_1 is the direct sum of V_0 and W_0. Assume there are $\psi^1, \ldots, \psi^{m-1} \in W_0$ such that $\{\psi_{0,k}^\mu : \mu = 1, \ldots, m-1, \ k \in \Gamma\}$ is a Riesz basis of W_0. The ψ^μ are called *wavelets*.

For $j \in \mathbb{Z}$, let W_j be the subspace with Riesz basis $\{\Psi_{j,k}^\mu, k \in \Gamma, \mu = 1, \ldots, m-1\}$. It follows that $V_{j+1} = V_j \oplus W_j$, a direct sum but not necessarily orthogonal.

DEFINITION A.4. Let V_j, $\tilde{V}_j$ be two multiresolution analyses corresponding to the same lattice Γ and dilation matrix M. Let ϕ and ψ^μ, for $\mu = 1, \ldots, m-1$, be the scaling function and wavelets corresponding to V_j. Let $\tilde{\phi}$ and $\tilde{\psi}^\mu$, for $\mu = 1, \ldots, m-1$, be the scaling function and wavelets corresponding to $\tilde{V}_j$. The multiresolution analyses are called *biorthogonal* if the following conditions hold for $j, j' \in \mathbb{Z}$, $\mu, \mu' = 1, \ldots, m-1$, and $k, k' \in \Gamma$:

(i) $\langle \psi_{j,k}^{\mu}, \tilde{\psi}_{j',k'}^{\mu'} \rangle = \delta(j,j')\delta(\mu,\mu')\delta(k,k')$.

(ii) $\langle \tilde{\psi}_{j,k}^{\mu}, \phi_{j,k'} \rangle = \langle \tilde{\phi}_{j,k}, \psi_{j,k'}^{\mu} \rangle = 0$.

(iii) $\langle \phi_{j,k}, \tilde{\phi}_{j,k'} \rangle = \delta(k,k')$.

For $n = 1$, $M = 2$, $\Gamma = \mathbb{Z}$ and $\phi = \tilde{\phi}$, multiresolution analysis becomes the familiar one-dimensional, orthonormal case.

Since $\phi \in V_0 \subset V_1$, and $\{\phi_{1,k}, k \in \Gamma\}$ is a basis for V_1, there are coefficients $F_0(k)$, $k \in \Gamma$ such that

$$\phi(t) = m^{1/2} \sum_{k \in \Gamma} F_0(k)\phi_{1,k}(t)$$

From condition (iii) above, it follows that $F_0(k) = m^{-1/2}\langle \phi, \tilde{\phi}_{1,k} \rangle$. Similarly, since $\tilde{\phi} \in \tilde{V}_0 \subset \tilde{V}_1$, and $\{\tilde{\phi}_{1,k}, k \in \Gamma\}$ is a basis for $\tilde{V}_1$,

$$\tilde{\phi}(t) = m^{1/2} \sum_{k \in \Gamma} H_0(k)\tilde{\phi}_{1,k}(t)$$

where $H_0(k) = m^{-1/2}\langle \tilde{\phi}, \phi_{1,k} \rangle$. The above equations are called dilation equations. Similarly, the wavelets ψ^{μ}, $\tilde{\psi}^{\mu}$ must satisfy so-called wavelet equations:

$$\psi^{\mu}(t) = m^{1/2} \sum_{k \in \Gamma} F_{\mu}(k)\phi_{1,k}(t), \qquad \tilde{\psi}^{\mu}(t) = m^{1/2} \sum_{k \in \Gamma} H_{\mu}(k)\tilde{\phi}_{1,k}(t),$$

where $\mu = 1, \ldots, m-1$, $F_{\mu}(k) = m^{-1/2}\langle \psi^{\mu}, \tilde{\phi}_{1,k} \rangle$, and $H_{\mu}(k) = m^{-1/2}\langle \tilde{\psi}^{\mu}, \phi_{1,k} \rangle$.

The following lemma and theorems show how to decompose a function f into its wavelet coefficients and how to reconstruct f if its wavelet coefficients are known.

LEMMA A.5. *For $j \in \mathbb{Z}$ and $\mu = 1, \ldots, m-1$, we have:*

$$\phi_{j,l}(t) = m^{1/2} \sum_{k \in \Gamma} F_0(k)\phi_{j+1,Ml+k}(t), \qquad \text{(A-2)}$$

$$\tilde{\phi}_{j,l}(t) = m^{1/2} \sum_{k \in \Gamma} H_0(k)\tilde{\phi}_{j+1,Ml+k}(t), \qquad \text{(A-3)}$$

$$\psi_{j,l}^{\mu}(t) = m^{1/2} \sum_{k \in \Gamma} F_{\mu}(k)\phi_{j+1,Ml+k}(t), \qquad \text{(A-4)}$$

$$\tilde{\psi}_{j,l}^{\mu}(t) = m^{1/2} \sum_{k \in \Gamma} H_{\mu}(k)\tilde{\phi}_{j+1,Ml+k}(t). \qquad \text{(A-5)}$$

PROOF. This follows directly from the dilation and wavelet equations. $\square$

THEOREM A.6 (FAST WAVELET TRANSFORM). *Let $j \in \mathbb{Z}$ and $f \in V_{j+1}$. For $k \in \Gamma$ and $\mu = 1, \ldots, m-1$ define the approximation coefficients as*

$$\tilde{A}_{j,k} = \langle f, \tilde{\phi}_{j,k} \rangle = (f * (\overline{\tilde{\phi}_{j,0}})^{\vee})(M^{-j}k)$$

and the detail coefficients as

$$\tilde{D}_{j,k}^{\mu} = \langle f, \tilde{\psi}_{j,k}^{\mu} \rangle = (f * (\overline{\tilde{\psi}_{j,0}^{\mu}})^{\vee})(M^{-j}k).$$

Then

$$\tilde{A}_{j,k} = m^{1/2} \sum_{l \in \Gamma} \overline{H_0(l - Mk)} \tilde{A}_{j+1,l} \qquad \text{(A-6)}$$

and

$$\tilde{D}^{\mu}_{j,k} = m^{1/2} \sum_{l \in \Gamma} \overline{H_\mu(l - Mk)} \tilde{A}_{j+1,l}. \qquad \text{(A-7)}$$

PROOF. Consider the following expansion of f:

$$f(t) = \sum_{l \in \Gamma} \tilde{A}_{j+1,l} \phi_{j+1,l}(t) \qquad \text{(A-8)}$$

For $\tilde{A}_{j,k}$, take an inner product of (A-8) with $\tilde{\phi}_{j,k}$. Use (A-3) and biorthogonality to get (A-6).
For $\tilde{D}^{\mu}_{j,k}$, take an inner product of (A-8) with $\tilde{\psi}^{\mu}_{j,k}$. Use (A-5) and biorthogonality to get (A-7). $\qquad \square$

THEOREM A.7 (INVERSE FAST WAVELET TRANSFORM). *Under the hypothesis of Theorem A.6*

$$\tilde{A}_{j+1,l} = m^{1/2} \sum_{k \in \Gamma} \left(F_0(l - Mk) \tilde{A}_{j,k} + \sum_{\mu=1}^{m-1} F_\mu(l - Mk) \tilde{D}^{\mu}_{j,k} \right) \qquad \text{(A-9)}$$

PROOF. Consider the following expansions of f:

$$f(t) = \sum_{k \in \Gamma} \tilde{A}_{j+1,k} \phi_{j+1,k}(t)$$

$$= \sum_{k \in \Gamma} \tilde{A}_{j,k} \phi_{j,k}(t) + \sum_{\mu=1}^{m-1} \sum_{k \in \Gamma} \tilde{D}^{\mu}_{j,k} \psi^{\mu}_{j,k}(t). \qquad \text{(A-10)}$$

The second expansion comes from writing $f \in V_{j+1} = V_j \oplus W_j$ as a sum of elements of V_j and W_j. To get $\tilde{A}_{j+1,l}$, take an inner product of (A-10) with $\tilde{\phi}_{j+1,l}$. Use (A-2), (A-4) and biorthogonality to get (A-9). $\qquad \square$

An easy way to obtain wavelets in $\mathbb{R}^n$ is to use a tensor product construction with the wavelets in $\mathbb{R}$. We will look specifically at the two-dimensional case. Define the lattice as $\Gamma = \mathbb{Z}^2$, and the dilation matrix $M = 2I$, where I denotes the identity matrix. Since $|\det M| = 4 = m$, one can expect 3 wavelets and 1 scaling function. Let the spaces V_j, W_j be the chosen one-dimensional multiresolution analysis with scaling function ϕ, and wavelet ψ. The coefficients for the dilation and wavelet equation are $F_\mu(k)$, $\mu = 0, 1$. Constructing the two-dimensional scaling function and wavelets by taking products of the one-dimensional functions leads to the following definition.

DEFINITION A.8. From a one-dimensional scaling function $\phi(x)$ and its corresponding wavelet $\psi(x)$, *two-dimensional separable wavelets* are defined for $(x, y) \in \mathbb{R}^2$:

$$\Phi(x, y) = \phi(x)\phi(y),$$

$$\Psi^1(x, y) = \phi(x)\psi(y), \quad \Psi^2(x, y) = \psi(x)\phi(y), \quad \Psi^3(x, y) = \psi(x)\psi(y).$$

Consider a biorthogonal pair of one-dimensional multiresolution analyses. Recall for separable, two-dimensional wavelets, $M = 2I$, $m = 4$ and $\Gamma = \mathbb{Z}^2$. Let V_j and W_j be the multiresolution analysis with scaling function ϕ, wavelet ψ, and coefficients F_0, F_1 for the dilation and wavelet equations. Let $\tilde{V}_j$ and $\tilde{W}_j$ be the multiresolution analysis with scaling function $\tilde{\phi}$ wavelet $\tilde{\psi}$ and coefficients H_0, H_1 for the dilation and wavelet equations.

We want to rewrite the fast wavelet transform and inverse fast wavelet transform for the case of two-dimensional separable wavelets. Let $\mathbf{F}_\mu(\mathbf{k})$, $\mu = 0, 1, 2, 3$ be the coefficients in the dilation and wavelet equations for the separable wavelets. It is easy to verify that

$$\mathbf{F}_0(\mathbf{k}) = F_0(k_1)F_0(k_2), \qquad \mathbf{F}_1(\mathbf{k}) = F_0(k_1)F_1(k_2),$$
$$\mathbf{F}_2(\mathbf{k}) = F_1(k_1)F_0(k_2), \qquad \mathbf{F}_3(\mathbf{k}) = F_1(k_1)F_1(k_2),$$

where $\mathbf{k} = (k_1, k_2)$. Similarly for the $\mathbf{H}_\mu(\mathbf{k})$, $\mu = 0, 1, 2, 3$. Thus, equation (A–6) becomes:

$$\tilde{A}_{j,k} = 2 \sum_{l \in \Gamma} \overline{\mathbf{H}_0(l - 2k)} \tilde{A}_{j+1,l}$$

and equation (A–7) becomes

$$\tilde{D}^\mu_{j,k} = 2 \sum_{l \in \Gamma} \overline{\mathbf{H}_\mu(l - 2k)} \tilde{A}_{j+1,l} \quad \text{for } \mu = 1, 2, 3.$$

The inverse wavelet transform, equation (A–9), becomes

$$\tilde{A}_{j+1,n} = m^{1/2} \sum_{k \in \Gamma} \left(\mathbf{F}_0(n - Mk)\tilde{A}_{j,k} + \sum_{\mu=1}^{3} \mathbf{F}_\mu(n - Mk)\tilde{D}^\mu_{j,k} \right)$$

$$= 2 \sum_{k_1, k_2 \in \mathbb{Z}} \left(F_0(n_1 - 2k_1)F_0(n_2 - 2k_2)\tilde{A}_{j,k} + F_0(n_1 - 2k_1)F_1(n_2 - 2k_2)\tilde{D}^1_{j,k} \right.$$

$$\left. + F_1(n_1 - 2k_1)F_0(n_2 - 2k_2)\tilde{D}^2_{j,k} + F_1(n_1 - 2k_1)F_1(n_2 - 2k_2)\tilde{D}^3_{j,k} \right).$$

Recall that the approximation coefficients are a convolution of f with $\overline{\tilde{\phi}_{j,0}}$: $\tilde{A}_{j,k} = (f * (\overline{\tilde{\phi}_{j,0}})^\vee)(M^{-j}k)$. We would like to consider, for sufficiently large j, the approximation coefficients as an approximation for f at a particular point. Notice that for $M = 2I$, $M^j x = 2^j x$, and $m = 2^n$. From Real Analysis, the following lemma holds.

LEMMA A.9. *Let f be continuous, $g \in L^1(\mathbb{R}^n)$ with $\int g\,dx = 1$, g is bounded and has compact support. Then for all $x \in \mathbb{R}^n$*

$$f(x) = \lim_{j \to \infty} 2^{jn} \int f(x+y)\overline{g(2^j y)}\,dy = \lim_{j \to \infty} 2^{j/2} f * (\overline{g_{j,0}})^\vee(x)$$

PROOF. Let $t = 2^{-j}$. Then we have a special case of [20, Theorem 8.15, p. 235]. $\square$

Thus, for j sufficiently large, we get an approximation for f:

$$2^{j/2} f * (\overline{\tilde{\phi}_{j,0}})^\vee(2^{-j}k) \approx f(2^{-j}k)$$

or, with $n = 2$,

$$2^{j/2} \tilde{A}_{j,k} \approx f(2^{-j}k).$$

Appendix B. Matlab Source Code

We have made available online at http://www.msri.org/publications/books/Book47/faridani a number of Matlab M-files related to the algorithms described here. *This source code is provided for illustrative purposes only and comes without warranties of any kind.*

First there is an implementation of the parallel-beam filtered backprojection algorithm for the standard lattice. The main file for this is fbp.m, and it is supplemented by three function M-files: Rad.m computes line integrals for a mathematical phantom consisting of ellipses, slkernel.m computes the discrete Shepp–Logan convolution kernel (7–7), and window3.m allows one to view the reconstructed image. It is automatically called at the end of the reconstruction. (However, with the example phantom — the well-known Shepp–Logan phantom — given in fbp.m the picture shown does not display the most interesting details. It is better to call window3 again with the parameters window3(−0.07,0.07,roi,P).

Also included at the same address are the files fdk.m (a simple implementation of the Feldkamp–Davis–Kress algorithm) and Divray.m (an implementation of the transform discussed in Problem 9 below).

Appendix C. Some Exercises

Problem 1: Let $f(x)$ be the characteristic function of an ellipse with center (x_0, y_0), half-axes of length a and b, respectively, such that the axis of length $2a$ makes an angle ψ with the x-axis when measured counterclockwise starting from the x-axis. Compute the x-ray transform $Pf(\theta, s)$. Parameterize the unit vector θ with its polar angle φ, i.e., $\theta = (\cos\varphi, \sin\varphi)$.

Problem 2: Compute the Fourier transform of the characteristic function of the unit disk in $\mathbb{R}^2$. Hint: Use polar coordinates and formula (3.16) in [62, p. 197].

Problem 3: Using the Matlab code described in Appendix B for the filtered backprojection algorithm (with the Shepp Logan phantom), perform the following experiments.

(a) Run the program for the following values of p and q, leaving the other parameters unchanged: $q = 16, p = 50$; $q = 32, p = 100$; $q = 64, p = 200$; $q = 128, p = 400$. For each case plot a crossection along the horizontal line of pixels closest to the line $y = -0.605$ which passes through the centers of the three small ellipses. Compare the resolution for the various parameter choices.

(b) Fix the parameter b in the program at the value 64π. Theory suggests that the choice $q = 64, p = 200$ is a good one. Compare the images for the following choices of p and q, again leaving the other parameters unchanged. $q = 128, p = 200$; $q = 64, p = 200$; $q = 32, p = 200$; $q = 16, p = 200$; $q = 64, p = 400$; $q = 64, p = 100$; $q = 64, p = 50$; $q = 64, p = 20$; $q = 64, p = 10$. Summarize your findings about the influence of chosing larger or smaller values of p or q than the ones suggested by theory.

Problem 4: Modify the filtered backprojection program so that it reconstructs the function $P^{\#} P f$. Compute an image of $P^{\#} P f$ for the Shepp–Logan phantom with $p = 200$, $q = 64$.

Problem 5: A fundamental question for image reconstruction is if the data uniquely determine the original image.

(a) Convince yourself that the x-ray transform is a linear operator, i.e., $P(\alpha f) = \alpha P f$ and $P(f + g) = P f + P g$. Show that for linear operators the question of uniqueness is equivalent to the question if there are nontrivial null-functions. I.e., $P f = P g$ implies $f = g$, if and only if $P f \equiv 0$ implies $f \equiv 0$.

(b) While the data $P f(\theta, s)$ for all s and infinitely many directions θ uniquely determine the function f, it was already known to the pioneers of tomography that this is not the case if $P f(\theta, s)$ is known for all s but for only finitely many directions θ. For example, in his 1963 paper "Representation of a function by its line integrals, with some radiological applications" (*Journal of Applied Physics*, **34** (1963), 2722–2727), A. M. Cormack, who later shared the Nobel prize in medicine for his contributions to tomography, claims that if the function $f(x)$ vanishes outside the unit disk and inside the unit disk is given by $f(x) = A \cos(n\psi)$, $n > 0$, where ψ is the polar angle of x, then the line integrals of f are zero along all lines perpendicular to the directions with polar angle $\varphi = (2m + 1)\pi/(2n)$, $m = 0, \ldots, 2n - 1$. Show that Cormack's claim is correct.

(c) What do you think may be the implications of the existence of such null functions (or "ghosts") for medical imaging?

Problem 6: (a) Modify the filtered backprojection program so that it reconstructs from fan-beam data. Test it for the Shepp–Logan phantom with $p = 200$, $q = 64$, $MX = MY = 128$, $R = 2.868$ and $c = \pi * q / \arcsin(1/R)$.

(b) Modify the program so that it can read the projection data from a file, using the fread command. Request from the author the data file pelvis.ctd. It contains real data from a hospital scanner. The data are stored in integer*2 format and correspond to a fan-beam geometry with $p = 360$, $q = 256$, and $R = 2.868$. The angle β in the data is incremented in reverse order compared to the lecture. Reconstruct an image (the best you can get) from these data.

Problem 7: Modify the filtered backprojection program so that it reconstructs from fan-beam data with detectors on a line as described in [67, pp. 93–95]. Test it for the Shepp–Logan phantom with $p = 200$, $q = 64$, $MX = MY = 128$, $R = 2.868$ and $b = \pi * q$.

Problem 8: Consider a crude method for so-called region-of-interest tomography, as follows.

(a) Modify the parallel-beam reconstruction program so that the data outside the circle inscribed in a square given by the parameter roi are set to zero. Test your program with the Shepp–Logan phantom and roi $= [-0.2, 0.2, -0.8, -0.4]$. (Set the parameter circle equal to 1.) Discuss the quality of the resulting image and compare with the reconstruction from complete data.

(b) Perform the same experiment as in part (a), only do not set the data to zero outside the region of interest (ROI) but set them to a constant equal to the nearest line integral intersecting the circle inscribed in the ROI.

Problem 9: (a) Consider the family of functions

$$f(x) = \left(1 - \|x\|^2\right)_+^m, \qquad x \in \mathbb{R}^3, \qquad m > -1.$$

(The case $m = 0$ gives the characteristic function of the unit ball. The larger m is, the smoother the function becomes.) Compute the transform

$$Df(z, \theta) = \int f(z + t\theta)\, dt, \qquad z \in \mathbb{R}^3, \qquad \theta \in S^2$$

for these functions. The following formula may be helpful:

$$\int_{-1}^{1} (1 - u^2)^m \, du = 2^{2m+1} \frac{(\Gamma(m+1))^2}{\Gamma(2m+2)}$$

(Gradshteyn and Ryzhik, p. 949, section 8.380, Formula 9).

(b) Show that for $g(x) = f(x - x_0)$, we have $Dg(z, \theta) = Df(z - x_0, \theta)$, and for $h(x) = f(Ax)$ with A a non-singular matrix,

$$Dh(z, \theta) = \|A\theta\|^{-1} Df(Az, \omega), \qquad \omega = \frac{A\theta}{\|A\theta\|}.$$

The M-file Divray.m provided in Appendix B implements this transform for the functions of part (a).

Problem 10: Use the source code for the FDK algorithm [18] provided in Appendix B and familiarize yourself with its use. Input the parameters for a phantom consisting of one ellipsoid with center at $(0.2, 0.3, 0.1)$ and half-axes of lengths $0.4, 0.2, 0.3$ along the directions $(1, 1, 0)/\sqrt{(2)}$, $(-1, 1, 0)/\sqrt{(2)}$ and $(0, 0, 1)$, respectively. In the code the rows of the orthonormal matrix OV indicate the directions of the principal axes. Produce reconstructions along the planes $y = 0.3$, $z = 0.1$, and $x + y + z = 0.6$, respectively. Use values $p = 20$ and $q = 64$ and indicate for each case the orthonormal vectors $n, w1, w2$ which you are using. Observe which boundaries are well reconstructed and which are blurred.

References

[1] J. Boman and E.T. Quinto, *Support theorems for real-analytic Radon transforms on line complexes in three-space*, Trans. Amer. Math. Soc., 335(1993), pp. 877-890.

[2] C. Berenstein and D. Walnut, *Local inversion of the Radon transform in even dimensions using wavelets*, in [22, pp. 45-69].

[3] A. M. Cormack, *Representation of a function by its line integrals, with some radiological applications*, J. Appl. Phys., 34 (1963), pp. 2722-2727.

[4] K. Buglione, *Pseudolocal tomography*, M.S. paper, Dept. of Mathematics, Oregon State University, Corvallis, OR 97331, U.S.A., (1998).

[5] A. M. Cormack, *Sampling the Radon transform with beams of finite width*, Phys. Med. Biol., 23 (1978), pp. 1141-1148.

[6] S. R. Deans, *The Radon transform and some of its applications*, Wiley, 1983.

[7] I. Daubechies, *Ten Lectures on Wavelets*, SIAM, Capital City Press, VT, 1992

[8] M. Defrise and R. Clack, *A cone-beam reconstruction algorithm using shift–variant filtering and cone–beam backprojection*, IEEE Trans. Med. Imag., MI-13 (1994), pp. 186-195.

[9] L. Desbat, *Efficient sampling on coarse grids in tomography*, Inverse Problems, 9(1993), pp. 251-269.

[10] A. Faridani, *An application of a multidimensional sampling theorem to computed tomography*, in [28, pp. 65-80].

[11] A. Faridani, *Reconstructing from efficiently sampled data in parallel-beam computed tomography*, in G.F. Roach (ed.), Inverse problems and imaging, Pitman Research Notes in Mathematics Series, Vol. 245, Longman, 1991, pp. 68-102.

[12] A. Faridani, *Results, old and new, in computed tomography*, in: Inverse Problems in Wave Propagation, G. Chavent et al. (editors), The IMA Volumes in Mathematics and its Applications, Vol. 90, Springer Verlag, New York, 1997, pp. 167-193.

[13] A. Faridani, *Sampling in parallel-beam tomography*, in: Inverse Problems and Imaging, A.G. Ramm (editor), Plenum, New York, 1998, pp. 33-53.

[14] A. Faridani, E. L. Ritman, and K. T. Smith, *Local tomography*, SIAM J. Appl. Math., 52 (1992), pp. 459-484. *Examples of local tomography*, SIAM J. Appl. Math., 52 (1992), pp. 1193-1198.

[15] A. Faridani, D.V. Finch, E. L. Ritman, and K. T. Smith, *Local tomography II*, SIAM J. Appl. Math., 57 (1997), pp. 1095-1127.

[16] A. Faridani and E. L. Ritman, *High-resolution computed tomography from efficient sampling*, Inverse Problems, 16(2000), pp. 635-650.

[17] A. Faridani, K. A. Buglione, P. Huabsomboon, O. D. Iancu, and J. McGrath, *Introduction to Local Tomography*, Contemporary Mathematics, Vol. 278 (2001), pp. 29–47.

[18] L. A. Feldkamp, L. C. Davis, and J. W. Kress, *Practical cone–beam algorithm*, J. Opt. Soc. Am. A, 1 (1984), pp. 612-619.

[19] D. V. Finch, *Cone beam reconstruction with sources on a curve*, SIAM J. Appl. Math., 45(1985), pp. 665-673.

[20] G. Folland, *Real Analysis–Modern Techniques and their Applications* John Wiley and Sons, Inc, (1984)

[21] I.M. Gelfand and S. G. Gindikin (eds.), *Mathematical problems of tomography*, Translations of Mathematical Monographs, Vol. 81, Amer. Math. Soc., 1990.

[22] S. Gindikin and P. Michor (eds.), *75 Years of Radon Transform*, Conference Proceedings and Lecture Notes in Mathematical Physics, Vol. 4, International Press, Boston, 1994.

[23] P. Grangeat, *Mathematical framework of cone beam 3D reconstruction via the first derivative of the Radon transform*, in [34, pp. 66-97].

[24] A. Greenleaf and G. Uhlmann, *Nonlocal inversion formulas for the X-ray transform*, Duke Math. J., 58(1989), pp. 205-240.

[25] A. Greenleaf and G. Uhlmann, *Estimates for singular Radon transforms and pseudodifferential operators with singular symbols*, J. Funct. Anal., 89(1990), pp. 202-232.

[26] A. Greenleaf and G. Uhlmann, *Composition of some singular Fourier integral operators and estimates for restricted X-ray transforms.*, Ann. Inst. Fourier, 40(1990), pp. 443-466.

[27] A. Greenleaf and G. Uhlmann, *Microlocal techniques in integral geometry.*, in [28, pp. 121-135].

[28] E. Grinberg and E.T. Quinto (eds.), *Integral Geometry and Tomography*, Contemporary Mathematics, Vol. 113, Amer. Math. Soc., Providence, R.I., 1990.

[29] C. W. Groetsch, *Inverse problems in the mathematical sciences*, Vieweg, Braunschweig, 1993.

[30] S. Gutmann, J .H. B. Kemperman, J. A. Reeds, and L. A. Shepp, *Existence of probability measures with given marginals*, Annals of Probability, 19 (1991), pp. 1781-1797.

[31] S. Helgason, *The Radon transform*, Birkhäuser, 1980.

[32] G. T. Herman, *Image reconstruction from projections: the fundamentals of computerized tomography*, Academic Press, New York, 1980.

[33] G. T. Herman (ed.), *Image reconstruction from projections: implementation and applications*, Springer, 1979.

[34] G.T. Herman, A.K. Louis, and F. Natterer (eds.), *Mathematical Methods in Tomography*, Lecture Notes in Mathematics, Vol. 1497, Springer, 1991.

[35] O. D. Iancu, *Contour reconstruction in 3D x-ray computed tomography*. M.S. paper, Dept. of Mathematics, Oregon State University, Corvallis, OR 97331, U.S.A., (1999).

[36] A. C. Kak and M. Slaney, *Principles of computerized tomographic imaging*, IEEE Press, New York, 1988.

[37] A. I. Katsevich, *Local Tomography for the generalized Radon transform*, SIAM J. Appl. Math. 57(1997), pp. 1128-1162.

[38] A. Katsevich, *Cone beam local tomography*, SIAM J. Appl. Math, 59(1999), pp. 2224-2246.

[39] A.I. Katsevich and A. G. Ramm, *A method for finding discontinuities of functions from the tomographic data*, in [76, pp. 115-123].

[40] A.I. Katsevich and A. G. Ramm, *New methods for finding jumps of a function from its local tomographic data*, Inverse Problems, 11 (1995), pp. 1005-1023.

[41] A.I. Katsevich and A. G. Ramm, *Pseudolocal tomography*, SIAM J. Appl. Math., 56, (1996), pp. 167-191.

[42] F. Keinert, *Inversion of k-plane transforms and applications in computer tomography*, SIAM Review, 31(1989), pp. 273-298.

[43] W. Klaverkamp, *Tomographische Bildrekonstruktion mit direkten algebraischen Verfahren*, Ph.D. thesis, Fachbereich Mathematik der Universität Münster, Münster, Germany, 1991.

[44] H. Kruse, *Resolution of reconstruction methods in computerized tomography*, SIAM J. Sci. Stat. Comput.,10(1989), pp. 447-474.

[45] P. Kuchment, K. Lancaster and L. Mogilevskaya, *On local tomography*, Inverse Problems, 11 (1995), pp. 571-589.

[46] A. V. Lakshminarayanan, *Reconstruction from divergent x-ray data*, Suny Tech. Report 32, Comp. Sci. Dept., State University of New York, Buffalo, NY, 1975.

[47] I. Lan, *On an operator associated to a restricted x-ray transform*, Ph.D. thesis, Dept. of Mathematics, Oregon State University, Corvallis, OR 97331, U.S.A., (1999).

[48] J. V. Leahy, K. T. Smith, and D. C. Solmon, *Uniqueness, nonuniqueness and inversion in the x-ray and Radon problems*, submitted to Proc. Internat. Symp. on Ill-posed Problems, Univ. Delaware Newark, 1979. The proceedings did not appear. Some results of this article have been published in [93].

[49] R. M. Lewitt, *Reconstruction algorithms: transform methods*, Proc. IEEE, 71 (1983), pp. 390-408.

[50] R. M. Lewitt, R. H. T. Bates, and T. M. Peters, *Image reconstruction from projections: II: Modified back-projection methods*, Optik, 50 (1978), pp. 85-109.

[51] B. F. Logan, *The uncertainty principle in reconstructing functions from projections*, Duke Math. J., 42 (1975), pp. 661-706.

[52] A. K. Louis, *Nonuniqueness in inverse Radon problems: the frequency distribution of the ghosts*, Math. Z. 185 (1984), pp. 429-440.

[53] A. K. Louis, *Orthogonal function series expansions and the null space of the Radon transform*, SIAM J. Math. Anal., 15(1984), pp. 621-633.

[54] A. K. Louis, *Incomplete data problems in x-ray computerized tomography I. Singular value decomposition of the limited angle transform*, Numer. Math. 48(1986), pp. 251-262.

[55] A. K. Louis, *Inverse und schlecht gestellte Probleme*, B.G. Teubner, Stuttgart, 1989.

[56] A. K. Louis, *Medical imaging: state of the art and future development*, Inverse Problems 8(1992), pp. 709-738.

[57] A. K. Louis, and P. Maass, *Contour reconstruction in 3-D x-ray CT*, IEEE Trans. Med. Imag., MI-12 (1993), pp. 764–769.

[58] A. K. Louis and F. Natterer, *Mathematical problems in computerized tomography*, Proc. IEEE, 71 (1983), pp. 379-389.

[59] W. R. Madych, *Summability and approximate reconstruction from Radon transform data*, in [28, pp. 189-219].

[60] P. Maass, *The x-ray transform: singular value decomposition and resolution*, Inverse Problems, 3 (1987), pp. 729-741.

[61] P. Maass, *The interior Radon transform*, SIAM J. Appl. Math. 52(1992), pp. 710-724.

[62] F. Natterer, *The Mathematics of Computerized Tomography*, Wiley, 1986.

[63] F. Natterer, *Sampling in fan-beam tomography*, SIAM J. Appl. Math. 53(1993), pp. 358-380.

[64] F. Natterer, *Recent developments in x-ray tomography*, in [76, pp. 177-198].

[65] F. Natterer, *Algorithms in Tomography*, Preprint, (1997). Available online at wwwmath.uni-muenster.de/math/inst/num

[66] F. Natterer, *Numerical methods in tomography*, in Acta Numerica, Vol. 8, 1999, Cambridge University Press, New York, pp. 107-141.

[67] F. Natterer and F. Wuebbeling, *Mathematical Methods in Image Reconstruction*, SIAM, Philadelphia, 2001.

[68] T. Olson and J. de Stefano, *Wavelet localization of the Radon transform*, IEEE Trans. Sig. Proc., 42 (1994), pp. 2055-2067 .

[69] D. P. Petersen and D. Middleton, *Sampling and reconstruction of wave-number-limited functions in N-dimensional euclidean space*, Inf. Control, 5(1962), pp. 279-323.

[70] R. M. Perry, *On reconstructing a function on the exterior of a disc from its Radon transform*, J. Math. Anal. Appl. 59 (1977), pp. 324-341.

[71] D. A. Popov, *On convergence of a class of algorithms for the inversion of the numerical Radon transform*, in [21, pp. 7-65].

[72] E.T. Quinto, *The dependence of the generalized Radon transform on defining measures*, Trans. Amer. Math. Soc. 257 (1980), pp. 331-346.

[73] E. T. Quinto, *Singular value decompositions and inversion methods for the exterior Radon transform and a spherical transform*, J. Math. Anal. Appl., 95 (1983), pp.437-448.

[74] E.T. Quinto, *Tomographic reconstructions from incomplete data – numerical inversion of the exterior Radon transform*, Inverse Problems, 4 (1988), pp. 867-876.

[75] E. T. Quinto, *Singularities of the x-ray transform and limited data tomography in* $\mathbb{R}^2$ *and* $\mathbb{R}^3$, SIAM J. Math. Anal., 24 (1993), pp. 1215-1225.

[76] E.T. Quinto, M. Cheney, and P. Kuchment (eds.), *Tomography, Impedance Imaging, and Integral Geometry*, Lectures in Applied Mathematics, Vol. 30, Amer. Math. Soc., 1994.

[77] E.T. Quinto, L. Ehrenpreis, A. Faridani, F. Gonzalez and E. Grimberg (eds.), *Radon Transforms and Tomography*, Contemporary Mathematics, Vol. 278, American Mathematical Society, Providence, RI, 2000.

[78] A. G. Ramm and A. I. Katsevich, *The Radon Transform and Local Tomography*, CRC Press, Boca Raton, 1996.

[79] A. G. Ramm and A.I. Zaslavsky, *Reconstructing singularities of a function given its Radon transform*, Math. and Comput. Modelling, 18(1993), pp. 109-138.

[80] F. Rashid-Farrokhi, K. J. R. Liu, C. A. Berenstein, and D. Walnut, *Wavelet-based multiresolution local tomography*, IEEE Transactions on Image Processing, 6 (1997), pp. 1412-1430.

[81] P. A. Rattey and A. G. Lindgren, *Sampling the 2-D Radon transform*, IEEE Trans. Acoust. Speech Signal Processing, ASSP-29(1981), pp. 994-1002.

[82] M.G. Raymer, M. Beck, and D.F. McAllister, *Complex wave-field reconstruction using phase-space tomography*, Phys. Rev. Lett., 72(1994), pp. 1137-1140.

[83] E. L. Ritman, J. H. Dunsmuir, A. Faridani, D. V. Finch, K. T. Smith, and P. J. Thomas, *Local reconstruction applied to microtomography*, in: Inverse Problems in Wave Propagation, G. Chavent et al. (editors), The IMA Volumes in Mathematics and its Applications, Vol. 90, Springer Verlag, New York, 1997, pp. 443-452.

[84] L. A. Shepp and J. B. Kruskal, *Computerized tomography: the new medical x-ray technology*, Amer. Math. Monthly, 85 (1978), pp. 420-439.

[85] E. A. Sivers, D. L. Halloway, W. A. Ellingson, and J. Ling, *Development and application of local 3-D CT reconstruction software for imaging critical regions in large ceramic turbine rotors*, in Rev. Prog. Quant. Nondest. Eval.:, D.O. Thompson and D.E. Chimenti (eds.), Plenum, New York, 1993, pp. 357-364.

[86] E. A. Sivers, D. L. Halloway, W. A. Ellingson, *Obtaining high-resolution images of ceramics from 3-D x-ray microtomography by region-of-interest reconstruction*, Ceramic Eng. Sci. Proc., 14, no. 7-8, (1993), pp. 463-472.

[87] J. Skaggs, *Region of interest tomography using biorthogonal wavelets*, M.S. paper, Dept. of Mathematics, Oregon State University, Corvallis, OR 97331, U.S.A., (1997).

[88] K. T. Smith, *Reconstruction formulas in computed tomography*, Proc. Sympos. Appl. Math., No. 27, L. A. Shepp, ed., AMS, Providence, RI, (1983), pp. 7-23.

[89] K. T. Smith, D.C. Solmon, and S. L. Wagner, *Practical and mathematical aspects of the problem of reconstructing objects from radiographs*, Bull. Amer. Math. Soc., 83(1977), pp. 1227-1270. Addendum in Bull. Amer. Math. Soc., 84(1978), p. 691.

[90] K. T. Smith and F. Keinert, *Mathematical foundations of computed tomography*, Appl. Optics 24 (1985), pp. 3950-3957.

[91] D. T. Smithey, M. Beck, M. G. Raymer, and A. Faridani, *Measurement of the Wigner distribution and the density matrix of a light mode using optical homodyne*

tomography: application to squeezed states and the vacuum, Phys. Rev. Lett., 70 (1993), pp. 1244-1247.

[92] D. C. Solmon, *The x-ray transform*, J. Math. Anal. Appl., 56(1976), pp. 61-83.

[93] D. C. Solmon, *Nonuniqueness and the null space of the divergent beam x-ray transform*, in [28, pp. 243-249].

[94] W. J. T. Spyra, A. Faridani, E. L. Ritman, and K. T. Smith, *Computed tomographic imaging of the coronary arterial tree - use of local tomography*, IEEE Trans. Med. Imag., 9 (1990), pp. 1-4.

[95] G. Strang, T. Nguyen, *Wavelets and Filter Banks*, Wellesley–Cambridge Press, Wellesley, MA, (1996).

[96] R. Strichartz, *Construction of Orthonormal Wavelets*, in *Wavelets: Mathematics and Applications*, pg. 23-50, Editors: L. Benedetto, M. Frazier, CRC Press, Inc, Baton Rouge, FL, (1994).

[97] H. K. Tuy, *An inversion formula for cone beam reconstruction*. SIAM J. Appl. Math. 43(1983), pp. 546-552.

[98] É. I. Vainberg, I. A. Kazak, and V. P. Kurozaev, *Reconstruction of the internal three-dimensional structure of objects based on real-time internal projections*, Soviet J. Nondestructive Testing, 17 (1981), pp. 415-423.

[99] É. I. Vainberg, I. A. Kazak, and M. L. Faingoiz, *X-ray computerized back projection tomography with filtration by double differentiation. Procedure and information features*, Soviet J. Nondestructive Testing, 21 (1985), pp. 106-113.

[100] R. C. Vaughn, *The parallel-beam filtered backprojection algorithm with reconstruction on a polar grid*, M.S. paper, Dept. of Mathematics, Oregon State University, Corvallis, OR 97331, USA, 1994.

ADEL FARIDANI
DEPARTMENTOF MATHEMATICS
OREGON STATE UNIVERSITY
CORVALLIS, OR 97331
UNITED STATES
    faridani@math.orst.edu
    http://oregonstate.edu/~faridana

The Attenuated X-Ray Transform: Recent Developments

DAVID V. FINCH

ABSTRACT. We survey recent work on the attenuated x-ray transform, concentrating especially on the inversion formulas found in the last few years.

1. Introduction

The attenuated x-ray transform is a variant of the classical x-ray transform in which functions are integrated over straight lines with respect to an exponential weight. It arises as a model in single photon emission computed tomography (SPECT) and in the study of the stationary linear single speed transport equation. Let a, f be continuous functions of compact support in $\mathbb{R}^n$ and let θ be a unit vector. We define the divergent beam x-ray transform of a at x in direction θ by

$$Da(x, \theta) = \int_0^\infty a(x + s\theta)\, ds,$$

where the integration is with respect to arc length. The attenuated x-ray transform of f is a function on the space of directed lines, whose value on the line l with direction θ is given by

$$P_a f(l) = \int_l f(y(\tau)) e^{-Da(y(\tau), \theta)}\, d\tau,$$

where $y(\tau)$ is an arc length parametrization of l. When the attenuation, a, is identically zero, the attenuated x-ray transform reduces to the ordinary x-ray transform. In the model of single photon emission tomography, the function f represents the spatial density of emitters which are assumed to emit photons isotropically. The function a is the linear attenuation coefficient, and so the attenuated x-ray transform is supposed to represent the photon intensity at a detector, collimated to accept only photons which have travelled along a specific line. A useful survey of the physics can be found in [9]. (The density of emitters is called the activity distribution, and so some authors denote it by a, whereas we use a for attenuation.)

In the plane, lines are often parametrized by their unit normal and directed distance from the origin, as are hyperplanes in higher dimensions. In that case, the attenuated x-ray transform is usually called the attenuated Radon transform and given by

$$R_a f(\omega, p) = \int_{x \cdot \omega = p} f(x) e^{-Da(x, \omega^{\perp})} dx, \qquad (1\text{--}1)$$

where $\omega^{\perp} = (-\sin \phi, \cos \phi)$ if $\omega = (\cos \phi, \sin \phi)$. The exponential Radon transform in $\mathbb{R}^2$ is given by

$$E_\mu f(\omega, p) = \int_{x \cdot \omega = p} f(x) e^{\mu x \cdot \omega^{\perp}} dx.$$

If the attenuation a is a constant on a convex set containing the support of f, then the attenuated Radon transform can be expressed in terms of the exponential Radon transform, and vice versa. The theory of the exponential Radon (or x-ray) transform is far more complete than that of the attenuated x-ray transform. The attenuated x-ray transform itself is a special case of the generalized x-ray transform, where the measure $e^{-Da} ds$ is replaced by a general measure $\mu(y, \theta) ds$. Even in this setting, Boman [7] has produced an example of a smooth measure on lines in the plane so that the associated generalized Radon transform has non-trivial kernel. One may further pass from lines in Euclidean space to curves on (or submanifolds of) a manifold. There has been a great deal of work done on such generalized Radon transforms, and many open questions remain, but we will only touch upon these extensions in this paper.

There are several inverse problems which can be posed for the attenuated x-ray transform. The simplest, and the one which has recently been solved, is the linear problem of recovering the activity f when the attenuation a is known. A much harder problem is to determine both f and a from $P_a f$. Given the resolution of the linear inverse problem, this amounts to determining a from $P_a f$. This is called the identification problem, and it is easy to see that there are rotationally invariant pairs of distinct a and f which give the same measurements, even when a is constant. Nonetheless, some progress has been made if one assumes that f has a special structure, e.g. a sum of delta functions, [28; 29] or a sum of point measures and an L^p function, [6]. In the special case of the exponential Radon transform in two dimensions, it has been proved that the identification problem has a unique solution if and only if f is not radial, [39; 17]. While we won't discuss the identification problem further in this article, it is worth saying that one of the tools used by Natterer is a set of consistency conditions for the range of the attenuated transform. The range is a topic discussed in Section 4, and its application to the identification problem has been an important motivation in its study.

The first uniqueness results which applied to the attenuated transform were of local nature. Local uniqueness for a generalized Radon transform R_μ means that each point x has a neighborhood U_x so that no non-trivial f supported in U_x lies in the kernel of R_μ. The size of U_x usually depends on some norms of the measure defining the generalized Radon transform, and of its derivatives. Examples of such results can be found in [23; 24]. A method yielding stronger local uniqueness results in the plane, as well as uniqueness for some problems of integral geometry, was introduced by Mukhometov [25] in the mid 70's. It was based on energy type estimates for a boundary value problem for a partial differential equation arising from a transport equation formulation of the integral transform. The method of Mukhometov was adapted by the author, [13], in the special case of the attenuated x-ray transform to prove uniqueness when the product of the diameter of the support of the activity and the supremum norm of the attenuation was not too large. Subsequently, Mukhometov's method was systematized and extended by Sharafutdinov and collaborators. An account may be found in his book [35]. Sharafutdinov also considered, see for example [36], the uniqueness problem for the attenuated x-ray transform on a class of Riemannian manifolds with boundary, where the integrals are taken over geodesics of the metric. The results are of the form that if the the integral over all geodesic segments joining boundary points of a weighted average of the attenuation and a geometric quantity depending on sectional curvatures is not too large, then the x-ray transform is injective. To our knowledge, these papers of Sharafutdinov are the only works on the attenuated x-ray transform on manifolds.

The theory for the exponential transform is fairly complete, but will not be much discussed in this paper. The first analytic inversion formulas were found in the late 1970's, [5; 42]. Recently, [31], an inversion procedure when the data is only collected for a range of 180° has been found (implementation is based on truncation of a Neumann series); this paper also has a good bibliography on inversion methods. The range was first characterized, in a complicated manner, by Kuchment and L'vin, [19], with later simplifications and extensions by Kuchment and coworkers appearing in [2; 1] and elsewhere. Their most recent contribution, [12], discusses a differential equation range characterization for a family of transforms which encompasses the exponential transform.

To the author's knowledge, most practical reconstruction in SPECT is done using iterative methods. In conventional x-ray tomography the greater speed and provable convergence properties of analytic methods have generally outweighed the benefits of iterative methods. In SPECT, where the photon flux is much smaller, the statistics of the emission process must be taken into account. The flexibility of iterative methods better allows them to account for these effects, to incorporate prior information, and to be adapted to incomplete sampling geometries. Of course, the price is that very little can be proved. The reader who wants to pursue this side of the subject might start by scanning some recent issues of *IEEE Transactions on Medical Imaging* or *Physics in Medicine and Biology.*

In the last five years, several exact inversion formulas have been found for the attenuated x-ray transform, as well as some results on characterization of the range of the transform. This paper is devoted to a survey of these results. In section 2 we introduce some standard notations and review some background results from complex analysis. In the next section, we sketch the methods of proof of the various inversion formulas. Section 4 is devoted to the range results, and the last section mentions some open problems.

2. Background and Preliminaries

Each of the inversion formulas makes use in some way of boundary values of analytic functions defined in a region in the complex plane. We recall a few results which we will need later. The first result is the Plemelj–Sokhozki formulas for the boundary values of an analytic function defined by a Cauchy integral. Suppose that L is a C^1 oriented simple path or simple closed curve in the complex plane, and that g is Hölder continuous of order α on L for some positive α. Let $G(z)$ be defined for $z \in \mathbb{C} \setminus L$ by the Cauchy integral

$$G(z) = \frac{1}{2\pi i} \int_L \frac{g(t)}{t - z} \, dt.$$

Then G is holomorphic in $\mathbb{C} \setminus L$ and the following formulas hold. For a proof we refer the reader to [26]. The existence of the principal value integrals is part of the assertion.

PROPOSITION 2.1. *Let g, G, and L be as above. If $t_0 \in L$ and is not either endpoint in the case when L is not closed, then the limit of G from the left of L exists and is given by*

$$G_+(t_0) = \tfrac{1}{2}g(t_0) + \frac{1}{2\pi i} \int_L \frac{g(t)}{t - t_0} \, dt,$$

where the integral on L is taken in principal value sense. Similarly the limit from the right exists and is given by

$$G_-(t_0) = -\tfrac{1}{2}g(t_0) + \frac{1}{2\pi i} \int_L \frac{g(t)}{t - t_0} \, dt.$$

COROLLARY 2.1. *Let g be Hölder continuous of order $\alpha > 0$ on the unit circle, and let G be Cauchy integral of g, as above, for L the unit circle oriented counterclockwise. Then for ω in the unit circle,*

$$G_+(-\omega^\perp) - G_+(\omega^\perp) = \tfrac{1}{2}(g(-\omega^\perp) - g(\omega^\perp)) + \tfrac{1}{2\pi i} \, \text{p.v.} \int_{S^1} \frac{1}{\omega \cdot \theta} g(\theta) \, d\theta. \quad (2\text{-}1)$$

PROOF. Let $\zeta = e^{i\sigma}$ where $\omega = (\cos\sigma, \sin\sigma)$ and $w = e^{i\psi}$ for $\theta = (\cos\psi, \sin\psi)$. Then, by the Plemelj–Sokhozki formula,

$$G_+(-i\zeta) - G_+(i\zeta) = \tfrac{1}{2}(g(-i\zeta) - g(i\zeta)) + \frac{1}{2\pi i}\,\text{p.v.}\int_{S^1}\left(\frac{1}{w + i\zeta} - \frac{1}{w - i\zeta}\right)g(w)\,dw.$$

Combining terms in the integral and using $dw = iw\,d\psi$, the integral becomes

$$\text{p.v.}\int_{S^1}\frac{2w\zeta}{w^2 + \zeta^2}\,g(e^{i\psi})\,d\psi.$$

But $\omega \cdot \theta = \tfrac{1}{2}(\bar{\zeta}w + \bar{w}\zeta)$, which in turn is equal to $\dfrac{w^2 + \zeta^2}{2w\zeta}$, since both w and ζ lie on the unit circle. $\square$

For f a smooth function with compact support on $\mathbb{R}$ the Hilbert transform of f is defined by the principal value integral

$$Hf(x) = \frac{1}{\pi}\,\text{p.v.}\int\frac{f(t)}{x - t}\,dt.$$

The Hilbert transform extends to a bounded operator on L^p, for $1 < p < \infty$.

We shall use $\mathcal{S}(\mathbb{R}^n)$ to denote the Schwartz space of infinitely differentiable functions f on $\mathbb{R}^n$ which satisfy $D^\alpha f$ is $O((1 + |x|)^{-k})$ for every natural number k and for every derivative D^α. The space of oriented lines in the plane can be parametrized by $\mathbb{S}^1 \times \mathbb{R}$ either in x-ray coordinates where (θ, s) corresponds to the line $s\theta^\perp + \mathbb{R}\theta$ or in Radon coordinates in which (θ, s) corresponds to $s\theta + \mathbb{R}\theta^\perp$. In either case, we define the Schwartz space of the space of lines to be the infinitely differentiable functions g on $\mathbb{S}^1 \times \mathbb{R}$ such that $\partial_\theta^k \partial_s^j g(\theta, s)$ is $O((1 + |s|)^{-n})$ for every $n \in \mathbb{N}$.

We shall frequently use the complex differential operators $\frac{\partial}{\partial z}$, also abbreviated ∂, and $\frac{\partial}{\partial \bar{z}}$, abbreviated $\bar{\partial}$, defined as follows. In $\mathbb{R}^2$ with standard coordinates x and y,

$$\frac{\partial}{\partial z} = \frac{1}{2}\left(\frac{\partial}{\partial x} - i\frac{\partial}{\partial y}\right), \qquad \frac{\partial}{\partial \bar{z}} = \frac{1}{2}\left(\frac{\partial}{\partial x} + i\frac{\partial}{\partial y}\right).$$

With this notation, the Cauchy–Riemann equations for $f = u + iv$ are written simply as $\bar{\partial}f = 0$. Let D be an open set in the plane with C^1 boundary. If $g \in C_0^1(R^2)$ and $\zeta \in D$ then

$$g(\zeta) = -\frac{1}{\pi}\int_D \frac{\partial g}{\partial \bar{z}}(x, y)\frac{1}{z - \zeta}\,dx\,dy + \frac{1}{2\pi i}\int_{\partial D}g(x, y)\frac{1}{z - \zeta}\,dz. \qquad (2\text{-}2)$$

There is an analogous formula with kernel $(\bar{z} - \bar{\zeta})^{-1}$ involving $\partial g/\partial z$ obtained by conjugating the preceding equation after g is replaced by $\bar{g}$. If D is the entire plane, this shows that $(\pi z)^{-1}$ is a fundamental solution of the $\bar{\partial}$ operator, and $(\pi\bar{z})^{-1}$ is a fundamental solution for ∂. For details see [18].

The function

$$h(\theta, s) = \tfrac{1}{2}(I + iH)Ra(\theta, s) \qquad (2\text{-}3)$$

plays a role in all of the inversion formulas. In this definition, I is the identity operator, R is the Radon transform, and the Hilbert transform H is applied to Ra in the second variable. Its importance was first discovered by Natterer [27; 28] in his work on consistency conditions for the range of the attenuated transform. We have adopted his Radon parametrization rather than the x-ray parametrization for that reason. Note that $h(\theta, x \cdot \theta)$ is constant on oriented lines, but not independent of orientation since $HRa(\theta, x \cdot \theta)$ is odd in θ. Natterer proved the following lemma, with a different proof. A proof similar to the one given here was found by Boman and Strömberg [8].

LEMMA 2.1. *The coefficients of the Fourier expansion in the angular variable of the function $h(\theta, x \cdot \theta) - Da(x, \theta^{\perp})$ are zero for negative or even index.*

PROOF. Since $Ra(\theta, x \cdot \theta) = Da(x, -\theta^{\perp}) + Da(x, \theta^{\perp})$, it needs to be shown that

$$\tfrac{1}{2}(Da(x, -\theta^{\perp}) - Da(x, \theta^{\perp})) + \tfrac{i}{2}HRa(\theta, x \cdot \theta)$$

has the desired property. Writing the Hilbert transform of the Radon transform as an iterated integral, and changing to polar coordinates yields

$$HRa(\theta, x \cdot \theta) = -\frac{1}{\pi} \, \text{p.v.} \int_{S^1} \frac{1}{\theta \cdot \omega} Da(x, \omega) \, d\omega.$$

By the corollary to the Plemelj–Sokhozki relations (2–1) the combination is the boundary value of an analytic function, and so has only non-negative Fourier coefficients. Since it is also an odd function, the result follows.   □

In fact more can be said: if $\sum_{k \in \mathbb{Z}} m_k(x) e^{ik\phi}$ is the Fourier series expansion of $Da(x, -\theta^{\perp})$, the Fourier expansion of $h(\theta, x \cdot \theta) - Da(x, \theta^{\perp})$ is

$$\sum_{d > 0} m_{2d+1}(x) e^{i(2d+1)\phi}.$$

This is the form (after a rotation) in which the expression enters in [4].

3. Uniqueness and Inversion

In this section we will present the inversion formulas found in the last five years. They are all formulated in two dimensions, but that is sufficient, since in higher dimensions one may restrict the full attenuated x-ray transform to lines in a family of planes whose union is the full space. (Whether in higher dimensions there exist other families of lines yielding inversion formulas is an open question.) We will present the formulas in order of discovery. Several of the authors have also stated uniqueness results for limited angle data, in which it is supposed that the attenuated x-ray transform is known only for lines with directions lying in a proper subset of the circle. The specific statements will be mentioned below.

3.1. The approach of Arbuzov, Bukhgeĭm, and Kazantsev. The first result, due to Arbuzov, Bukhgeĭm, and Kazantsev [4], is an application of the theory of A-analytic functions developed by Bukhgeĭm and collaborators. A summary of this theory may be found in [11]. The analysis begins with a formulation of the attenuated transform as a transport problem. Let f be the activity distribution and a the attenuation, and let Ω be a bounded convex set in the plane with smooth boundary. For a point $x \in \Omega$, let $\tau_\pm(x, \theta)$ be the point of intersection of the boundary and the ray from x in the direction $\pm\theta$. Let ∇ denote the gradient in space and consider the stationary transport equation

$$\theta \cdot \nabla u(x, \theta) + a(x)u(x, \theta) = f(x). \tag{3-1}$$

This equation may be integrated along lines in direction θ to obtain

$$e^{-Da(y,\theta)}u(y, \theta)\Big|_{\tau_-(x,\theta)}^{\tau_+(x,\theta)} = \int_{\tau_-}^{\tau_+} e^{-Da(y,\theta)} f(y)\, ds(y), \tag{3-2}$$

where the integral extends over the segment of the line through x in direction θ lying in Ω. The right hand side is the attenuated x-ray transform of f and a restricted to Ω. Thus knowing the boundary values of a solution of (3–1) determines the attenuated x-ray transform (since a is assumed known). Since (3–1) is a parametrized family of ordinary differential equations, the forward problem does not have a unique solution without some specification of initial conditions. For example, if the incoming flux is given, $u(x, \theta) = u_0(x, \theta)$ for $x \in \partial\Omega$ and $\theta \cdot \nu(x) < 0$, for the $\nu(x)$ the outer normal at x, then the solution is unique. Then supposing the attenuated transform is known, (3–2) completes the specification of the boundary values of u. The questions of uniqueness and inversion of the attenuated x-ray transform are then transferred to the questions of uniqueness and inversion for f from boundary values for the transport equation.

We will now present a proof of the inversion formula of Arbuzov, Bukhgeĭm, and Kazantsev. Their proof is an application of the theory of A-analytic functions, but we have chosen to avoid them by working with Fourier series expansions directly. This loses the elegance and some of the power of the approach taken by these authors, but it shows clearly how easily the result may be attained. One word of caution: we have stayed with the conventions of Fourier analysis and write $g(\theta) \sim \sum_{k \in \mathbb{Z}} g_k e^{ik\phi}$ whereas Arbuzov, Bukhgeĭm, and Kazantsev write $\sum g_k e^{-ik\phi}$.

Returning to the transport equation, it is clear that one may use any nonzero multiple of $e^{-Da(x,\theta)}$ as an integrating factor. Let $b(x, \theta) = h(-\theta^\perp, -x \cdot \theta^\perp)$, where h is given in (2–3). Although it is not needed in what follows, a calculation similar to that in (2.1) shows that $b(x, \theta) = \frac{1}{2}(I - iH)Pa(\theta, x \cdot \theta^\perp)$, where Pa is the parallel beam transform of a. Then b is constant on lines in direction θ and so $e^{b(x,\theta)-Da(x,\theta)}$ is also an integrating factor. Let $v(x, \theta) = e^{b(x,\theta)-Da(x,\theta)}u(x, \theta)$. Then v satisfies the equation $\theta \cdot \nabla v = f(x)e^{b(x,\theta)-Da(x,\theta)}$, which may be written in complex form, identifying $(x_1, x_2) \in \mathbb{R}^2$ with $z = x_1 + ix_2$ and setting $\theta = (\cos\phi, \sin\phi)$, as

$$e^{-i\phi}\frac{\partial v}{\partial \bar{z}} + e^{i\phi}\frac{\partial v}{\partial z} = f(z)e^{b(z,\theta)-Da(z,\theta)}. \tag{3-3}$$

By Lemma 2.1, the Fourier coefficients of $b(x,\theta) - Da(x,\theta)$ are zero for negative (or even) index, and so the Fourier coefficients of $e^{b(x,\theta)-Da(x,\theta)}$ are zero for negative index. Thus if $v = \sum v_k e^{ik\phi}$ is substituted in (3–3) and Fourier coefficients are equated, there results the system of equations

$$\frac{\partial v_{k+1}}{\partial \bar{z}} + \frac{\partial v_{k-1}}{\partial z} = \begin{cases} 0 & \text{for } k < 0, \\ f(z)\gamma_k & \text{for } k \geq 0, \end{cases} \tag{3-4}$$

where the expansion of e^{-G}, with $G = Da(x,\theta) - b(x,\theta)$, is $\sum_{k\geq 0}\gamma_k e^{ik\phi}$.

The aim of the following calculations is to show that each v_k for $k \leq 0$, can be expressed in terms of the boundary values of v_j for $j \leq k$. Since these are given in terms of u and e^{-G} on the boundary, and since for $k \leq 0$, the Fourier coefficient u_k of u is the k-th Fourier coefficient of $e^G v$ which is expressible in terms of the γ_l and the Fourier coefficients v_j for $j \leq k$, we can determine everywhere u_{-1} and u_0. These determine f by

$$2\,\mathrm{Re}\,\frac{\partial u_{-1}}{\partial z} + a(z)u_0 = f(z),$$

which results from writing the transport equation $\theta \cdot \nabla u + au = f$ in complex form, and separating Fourier coefficients, as was worked out above for v. Here it is also used that u is real valued, and so $\partial u_{-1} + \bar{\partial}u_1 = 2\,\mathrm{Re}\,\partial u_{-1}$.

Let

$$\rho_k(x,\phi) = \sum_{j=0}^{\infty} v_{k-2j}(z)e^{i(k-2j)\phi}.$$

We assume that Ω is an open bounded convex set with C^1 boundary, and that ρ_k is C^1 in Ω and continuous on the closure. Let $\zeta \in \Omega$ and for each ϕ let $l(\phi)$ be the length of the ray from ζ to the boundary in direction $e^{i\phi}$. Denote by $w(\phi) = \zeta + l(\phi)e^{i\phi}$ the point where the ray meets the boundary. Then, for $k \leq 0$,

$$\rho_k(w(\phi),\phi) - \rho_k(\zeta,\phi) = \int_0^l \frac{\partial\rho_k}{\partial s}(\zeta + se^{i\phi},\phi)\,ds$$

$$= \int_0^l \left(\frac{\partial\rho_k}{\partial\bar{z}}e^{-i\phi} + \frac{\partial\rho_k}{\partial z}e^{i\phi}\right)ds$$

$$= \int_0^l \sum_{j=0}^{\infty}\frac{\partial v_{k-2j}}{\partial\bar{z}}e^{i(k-2j-1)\phi} + \sum_{j=0}^{\infty}\frac{\partial v_{k-2j}}{\partial z}e^{i(k-2j+1)\phi}\,ds$$

$$= \int_0^l \frac{\partial v_k}{\partial z} e^{i(k+1)\phi} + \sum_{j=0}^{\infty} \left(\frac{\partial v_{k-2j}}{\partial \bar{z}} + \frac{\partial v_{k-2j-2}}{\partial z} \right) e^{i(k-2j-1)\phi} \, ds$$

$$= \int_0^{l(\phi)} \frac{\partial v_k}{\partial z} e^{i(k+1)\phi} \, ds,$$

where the series in the second to last integral is zero by (3–4). From this $v_k(\zeta)$ is found:

$$v_k(\zeta) = \frac{1}{2\pi} \int_0^{2\pi} \rho_k(\zeta, \phi) e^{-ik\phi} \, d\phi$$

$$= \frac{1}{2\pi} \int_0^{2\pi} \left(\rho_k(w(\phi), \phi) - \int_0^{l(\phi)} \frac{\partial v_k}{\partial z} e^{i(k+1)\phi} \, ds \right) e^{-ik\phi} \, d\phi$$

$$= \frac{1}{2\pi} \int_0^{2\pi} \rho_k(w(\phi), \phi) e^{-ik\phi} \, d\phi - \frac{1}{2\pi} \int_0^{2\pi} \int_0^{l(\phi)} \frac{\partial v_k}{\partial z} \frac{1}{se^{-i\phi}} s \, ds \, d\phi$$

$$= \frac{1}{2\pi} \int_0^{2\pi} \sum_{j=0}^{\infty} v_{k-2j}(w(\phi), \phi) e^{-2ij\phi} \, d\phi - \frac{1}{2\pi} \int_{\Omega} \frac{\partial v_k}{\partial z} \frac{1}{\bar{z}} \frac{1}{\zeta} \, dA. \tag{3–5}$$

By the conjugate form of (2–2),

$$-\frac{1}{2\pi} \int_{\Omega} \frac{\partial v_k}{\partial z} \frac{1}{z - \zeta} \, dA = \tfrac{1}{2} v_k(\zeta) + \frac{1}{4\pi i} \int_{\partial \Omega} v_k(w) \frac{1}{\bar{w} - \bar{\zeta}} \, d\bar{w}.$$

From $w(\phi) = \zeta + l(\phi)e^{i\phi}$,

$$e^{-i2\phi} = \frac{\overline{w - \zeta}}{w - \zeta}, \qquad d\phi = \frac{1}{2i} \left(\frac{1}{w - \zeta} \, dw - \frac{1}{\bar{w} - \bar{\zeta}} \, d\bar{w} \right).$$

Substituting these into (3–5) and gathering terms gives

$$v_k(\zeta) = \frac{1}{2\pi i} \int_{\partial \Omega} \left(dw \frac{1}{w - \zeta} \sum_{j=0}^{\infty} v_{k-2j}(w) \left(\frac{\overline{w - \zeta}}{w - \zeta} \right)^j \right.$$

$$\left. - d\bar{w} \frac{1}{\bar{w} - \bar{\zeta}} \sum_{j=1}^{\infty} v_{k-2j}(w) \left(\frac{\overline{w - \zeta}}{w - \zeta} \right)^j \right). \tag{3–6}$$

Recalling that $v = e^{-G}u$ and that we have reversed the indexing of [4], this is the k-th component of the equation in Theorem 4.3 of [4].

3.2. Novikov's inversion formula.

In the late spring of 2000, Novikov circulated a manuscript with an inversion formula for the attenuated transform. A revised version with some additional results was written in the fall and has now appeared in [33]. A published announcement with an outline of the proof appears in [32]. The paper makes heavy use of the boundary value distributions $(x \pm i0)^{-1}$ and becomes notationally dense when operators are dressed with $\pm$ and direction subscripts. We will first present Novikov's formula in his own notation, and then modify the notation to accord with that used in this paper. Suppose that a and f are Hölder continuous of order α, for some $\alpha \in (0, 1)$, that there is an $\varepsilon > 0$ such that they are $O(|x|^{-1-\varepsilon})$ as $|x| \to \infty$ and that $\sup_{0<|y|\leq 1} |y|^{-\alpha} |f(x+y) - f(x)|$ is also $O(|x|^{-1-\varepsilon})$, and similarly for a. Then, using the notations $D_\theta u(x) := Du(x,\theta)$, $P_\theta^\perp u(s) := Pu(\theta, s\theta^\perp)$, $P_{a,\theta}^\perp u(s) := P_a u(\theta, s\theta^\perp)$ and

$$H_\pm v(s) = \frac{1}{\pi} \int_{\mathbb{R}} \frac{v(t)}{s \pm i0 - t}\, dt$$

and

$$\exp\big(\pm (2i)^{-1} H_\mp P_\theta^\perp a\big) v(s) = \exp\big(\pm (2i)^{-1} H_\mp P_\theta^\perp a(s)\big) v(s),$$

Novikov's formula reads

$$f(x) = -\frac{1}{4\pi} \left(\frac{\partial}{\partial x_1} - i \frac{\partial}{\partial x_2} \right) \int_{S^1} \varphi(x,\theta)(\theta_1 + i\theta_2)\, d\theta,$$

where

$$\varphi(x,\theta) = \exp\big(-D_{-\theta} a(x)\big) m(x \cdot \theta^\perp, \theta),$$

with m given by

$$m(s,\theta) = m_+(s,\theta) - m_-(s,\theta)$$

$$= (2i)^{-1} \exp\big(-(2i)^{-1} H_+ P_\theta^\perp a(s)\big) H_+ \big(\exp((2i)^{-1} H_- P_\theta^\perp a) P_{a,\theta}^\perp f\big)(s)$$

$$-(-2i)^{-1} \exp\big((2i)^{-1} H_- P_\theta^\perp a(s)\big) H_- \big(\exp(-(2i)^{-1} H_+ P_\theta^\perp a) P_{a,\theta}^\perp f\big)(s).$$

$$(3\text{-}7)$$

It may be easily shown that

$$-\frac{1}{2i} H_+ = \tfrac{1}{2}(I + iH), \qquad \frac{1}{2i} H_- = \tfrac{1}{2}(I - iH),$$

which is useful when trying to compare Novikov's development with the work of others. Further since a and f are real, m_+ and m_- are conjugate, so m is pure imaginary.

Novikov's proof is also based on reformulation of the problem as a scattering problem for the transport equation

$$\theta \cdot \nabla u + au = f. \tag{3-8}$$

Let $\psi^+ = \psi^+(x,\theta)$ be the solution of the transport equation satisfying

$$\lim_{s \to -\infty} \psi^+(x + s\theta, \theta) = 0.$$

Then $\lim_{s \to \infty} \psi^+(x + s\theta, \theta) = P_a f(\theta, x \cdot \theta^\perp)$ which identifies the attenuated transform as scattering data. Let Σ be the complex quadric in $\mathbb{C}^2$ given by $\Sigma = \{(\theta_1, \theta_2) \in \mathbb{C}^2 : \theta_1^2 + \theta_2^2 = 1\}$. (This intersects the real space in the unit circle.) Novikov shows that for $\theta = (\theta_1, \theta_2) \in \Sigma \setminus S^1$ there is a unique solution ψ of the complex transport equation

$$\theta \cdot \nabla \psi(x,\theta) + a(x)\psi(x,\theta) = f(x), \qquad x \in \mathbb{R}^2,$$

which satisfies $\psi(x,\theta) \to 0$ as $|x| \to \infty$. An explicit formula is given (see (3–11) below). It is not remarked by Novikov, but simplifies some of his later results, to observe that

$$\bar{\psi}(x,\theta) = \psi(x,\bar{\theta}), \qquad (3\text{–}9)$$

which follows directly from conjugating the differential equation and using the uniqueness assertion. The next step is to study the limit of these solutions as θ tends to the real space. (This is the most delicate part of the analysis.) It is proved that for $\theta \in \mathbb{S}^1$, the limits

$$\psi_\pm(x,\theta) = \lim_{0 < \tau \to 0} \psi(x, \omega(\pm\tau))$$

for $\omega(\tau) = \sqrt{1+\tau^2}\theta + i\tau\theta^\perp$ with τ real and the positive square root, exist and are continuous, and satisfy the real transport equation (3–8), with initial conditions

$$\lim_{s \to -\infty} \psi_\pm(x + s\theta, \theta) = m_\pm(x \cdot \theta^\perp, \theta),$$

where $m_\pm$ were defined in (3–7). It follows from (3–9) that ψ_+ and ψ_- are conjugate. The difference $\psi_+ - \psi_-$ is then a solution of the homogeneous transport equation with value at $-\infty$ equal to $m_+ - m_-$. Since

$$\varphi(x,\theta) = e^{-Da(x,-\theta)}(m_+ - m_-)$$

is the unique such solution, it must hold that

$$\varphi(x,\theta) = \psi_+ - \psi_-.$$

(This is the φ of the inversion formula.) Next the quadric Σ is seen to have the holomorphic parametrization $\theta(\lambda)$, given by

$$\theta_1 = \frac{\lambda + \lambda^{-1}}{2}, \qquad \theta_2 = \frac{\lambda - \lambda^{-1}}{2i},$$

for $\lambda \in \mathbb{C} \setminus \{0\}$. Moreover, the unit circle $T \subset \mathbb{C}$ corresponds to the unit circle $S^1 \subset \mathbb{R}^2$, with the interior of the unit circle in $\mathbb{C}$ mapping to the the subset of Σ parametrized above by $\omega(\tau)$, for $\tau > 0$ (and all θ). It is easy to check that $\theta(\lambda)$ also satisfies

$$\theta(1/\bar{\lambda}) = \bar{\theta}(\lambda). \qquad (3\text{–}10)$$

It is then shown that for each $x \in \mathbb{R}^2$, $\psi(x,\theta(\lambda))$ is holomorphic in λ for $\lambda \in \mathbb{C} \setminus \{0 \cup T\}$, and that the limit from inside the circle (resp. from outside the circle) correspond to the boundary values $\psi_\pm$, respectively. The analyticity results from the specific form of the solution of the complex transport equation, referred to above. Here it is appropriate to indicate the form: for $\theta = \theta(\lambda)$ one has

$$\psi(x,\theta(\lambda)) = \int_{\mathbb{R}^2} e^{-G_{\theta(\lambda)}a(x)} G(x-y, \theta(\lambda)) e^{G_{\theta(\lambda)}a(y)} f(y)\, dy, \qquad (3\text{–}11)$$

where $(x_1, x_2) \in \mathbb{R}^2$ is identified with $z = x_1 + ix_2 \in \mathbb{C}$,

$$G(z, \theta(\lambda)) = \frac{\operatorname{sgn}(1 - |\lambda|)}{2\pi i(i/2)(\lambda \bar{z} - z/\lambda)},$$

for $\lambda \neq 0, \lambda \notin T$, and $G_\theta a(x)$ is the convolution $\int G(x-w, \theta)a(w)\, dw$. Moreover, from (3–9) and (3–10),

$$\psi(x, \theta(\lambda)) = \bar{\psi}(x, \theta(\bar{\lambda}^{-1})),$$

so the Laurent expansion in the punctured circle determines that in the exterior of the circle, and vice versa. Looking at the kernel, $G(x, \theta(\lambda))$, it is clear that the expansion around $\lambda = 0$ has the form, with $\zeta = y_1 + iy_2$ and $z = x_1 + ix_2$

$$\psi(x, \theta(\lambda)) = \lambda \int \frac{f(\zeta)}{\pi(z - \zeta)}\, dy + O(\lambda^2),$$

and so the expansion at infinity is

$$\psi(x, \theta(\lambda)) = \bar{\psi}(x, \theta(\bar{\lambda}^{-1})) = \lambda^{-1} \int \frac{f(\zeta)}{\pi(\overline{z - \zeta})}\, dy + O(\lambda^{-2}).$$

Using this last relation and taking the limit as the contour shrinks to the unit circle we get

$$\int \frac{f(\zeta)}{\pi(\overline{z - \zeta})}\, dy = \frac{1}{2\pi i}\int_T \psi_-(x, \theta(\lambda))\, d\lambda$$
$$= -\frac{1}{2\pi i}\int_T (\psi_+(x, \theta(\lambda)) - \psi_-(x, \theta(\lambda)))\, d\lambda$$
$$= -\frac{1}{2\pi}\int_{S^1} (\psi_+(x, \theta) - \psi_-(x, \theta))(\theta_1 + i\theta_2)\, d\theta,$$

which finishes the proof of the formula, since $(\pi \bar{z})^{-1}$ is a fundamental solution for ∂.

Since the constant terms of the Laurent expansions inside and outside the circle are both zero, a similar chain of equalities shows that

$$0 = \frac{1}{2\pi i}\int_T (\psi_+ - \psi_-)\frac{d\lambda}{\lambda} = \frac{1}{2\pi}\int_{S^1} (\psi_+(x, \theta) - \psi_-(x, \theta))\, d\theta. \qquad (3\text{–}12)$$

This is a consistency condition which will be used later.

Novikov proves a limited angle theorem assuming that f is continuous with compact support. If $P_a f(\theta, s)$ is known for all $s \in \mathbb{R}$ and θ in a set of positive length, then f is uniquely determined.

The inversion formula of Novikov has been implemented by Kunyansky, [22], with further particulars in [21]. Kunyansky also shows that in the case of constant attenuation Novikov's formula reduces to the inversion formula for the exponential transform given by Tretiak and Metz, [42].

3.3. The results of Natterer and of Boman and Strömberg.

Natterer, [30], works with the attenuated Radon transform, (1–1), and proves the inversion formula

$$f(x) = \frac{1}{4\pi} \operatorname{Re} \operatorname{div} \int_{S^1} \theta e^{Da(x,\theta^\perp)} e^{-h(\theta, x\cdot\theta)} H(e^h R_a f)(\theta, x \cdot \theta) \, d\theta,$$

where h is given by (2–3) and the Hilbert transform H is applied to $e^h R_a f$ in the second variable. This formula can be shown to be equivalent to Novikov's formula after changing from x-ray to Radon coordinates and taking account of the parities of the constituent functions under change of sign in the argument (these occur since $R_a f(\theta, s) = P_a f(\theta^\perp, -s)$). Natterer's proof is very economical: our prose description will be nearly as long as his full exposition.

It begins with a few lemmas. The first is the result given above as (2.1) on the Fourier coefficients of $u(x, \theta) = h(\theta, x\cdot\theta) - Da(x, \theta^\perp)$, and the second, which evaluates the integrals $\int_0^{2\pi} \frac{\theta}{x\cdot\theta} e^{il\phi} \, d\phi$ (with $\theta = (\cos\phi, \sin\phi)$), is easily derivable from (2–1). The proof of the main theorem has two steps. The first expresses the integrand in the inversion formula as

$$\theta e^{-u(x,\theta)} H(e^h R_a f)(\theta, x \cdot \theta) = \frac{1}{\pi} \int_{R^2} f(y) \frac{\theta}{(x-y)\cdot\theta} e^{u(y,\theta) - u(x,\theta)} \, dy$$

(there are two minor misprints in his equation (2.4)) and then shows that

$$\operatorname{Re} \int_{S^1} \frac{\theta}{(x-y)\cdot\theta} e^{u(y,\theta) - u(x,\theta)} \, d\theta = 2\pi \frac{x-y}{|x-y|^2}$$

which is a multiple of a fundamental solution of the divergence operator. The proof of this formula follows by expanding $e^{u(y,\cdot) - u(x,\cdot)}$ in Fourier series, applying the lemma on the evaluation of the integrals $\int \frac{\theta}{x\cdot\theta} e^{il\phi} \, d\phi$, and observing that only the $l = 0$ term of the Fourier expansion contributes to the real part. Natterer does not formulate a precise theorem on the necessary regularity of f, but notes that the formula does hold pointwise for $f \in C_0^1(\mathbb{R}^2)$. The conditions on a are even less specific: only that it is sufficiently smooth and of sufficiently rapid decay at infinity. Natterer has implemented the formula and presents an example reconstruction.

The work of Boman and Strömberg, [8], does not yet have its final form, so we can only indicate the preliminary results given by Boman in lecture. First they prove an inversion formula for continuous functions with compact support in an open set Ω for the generalized Radon transform

$$R_\rho f(\theta, p) = \int_{x\cdot\theta = p} f(x) \rho(\theta, x) \, ds, \qquad (\theta, p) \in \mathbb{S}^1 \times \mathbb{R},$$

for complex measures $\rho(x, \theta)$ such that for each $x \in \Omega$, $\rho(x, \cdot)$ extends to a continuous nowhere zero function on the unit disk which is analytic on the open disk, ρ is Hölder continuous on $\Omega \times T$, and such that $\arg \rho(x, \theta)$ is constant on each oriented line $x \cdot \theta = p$ for $x \in \Omega$. With $m(x) = \frac{1}{2\pi} \int \rho(x, \theta) \, d\theta$ the (angular) mean value of ρ, assumed to be real, the inversion formula takes the form

$$f(x) = \frac{1}{4\pi m(x)} \operatorname{div} \left(m(x) \operatorname{Re} \int_{S^1} \theta (H R_\rho f)(\theta, x \cdot \theta) \frac{1}{\rho(x, \theta)} \, d\theta \right).$$

They then observe that their argument can be applied to any measure ρ_0 for which there is a nowhere zero function τ, constant on oriented lines, such that $\rho = \tau \rho_0$ satisfies the conditions above. They prove that if $\rho_0 = e^q$ is real, and $q(x, \theta) = w(x, \theta) + u(x, \theta)$ for real u, w where u and the conjugate function $\tilde{w}$ are constant on all oriented lines $x \cdot \theta = p$ in Ω, then the trick applies with $\tau = e^{-u + i\tilde{w}}$. It is further shown that for the attenuated transform, where $q = -Da(x, \theta^\perp)$, this kind of decomposition holds, so their inversion formula applies. They also prove a limited angle theorem which states that for the generalized Radon transform R_ρ, with ρ satisfying the conditions above, if $R_\rho f(\theta, p) = 0$ for all $p \in \mathbb{R}$ and θ in an non-empty open subset of the circle for some compactly supported continuous function f, then f must be identically zero.

3.4. Additional remarks. We add a few remarks on the preceding formulas and cite a new paper that has just come to our attention.

REMARK 3.1. The analysis of the stationary transport equation (3–1) is an important ingredient in the study of inverse problems for the stationary transport equation with scattering. This seems to have the motivation for much of the work by the Novosibirsk group. Indeed, it was in this context that (to our knowledge) it was first observed, see [3], that local uniqueness for the attenuated transform in dimension two implies global uniqueness for compactly supported activity distributions in higher dimensions. We refer the reader to the recent thesis and paper by Tamasan, [41; 40], for applications of the recent inversion formulas for the attenuated transform to inverse problems for the more general transport equations, and for further references.

REMARK 3.2. The results of Natterer and of Boman and Strömberg provide more direct proofs of inversions formulas of Novikov type, but they clearly owe their formulation to the work of Novikov. It is not hard to modify Natterer's Fourier series expansion to arrive at the fundamental solution of ∂ instead of working with the real part of the vector adjoint to arrive at a fundamental solution for the divergence operator.

REMARK 3.3. The inversion procedure given by Arbuzov, Bukhgeĭm, and Kazantsev is fairly complicated, because it requires evaluating u_0 and u_{-1}, both of which require the full sequence of $\{v_k\}$ for $k \leq 0$. One explanation may be that the method is underspecified. The derivation leading up to (3–6) holds no matter what choices are made in (3–2) to complete the boundary values (e.g. zero incoming or some other), and it would be no surprise if some choice might lead to a more efficient inversion formula.

REMARK 3.4. In a recent work, [10], which came to the author too late for a full treatment in this paper, A. A. Bukhgeim and Kazantsev have found an inversion formula adapted to divergent beam geometry, where it is supposed that the attenuation and activity are supported in the unit disk, and lines are parametrized by direction and point of intersection with the unit circle. The formula makes use of the circular Hilbert transform Γ, given by

$$\Gamma q(\phi) = \frac{1}{2\pi} \int_0^{2\pi} \frac{\cot(\phi - \alpha)}{2} q(\alpha)\, d\alpha,$$

with respect to both direction at points on the measurement circle, and with respect to position on the measurement circle. Assuming that f is square integrable on the unit disk and that a is C^2 on the closed unit disk, they state and outline the proof of the following:

$$f(z) = \frac{\partial}{\partial \bar{z}} \frac{i}{\pi} \int_0^{2\pi} e^{-i\phi} e^{-m(z,\phi)} \operatorname{Im}\left(e^{-2(Da)^+(\gamma(z,\phi),\phi)} v^*(\gamma(z,\phi),\phi)\right) d\phi,$$

where for t in the unit circle,

$$v^*(t,\phi) = (I + i\Gamma)(v(\,\cdot\,,\phi) - \tfrac{1}{2}v(\gamma(\,\cdot\,,\phi+\pi),\phi))(t)$$

with $v(t,\phi) = e^{-2(Da)^+(t,\phi)} D_a f(t,\phi)$. Here the notations are $\theta = (\cos\phi, \sin\phi)$, $m(z,\phi) = Da(z,-\theta)$, $\gamma(z,\phi)$ is the point of intersection of the unit circle with the ray from z in the direction $-\theta$, $D_a f(z,\phi) = \int_{-\infty}^0 f(z + s\theta) \exp(-Da(z + s\theta, -\theta))\, ds$, and $(Da)^+(z,\phi) = \tfrac{1}{2}(I - i\Gamma)m^{\mathrm{odd}}(z,\cdot)(\phi)$, where m^{odd} is the odd part of $m(z,\phi)$ with respect to the angular variable.

4. Range Characterization

Prior to presenting what is known about range characterization for the attenuated x-ray transform, it is valuable to recall what is known for the classical x-ray and Radon transforms. The results are most easily stated for the Radon transform. The chief theorem, which was first proved fully by Helgason, [16], gives a characterization of the range of the Radon transform on $S(\mathbb{R}^n)$, the Schwartz space of smooth rapidly decreasing functions on $\mathbb{R}^n$.

THEOREM 4.1. *A function* $g \in S(\mathbb{S}^{n-1} \times \mathbb{R})$ *is in the range of the Radon transform on* $S(\mathbb{R}^n)$ *if and only if*

(i) g *is even, that is,* $g(\theta,p) = g(-\theta,-p)$ *for all* $(\theta,p) \in \mathbb{S}^{n-1} \times \mathbb{R}$, *and*
(ii) *for each natural number* m *the function* $p_m(\theta) = \int_{\mathbb{R}} g(\theta,p)p^m\, dp$ *is the restriction to* $\mathbb{S}^{n-1}$ *of homogeneous polynomial on* $\mathbb{R}^n$.

The last set of conditions are called the moment conditions. They also enter what are called the Paley–Wiener type theorems for the Radon transform, which characterize the range of the transform on various spaces of compactly supported functions. There are analogous results on the range of the classical x-ray transform in higher dimensions, [37], but also characterizations of the range as the solution space of certain differential equations.

In the mid-80's, Solmon, [38], gave a very nice extension of the range theorem, which said that if g is an even smooth function of rapid decay on $\mathbb{S}^{n-1} \times \mathbb{R}$, then g is the Radon transform of a function f which is $O(|x|^{-n})$, moreover f is $O(|x|^{-n-m-1})$, $m \geq 0$, as $|x| \to \infty$ if and only if g satisfies the moment conditions through order m. Furthermore, any k-th order derivative of f is $O(|x|^{-n-1-m-k})$ as $|x| \to \infty$.

For the attenuated Radon transform, the first result found was an analogue of the moment conditions. Natterer, [27; 28], showed the following,

THEOREM 4.2. *Let $k > m \geq 0$ be integers. If $g = R_a f$ for $a, f \in \mathcal{S}(\mathbb{R}^2)$ then*

$$\int_{\mathbb{R}} \int_0^{2\pi} p^m e^{ik\phi + 1/2(I+iH)Ra} g(\omega, p) \, dp \, d\omega = 0,$$

where as usual, $\omega = (\cos\phi, \sin\phi)$. (Additional conditions result from taking the conjugate, since g is real.)

The proof follows from writing out the integral defining the attenuated Radon transform, applying Fubini's theorem, and (2.1) on the vanishing of the negative Fourier coefficients of $e^{h(x, x \cdot \theta) - Da(x, \theta^\perp)}$. About ten years later, Kuchment and L'vin gave a characterization of the range of the exponential Radon transform, [19]. One consequence was that Natterer's conditions are not sufficient to characterize the range, even for the case of constant attenuation. Something which was obviously lacking was the analogue of the evenness condition in the classical range theorem. Novikov, [33; 32] found such a condition, and has proved, [34], the following theorem.

THEOREM 4.3. *Let a, g be in the Schwartz space $\mathcal{S}(\mathbb{S}^1 \times \mathbb{R})$ and let g satisfy*

$$\text{Re} \int_{S^1} e^{-Da(x,-\theta)} e^{1/2(I+iH)Pa(\theta, x \cdot \theta^\perp)} H_+ (e^{1/2(I-iH)Pa(\theta, \cdot)} g(\theta, \cdot)) (\theta^\perp \cdot x) \, d\theta = 0$$
$$\text{for } x \in \mathbb{R}^2. \quad (4\text{-}1)$$

Then there is a C^∞ function f such that f and all its derivatives are $O(|x|^{-2})$ as $|x| \to \infty$ with $P_a f = g$.

The necessity of (4-1) was established in (3-12). The scheme of the proof is to define f by Novikov's inversion formula, using g in place of $P_a f$, and then to prove that the resulting function has the required decay and that its image under the attenuated transform is g. In the case of the classical x-ray transform, (4-1) takes the form

$$0 = \int_{S^1} Hg(\theta, x \cdot \theta^\perp) \, d\theta \text{ for all } x \in \mathbb{R}^2.$$

It is obvious that if $g = Pf$ then the conditions are satisfied, for then Pf is even and so $HPf(\theta, x \cdot \theta^{\perp})$ is odd in θ, but the converse takes some work. It suffices to prove that under the hypothesis $0 = \int_{S^1} Hg(\theta, x \cdot \theta^{\perp}) \, d\theta$ the function g must be even, for one can then apply Solmon's theorem, but the author has not found a simple proof of this.

Arbuzov, Bukhgeĭm, and Kazantsev give a different range result, whose consequence and interpretation in the context of the attenuated transform have not, to the author's knowledge, been fully worked out. (Kuchment has some ideas on the matter. A discussion, along with many other matters related to the present paper, can be found in [20].) Recalling the Plemelj–Sokhozki relations from Section 2, one sees that a Hölder continuous function g on the boundary of a simply connected domain with smooth boundary is the boundary value of a function analytic in the domain (its Cauchy integral) if and only if the principal value integral over the boundary is one-half the value of the function on the boundary. The same result holds for Hölder continuous functions on the boundary, taking values in X^{m+2} of the scale of Banach space defining A-analyticity. (See [4] for further explanation of the following.) Since the transmuted function $e^{-G}u$ is A-analytic, if the boundary values are Hölder continous, they must be equal to twice the principal value integral over the boundary.

5. Open Problems

In the author's opinion, the "most wanted" of the open problems is to find an explicit set of consistency conditions which characterize the range of the attenuated transform on functions of compact support or rapid decay. As mentioned above, Novikov has proved the necessity of (4–1), and its sufficiency for a rapidly decaying smooth function on the space of lines to be the attenuated transform of a function with quadratic decay. Natterer's conditions, (4.2), are also necessary range conditions for functions of rapid decay. It is unknown whether the union of these conditions is sufficient. There are some implicit conditions, such as the function produced by the inversion formula has the desired property, but these do not seem useful for application elsewhere, such as to the identification problem. An allied question would be the existence of a support theorem for the attenuated transform. If f is continuous and decays faster than any reciprocal power of $|x|$ and $P_a f(l) = 0$ for every directed line l disjoint from some compact convex set K, is f supported in K?

Inversion from partial data. In the plane, Novikov and Boman–Strömberg proved uniqueness for compactly supported activity distributions when the line directions are restricted to a non-empty open subset of the circle. Natterer has posed the question of whether there exists a stable inversion formula when the directions comprise half the circle. If such could be found, it could help make analytic methods more competitive with iterative methods in clinical applications.

The analogous problem for the exponential transform has been treated by Noo and Wagner in [31].

In higher dimensions, for what submanifolds of the manifold of lines can inversion formulas be derived? Gel'fand, Gindikin, and Shapiro [14] have given conditions on curve families with densities in the plane for admissibility, and for curve families in higher dimensions this was extended in [15]. These papers are concerned with a geometric condition (admissibility) which corresponds to a certain differential form being closed. In complex space this can lead to local inversion formulae, though not in real space. For the attenuated transform one might start by making a similar analysis on submanifolds of oriented curves.

As mentioned in the introduction, there has been a lot of work on generalized Radon transforms on manifolds, but very little specifically about the attenuated x-ray transform. In particular, the uniqueness problem remains open. One might hope also for an inversion formula, perhaps under more stringent hypotheses.

Acknowledgments

The author would like to thank Jan Boman, Alexander Bukhgeĭm, Frank Natterer, Alex Tamasan and Gunther Uhlmann for helpful discussions at MSRI. He also thanks the Mathematical Sciences Research Institute, MSRI, of Berkeley, CA for enabling his participation in the program in Inverse Problems during fall 2001. He would also like to thank Professor R. Novikov for sending him preprints of the papers [34; 33] and R. Clackdoyle for stimulating his interest in these problems.

References

[1] V. Aguilar, L. Ehrenpreis, and P. Kuchment, *Range conditions for the exponetial Radon transform*, J. d'Anal. Math. **68** (1996), 1–13.

[2] V. Aguilar and P. Kuchment, *Range conditions for the multidimensional exponential X-ray transform*, Inverse Problems **11** (1995), 977–982.

[3] D. S. Anikonov, *Multidimensional inverse problems for the transport equation*, Differentsial'nye Uravneniya **20** (1984), no. 5, 817–824, English trans. Differential Equations, 20, 608–614. MR 86f:35171

[4] E. V. Arbuzov, A. L. Bukhgeĭm, and S. G. Kazantsev, *Two-dimensional tomography problems and the theory of A-analytic functions*, Siberian Advances in Mathematics **8** (1998), 1–20.

[5] S. Bellini, M. Pacentini, C. Cafforio, and F. Rocca, *Compensation of tissue absorption in emission tomography*, IEEE Trans. Acoust. Speech Signal Process. **27** (1979), 213–218.

[6] J. Boman, *On generalized Radon transforms with unknown measures*, Integral Geometry and Tomography (Arcata, CA) (E. Grinberg and E. T. Quinto, eds.), Contemporary Mathematics, vol. 113, AMS, Providence R.I., 1990, pp. 5–15.

[7] ———, *An example of nonuniqueness for a generalized Radon transform*, J. Anal. Math. **61** (1993), 395–401. MR 94j:44004

[8] J. Boman and J.-O. Strömberg, *Novikov's inversion formula for the attenuated Radon transform – a new approach*, In preparation.

[9] T. F. Budinger, G. T. Gullberg, and R. H. Huesman, *Emission computed tomography*, Image Reconstruction from Projections (G. T. Herman, ed.), Topics in Applied Physics, vol. 32, Springer, Berlin - New York, 1979, pp. 147–246.

[10] A. A. Bukhgeim and S. G. Kazantsev, *The attenuated Radon transform inversion formula for divergent beam geometry*, preprint, 6 pp., 2002.

[11] A. L. Bukhgeĭm, *Supplement: Inversion formulas in inverse problems*, pp. 323–376, in M. M. Lavrent'ev and L. Ya. Savel'ev, *Linear Operators and Ill-Posed Problems*, Consultants Bureau, N.Y., 1995.

[12] L. Ehrenpreis, P. Kuchment, and A. Panchenko, *Attenuated Radon transform and F. John's equation I: Range conditions*, Analysis, Geometry, Number Theory: The Mathematics of Leon Ehrenpreis (E. L. Grinberg, S. Berhanu, M. Knopp, G. Mendoza, and E.T. Quinto, eds.), Contemporary Mathematics, vol. 251, Amer. Math. Soc., 2000, pp. 173–188.

[13] D. V. Finch, *Uniqueness for the attenuated x-ray transform in the physical range*, Inverse Problems (1986), 197–203.

[14] I. M. Gel'fand, S. G. Gindikin, and Z. Ya. Shapiro, *A local problem of integral geometry*, Functional Analysis and its Applications **13** (1979), 87–102.

[15] I. M. Gel'fand and M. I. Graev, *Admissible n-dimensional complexes of curves in* **R**^{n}, Funct. Anal. Appl. **14** (1980), 274–281.

[16] S. Helgason, *The Radon transform on Euclidean spaces, compact two-point homogeneous spaces and Grassmann manifolds*, Acta Math. **113** (1965), 153–180.

[17] A. Hertle, *The identification problem for the constantly attenuated Radon transfrom*, Math. Z. **197** (1988), 13–19.

[18] L. Hörmander, *The analysis of linear partial differential operators I*, Springer-Verlag, Berlin, Heidelberg, New York, Tokyo, 1983.

[19] P. A. Kuchment and S. Ya. L'vin, *Paley-Wiener theorem for the exponential Radon transform*, Acta Appl. Math. **18** (1990), 251–260.

[20] P. A. Kuchment and E. T. Quinto, *Some problems of integral geometry arising in tomography*, chapter 11 in L. Ehrenpreis' forthcoming book on the Radon Transform.

[21] L. A. Kunyanksy, *Analytic reconstruction algorithms in emission tomography with variable attenuation*, J. Comp. Math. in Sci. and Engr., to appear.

[22] L. A. Kunynasky, *A new SPECT reconstruction algorithm based on the Novikov explict inversion formula*, Inverse Problems **17** (2001), 293–306.

[23] M. M. Lavrent'ev and A. L. Bukhgeĭm, *On a class of problems of integral geometry*, Dokl. Akad. Nauk SSSR **211** (1973), 38–39, Engl. trans. in Soviet Math Dokl., **14** (1973).

[24] A. Markoe and E. T. Quinto, *An elementary proof of local invertibility for generalized and attenuated Radon transforms*, SIAM J. Math. Anal. **16** (1985), 1114–1119.

[25] R. G. Mukhometov, *The problem of recovery of a two-dimensional Riemannian metric and integral geometry*, Dokl. Akad. Nuak SSSR **232** (1977), 32–35, English trans. in Soviet Math. Dokl. **18** (1977), pp. 27–31.

[26] N. I. Muskhelishvili, *Singular integral equations*, Noordhoff, Groningen, 1953.

[27] F. Natterer, *The ill-posedness of Radon's integral equation*, Symposium on Ill-Posed Problems: Theory and Practice, Newark, DE, 1979.

[28] _____, *The identification problem in emission computed tomography*, Mathematical aspects of computerized tomography (Oberwolfach, 1980), Springer, Berlin, 1981, pp. 45–56. MR 84m:92008

[29] _____, *Computerized tomography with unknown sources*, SIAM J. Appl. Math. **43** (1983), no. 5, 1201–1212. MR 86i:92011

[30] _____, *Inversion of the attenuated Radon transform*, Inverse Problems **17** (2001), 113–119.

[31] F. Noo and J.-M. Wagner, *Image reconstruction in 2D SPECT with 180° acquisition*, Inverse Problems **17** (2001), 1357–1371.

[32] R. G. Novikov, *Une formule d'inversion pour la transformation d'un rayonnement X atténué*, C. R. Acad. Sci. Paris Sér. I Math. **332** (2001), no. 12, 1059–1063. MR 2002e:44003

[33] _____, *An inversion formula for the attenuated X-ray transformation*, Ark. Mat. **40** (2002), 145–167.

[34] _____, *On the range characterization for the two-dimensional attenuated X-ray transformation*, Inverse Problems **18** (2002), 677–700.

[35] V. A. Sharafutdinov, *Integral geometry of tensor fields*, VSP, Utrecht, 1994. MR 97h:53077

[36] _____, *On emission tomography of inhomogeneous media*, SIAM J. Appl. Math. **55** (1995), no. 3, 707–718. MR 96e:44004

[37] D. C. Solmon, *The x-ray transform*, J. Math. Anal. Appl. **56** (1976), 61–83.

[38] _____, *Asymptotic formulas for the dual Radon transform and applications*, Math. Z. **195** (1987), 321–343.

[39] _____, *The identification problem for the exponential Radon transform*, Math. Meth. Appl. Sci. **18** (1995), 687–695.

[40] A. Tamasan, *In inverse boundary value problem in two-dimensional transport*, Inverse Problems **18** (2002), 209–219.

[41] _____, *A two dimensional inverse boundary value problem in radiation transport*, Ph.D. thesis, U. Washington, 2002.

[42] O. J. Tretiak and C. Metz, *The exponential Radon transform*, SIAM J. Appl. Math. **39** (1980), 341–354.

DAVID V. FINCH
DEPARTMENT OF MATHEMATICS
OREGON STATE UNIVERSITY
CORVALLIS, OR 97331, USA
   finch@math.orst.edu

Inverse Acoustic and Electromagnetic Scattering Theory

DAVID COLTON

ABSTRACT. This paper is a survey of the inverse scattering problem for time-harmonic acoustic and electromagnetic waves at fixed frequency. We begin by a discussion of "weak scattering" and Newton-type methods for solving the inverse scattering problem for acoustic waves, including a brief discussion of Tikhonov's method for the numerical solution of ill-posed problems. We then proceed to prove a uniqueness theorem for the inverse obstacle problems for acoustic waves and the linear sampling method for reconstructing the shape of a scattering obstacle from far field data. Included in our discussion is a description of Kirsch's factorization method for solving this problem. We then turn our attention to uniqueness and reconstruction algorithms for determining the support of an inhomogeneous, anisotropic media from acoustic far field data. Our survey is concluded by a brief discussion of the inverse scattering problem for time-harmonic electromagnetic waves.

1. Introduction

The field of inverse scattering, at least for acoustic and electromagnetic waves, can be viewed as originating with the invention of radar and sonar during the Second World War. Indeed, as every viewer of World War II movies knows, the ability to use acoustic and electromagnetic waves to determine the location of hostile objects through sea water and clouds played a decisive role in the outcome of that war. Inspired by the success of radar and sonar, the prospect was raised of the possibility of not only determining the range of an object from the transmitter, but to also image the object and thereby identify it, i.e. to distinguish between a whale and a submarine or a goose and an airplane. However, it was soon realized that the problem of identification was considerably more difficult than that of simply determining the location of a target. In particular, not only was the identification problem computationally extremely expensive, and indeed beyond the capabilities of post-war computing facilities, but the problem was also ill-posed in the sense that the solution did not depend continuously on the measured data. It was not until the 1970's with the development of the mathematical theory of ill-posed problems by Tikhonov and his school in the Soviet Union and Keith Miller and others in the United States, together with the rise of high speed computing facilities, that the possibility of imaging began to appear as a realistic possibility. Since that time, the mathematical basis of the acoustic and electromagnetic inverse scattering problem has reached a level of maturity that the imaging hopes expressed in the early post-war years have to a certain extent been realized, at least in the case of electromagnetic waves with the invention of synthetic aperature radar [3], [8]. However, as the imaging demands have increased so have the mathematical and computational expectations and hence at this time it seems appropriate to make an attempt at describing the state of the art in the mathematical foundations of acoustic and electromagnetic inverse scattering theory. This article is directed towards that goal.

Before proceeding, a few caveats are perhaps in order. The first one is obvious: we are not proposing to survey the entire field of inverse scattering theory in a few pages. In particular, we will restrict our attention to a specific area, that is inverse scattering in the frequency domain and deterministic models. This means such important topics as time-reversal and scattering by random media are ignored. Even within this restrictive framework we will be selective and hence opinionated. In particular, our view is that the mathematical field of inverse scattering theory should remain close to the applications and in particular should have the numerical solution of practical imaging problems in "real time" as a high priority. Uniqueness theorems are important since they indicate what is possible to image in an ideal noise-free world but not all reconstruction algorithms are equally valuable from this point of view. Proceeding with such judgments, since the inverse scattering problem is ill-posed, restoring stability is clearly of central importance, but again not all stability results are of equal value. In particular, in order to restore stability some type of a priori information is needed and an estimate on the noise level is in general more realistic than the knowledge of, for example, an a priori bound on the curvature of the scattering object. It is freely acknowledged that points of view other than our own are both reasonable and legitimate and we are only emphasizing our own view here in order to warn the reader of what to expect in the pages that follow.

Before proceeding to a discussion of the inverse scattering problem and methods for its numerical solution we need to be clear on what inverse scattering problem we are talking about since depending on what a priori information is available there are many inverse scattering problems! For example, in using ultrasound to image the human body it is not unreasonable to assume that the density is known and equal to the density of water. In this case incident waves at a single fixed frequency are sufficient for imaging purposes whereas this is not the case if the density is not known a priori. On the other hand in imaging a target that has been (partially) coated by an unknown material, it is not reasonable to assume that the boundary condition on the surface of the scatterer is known.

Indeed, in my opinion, this last example is more typical in the sense that one usually knows neither the shape nor the material properties of a scattering object and hence neither the shape nor boundary conditions are known. Of course if a priori information on the material properties of the scatterer are known (as, for example, in the case of ultrasound imaging of the human body) it is usually beneficial to use an algorithm which makes use of this information.

The plan of this paper is as follows. Except for the final section, we shall concentrate on the scattering of time-harmonic acoustic waves at a fixed frequency. Hence in Section 2 we shall formulate the acoustic scattering problem and discuss various inverse scattering problems and their solution by either "weak-scattering" or Newton-type methods. These two methods are the work horses of inverse scattering and typically lead to the problem of solving linear integral equations of the first kind arising in either the weak-scattering approximation or in the computation of the Fréchet derivative of a nonlinear operator. With this as motivation we shall give a brief introduction to Tikhonov's method for the numerical solution of ill-posed problems.

The methods presented in Section 2 for solving acoustic inverse scattering problems rely on rather strong a priori information on the scattering object. In Section 3 we shall turn to more recent methods which avoid such strong assumptions but at the expense of needing more data. In particular, we shall concentrate on the case of obstacle scattering and prove the Kirsch–Kress uniqueness theorem [46] which in turn serves as motivation for the linear sampling method for determining the shape of the scatterer [12],[42]. We shall in addition present a recent optimization scheme of Kirsch which has certain attractive characteristics and is closely related to the linear sampling method [44].

In Section 4 we will first consider acoustic inverse scattering problems associated with an isotropic inhomogeneous medium and begin with the uniqueness theorems of Nachman [53], Novikov [57] and Ramm [66]. In this case special problems occur in the case of scattering in R^2. We will again discuss the linear sampling method for determining the support of the inhomogeneous scattering object, leading to an investigation of the existence, uniqueness and spectral properties of the interior transmission problem [13], [14], [19], [67]. We will then proceed to an extension of these results to the case of anisotropic media. In contrast to the case of isotropic media, variational methods rather than integral equation techniques are a more convenient tool in this case [6], [7], [31].

Finally, in Section 5, we consider the inverse scattering problem for Maxwell's equations and extend some of the results in the previous sections to this situation. However, much of what is known for the scalar case of acoustic waves remains unknown in the vector case and hence is a rich area for future study. To this end, we conclude our survey with a list of open problems for the electromagnetic inverse scattering problem.

2. The Inverse Scattering Problem for Acoustic Waves

We now consider the scattering of a time harmonic acoustic wave of frequency ω by an inhomogeneous medium of compact support D having density $\rho_D(x)$ and sound speed $c_D(x)$, $x \in D \subset R^3$. We assume that the boundary ∂D is of class C^2 having unit outward normal ν (although much of the analysis which follows is also valid for Lipschitz domains-see [4],[6]) and that $\rho_D, c_D \in C^2(\bar{D})$. Then if the host medium is homogeneous with density ρ and sound speed c, the wave number k is defined by $k = \omega/c$,

$$n(x) = c/c(x), \quad x \in D$$

and the pressure $p(x, t)$ is given by

$$p(x,t) = \begin{cases} \mathrm{Re}(u(x)e^{-i\omega t}) & \text{if } x \in R^3 \setminus D, \\ \mathrm{Re}(v(x)e^{-i\omega t}) & \text{if } x \in D, \end{cases}$$

then $u \in C^2(R^3 \setminus \bar{D}) \cap C^1(R^3 \setminus D)$ and $v \in C^2(D) \cap C^1(\bar{D})$ satisfy the acoustic transmission problem

$$\Delta u + k^2 u = 0 \quad \text{in } R^3 \setminus \bar{D}, \tag{2--1a}$$

$$\rho_D(x)\nabla\left(\frac{1}{\rho_D(x)}\nabla v\right) + k^2 n(x)v = 0 \quad \text{in } D, \tag{2--1b}$$

$$u = u^i + u^s, \tag{2--1c}$$

$$\lim_{r\to\infty} r\left(\frac{\partial u^s}{\partial r} - iku^s\right) = 0, \tag{2--1d}$$

$$\begin{aligned} u &= v & \text{on } \partial D, \\ \frac{1}{\rho}\frac{\partial u}{\partial \nu} &= \frac{1}{\rho_D}\frac{\partial v}{\partial \nu} & \text{on } \partial D, \end{aligned} \tag{2--1e}$$

where u^i is the incident field, which we assume is given by

$$u^i(x) = e^{ikx\cdot d}, \quad |d| = 1,$$

and the Sommerfeld radiation condition (2–1d) holds uniformly in $\hat{x} = x/|x|$, $r = |x|$.

To allow the possibility of absorption in D we allow n to possibly have a positive imaginary part; that is, in addition to $\mathrm{Re}\, n(x) > 0$ for $x \in D$ we require that

$$\mathrm{Im}\, n(x) \geq 0$$

for $x \in D$. The existence of a unique solution to (2–1) has been established by Werner [73].

For the sake of simplicity, we shall be concerned in this and the next two sections with certain special cases of the above transmission problem. In particular, if $\rho_D \to \infty$ we are led to the exterior Neumann problem for $u \in$

$C^2(R^3 \setminus \bar{D}) \cap C^1(R^3 \setminus D),$

$$\Delta u + k^2 u = 0 \qquad \text{in } R^3 \setminus \bar{D}, \tag{2-2a}$$

$$u = u^i + u^s, \tag{2-2b}$$

$$\lim_{r \to \infty} r \left(\frac{\partial u^s}{\partial r} - iku^s \right) = 0, \tag{2-2c}$$

$$\frac{\partial u}{\partial \nu} = 0 \qquad \text{on } \partial D; \tag{2-2d}$$

if $\rho_D \to 0$ we are led to the exterior Dirichlet problem for $u \in C^2(R^3 \setminus \bar{D}) \cap C(R^3 \setminus D),$

$$\Delta u + k^2 u = 0 \qquad \text{in } R^3 \setminus \bar{D}, \tag{2-3a}$$

$$u = u^i + u^s, \tag{2-3b}$$

$$\lim_{r \to \infty} r \left(\frac{\partial u^s}{\partial r} - iku^s \right) = 0, \tag{2-3c}$$

$$u = 0 \qquad \text{on } \partial D; \tag{2-3d}$$

and if $\rho = \rho_D$ we are led to the inhomogeneous medium problem for $u \in C^1(R^3) \cap C^2(R^3 \setminus \partial D),$

$$\Delta u + k^2 n(x) u = 0 \qquad \text{in } R^3, \tag{2-4a}$$

$$u = u^i + u^s, \tag{2-4b}$$

$$\lim_{r \to \infty} r \left(\frac{\partial u^s}{\partial r} - iku^s \right) = 0, \tag{2-4c}$$

where $n(x) = 1$ in $R^3 \setminus \bar{D}$.

For the purpose of exposition, in the sequel we shall restrict our attention to the exterior Dirichlet problem and inhomogeneous medium problem (the exterior Neumann problem can be treated in essentially the same way as the exterior Dirichlet problem).

We can now be more explicit about what we mean by the acoustic inverse scattering problem. In particular, using Green's theorem and the radiation condition it is easy to show that the scattered field u^s has the representation

$$u^s(x) = \int_{\partial D} \left(u^s(y) \frac{\partial \Phi(x,y)}{\partial \nu(y)} - \frac{\partial u^s}{\partial \nu}(y) \Phi(x,y) \right) ds(y) \tag{2-5}$$

for $x \in R^3 \setminus \bar{D}$ where Φ is the radiating fundamental solution to the Helmholtz equation (2-1a) defined by

$$\Phi(x,y) := \frac{e^{ik|x-y|}}{4\pi|x-y|}, x \neq y. \tag{2-6}$$

From (2-5) and (2-6) we see that u^s has the asymptotic behavior

$$u^s(x) = \frac{e^{ikr}}{r} u_\infty(\hat{x}, d) + O\left(\frac{1}{r^2}\right) \tag{2-7}$$

as $r \to \infty$ where u_∞ is the *far field pattern* of the scattered field u^s. In the case of the exterior Dirichlet problem, the inverse scattering problem we will be concerned with is to determine D from a knowledge of $u_\infty(\hat{x}, d)$ for $\hat{x}$ and d on the unit sphere $\Omega := \{x : |x| = 1\}$ and fixed wave number k. For the inhomogeneous medium problem we will consider two inverse scattering problems, that of either determining D from $u_\infty(\hat{x}, d)$ or $n(x)$ from $u_\infty(\hat{x}, d)$, again assuming that k is fixed. In all cases, we will always assume (except in discussing uniqueness) that u_∞ is not known exactly but is determined by measurements that by definition are inexact.

The inverse scattering problems defined above are particularly difficult to solve for two reasons: they are 1) nonlinear and 2) ill-posed. Of these two reasons, it is the latter that creates the most difficulty. In particular, it is easily verified that u_∞ is an analytic function of both $\hat{x}$ and d on the unit sphere and hence, for a given measured far field pattern (i.e. "noisy data"), in general no solution exists to the inverse scattering problem under consideration. On the other hand, if a solution does exist it does not depend continuously on the measured data in any reasonable norm. Hence, before we can begin to construct a solution to an inverse scattering problem we must explain what we mean by a solution. In order to do this it is necessary to introduce "nonstandard" information that reflects the physical situation we are trying to model. Various ideas for doing this have been introduced, ranging from a priori bounds on the curvature of ∂D or the derivatives of $n(x)$ to having an a priori estimate of the noise level. The latter approach leads to what is called the Morozov discrepancy principle and will be discussed at the end of this section.

The two most popular methods for solving inverse scattering problems such as those described above are based on either what is called the "weak-scattering" approximation or on nonlinear optimization techniques. For a comprehensive discussion of such methods we refer the reader to Langenberg [51] and Biegler, et.al. [2] respectively. Here we shall content ourselves with only a brief description of these two approaches. We begin with the weak-scattering approximation, in particular the physical optics approximation for the case of the exterior Dirichlet problem.

The physical optics approximation is valid for a convex obstacle and large values of the wave number k. In particular, it is assumed that in a first approximation a convex object D may locally be considered at each point x of ∂D as a plane with normal $\nu(x)$. For the exterior Dirichlet problem, this means that not only does the total field u satisfy $u = 0$ on ∂D but also

$$\frac{\partial u}{\partial \nu} = 2\frac{\partial u^i}{\partial \nu} \tag{2-8}$$

in the illuminated region $\partial D_- := \{x \in \partial D : \nu(x) \cdot d < 0\}$ and

$$\frac{\partial u}{\partial \nu} = 0 \tag{2-9}$$

in the shadow region $\partial D_+ := \{x \in \partial D : \nu(x) \cdot d > 0\}$. Hence, using the identity

$$0 = \int\limits_{\partial D} \left(u^i(y) \frac{\partial \Phi(x,y)}{\partial \nu(y)} - \frac{\partial u^i}{\partial \nu}(y) \Phi(x,y) \right) ds(y)$$

for $x \in R^3 \setminus \bar{D}$ we see from (2–5) that under the *physical optics approximation* (2–8), (2–9)

$$u_\infty(\hat{x}, d) = -\frac{1}{2\pi} \int\limits_{\partial D_-} \frac{\partial}{\partial \nu} e^{iky \cdot d} e^{-ik\hat{x} \cdot y} ds(y) = \frac{-ik}{4\pi} \int\limits_{\partial D_1} \nu(y) \cdot d e^{ik(d-\hat{x}) \cdot y} ds(y).$$

Hence, setting $\hat{x} = -d$, replacing d by $-d$ and adding yields the *Bojarski identity*

$$u_\infty(-d, d) + \overline{u_\infty(d, -d)} = -\frac{1}{4\pi} \int\limits_{\partial D} \frac{\partial}{\partial \nu(y)} e^{2ikd \cdot y} ds(y)$$

$$= \frac{k^2}{4\pi} \int\limits_{R^3} \chi(y) e^{2ikd \cdot y} dy, \qquad (2\text{–}10)$$

where χ is the characteristic function of D. Hence, under the assumption that k is large, D is convex and $u = 0$ on ∂D, (2–10) is a linear integral equation which in principle yields D from a knowledge of u_∞. However, the kernel of this integral equation is analytic and hence solving (2–10) is a severely ill-posed problem! We shall indicate possible methods for solving such problems at the end of this section. Note that in order to ensure injectivity we must assume that (2–10) holds for an interval of k values.

An analogous procedure to the above method for attempting to solve the inverse obstacle problem can also be carried out for the inverse inhomogeneous medium problem, this time under the assumption that the wave number k is small. To derive the desired integral equation we reformulate the inhomogeneous medium problem (2–4) as the *Lippmann–Schwinger integral equation*

$$u(x) = u^i(x) - k^2 \int\limits_{R^3} \Phi(x,y) m(y) u(y) dy, \qquad x \in R^3, \qquad (2\text{–}11)$$

where $m := 1 - n$. If k is small, we can solve (2–11) by successive approximations and, if we replace u by the first term in this iterative process and let $r = |x| \to \infty$, we obtain the *Born approximation*

$$u_\infty(\hat{x}, d) = -\frac{k^2}{4\pi} \int\limits_{R^3} e^{-ik\hat{x} \cdot y} m(y) u^i(y) dy. \qquad (2\text{–}12)$$

(2–12) is again a linear integral equation of the first kind for the determination of m from u_∞ under the assumption that k is sufficiently small and $\rho = \rho_D$ in (2–1). In order to ensure injectivity we must again assume that (2–12) is valid for an interval of k values. For further developments in this direction see [22].

Although the weak scattering models discussed above have had considerable success, particularly in their extensions to the electromagnetic case and use in the development of synthetic aperature radar, they suffer in more complicated imaging problems where multiple scattering effects can no longer be ignored. In order to treat such problems, a considerable effort has been put into the derivation of robust nonlinear optimization schemes. The advantage of such an approach is that u_∞ need only be known for a single fixed value of k and multiple scattering effects are no longer ignored, although it is still necessary to have some a priori knowledge of the physical properties of the scattering object (e.g. $u = 0$ on ∂D or $\rho = \rho_D$ as in the above examples). A difficulty with nonlinear optimization techniques is that they are often computationally very expensive.

We begin our discussion of nonlinear optimization methods for solving the inverse scattering problem by considering the exterior Dirichlet problem (2–3). To this end we note the solution to the direct scattering problem with a fixed incident plane wave u^i defines an operator $\mathcal{F} : \partial D \to U_\infty$ which maps the boundary ∂D onto the far field pattern u_∞ of the scattered field. In terms of this operator, the inverse problem consists in solving the nonlinear equation $\mathcal{F}(\partial D) = u_\infty$. Having in mind that for ill-posed problems the norm in the data space has to be suitable for describing the measurement error, we make the assumption that u_∞ is in the Hilbert space $L^2(\Omega)$. For ∂D we need to choose a class of admissible surfaces described by some suitable parameterization and equipped with an appropriate norm. For the sake of simplicity, we restrict ourselves to the class of domains D that are star-like with respect to the origin with C^2 boundary ∂D, i.e. we assume that ∂D is represented in its parametric form

$$x = r(\hat{x})\hat{x}, \quad \hat{x} \in \Omega,$$

for a positive function $r \in C^2(\Omega)$. We now view the operator $\mathcal{F}$ as a mapping from $C^2(\Omega)$ into $L^2(\Omega)$ and write $\mathcal{F}(\partial D) = u_\infty$ as

$$\mathcal{F}(r) = u_\infty. \tag{2–13}$$

The following basic theorem was first proved by Kirsch [41] using variational methods and subsequently by Potthast [63] using a boundary integral equation approach (see also [34] and [48]). We note that the validity of the following theorem for the case of the exterior Neumann problem remains an open question [50].

THEOREM 2.1. *The boundary to far field map* $\mathcal{F} : C^2(\Omega) \to L^2(\Omega)$ *has a Fréchet derivative* $\mathcal{F}'$. *The linear operator* $\mathcal{F}'$ *is compact and injective with dense range.*

Theorem 2.1 now allows us to apply Newton's method to solve (2–13). In particular, given a far field pattern u_∞ and initial guess r_0 to r, the nonlinear equation (2–13) is replaced by the linearized equation

$$\mathcal{F}(r_0) + \mathcal{F}'q = u_\infty,$$

which is then solved for q to yield the new approximation r, given by $r_1 = r_0 + q$. Newton's method than consists in iterating this procedure. Note that since $\mathcal{F}'$ is compact each step of the iteration procedure is ill-posed. Alternate optimization strategies for determining D have been proposed by numerous people in particular Kirsch and Kress [45], Angell, Kleinman and Roach [1] and Maponi et al. [52].

Newton's method can also be used to determine the coefficient $n(x)$ in the inverse inhomogeneous medium problem [32] In this case the nonlinear operator $\mathcal{F}$ is defined by means of the Lippmann–Schwinger integral equation (2–11). Other methods for determining $n(x)$ have been proposed by Colton and Monk (see Chapter 10 of [14])who use an averaging procedure to reduce the number of unknowns, Gutman and Klibanov [26], who confine themselves to reconstructing a fixed number of Fourier coefficients of n where the number depends on the wave number k, Kleinman and Van den Berg [72], who use a modified gradient method for an output least squares formulation of the problem, and Natterer and Wübbeling [54], [55] who employ an algebraic reconstruction technique (ART) to determine $n(x)$. We shall conclude our brief discussion of nonlinear optimization schemes to solve inverse scattering problems by describing the method of Natterer and Wübbeling.

Our aim is to reconstruct the coefficient $n(x)$ from a knowledge of the far field pattern corresponding to the inhomogeneous medium scattering problem (2–4). We assume that $D \subset \{x : |x| < \rho\}$ and that our data is p far field patterns $u_\infty(\hat{x}, d_j)$, $j = 1, \ldots, p$, corresponding to p distinct incident plane waves. From each $u_\infty(\hat{x}, d_j)$ we can determine the Cauchy data on the planes $\Gamma_j^\pm$ perpendicular to d_j for the solution u_j of (2–4) corresponding to $u^i(x) = \exp(ikx \cdot d_j)$ (see figure). Assuming to begin with that $n(x)$ is known, we want to determine u_j

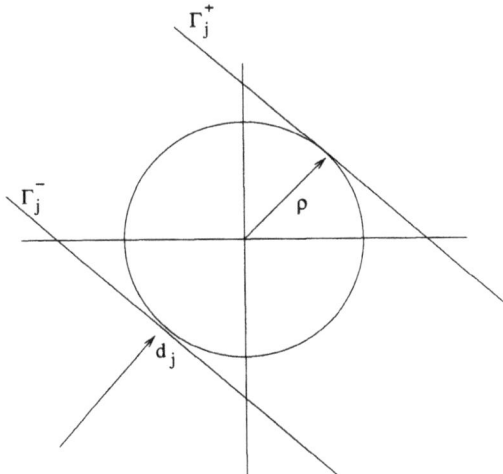

on Γ_j^+ from the ill-posed Cauchy problem

$$\Delta u_j + k^2 n(x) u_j = 0 \qquad \text{in } R^3, \tag{2-14a}$$

$$u = f \qquad \text{on } \Gamma_j^-,$$
$$\nabla u \cdot d_j = g \qquad \text{on } \Gamma_j^- \tag{2-14b}$$

This can be done in a stable fashion by finite difference methods if we first filter out frequencies greater than κ where $\kappa < k$. We now define the nonlinear operator $R_j : L^2(|x| < \rho) \to L^2(\Gamma_j^+)$ by

$$R_j(n) = u_\kappa^j|_{\Gamma_j^+}, \tag{2-15}$$

where u_κ^j is the filtered solution of (2-14). Our aim is to now solve the inverse scattering problem by using an ART-type procedure to solve (2-15) for $j = 1, \ldots, p$.

To solve (2-15) for n we set $g_j = u_\kappa^j|_{\Gamma_j^+}$ and solve this equation iteratively by first determining n_p from

$$n_0 = n^0,$$
$$n_j = n_{j-1} + \omega R_j'(n_{j-1})^* C_j^{-1}(g_j - R_j(n_{j-1})),$$

where n^0 is an initial guess, ω is a relaxation parameter, R_j' is the Frechet derivative of R_j, $C_j = R_j'(0)(R_j'(0))^*$ where $*$ denotes the adjoint operator and the operator C_j^{-1} can be applied through the use of Fourier transforms (see [54]). The first approximation is now defined to be n_p and the procedure is repeated. For details we refer the reader to Natterer and Wübbeling [54] [55] where the computational advantages of using such an approach are discussed. An extension of this method (which is sometimes called the "adjoint field method") to the case of time-harmonic electromagnetic waves has been done by Dorn, et.al.[23].

In both the weak scattering and Newton-type methods for solving the inverse scattering problem we are faced with the problem of solving a linear operator equation of the form

$$A\varphi = f$$

where $A : X \to Y$ is compact and X and Y are infinite dimensional normed spaces. We shall also encounter such equations in the sequel when we consider linear sampling methods for solving the inverse scattering problem. Hence it is appropriate to conclude this section of our paper by giving some idea of how such equations can be solved numerically. The problem in doing this is that since A is compact solving $A\varphi = f$ is an ill-posed problem in the sense that A^{-1}, if it exists, is unbounded. This follows immediately from the fact that if A^{-1} were bounded then $I = A^{-1}A$ is compact, a contradiction since X is infinite dimensional. Our discussion will purposefully be brief and for more information on the solution of ill-posed problems we refer the reader to Engl, Hanke and Neubauer [24], Kirsch [40] and Kress[47].

We restrict our attention to the case when X and Y are infinite dimensional Hilbert spaces. We denote by $(\sigma_n, \varphi_n, g_n)$ a singular system for the compact operator $A : X \to Y$, so that

$$A\varphi_n = \sigma_n g_n, \qquad A^* g_n = \sigma_n \varphi_n,$$

and we denote the null space of A by $N(A)$.

PICARD'S THEOREM. $A\varphi = f$ is solvable if and only if $f \in N(A^*)^\perp$ and

$$\sum_{n=1}^\infty \frac{1}{\sigma_n^2} |(f, g_n)|^2 < \infty.$$

In this case a solution is given by

$$\varphi = \sum_{n=1}^\infty \frac{1}{\sigma_n} (f, g_n) \varphi_n.$$

DEFINITION 2.2. The equation $A\varphi = f$ is *mildly ill-posed* if $\sigma_n = O(n^{-\beta})$, for $\beta \in R^+$, and *severely ill-posed* if the σ_n decay faster than this.

We note that the equations appearing in inverse scattering theory are typically severely ill-posed.

For severely ill-posed problems we must use regularization methods to arrive at a solution.

DEFINITION 2.3. Let $A : X \to Y$ be an injective compact linear operator. Then a family of bounded linear operators $R_\alpha : Y \to X, \alpha > 0$, such that

$$\lim_{\alpha \to 0} R_\alpha A\varphi = \varphi$$

for all $\varphi \in X$ is called a *regularization scheme* for A with *regularization parameter* α.

Suppose the solution φ of $A\varphi = f$ is approximated by

$$\varphi_\alpha^\delta := R_\alpha f^\delta$$

where $\|f - f^\delta\| \leq \delta$. Then

$$\varphi_\alpha^\delta - \varphi = R_\alpha f^\delta - R_\alpha f + R_\alpha A\varphi - \varphi$$

and hence

$$\|\varphi_\alpha^\delta - \varphi\| \leq \delta \|R_\alpha\| + \|R_\alpha A\varphi - \varphi\|.$$

The first term in the above equation increases as α tends to zero (since A^{-1} is not bounded) whereas the second term is only small when α tends to zero. How should $\alpha = \alpha(\delta)$ be chosen?

DEFINITION 2.4. The choice of $\alpha = \alpha(\delta)$ is called *regular* if for all $f \in A(X)$ and all $f^\delta \in Y$ with $\|f - f^\delta\| \le \delta$ we have

$$R_{\alpha(\delta)} f^\delta \to A^{-1} f$$

as δ tends to zero.

We shall now describe a regular regularization scheme due to Tikhonov and Morozov for solving the ill-posed equation $A\varphi = f$.

Assume once again that $A : X \to Y$ is compact. Then, since $A^* A \ge 0$, for every $\alpha > 0$ the operator $\alpha I + A^* A : X \to X$ is bijective with bounded inverse. The *Tikhonov regularization method* for solving $A\varphi = f$ is to set

$$R_\alpha := (\alpha I + A^* A)^{-1} A^*.$$

Then the regularized solution φ_α of $A\varphi = f$ is the unique solution of

$$\alpha \varphi_\alpha + A^* A \varphi_\alpha = A^* f.$$

In particular, if A is injective, then

$$\varphi_\alpha = \sum_{n=l}^{\infty} \frac{\sigma_n}{\alpha + \sigma_n^2} (f, g_n) \varphi_n = R_\alpha f$$

and hence as α tends to zero we have $R_\alpha A\varphi \to \varphi$ for all $\varphi \in X$. The function φ_α can also be obtained by minimizing the *Tikhonov functional*

$$\|A\varphi - f\|^2 + \alpha \|\varphi\|^2.$$

Note that if A is injective with dense range then $\|\varphi_\alpha\| \to \infty$ as α tends to zero if and only if $f \notin A(X)$.

We now turn to the choice of the regularization parameter α. If A is injective with dense range then a regular method for choosing α is the *Morozov discrepancy principle*. In particular, assume that we want to solve

$$A_\delta \varphi = f_\varepsilon$$

where $\|A - A_\delta\| \le \delta$ and $\|f - f_\varepsilon\| \le \varepsilon$ with known δ and ε, i.e. an estimate of the noise level is known a priori. We require that the residual be commensurate with the accuracy of the measurements of A and f, i.e.

$$\|A_\delta \varphi_\alpha - f_\varepsilon\| \approx \varepsilon + \delta \|\varphi_\alpha\|.$$

In applications to the linear sampling method described in the sequel we have $\varepsilon \ll \delta$. Hence, in this case, the Morozov discrepancy principle is to choose $\alpha = \alpha(\delta)$ such that $\mu(\alpha) = 0$, where

$$\mu(\alpha) := \|A_\delta \varphi_\alpha - f_\varepsilon\|^2 - \delta^2 \|\alpha_\alpha\|^2 = \sum_{n=1}^{\infty} \frac{\alpha^2 - \delta^2 \sigma_n^2}{(\sigma_n^2 + \alpha)^2} |(f_\varepsilon, g_n)|^2$$

and $(\sigma_n, \varphi_n, g_n)$ is now a singular system for the operator A_δ. We note that $\mu(\alpha)$ is monotonously increasing and in practice only a rough approximation to the root of $\mu(\alpha) = 0$ is necessary.

In closing this section, we make a few comments on the use of Tikhonov regularization and the Morozov discrepancy principle in solving ill-posed problems arising in inverse scattering theory. In particular, we note that in general one has no idea if the noise level is small enough so that the regularized solution of the equation with noisy data is in fact a good approximation to the solution of the noise free equation. Without further a priori information the only statement that can be made is what happens if the noise tends to zero. However, since the noise is fixed and nonzero, in general all that can be said is that there is a "nearby" equation (i.e. noisy A and f) whose solution can be obtained and if this nearby equation is "close enough" to the noise free equation then one expects the regularized solution to behave like the true solution, assuming it exists. In particular, since without severe a priori assumptions, which are in general not available, error estimates are not known for the dependency of the regularized solution on the noise level, and the remark of Lanczos is valid: "A lack of information cannot be remedied by any mathematical trickery." Nevertheless, a regularized solution based on Tikhonov regularization and the Morozov discrepancy principle provides a rational approach for arriving at a candidate for a solution to an ill-posed problem when an a priori estimate of the noise level is available.

3. The Inverse Dirichlet Problem for Acoustic Waves

In the previous section we discussed two of the most popular methods for solving the inverse scattering problem, i.e. the weak scattering approximation and nonlinear optimization techniques, as well as regularization methods that can be used for their numerical solution. However, as previously mentioned, both of these methods rely on some a priori knowledge of the physical properties of the scattering object D in order to know the boundary conditions on ∂D. Furthermore, uniqueness theorems were not discussed, in particular how much information is needed in principle to determine D or, more importantly, can D be determined from the far field data if the boundary conditions are not known? We view these issues as particularly important since in many practical inverse scattering problems both the material properties of the scatterer as well as its shape are unknown. In this and the sections that follow we will be paying particular attention to inverse scattering problems such as these.

In this section we will consider the inverse scattering problem associated with the exterior Dirichlet problem (2–3) and, when relevant, point out what results are in fact independent of the boundary condition on ∂D. We will always assume the existence of a solution $u \in C^2(R^3 \setminus \bar{D}) \cap C(R^3 \setminus D)$ to (2–3) as well as the fact that since ∂D is in class C^2 we have $u \in C^1(R^3 \setminus D)$ [14]. We begin with

Rellich's lemma which forms the basis of the entire field of acoustic scattering theory.

THEOREM 3.1 (RELLICH'S LEMMA). *Let u^s be a solution of the Helmholtz equation in the exterior of D satisfying the Sommerfeld radiation condition such that the far field pattern u_∞ of u^s vanishes. Then $u^s = 0$ in $R^3 \setminus \bar{D}$.*

PROOF. For sufficiently large $|x|$ we have a Fourier expansion

$$u^s(x) = \sum_{n=0}^{\infty} \sum_{m=-n}^{n} a_n^m(r) Y_n^m(\hat{x})$$

with respect to the spherical harmonics Y_n^m where the coefficients are given by

$$a_n^m(r) = \int_\Omega u^s(r\hat{x}) \overline{Y_n^m(\hat{x})}\, ds(\hat{x}).$$

Since $u^s \in C^2(R^3 \setminus \bar{D})$ and the radiation condition holds uniformly in $\hat{x}$, we can differentiate under the integral sign and integrate by parts to conclude that a_n^m is a solution of the spherical Bessel equation

$$\frac{d^2 a_n^m}{dr^2} + \frac{2}{r}\frac{da_n^m}{dr} + (k^2 - \frac{n(n+1)}{r^2})a_n^m = 0$$

satisfying the radiation condition, i.e.

$$a_n^m(r) = \alpha_n^m h_n^{(1)}(kr)$$

where $h_n^{(1)}$ is a spherical Hankel function of the first kind of order n and the α_n^m are constants depending only on n and m. From (2–7) we have that, since $u_\infty = 0$,

$$\lim_{r\to\infty} \int_{|x|=r} |u^s(x)|^2 ds = \int_\Omega |u_\infty(\hat{x})|^2 ds = 0.$$

But by Parseval's equality

$$\int_{|x|=r} |u^s(x)|^2 ds = r^2 \sum_{n=0}^{\infty} \sum_{m=-n}^{n} |a_n^m(r)|^2.$$

Substituting the above expression for a_n^m into this identity, letting r tend to infinity, and using the asymptotic behavior of the spherical Hankel functions now yields $\alpha_n^m = 0$ for all n and m. Hence $u^s = 0$ outside a sufficiently large sphere. By the representation formula (2–5) we see that u^s is an analytic function of x, and hence we can now conclude that $u^s = 0$ in $R^3 \setminus \bar{D}$ by analyticity. $\square$

In the sequel we will need two *reciprocity relations* which can be proved by a straightforward application of Green's theorem. The first of these is for scattering by a plane wave, i.e. $u^i(x) = e^{ik\hat{x}\cdot d}$ in (2–3), and is given by [14]

$$u_\infty(\hat{x}, d) = u_\infty(-d, -\hat{x}) \tag{3–1}$$

and the second of these is for scattering due to the point source $u^i(x) = \Phi(x, z)$ having far field pattern $u_\infty(\hat{x}, z)$ and is given by [62]

$$4\pi u_\infty(-d, z) = u^s(z, d) \tag{3-2}$$

for $z \in R^3 \setminus \bar{D}, d \in \Omega$, where u^s is the scattered field in (2–3) corresponding to $u^i(x) = e^{ikx \cdot d}$.

We now consider the scattering problem (2–3) and let u_∞ be the far field pattern of the scattered field. The *far field operator* $F : L^2(\Omega) \to L^2(\Omega)$ for this problem is defined by

$$(Fg)(\hat{x}) := \int_\Omega u_\infty(\hat{x}, d) g(d) \, ds(d) \tag{3-3}$$

and is easily seen to be the scattered field v^s_g corresponding to the *Herglotz wave function*

$$v^i_g(x) := \int_\Omega e^{ikx \cdot d} g(d) \, ds(d), \quad x \in R^3$$

as incident field. The function $g \in L^2(\Omega)$ is known as the kernel of the Herglotz wave function. Of basic importance to us is the following theorem [14].

THEOREM 3.2. *The far field operator F is injective with dense range if and only if there does not exist a Dirichlet eigenfunction for D which is a Herglotz wave function.*

PROOF. For the L^2 adjoint $F^* : L^2(\Omega) \to L^2(\Omega)$ the reciprocity relation (3–1) implies that

$$F^* g = \overline{RFR\bar{g}} \tag{3-4}$$

where $R : L^2(\Omega) \to L^2(\Omega)$ is defined by

$$(Rg)(d) := g(-d).$$

Hence, the operator F is injective if and only if its adjoint F^* is injective. Recalling that the denseness of the range of F is equivalent to the injectivity of F^* we therefore must only show the injectivity of F. To this end, we note that $Fg = 0$ with $g \neq 0$ is equivalent to the existence of a nontrivial Herglotz wave function v^i_g with kernel g for which the far field pattern of the corresponding scattered field v^s is $v_\infty = 0$. By Rellich's lemma this implies $v^s = 0$ in $R^3 \setminus D$ and the boundary condition $v^i_g + v^s = 0$ on ∂D now shows that $v^i_g = 0$ on ∂D. The proof is finished.   □

We now want to show that the far field F is normal. To this end, we need the following lemma [15].

LEMMA 3.3. *The far field operator F satisfies*

$$2\pi\big((Fg, h) - (g, Fh)\big) = ik(Fg, Fh).$$

PROOF. If v^s and w^s are radiating solutions to the Helmholtz equation with far field patterns v_∞ and w_∞ then from the radiation condition and Green's theorem we obtain

$$\int_{\partial D} \left(v^s \frac{\partial \overline{w^s}}{\partial \nu} - \overline{w^s} \frac{\partial v^s}{\partial \nu} \right) ds = -2ik \int_{\Omega} v_\infty \overline{w_\infty} \, ds.$$

From the representation formula (2–5) and letting $x \to \infty$ we see that, if w_h^i is a Herglotz wave function with kernel h, then

$$\int_{\partial D} \left(v^s(x) \frac{\partial \overline{w_h^i}}{\partial \nu}(x) - \overline{w_h^i(x)} \frac{\partial v^s}{\partial \nu}(x) \right) ds(x)$$

$$= \int_{\Omega} \overline{h(d)} \int_{\partial D} \left(v^s(x) \frac{\partial e^{-ikx \cdot d}}{\partial \nu} - e^{-ikx \cdot d} \frac{\partial v^s}{\partial \nu}(x) \right) ds(x) \, ds(d)$$

$$= 4\pi \int_{\Omega} \overline{h(d)} v_\infty(d) \, ds(d).$$

Now let v_g^i and v_h^i be Herglotz wave functions with kernels $g, h \in L^2(\Omega)$, respectively, and let v_g, v_h be the solutions of (2–3) with u^i replaced by v_g^i and v_h^i, respectively. Let $v_{g,\infty}$ and $v_{h,\infty}$ be the corresponding far field patterns. Then we can combine the two previous equations to obtain

$$-2ik(Fg, Fh) + 4\pi(Fg, h) - 4\pi(g, Fh)$$

$$= -2ik \int_{\Omega} v_{g,\infty} \overline{v_{h,\infty}} \, ds + 4\pi \int_{\Omega} v_{g,\infty} \overline{h} \, ds - 4\pi \int_{\Omega} g \overline{v_{h,\infty}} \, ds$$

$$= \int_{\partial D} \left(v_g \frac{\partial \overline{v_h}}{\partial \nu} - \overline{v_h} \frac{\partial v_g}{\partial \nu} \right) ds$$

and the lemma follows from the Dirichlet boundary condition satisfied by v_g and v_h. $\square$

THEOREM 3.4. *The far field operator F is compact and normal.*

PROOF. Since F is an integral operator on Ω with a continuous kernel, it is compact. From Lemma 3.3 we have

$$(g, ikF^*Fh) = 2\pi\big((g, Fh) - (g, F^*h)\big)$$

for all $g, h \in L^2(\Omega)$ and hence

$$ikF^*F = 2\pi(F - F^*). \qquad (3\text{–}5)$$

Using (3–4) we can deduce that

$$(F^*g, F^*h) = (FR\overline{h}, FR\overline{g})$$

and hence from Lemma 3.3 again it follows that

$$ik(F^*g, F^*h) = 2\pi\big((g, F^*h) - (F^*g, h)\big)$$

for all $g, h \in L^2(\Omega)$. If we now proceed as in the derivation of (3–6) we find that

$$ikFF^* = 2\pi(F - F^*) \tag{3-6}$$

and the proof is finished. □

The proof of Theorem 3.4 carries over to the case of Neumann boundary data. However for the impedance boundary condition

$$\frac{\partial u}{\partial \nu} + ik\lambda u = 0$$

where $\lambda > 0$ the operator F is no longer normal since Lemma 3.3 is not valid, i.e. absorption destroys normality. Finally, returning to the case of Dirichlet boundary data, if we define the *scattering operator* $S : L^2(\Omega) \to L^2(\Omega)$ by

$$S = I + \frac{ik}{2\pi}F$$

then from (3–5) and (3–6) we see that $SS^* = S^*S = I$, i.e., S is unitary.

Having established the basic properties of the far field pattern and far field operator, we now turn our attention to the uniqueness of a solution to the inverse scattering problem, basing our analysis on the approach of Kirsch and Kress [46] with a subsequent simplification of the proof by Potthast [62].

THEOREM 3.5. *Assume that D_1 and D_2 are two obstacles such that the far field patterns corresponding to the exterior Dirichlet problem (2–3) for D_1 and D_2 coincide for all incident directions d. Then $D_1 = D_2$.*

PROOF. By analyticity and Rellich's lemma the scattered fields $u_1^s(\,\cdot\,, d) = u_2^s(\,\cdot\,, d)$ for the incident fields $u^i(x, d) = e^{ikx\cdot d}$ coincide in the unbounded component G of the complement of $\bar{D}_1 \cup \bar{D}_2$ for all $d \in \Omega$. Then from the reciprocity relation (3–2) we can conclude that the far field patterns $u_{1,\infty}(\,\cdot\,, z) = u_{2,\infty}(\,\cdot\,, z)$ for the scattering of point sources $\Phi(\,\cdot\,, z)$ coincide for all point sources located at $z \in G$. Again by Rellich's lemma, this implies that the corresponding scattered fields satisfy $u_1^s(x, z) = u_2^s(x, z)$ for all $x, z \in G$.

Now assume that $D_1 \neq D_2$. Then, without loss of generality, there exists $x^* \in \partial G$ such that $x^* \in \partial D_1$ and $x^* \notin \bar{D}_2$. In particular, we have

$$z_n := x^* + \frac{1}{n}\nu(x^*)$$

is in G for integers n sufficiently large. Then, on the one hand, we have

$$\lim_{n\to\infty} u_2^s(x^*, z_n) = u_2^s(x^*, x^*)$$

since $u_2^s(x^*, \cdot)$ is continuous in a neighborhood of $x^* \notin \bar{D}_2$ due to the well-posedness of the direct scattering problem. On the other hand, we have

$$\lim_{n \to \infty} u_1^s(x^*, z_n) = \infty$$

because of the boundary condition $u_1^s(x^*, z_n) + \Phi(x^*, z_n) = 0$. This contradicts the fact that $u_1^s(x^*, z_n) = u_2^s(x^*, z_n)$ for n sufficiently large and the proof in complete. □

A closer examination of the proof of Theorem 3.5 shows that the boundary condition $u = 0$ is not used explicitly but rather only the well-posedness of the direct scattering problem. Hence it is not necessary to know the boundary condition (2–3d) a priori in order to conclude that the far field pattern uniquely determines the scatterer. In fact it is not even necessary to know if u_∞ is the far field pattern of (2–2), (2–3) or (2–4) in order to conclude that D is uniquely determined [41], [42]. In a related direction, Potthast [62], [64] has considered the important case of finite data. In particular, if $\Omega_n \subset \Omega$ is a set of n uniformly distributed unit vectors such that if

$$d(\hat{x}, \Omega_n) := \inf_{d \in \Omega_n} |\hat{x} - d|$$

then (1) $d(\hat{x}, \Omega_n) \to 0$ as $n \to \infty$; (2) $d \in \Omega_n \implies -d \in \Omega_n$ if n is even; and (3) $\Omega_{n'} \subset \Omega_n$ for $n > n'$ then the following theorem is valid.

THEOREM 3.6. *Let u_1^∞ and u_2^∞ be the far field patterns corresponding to one of (2–2), (2–3) or (2–4). Given $\varepsilon > 0$ there exists integers n_0 and n_i such that if $u_1^\infty(\hat{x}, d) = u_2^\infty(\hat{x}, d)$ for $\hat{x} \in \Omega_{n_0}, d \in \Omega_{n_i}$ then*

$$d(D_1, D_2) \leq \varepsilon,$$

where $d(D_1, D_2)$ denotes the Hausdorff distance between D_1 and D_2.

An open problem is to determine if one incident plane wave at a fixed wave number k is sufficient to uniquely determine the scatterer D. If it is known a priori that the boundary condition (2–3d) is satisfied and that furthermore D is contained in a ball of radius R such that $kR < \pi$ then it was shown by Colton and Sleeman ([20] and Corollary 5.3 of [14]) that D is uniquely determined by its far field pattern for a single incident direction d and fixed wave number k.

We now turn our attention to a method for reconstructing D from an inexact knowledge of the far field pattern u_∞ of the scattering problem (2–3) that is closely related to the ideas of the proof of the uniqueness Theorem 3.5. Indeed, as with Theorem 3.5, this method can be implemented without knowing a priori which of the scattering problems (2–2), (2–3) or (2–4) is associated with u_∞ and in this sense has a clear advantage over the reconstruction methods discussed in the previous section. On the other hand, the implementation requires a knowledge of $u_\infty(\hat{x}, d)$ for $\hat{x}, d$ on open subsets of Ω whereas for obstacle scattering Newton's method only requires a single incident direction d. Furthermore, for the case of the scattering problem (2–4), only the support D is obtained rather than the coefficient $n(x)$. The method we have on mind was first introduced by Colton and Kirsch in [12], with a subsequent second version being given by Kirsch in [42] and [43] and has become known as the *linear sampling method* (related methods have been considered by Ikehata [35], Norris [56] and Potthast [65]). Here, for the sake of simplicity, we will assume that $u_\infty(\hat{x}, d)$ is known for all $\hat{x}, d \in \Omega$ rather than only on a subset of Ω. For the case of the first version of the linear sampling method, this latter case can be easily handled by appealing to the result that a Herglotz wave function and its first derivatives can be approximated on compact subsets of a ball B by another Herglotz wave function having a kernel that is compactly supported on Ω. (This can easily be shown by assuming without loss of generality that k^2 is not a Dirichlet eigenvalue for B and then using the ideas of the proof of Theorem 5.5 of [14] to show that Herglotz wave functions with compactly supported kernels are dense in $L^2(\partial B)$).

To describe the basic idea behind the linear sampling method, assume that for every $z \in D$ there exists a unique solution $g = g(\,\cdot\,, z) \in L^2(\Omega)$ to the *far field equation*

$$\int_\Omega u_\infty(\hat{x}, d)g(d)\, ds(d) = \Phi_\infty(\hat{x}, z), \tag{3–7}$$

where

$$\Phi_\infty(\hat{x}, z) = \frac{1}{4\pi}e^{-ik\hat{x}\cdot z}$$

and u_∞ is the far field pattern corresponding to the scattering problem (2–3). Then, since the right hand side of (3–7) is the far field pattern of the fundamental solution $\Phi(x, z)$, it follows from Rellich's lemma that

$$\int_\Omega u^s(x, d)g(d)\, ds(d) = \Phi(x, z)$$

for $x \in R^3 \setminus D$. From the boundary condition $u = 0$ on ∂D it now follows that

$$v_g^i(x) + \Phi(x, z) = 0 \quad \text{for } x \in \partial D, \tag{3–8}$$

where v_g^i is the Herglotz wave function with kernel g. We now see from (3–8) that v_g^i becomes unbounded as $z \to x \in \partial D$ and hence

$$\lim_{\substack{z \to \partial D \\ z \in D}} \|g(\,\cdot\,, z)\| = \infty,$$

that is, ∂D is characterized by points z where the solution of (3–7) becomes unbounded.

Unfortunately, in general the far field equation

$$Fg = \Phi_\infty(\,\cdot\,, z)$$

does not have a unique solution, nor does the above analysis say anything about what happens when $z \in R^3 \setminus D$. However, using on the one hand the fact that

Herglotz wave functions are dense in the space of solutions to the Helmholtz equation in D with respect to the norm in the Sobolev space $H^1(D)$ (see [16], [21]) and on the other the factorization of the far field operator F as

$$(Fg) = -\frac{1}{4\pi}\mathcal{F}S^{-1}(Hg),$$

where $S : H^{-1/2}(\partial D) \to H^{1/2}(\partial D)$ is the single layer potential

$$(S\varphi)(x) := \int_{\partial D} \varphi(y)\Phi(x,y)\,ds(y), \qquad (3\text{-}9)$$

Hg is the trace on ∂D of the Herglotz wave function, and $\mathcal{F} : H^{-1/2}(\partial D) \to L^2(\Omega)$ is defined by

$$(\mathcal{F}\varphi)(\hat{x}) := \int_{\partial D} \varphi(y)\varphi^{-ik\hat{x}\cdot y}\,ds(y),$$

we can prove the following result [4].

THEOREM 3.7. *Assume that k^2 is not a Dirichlet eigenvalue for D. Then*

(1) *if $z \in D$ for every $\varepsilon > 0$ there exists a solution $g(\,\cdot\,,z) \in L^2(\Omega)$ of the inequality*

$$\left\|Fg(\,\cdot\,,z) - \Phi_\infty(\,\cdot\,,z)\right\| < \varepsilon$$

such that

$$\lim_{z\to\partial D}\left\|g(\,\cdot\,,z)\right\|_{L^2(\Omega)} = \infty, \qquad \lim_{z\to\partial D}\left\|v_g^i(\,\cdot\,,z)\right\|_{H^1(D)} = \infty,$$

(2) *if $z \in R^3\setminus\bar{D}$ for every $\varepsilon > 0$ and $\gamma > 0$ there exists a solution $g(\,\cdot\,,z) \in L^2(\Omega)$ of the inequality*

$$\left\|Fg(\,\cdot\,,z) - \Phi_\infty(\,\cdot\,,z)\right\| < \varepsilon + \gamma$$

such that

$$\lim_{\gamma\to 0}\left\|g(\,\cdot\,,z)\right\|_{L^2(\Omega)} = \infty, \qquad \lim_{\gamma\to 0}\left\|v_g^i(\,\cdot\,,z)\right\|_{H^1(D)} = \infty.$$

We note that the difference between cases (1) and (2) of this theorem is that, for $z \in D$, $\Phi_\infty(\,\cdot\,,z)$ is in the range of $\mathcal{F}$ whereas for $z \in R^3 \setminus \bar{D}$ this is no longer true.

The above theorem now suggests a numerical procedure for determining ∂D from noisy far field data. In particular, let u_∞^δ be the measured far field data, i.e. $\|u_\infty^\delta - u_\infty\| < \delta$ and assume g is such that $\|Fg - \Phi_\infty(\,\cdot\,,z)\| < \varepsilon$. If F_δ is the operator F with kernel u_∞ replaced by u_∞^δ, then we want to find an approximation to g by solving $F_\delta\varphi = \Phi_\infty(\,\cdot\,,z)$; that is, we view both the operator and the right hand side as being inexact. This equation is now solved using Tikhonov regularization and the Morozov discrepancy principle. The unknown boundary ∂D is now determined by looking for those points z where $\|\varphi(\,\cdot\,,z)\|$ begins to sharply increase. Numerical examples using this procedure can be found in [9], [10] and [74].

The analogue of Theorem 3.7 for the exterior Neumann problem is established in exactly the same way as Theorem 3.7 where now it is assumed that k^2 is not a Neumann eigenvalue for D. It is also possible to treat mixed boundary value problems [4], [6]. As will be seen in the next section, a similar result also holds for the inhomogeneous medium problem (2–4) as well as the more general problem (2–1) provided k^2 is not a transmission eigenvalue (to be defined in the following section of this paper). In particular, as with Theorems 3.5 and 3.6, it is not necessary to know the material properties of the scatterer (e.g., the boundary condition) in order to determine D from a knowledge of the regularized solution of the far field equation. it is also possible to treat the case when the background medium is piecewise homogeneous by appropriately modifying the far field equation [9]. The possibility of doing this is particularly important in numerous applications, e.g. the detection of buried objects or structures under foliage.

Theorem 3.7 is complicated by the fact that in general $\Phi_\infty(\cdot, z)$ is not in the range of F for neither $z \in D$ nor $z \in R^3 \setminus \bar{D}$. For the case when F is normal (say nonabsorbing media and data on all of Ω rather than some subset of Ω), this problem was resolved by Kirsch [42], who proposed replacing the equation $Fg = \Phi_\infty(\cdot, z)$ by $(F^*F)^{\frac{1}{4}}g = \Phi_\infty(\cdot, z)$ where F^* is again the adjoint of F in $L^2(\Omega)$. He was then able to show that $\Phi(\cdot, z)$ is in the range of $(F^*F)^{\frac{1}{4}}$ if and only if $z \in D$. We will now outline the main ideas of Kirsch's proof of this result. In what follows, $S : L^2(\partial D) \to L^2(\partial D)$ is the single layer potential defined by (3–9) and $G : L^2(\partial D) \to L^2(\Omega)$ is defined by $Gh = v_\infty$ where v_∞ is the far field pattern of the solution to the radiating exterior Dirichlet problem with boundary data $h \in L^2(\partial D)$. The relation among the operators F, G and S is given by the following lemma.

LEMMA 3.8. *The relation*

$$F = -4\pi GS^*G^*$$

is valid where $G^* : L^2(\Omega) \to L^2(\partial D)$ and $S^* : L^2(\partial D) \to L^2(\partial D)$ are the L^2 adjuncts of G and S respectively.

PROOF. Define the operator $H : L^2(\Omega) \to L^2(\partial D)$ by

$$(Hg)(x) := \int_\Omega g(d)e^{ikx\cdot d}\, ds(d).$$

Note that Hg is the Herglotz wave function with density g. The adjoint operator $H^* : L^2(\partial D) \to L^2(\Omega)$ is given by

$$(H^*\varphi)(\hat{x}) = \int_{\partial D} \varphi(y)e^{-ik\hat{x}\cdot y}\, ds(y).$$

and we note that $\frac{1}{4\pi}H^*\varphi$ is the far field pattern of the single layer potential (3–9). The single layer potential with continuous density φ is continuous in R^3 and thus $\frac{1}{4\pi}H^*\varphi = GS\varphi$, i.e. by a denseness argument

$$H = 4\pi S^*G^* \tag{3–10}$$

on $L^2(\partial D)$. We now observe that Fg is the far field pattern of the solution to the radiating exterior Dirichlet problem with boundary data $-(Hg)(x), x \in \partial D$, and hence

$$Fg = -GHg. \tag{3–11}$$

Substituting (3–10) into (3–11) now yields the lemma.   □

We now assume that k^2 is not a Dirichlet eigenvalue for D. Then by Theorems 3.2 and 3.4 the far field operator F is normal and injective. In particular, there exists eigenvalues $\lambda_j \in \mathbb{C}$ of $F, j = 1, 2, \ldots$, with $\lambda_j \neq 0$, and the corresponding eigenfunctions $\psi_j \in L^2(\omega)$ form a complete orthonormal system in $L^2(\Omega)$. From Lemma 3.3 we can deduce the fact that the λ_j all lie on the circle of radius $\frac{2\pi}{k}$ and center $\frac{2\pi i}{k}$. We also note that $\{|\lambda_j|, \psi_j, \text{sign}(\lambda_j)\psi_j\}$ is a singular system for F. By the preceding lemma,

$$-4\pi GS^*G^*\lambda_j = \lambda_j\psi_j.$$

If we define the functions $\varphi_j \in L^2(\partial D)$ by

$$G^*\psi_j = -\sqrt{\lambda_j}\varphi_j,$$

where we choose the branch of $\sqrt{\lambda_j}$ such that $\text{Im}\sqrt{\lambda_j} > 0$, we see that

$$GS^*\varphi_j = \frac{\sqrt{\lambda_j}}{4\pi}\psi_j. \tag{3–12}$$

A central result of Kirsch is that the functions φ_j form a Riesz basis in the Sobolev space $H^{-1/2}(\partial D)$, i.e. $H^{-1/2}(\partial D)$ consists exactly of functions φ of the form

$$\varphi = \sum_{j=1}^{\infty} \alpha_j\varphi_j \quad \text{with} \quad \sum_{j=1}^{\infty} |\alpha_j|^2 < \infty.$$

The proof of this result relies in a fundamental way on the normality of F. Using these results we can now prove the main result of [42] where in the proof of the theorem $R(A)$ denotes the range of the operator A.

THEOREM 3.9. *Assume that k^2 is not a Dirichlet eigenvalue for D. Then the ranges of $G : H^{1/2}(\partial D) \rightarrow L^2(\Omega)$ and $(F^*F)^{\frac{1}{4}} : L^2(\Omega) \rightarrow L^2(\Omega)$ coincide.*

PROOF. We use the fact that $S^* : H^{-1/2}(\partial D) \rightarrow H^{1/2}(\partial D)$ is an isomorphism. Suppose $G\varphi = \psi$ for some $\varphi \in H^{1/2}(\partial D)$. Then $(S^*)^{-1}\varphi \in H^{-1/2}(\partial D)$ and thus $(S^*)^{-1}\varphi = \sum_{j=1}^{\infty} \alpha_j\varphi_j$ with $\sum_{j=1}^{\infty} |\alpha_j|^2 < \infty$. Therefore, by (3–12), we have

$$\psi = G\varphi = GS^*(S^*)^{-1}\varphi = \frac{1}{4\pi}\sum_{j=1}^{\infty} \alpha_j\sqrt{\lambda_j}\psi_j = \sum_{j=1}^{\infty} \rho_j\psi_j$$

with $\rho_j = \frac{1}{4\pi}\alpha_j\sqrt{\lambda_j}$ and thus

$$\sum_{j=1}^{\infty} \frac{|\rho_j|^2}{|\lambda_j|} = \frac{1}{(4\pi)^2}\sum_{j=1}^{\infty}|\alpha_j|^2 < \infty. \tag{3-13}$$

On the other hand, let $\psi = \sum_{j=1}^{\infty}\rho_j\psi_j$ with the ρ_j satisfying (3–13) and define $\varphi := \sum_{j=1}^{\infty}\alpha_j\varphi_j$ with $\alpha_j = 4\pi\rho_j/\sqrt{\lambda_j}$. Then $\sum_{j=1}^{\infty}|\alpha_j|^2 < \infty$ and hence $\varphi \in H^{-1/2}(\partial D), S^*\varphi \in H^{1/2}(\partial D)$, and

$$G(S^*\varphi) = \frac{1}{4\pi}\sum_{j=1}^{\infty}\alpha_j\sqrt{\lambda_j}\psi_j = \sum_{j=1}^{\infty}\rho_j\psi_j = \psi.$$

Since $\sqrt{|\lambda_j|}$ and ψ_j are the eigenvalues and eigenfunctions, respectively, of the self-adjoint operator $(F^*F)^{\frac{1}{4}}$, we have

$$R(F^*F)^{\frac{1}{4}} = \left(\sum_{j=1}^{\infty}\rho_j\psi_j : \sum_{j=1}^{\infty}\frac{|\rho_j|^2}{|\lambda_j|} < \infty\right),$$

and, as we have shown above, this is precisely $R(G)$. $\qquad\square$

Since $\Phi_{\infty}(\hat{x}, z) = \frac{1}{4\pi}e^{-ik\hat{x}\cdot z}$ is the far field pattern of the fundamental solution $\Phi(x, z)$, it is easy to verify that $\Phi_{\infty}(\cdot, z)$ is in the range of G and if and only if $z \in D$, i.e. $(F^*F)^{\frac{1}{4}}g = \Phi_{\infty}(\cdot, z)$ is solvable if and only if $z \in D$. In particular, if regularization methods are used to solve $(F^*F)^{\frac{1}{4}}g = \Phi_{\infty}(\cdot, z)$ then from Section 2 of this paper we see that as the noise level on u_{∞} tends to zero the norm of the regularized solution remains bounded if and only if $z \in D$. Numerical examples using this procedure can be found in [10] and [42].

Under the assumption that k^2 is not a Neumann eigenvalue, the equation $(F^*F)^{\frac{1}{4}}g = \Phi_{\infty}(\cdot, z)$ can also be derived for the determination of D, i.e. it is not necessary to know a priori whether or not the boundary data is of Dirichlet or Neumann type. However, in both cases the derivation of this equation depends on F being a normal operator. In particular this excludes the limited aperature case when Ω is replaced by a subset Ω_0 of Ω as well as the case of impedance boundary data. In an effort to avoid this problem, Kirsch has introduced a simple nonlinear optimization scheme which preserves some of the advantages of the second version of the linear sampling method while at the same time avoiding the assumption of normality of F [44]. We will conclude this section of our paper by describing Kirsch's optimization scheme.

The optimization scheme of Kirsch is based on the following theorem.

THEOREM 3.10. *Let X_1 be a (complex) reflexive Banach space with dual X_1^* and dual form $\langle\cdot, \cdot\rangle_1$. Let X_2 be a (complex) Hilbert space with inner product $\langle\cdot, \cdot\rangle_2$ and $F : X_2 \to X_2, B : X_1 \to X_2$, compact linear operators such that B is injective. Suppose there exists a bounded linear operator $A : X_1^* \to X_1$ such that $F = BAB^*$ and*

$$c_1\|A\varphi\|_1^2 \leq |\langle\varphi, A\varphi\rangle_1| \leq c_2\|A\varphi\|_1^2 \tag{3-14}$$

for all $\varphi \in X_1^*$ where c_1 and c_2 are positive constants. Then for any $\varphi \in X_2, \varphi \neq 0, \varphi \in R(BA^*)$ if and only if

$$\inf\left\{\left|\langle \psi, F\psi \rangle_2\right| : \psi \in X_2, \langle \psi, \varphi \rangle_2 = 1\right\} > 0.$$

PROOF. From (3–14), we have

$$\left|\langle \psi, F\psi \rangle_2\right| = \left|\langle B^*\psi, AB^*\psi \rangle_1\right| \geq c_1 \|AB^*\psi\|^2,$$

for all $\psi \in X_2$. Let $\varphi = BA^*\varphi_0$ for some $\varphi_0 \in X_1^*$. Then for $\psi \in X_2$ such that $\langle \psi, \varphi \rangle_2 = 1$ we have

$$\left|\langle \psi, F\psi \rangle_2\right| \geq c_1 \|AB^*\psi\|_1^2 = \frac{c_1}{\|\varphi_0\|_1^2} \|AB^*\psi\|_1^2 \|\varphi_0\|_1^2$$

$$\geq \frac{c_1}{\|\varphi_0\|_1^2} |\langle \varphi_0, AB^*\psi \rangle_1|^2 = \frac{c_1}{\|\varphi_0\|_1^2} |\langle BA^*\varphi_0, \psi \rangle_2|^2 = \frac{c_1}{\|\varphi_0\|_1^2} > 0.$$

Now assume that $\varphi \neq R(BA^*)$ and define the closed subspace $V := \{\psi \in X_2 : \langle \psi, \varphi \rangle_2 = 0\}$. We will show that $AB^*(V)$ is dense in $\overline{R(A)} = N(A^*)^\perp$. To see this, let $\varphi \in X_1^*$ such that $\langle \varphi, AB^*\psi \rangle_1 = 0$ for all $\psi \in V$. Then $\langle BA^*\varphi, \psi \rangle_2 = 0$ for all $\psi \in V$, i.e., $BA^*\varphi \in V^\perp = \mathrm{span}\{\varphi\}$. Since $\varphi \in R(BA^*)$ this implies that $BA^*\varphi = 0$ and hence $A^*\varphi = 0$ by the injectivity of B. Therefore $\varphi \in N(A^*) = R(A)^\perp$ and hence $AB^*(V)$ is dense in $\overline{R(A)}$. We can therefore find a sequence $\{\hat{\psi}_n\}$ in V such that

$$AB^*\hat{\psi}_n \to -\frac{1}{\|\varphi\|_2^2} AB^*\varphi$$

as $n \to \infty$. We now set $\psi_n = \hat{\psi}_n + \varphi / \|\varphi\|_2^2$. Then $\langle \psi_n, \varphi \rangle_2 = 1$ and $AB^*\psi_n \to 0$. From (3–14) we have

$$\left|\langle \psi_n, F\psi_n \rangle_2\right| = \left|\langle B^*\psi, AB^*\psi \rangle_1\right| \leq c_2 \|AB^*\psi_n\|_1^2$$

and hence $\langle \psi_n, F\psi_n \rangle_2 \to 0$ as $n \to \infty$, i.e.,

$$\inf\left\{\left|\langle \psi, F\psi \rangle_2\right| : \psi \in X_2, \langle \psi, \varphi \rangle_2 = 1\right\} = 0. \qquad \square$$

In order to make use of Theorem 3.10, Kirsch defines $G : H^{1/2}(\partial D) \to L^2(\Omega)$ and $S : H^{-1/2}(\partial D) \to H^{1/2}(\partial D)$ as in Lemma 3.8 (with the indicated changes in ranges and domains) and proves that if $F : L^2(\Omega) \to L^2(\Omega)$ is the far field operator corresponding to the exterior Dirichlet problem (2–3) then we again have the factorization $F = 4\pi GS^*G^*$. After showing that S^* satisfies the coercivity condition (3–14) if k^2 is not a Dirichlet eigenvalue, and using the fact that in this case S is an isomorphism, it is then possible to use Theorem 3.10 to conclude that if k^2 is not a Dirichlet eigenvalue we have

$$\varphi \in R(G) \iff \inf\left\{\left|\langle \psi, F\psi \rangle_{L^2(\Omega)}\right| : \psi \in L^2(\Omega), \langle \varphi, \psi \rangle_{L^2(\Omega)} = 1\right\} > 0.$$

Since $z \in D$ if and only if $\Phi_\infty(\cdot, z) \in R(G)$ we now have the following theorem [44].

THEOREM 3.11. *Assume that k^2 is not a Dirichlet eigenvalue. Then*

$$z \in D \iff \inf\{|\langle \psi, F\psi \rangle| : \psi \in L^2(\Omega), \ \langle \Phi_\infty(\cdot, z), \psi \rangle_{L^2(\Omega)} = 1\} > 0.$$

Theorem 3.11 leads in an obvious manner to a constrained optimization scheme for determining when a point $z \in R^3$ is in D [44]. Note that the proof of Theorem 3.11 does not rely on the normality of F.

4. The Inverse Medium Problem for Acoustic Waves

We will now turn our attention to inverse scattering problems associated with the inhomogeneous medium problem (2–4) and ultimately the acoustic transmission problem (2–1). As in Section 3, we will focus our attention on the situation where neither the material properties of the scatterer nor its shape are known. Then it follows from the uniqueness theorems of Nachman [53] and Isakov [36], [37]that for a single frequency the best that we can hope for is to determine the shape D of the scatterer. In particular, in order to determine the coefficients in (2–1b), either multi-frequency data is needed or an a priori knowledge of either $\rho_D(x)$ or $n(x)$ is required. If such information is available then nonlinear optimization techniques such as those described in Section 2 of this paper can be used to determine the coefficients. Here we will restrict ourselves to a fixed frequency and prove uniqueness theorems associated with the direct scattering problems (2–1) and (2–4) as well as reconstruction algorithms for determining D from the far field pattern u_∞.

In the previous section we presented three different methods for determining the shape D of the scatterer from a knowledge of the far field pattern associated with the exterior Dirichlet problem (2–3), in particular the two versions of the linear sampling method based on F and $(F^*F)^{\frac{1}{4}}$ respectively and the constrained optimization method based on Theorem 3.11. Each of these methods, under appropriate assumptions, can be extended to the inverse scattering problem associated with the inhomogeneous medium problems (2–4) [12], [19], [43], [44]. However, at the time of writing, only the linear sampling method associated with the far field operator F has been extended to the general acoustic transmission problem (2–1) [6], [7] and the case of Maxwell's equations [11], [29], [49]. Hence, in the interest of developing a unifying theme to our paper, we will restrict our attention to the first version of the linear sampling method in order to determine D.

We begin our discussion by considering the inhomogeneous medium problem (2–4) and again define the far field operator $F : L^2(\Omega) \to L^2(\Omega)$ by

$$(Fg)(\hat{x}) := \int_\Omega u_\infty(\hat{x}, d) g(d) \, ds(d), \tag{4–1}$$

where u_∞ is the far field pattern of the scattered field u^s defined in (2–4). It is again possible to establish the reciprocity relations (3–1) and (3–2). However, since $\operatorname{Im} n(x) \geq 0$, we cannot expect normality of F except in the case when $\operatorname{Im} n(x) = 0$. The question of when F is injective with dense range is addressed by the following theorem, where the role of the interior Dirichlet problem in Theorem 3.2 is now replaced by a new type of boundary value problem called the homogeneous *interior transmission problem*.

THEOREM 4.1. *The far field operator F defined by (4–1) is injective with dense range if and only if there does not exist $w \in C^2(D) \cap C^2(\bar{D})$ and a Herglotz wave function v with kernel $g \neq 0$ such that v, w is a solution to the homogeneous interior transmission problem*

$$\Delta v + k^2 v = 0 \qquad in \ D,$$
$$\Delta w + k^2 n(x) w = 0 \qquad in \ D, \tag{4–2a}$$
$$v = w \qquad on \ \partial D,$$
$$\frac{\partial w}{\partial \nu} \qquad on \ \partial D. \tag{4–2b}$$

PROOF. As in the case of Theorem 3.2, it suffices to establish conditions for when the far field operator is injective. To this end, we note that $Fg = 0$ with $g \neq 0$ is equivalent to the vanishing of the far field pattern of w^s where w is the solution of (2–4) with u^i a Herglotz wave function v with kernel g. By Rellich's lemma, $w^s = 0$ in $R^3 \setminus D$, and hence if $w = v + w^s$ we have

$$w = v \quad on \ \partial D, \qquad \frac{\partial w}{\partial \nu} = \frac{\partial v}{\partial \nu} \quad on \ \partial D. \qquad \square$$

An elementary application of Green's theorem and the unique continuation principle for elliptic equations shows (Theorem 8.12 of [14]) that if $\operatorname{Im} n(x_0) > 0$ for some $x_0 \in D$ then the only solution of (4–2) is $v = w = 0$, i.e., in this case F is injective with dense range. Knowing that the values of k for which the far field operator is not injective form a discrete set is of considerable importance in the inverse scattering problem associated with (2–4), just as it is in the case of the obstacle problem (2–3) where it is known that the set of Dirichlet eigenvalues forms a discrete set. In the case of the linear sampling method, for example, this enables us to conclude that the method can fail only for a discrete set of values of k. From Theorem 4.1 we see that F is injective if there does not exist a nontrivial solution to the homogeneous interior transmission problem. Values of k for which there exists a nontrivial solution to the homogeneous interior transmission problem are called *transmission eigenvalues*. It was shown by Colton, Kirsch and Päivärinta ([13] and Section 8.6 of [14]) and by Rynne and Sleeman [67] that if there exists $\varepsilon > 0$ such that either $n(x) \geq 1 + \varepsilon$ for $x \in \bar{D}$ or $0 < n(x) \leq 1 - \varepsilon$ for $x \in \bar{D}$ then the set of transmission eigenvalues is discrete.

We now turn to the problem of the unique determination of $n(x)$ in (2–4) from a knowledge of the far field pattern $u_\infty(\hat{x}, d)$ for $\hat{x}, d \in \Omega$. The original proof of this result is due to Nachman [53], Novikov [57] and Ramm [66] and is

based on the fundamental paper of Sylvester and Uhlmann [71]. Here we follow
a modification of the original proof due to Hähner [30] which is based on the
following two lemmas, where $H^2(B)$ denotes a Sobolev space.

LEMMA 4.2. *Let B be an open ball centered at the origin and containing the
support of $m := 1 - n$. Then there exists a positive constant C such that for each
$z \in \mathbb{C}^3$ with $z \cdot z = 0$ and $|\operatorname{Re} z| \geq 2k^2\|n\|_\infty$ there exists a solution $w \in H^2(B)$
to $\Delta w + k^2 n w = 0$ in B of the form*

$$w(x) = e^{iz \cdot x}\left(1 + r(x)\right),$$

where

$$\|r\|_{L^2(B)} \leq \frac{C}{|\operatorname{Re} z|}.$$

LEMMA 4.3. *Let B_1 and B_2 be two open balls entered at the origin and containing
the support of $m := 1 - n$ such that $\bar{B}_1 \subset B_2$. Then the set of total fields
$\{u(\,\cdot\,, d), d \in \Omega\}$ satisfying (2–4) is complete in the closure of*

$$H := \{w \in C^2(\bar{B}_2) : \Delta w + k^2 n w = 0 \text{ in } B_2\}$$

with respect to the norm in $L^2(B_1)$.

We are now ready to prove the following uniqueness result for the inverse inhomogeneous medium problem (2–4).

THEOREM 4.4. *The coefficient $n(x)$ in (2–4) is uniquely determined by a knowledge of the far field pattern $u_\infty(\hat{x}, d)$ for $\hat{x}, d \in \Omega$.*

PROOF. Assume that n_1 and n_2 are such that $u_{1,\infty}(\,\cdot\,, d) = u_{2,\infty}(\,\cdot\,, d), d \in \Omega$,
and let B_1 and B_2 be two open balls centered at the origin and containing the
supports of $1 - n_1$ and $1 - n_2$ such that $\bar{B}_1 \subset B_2$. Then by Rellich's lemma we
have $u_1(\,\cdot\,, d) = u_2(\,\cdot\,, d)$ in $R^3 \setminus \bar{B}_1$ for all $d \in \Omega$. Hence $u = u_1 - u_2$ satisfies
$u = \partial u/\partial \nu$ on ∂B_1 and the differential equation

$$\Delta u + k^2 n_1 u = k^2(n_2 - n_1)u_2$$

in B_1. From this and the differential equation for $\tilde{u}_1 = u_1(\,\cdot\,, \tilde{d}), \tilde{d} \in \Omega$, we obtain

$$k^2 \tilde{u}_1 u_2 (n_2 - n_1) = \tilde{u}_1(\Delta u + k^2 n_1 u) = \tilde{u}_1 \Delta u - u \Delta \tilde{u}_1.$$

From Green's theorem and the fact that the Cauchy data for u vanishes on ∂B_1
we now have

$$\iint_{B_1} u_1(\,\cdot\,, \tilde{d}) u_2(\,\cdot\,, d)(n_1 - n_2)\, dx = 0$$

for all $d, \tilde{d} \in \Omega$. It now follows from Lemma 4.3 that

$$\iint_{B_1} w_1 w_2 (n_1 - n_2)\, dx = 0 \qquad (4\text{–}3)$$

for all solutions $w_1, w_2 \in C^2(\bar{B}_2)$ of $\Delta w_1 + k^2 n_1 w_1 = 0$ and $\Delta w_2 + k^2 n_2 w_2 = 0$ in B_2.

Given $y \in R^3 \setminus \{0\}$ and $\rho > 0$, we now choose vectors $a, b \in R^3$ such that $\{y, a, b\}$ is an orthogonal basis in R^3 with the properties that $|a| = 1$ and $|b|^2 = |y|^2 + \rho^2$. Then for $z_1 := y + \rho a + ib$, $z_2 := y - \rho a - ib$ we have

$$z_j \cdot z_j = |\operatorname{Re} z_j|^2 - |\operatorname{Im} z_j|^2 + 2i \operatorname{Re} z_j \cdot \operatorname{Im} z_j = |y|^2 + \rho^2 - |b|^2 = 0$$

and $|\operatorname{Re} z_j|^2 = |y|^2 + \rho^2 \geq \rho^2$. In (4–3) we now substitute the solutions w_1 and w_2 from Lemma 4.2 for the coefficients n_1 and n_2 and vectors z_1 and z_2, respectively. Since $z_1 + z_2 = 2y$ this gives

$$\iint\limits_{B_1} e^{2iy \cdot x} \big(1 + r_1(x)\big)\big(1 + r_2(x)\big)\big(n_1(x) - n_2(x)\big)\, dx = 0$$

and, passing to the limit as $\rho \to \infty$, gives

$$\iint\limits_{B_1} e^{2iy \cdot x} \big(n_1(x) - n_2(x)\big)\, dx = 0.$$

Since this equation is true for arbitrary $y \in R^3$, by the Fourier integral theorem we have $n_1(x) = n_2(x)$ in B_1 and the proof is finished. $\square$

Before proceeding to reconstruction algorithms for determining the support of $m = 1 - n$, we note that at the time of writing a uniqueness theorem for the inverse inhomogeneous medium problem (2–4) in R^2 analogous to Theorem 4.4 for the case of R^3 is unknown. The problem in R^2 is more difficult than the case in R^3 due to the fact that the inverse scattering problem for fixed frequency in R^2 is not overdetermined as in the R^3 case. i.e., in R^2, $u_\infty(\hat{x}, d)$ is a function of two variables and $n(x)$ is also a function of two variables. Nevertheless, there have been numerous partial results in this case due to Novikov [58], Sun and Uhlmann [68], Isakov and Nachman [38], Isakov and Sun [39] and Eskin [25] among others. We content ourselves here by stating a single result in this direction due to Sun and Uhlmann [70] (see also [64]) which shows that the discontinuities of n are uniquely determined from the far field pattern u_∞. We note that in the R^2 case the radiation condition (2–4c) is replaced by

$$\lim_{r \to \infty} r^{1/2}\Big(\frac{\partial u^s}{\partial r} - iku^s\Big) = 0$$

and the asymptotic behavior (2–7) is replaced by

$$u^s(x) = \frac{e^{ikr}}{\sqrt{r}} u_\infty(\hat{x}, d) + O(r^{-3/2}).$$

THEOREM 4.5. *Let n_1 and n_2 be in $L^\infty(R^2)$ and suppose $m_1 := 1 - n_1$ and $m_2 := 1 - n_2$ have compact support. Then if u^j_∞ is the far field pattern corresponding to n_j for $j = 1, 2$ and $u^1_\infty(\hat{x}, d) = u^2_\infty(\hat{x}, d)$ for all $\hat{x}, d$ on the unit circle Ω, then $n_1 - n_2 \in C^{0,\alpha}(R^2)$ for every $\alpha, 0 \leq \alpha < 1$.*

We now return to the three dimensional inverse scattering problem associated with (2–4). Given the fact that $n(x)$ is uniquely determined from u_∞, we can now attempt to reconstruct $n(x)$ by using nonlinear optimization techniques as discussed in Section 2 of this paper. (A reconstruction procedure for determining n based on the techniques used in the uniqueness Theorem 4.4 has been given by Nachman [53] and Novikov [57] although it is not clear whether or not this leads to a viable numerical procedure.) However, a reconstruction of $n(x)$ is often more than is necessary. Indeed, it is frequently sufficient to determine the support of $m = 1 - n$ and, as mentioned at the beginning of this section, for fixed frequency and the more general acoustic transmission problem this is essentially all the information that can be extracted from the far field data u_∞. We will now proceed to show how the linear sampling method can be used to determine the support D of $m := 1 - n$ basing our analysis on the ideas of Colton and Kirsch [12] and Colton, Piana and Potthast [19]. In order to avoid the problem of transmission eigenvalues, we will limit our attention to the case when there exists a positive constant c such that

$$\operatorname{Im} n(x) \geq c \qquad\qquad (4\text{–}4)$$

for $x \in D$ where $\bar{D}$ is the support of $m = 1 - n$. If instead of (4–4) we have $\operatorname{Im} n(x) = 0$ for $x \in D$, the analysis that follows remains valid if we assume that k is not a transmission eigenvalue.

The derivation of the linear sampling method for the inverse scattering problem associated with (2–4) is based on a projection theorem for Hilbert spaces where the inner product is replaced by a bounded sesquilinear form together with an analysis of a special inhomogeneous interior transmission problem. We begin with the projection theorem. Let X be a Hilbert space with the scalar product $(\,\cdot\,,\,\cdot\,)$ and norm $\|\cdot\|$ induced by $(\,\cdot\,,\,\cdot\,)$ and let $\langle\,\cdot\,,\,\cdot\,\rangle$ be a bounded sesquilinear form on X such that

$$|\langle\varphi,\varphi\rangle| \geq C\|\varphi\|^2$$

for all $\varphi \in X$ where C is a positive constant. Then, using the Lax–Milgram theorem, it is easy to prove the following theorem ([19], Theorem 10.22 in [14]) where $\oplus_s$ is the orthogonal decomposition with respect to the sesquilinear form $\langle\,\cdot\,,\,\cdot\,\rangle$ and $H^{\perp_s}$ is the orthogonal complement of H with respect to $\langle\,\cdot\,,\,\cdot\,\rangle$.

THEOREM 4.6. *For every closed subspace $\bar{H} \subset X$ we have the orthogonal decomposition*

$$X = H^{\perp_s} \oplus_s \bar{H}.$$

The projection operator $P : X \to H^{\perp_s}$ defined by this decomposition is bounded in X.

We now turn our attention to the problem of showing the existence of a unique weak solution v, w of the interior transmission problem

$$\Delta v + k^2 v = 0 \qquad \text{in } D, \qquad\qquad (4\text{–}5)$$

$$\Delta w + k^2 n(x) w = 0 \qquad \text{in } D,$$

$$w - v = \Phi(\,\cdot\,, z) \qquad \text{on } \partial D,$$

$$\frac{\partial w}{\partial \nu} - \frac{\partial v}{\partial \nu} = \frac{\partial}{\partial \nu} \Phi(\,\cdot\,, z) \qquad \text{on } \partial D,$$

where $z \in D$, n is assumed to satisfy (4–4) and Φ as usual is defined by (2–6). To motivate the following definition of a weak solution of (4–5), we note that if a solution $v, w \in C^2(D) \cap C^1(\bar{D})$ to (4–5) exists, then from Green's formula and Rellich's lemma we have

$$w(x) + k^2 \iint\limits_D \Phi(x, y) m(y) w(y) \, dy = v(x) \qquad \text{for } x \in D,$$

$$-k^2 \iint\limits_D \Phi(x, y) m(y) w(y) \, dy = \Phi(x, z) \qquad \text{for } x \in \partial B,$$

where B is a ball centered at the origin with $\bar{D} \subset B$.

DEFINITION 4.7. Let H be the linear space of all Herglotz wave functions and $\bar{H}$ the closure of H in $L^2(D)$. For $\varphi \in L^2(D)$ define the volume potential by

$$(T\varphi)(x) := \iint\limits_D \Phi(x, y) m(y) \varphi(y) \, dy, \ x \in R^3.$$

Then a pair v, w with $v \in \bar{H}$ and $w \in L^2(D)$ is said to be a weak solution of the interior transmission problem (4–5) with source point $z \in D$ if v and w satisfy the integral equation

$$w + k^2 T w = v$$

and the boundary condition

$$-k^2 T w = \Phi(\,\cdot\,, z) \qquad \text{on } \partial B.$$

The uniqueness of a weak solution to the interior transmission problem follows from a limiting argument using (4–4) and a simple application of Green's theorem [19], [14, Theorem 10.24]. To prove existence we will use Theorem 4.6 applied to the sesquilinear form in $L^2(D)$ defined by

$$\langle \varphi, \psi \rangle := \iint\limits_D m(y) \varphi(y) \overline{\psi(y)} \, dy$$

and H as defined in the above definition.

THEOREM 4.8. *For every source point $z \in D$ there exists a weak solution to the interior transmission problem.*

PROOF. By a translation we can assume without loss of generality that $z = 0$. We consider the space

$$H_1^0 = \text{span}\{j_p(k|x|) Y_p^q(\hat{x}), \ p = 1, 2, \ldots, \ -p \leq q \leq p\}$$

and the closure H_1 of H_1^0 in $L^2(D)$, where j_p is a spherical Bessel function and Y_p^q is a spherical harmonic. It can be shown that there exists a nontrivial $\psi \in H_1^{\perp s} \cap \bar{H}$ such that $\langle j_0, \psi \rangle \neq 0$.

Now let P be the projection operator from $L^2(D)$ onto $H^{\perp s}$ as defined by Theorem 4.6. We first consider the integral equation

$$u + k^2 PTu = k^2 PT\psi \qquad (4\text{-}6)$$

in $L^2(D)$. Since T is compact and P is bounded, the operator PT is compact in $L^2(D)$. Hence to establish the existence of a solution to (4–6) we must prove uniqueness for the homogeneous equation. To this end, assume that $w \in L^2(D)$ satisfies

$$w + k^2 Tw = v.$$

Since $\langle w, \varphi \rangle = 0$ for all $\varphi \in H$, from the addition formula for Bessel functions we conclude that $Tw = 0$ on ∂B. Hence, by uniqueness for the weak interior transmission problem we have $v = w = 0$, and we obtain the continuous invertibility of $I + k^2 PT$ in $L^2(D)$.

Now let u be the unique solution to (4–6) and note that $u \in H^{\perp s}$. We define the constant c and function $w \in L^2(D)$ by

$$c := -\frac{1}{k^2 \langle j_0, \psi \rangle}, \qquad w := c(u - \psi).$$

Then we compute

$$w + k^2 PTw = -c\psi$$

and hence

$$w + k^2 Tw = v$$

where $v := k^2(I - P)Tw - c\psi \in \bar{H}$. Since

$$\langle h, w \rangle = c \langle h, u - \psi \rangle = 0$$

for all $h \in H_1$ and

$$\langle j_0, w \rangle = c \langle j_0, u - \psi \rangle = -\frac{1}{k^2},$$

we have from the addition formula for Bessel functions that

$$-k^2(Tw)(x) = ikh_0^{(1)}(k|x|) = \Phi(x, 0), \qquad x \in \partial B,$$

where $h_0^{(1)}$ is a spherical Hankel function of the first kind of order zero, and the proof is complete. $\qquad \square$

Having Theorem 4.8 at our disposal, we can now establish the linear sampling method for determining D. In particular, we again consider the far field equation $Fg = \Phi_\infty(\,\cdot\,, z)$, that is,

$$\int_\Omega u_\infty(\hat{x}, d)g(d)\, ds(d) = \Phi_\infty(\hat{x}, z), \qquad (4\text{-}7)$$

where $\Phi_\infty(\,\cdot\,,z)$ is the far field pattern of the fundamental solution $\Phi(\,\cdot\,,z)$. Following the proof of Theorem 4.1 we see that (4–7) has a solution if and only if $z \in D$ and the interior transmission problem (4–5) has a solution $v, w \in C^2(D) \cap C^1(\bar{D})$ such that v is a Herglotz wave function with kernel g. This is only true in very special cases. However, by Theorem 4.8 we know there exists a (unique) weak solution v, w to the interior transmission problem and that v can be approximated in $L^2(D)$ by a Herglotz wave function. This fact then enables us to establish a result for the far field equation (4–7) that is analogous to Theorem 3.7 for the far field equation (3–7) corresponding to the exterior Dirichlet problem [4] (Later on in this section we shall outline how this is done for the general case of an anisotropic medium). Note that the far field equations (3–7) and (4–7) are exactly the same except of course that the far field pattern u_∞ appearing in the kernel of F come from different scattering problems. This means that in order to determine the support D of the scatterer it is not necessary to know a priori whether the direct scattering problem is (2–3) or (2–4) or, as previously noted, (2–2). In particular, one can determine the support of the scatterer without a priori knowledge on whether or not the scattering object is penetrable or impenetrable, at least in the context of the three scattering problems (2–2), (2–3) and (2–4). In the remaining part of this section we will extend this observation to include the general acoustic transmission problem (2–1) and in fact consider the even more general case of anisotropic media. As with the case of the inhomogeneous medium problem (2–4), the basic ingredient will again be an analysis of an interior transmission problem, this time for anisotropic media.

Let $D \subset R^3$ be a bounded domain having C^2 boundary ∂D with unit outward normal ν. Let A be a 3×3 matrix-valued function whose entries a_{jk} ($j = 1, 2, 3$, $k = 1, 2, 3$) are continuously differentiable functions in $\bar{D}$, such that A is symmetric and satisfies

$$\bar{\xi} \cdot (\operatorname{Im} A)\xi \leq 0, \qquad \bar{\xi} \cdot (\operatorname{Re} A)\xi \geq \gamma |\xi|^2 \tag{4–8}$$

for all $\xi \in \mathbb{C}^3$ and $x \in \bar{D}$, where γ is a positive constant. For a function $u \in C^1(\bar{D})$ we define the conormal derivative by

$$\frac{\partial u}{\partial \nu_A}(x) := \nu(x) \cdot A(x)\nabla u(x) \quad \text{for } x \in \partial D$$

and let $k > 0$ again be the wave number and let $n \in C(\bar{D})$ satisfy $\operatorname{Re} n > 0$ and $\operatorname{Im} n \geq 0$. The anisotropic acoustic transmission problem, for which (2–1) is the special case of an isotropic medium, is to find $u \in C^2(R^3 \setminus \bar{D}) \cap C^1(R^3 \setminus D)$ and $v \in C^2(D) \cap C^1(\bar{D})$ such that

$$\Delta u + k^2 u = 0 \quad \text{in } R^3 \setminus \bar{D}, \tag{4–9a}$$

$$\nabla \cdot A\nabla v + k^2 n(x)v = 0 \quad \text{in } D, \tag{4–9b}$$

$$u = u^i + u^s, \tag{4–9c}$$

$$\lim_{r \to \infty} r\left(\frac{\partial u^s}{\partial r} - iku^s\right) = 0, \qquad\qquad\qquad (4\text{--}9\text{d})$$

$$u = v \qquad \text{on } \partial D, \qquad\qquad (4\text{--}9\text{e})$$

$$\frac{\partial u}{\partial \nu} = \frac{\partial v}{\partial \nu_A} \qquad \text{on } \partial D, \qquad\qquad (4\text{--}9\text{f})$$

where again $u^i(x) = e^{ikx \cdot d}$. The existence of a unique solution to (4–9) has been established by Hähner [31].

Since u satisfies the radiation condition, we can again conclude that u^s has the asymptotic behavior

$$u^s(x) = \frac{e^{ikr}}{r} u_\infty(\hat{x}, d) + O(r^{-2}).$$

The inverse scattering problem we are concerned with is to determine D from a knowledge of the far field pattern $u_\infty(\hat{x}, d)$ for $\hat{x}, d \in \Omega$. We note that the matrix A is not uniquely determined by u_∞ (see [27], [61]) and hence determinning D is the most that can be hoped for. To this end we have the following theorem due to Hähner [31] (see also [17] and [61]).

THEOREM 4.9. *Assume $\gamma > 1$. Then D is uniquely determined by $u_\infty(\hat{x}, d)$ for $\hat{x}, d \in \Omega$.*

The proof of this theorem uses the ideas of Theorem 3.5 together with a continuous dependence result for an associated interior transmission problem. It follows from the results of Cakoni and Haddar [7] that Theorem 4.9 remains valid if the condition $\gamma > 1$ is replaced by the condition

$$\bar{\xi} \cdot (\operatorname{Re} A^{-1})\xi \geq \mu|\xi|^2 \qquad\qquad (4\text{--}10)$$

for all $\xi \in \mathbb{C}^3$ and $x \in \bar{D}$ where μ is a positive constant such that $\mu > 1$. The isotropic case when $A = I$ is handled by Theorem 4.4.

Given the uniqueness Theorem 4.9 (and the variations on this theorem indicated above) we now want to establish the linear sampling method for determining D. In particular, we look for a (regularized) solution $g \in L^2(\Omega)$ of the far field equation

$$(Fg)(\hat{x}) := \int_\Omega u_\infty(\hat{x}, d)g(d)\, ds(d) = \Phi_\infty(\hat{x}, z) \qquad (4\text{--}11)$$

where $z \in R^3$ is an artificially introduced parameter point and u_∞ is the far field pattern of the scattered field defined by (4–9). Following the proof of Theorem 4.1 it is easily verified that (4–11) is solvable if and only if $z \in D$ and $v, w \in C^2(D) \cap C^1(\bar{D})$ is a solution of the interior transmission problem

$$\Delta v + k^2 v = 0 \qquad \text{in } D,$$
$$\nabla \cdot A\nabla w + k^2 n w = 0 \qquad \text{in } D, \qquad\qquad (4\text{--}12\text{a})$$

$$w - v = \Phi(\cdot, z) \qquad \text{on } \partial D,$$

$$\frac{\partial w}{\partial \nu_A} - \frac{\partial v}{\partial \nu} = \frac{\partial}{\partial \nu} \Phi(\cdot, z) \qquad \text{on } \partial D, \qquad (4\text{--}12b)$$

such that v is a Herglotz wave function. Values of k for which a nontrivial solution to the homogeneous interior transmission problem ($\Phi = 0$) exists are again called transmission eigenvalues. As in the case of an isotropic medium, our aim is to now study the interior transmission problem (4–12) with the aim of showing that, roughly speaking, D can be characterized as the set of points $z \in R^3$ where an arbitrarily good approximation of the solution to the far field equation (4–4) remains bounded (see Theorem 3.7).

We begin with uniqueness. The following theorem follows easily by an application of Green's theorem.

THEOREM 4.10. *If either* $\operatorname{Im} n > 0$ *or* $\bar{\xi} \cdot (\operatorname{Im} A)\xi < 0$ *in a neighborhood of a point* $x_0 \in D$, *then* (4–12) *has at most one solution.*

In order to study the solvability of (4–12) we first consider a modified interior transmission problem which is a compact perturbation of (4–12). In particular, let $m \in C(\bar{D})$ satisfy $m(x) > 0$ for $x \in \bar{D}$ and for $l_1, l_2 \in L^2(D)$, $f \in H^{1/2}(\partial D)$, $h \in H^{-1/2}(\partial D)$ we want to find $v, w \in H^1(D)$ such that

$$\Delta v + k^2 v = l_1 \qquad \text{in } D,$$

$$\nabla \cdot A \nabla w - m w = l_2 \qquad \text{in } D, \qquad (4\text{--}13a)$$

$$w - v = f \qquad \text{on } \partial D,$$

$$\frac{\partial w}{\partial \nu_A} - \frac{\partial v}{\partial \nu} = h \qquad \text{on } \partial D. \qquad (4\text{--}13b)$$

In [6], (4–13) is reformulated as a variational problem for $(w, \mathbf{v}) \in H^1(D) \times W(D)$ where $\mathbf{v} = \nabla v$ and

$$W(D) := \{\mathbf{v} \in (L^2(D))^3 : \nabla \cdot \mathbf{v} \in L^2(D) \quad \text{and curl } \mathbf{v} = 0\}.$$

The variational problem is then solved by appealing to the Lax–Milgram theorem under the assumption that in (4–8) we have $\gamma > 1$ and $m > \gamma$. Having established the existence of a unique solution to (4–13), and using the fact that (4–13) is a compact perturbation of the interior transmission problem (4–12), we can now appeal to Theorem 4.10 to deduce the following theorem [5].

THEOREM 4.11. *Assume that either* $\operatorname{Im} n > 0$ *or* $\bar{\xi} \cdot (\operatorname{Im} A)\xi < 0$ *in a neighborhood of a point* $x_0 \in D$ *and that* $\gamma > 1$. *Then* (4–12) *has a unique solution* $v, w \in H^1(D) \times H^1(D)$ *where the boundary data* (4–12b) *is assumed in the sense of the trace operator.*

It was shown in [7] that Theorem 4.11 remains valid if the condition $\gamma > 1$ is replaced by the condition (4–10) where $\mu > 1$ (In this case m is a constant restricted to satisfy $\mu^{-1} \leq m < 1$).

If A and n do not satisfy one of the above assumptions ($\gamma > 1$ in (4–8) or $\mu > 1$ in (4–10)) then in general we cannot conclude the solvability of the interior transmission problem (4–12). In particular, (4–12) is uniquely solvable if and only if k is not a transmission eigenvalue and if $\operatorname{Im} n = 0$ and $\operatorname{Im} A = 0$ it is not possible to exclude this possibility. However, from the point of view of applying regularization techniques to the far field equation (4–11), it is important to have F injective with dense range and this is true if k is not a transmission eigenvalue. To this end the following theorem is important [5], [7].

THEOREM 4.12. *Assume that* $\operatorname{Im} n = 0$ *and* $\operatorname{Im} A = 0$ *in* D *and that one of the following conditions is satisfied*:

(1) $\gamma > 1$ *in* (4–8) *and* $n(x) \geq \gamma$ *for* $x \in D$, *or*
(2) $\mu > 1$ *in* (4–10) *and* n *is a constant such that* $\mu^{-1} \leq n < 1$.

Then the set of transmission eigenvalues forms a discrete set.

The proof of Theorem 4.12 is based on the uniqueness of a solution to the modified interior transmission problem (4–13) together with the fact that the spectrum of a compact operator is discrete. At the time of writing it is not known whether or not transmission eigenvalues exist except for the special case when $A = I$ (isotropic media) and $n(x) = n(r)$ is spherically symmetric; see [14, Theorem 8.13].

We now turn our attention to showing how Theorems 4.11 and 4.12 lead to a justification of the linear sampling method for anisotropic media that is analogous to Theorem 3.7 for the case of the exterior Dirichlet problem. To this end we let B be the bounded linear operator from $H^{1/2}(\partial D) \times H^{-1/2}(\partial D)$ into $L^2(\Omega)$ which maps $(f, h) \in H^{1/2}(\partial D) \times H^{-1/2}(\partial D)$ onto the far field data $u_\infty \in L^2(\Omega)$ of the solution of the transmission problem

$$\Delta u^s + k^2 u^s = 0 \quad \text{in } R^3 \setminus \bar{D}, \tag{4–14a}$$

$$\nabla \cdot A\nabla v + k^2 n(x)v = 0 \quad \text{in } D, \tag{4–14b}$$

$$\lim_{r \to \infty} r\left(\frac{\partial u^s}{\partial r} - iku^s\right) = 0, \tag{4–14c}$$

$$v - u^s = f \quad \text{on } \partial D, \tag{4–14d}$$

$$\frac{\partial v}{\partial \nu_A} - \frac{\partial u^s}{\partial \nu} = h \quad \text{on } \partial D, \tag{4–14e}$$

where $u^s \in H^1_{\mathrm{loc}}(R^3 \setminus \bar{D})$ and $v \in H^1(D)$. It is shown in [5] that the range of B is dense in $L^2(\Omega)$. However, B is not injective. To remedy this problem, we define the subset $H(\partial D)$ of $H^{1/2}(\partial D) \times H^{-1/2}(\partial D)$ by

$$H(\partial D) := \left\{ \left(v|_{\partial D}, \frac{\partial v}{\partial \nu}|_{\partial D}\right) : v \in H \right\},$$

where $H := \{v \in H^1(D) : \Delta v + k^2 v = 0 \text{ in } D\}$. Then $H(\partial D)$ equipped with the induced norm from $H^{1/2}(\partial D) \times H^{-1/2}(\partial D)$ is a Banach space and if B_0 is the restriction of B to $H(\partial D)$ we conclude that for k not a transmission eigenvalue $B_0 : H(\partial D) \to L^2(\Omega)$ is injective, compact and has dense range [5].

We now write the far field equation (4–11) in the form

$$-B_0 Hg = \Phi_\infty(\,\cdot\,, z)$$

where Hg denotes the traces $(v_g|_{\partial D}, \partial v_g/\partial \nu|_{\partial D})$ for v_g a Herglotz wave function with kernel g. Using the facts that 1) Herglotz wave functions with kernels $g \in L^2(\Omega)$ are dense in H with respect to the norm in $H^1(D)$ and 2) $\Phi(\,\cdot\,, z)$ is in the range of B_0 if and only if $z \in D$, we can now deduce the analogue of Theorem 3.7 for anisotropic media under the assumption that either $\gamma > 1$ in (4–8) or $\mu > 1$ in (4–10) and k is not a transmission eigenvalue [5], [7]. As in the case of Theorem 3.7 for the exterior Dirichlet problem, this result now yields a numerical procedure for determining the support D of an anisotropic object from noisy far field data.

Partial results related to the above for the case when $A = I$ on ∂D but $A \neq I$ in D can be found in Chapter 7 of [62].

5. The Inverse Scattering Problem for Electromagnetic Waves

In this final section of our survey paper on inverse scattering problems for acoustic and electromagnetic waves we consider the scattering of a time harmonic electromagnetic wave by either a perfectly conducting obstacle or an isotropic inhomogeneous medium of compact support. We begin with the scattering of a time harmonic electromagnetic plane wave by a perfectly conducting obstacle. Let D be a bounded domain in R^3 with connected complement such that ∂D is in class C^2 and ν is the unit outward normal to ∂D. Then the direct scattering problem we are concerned with can be formulated as the problem of finding an electric field E and a magnetic field H such that $E, H \in C^1(R^3 \setminus \bar{D}) \cap C(R^3 \setminus D)$ and

$$\begin{aligned}
\operatorname{curl} E - ikH &= 0 \qquad \text{in } R^3 \setminus \bar{D}, \\
\operatorname{curl} H + ikE &= 0 \qquad \text{in } R^3 \setminus \bar{D},
\end{aligned} \tag{5–1a}$$

with

$$\nu \times E = 0 \qquad \text{on } \partial D \tag{5–1b}$$

and

$$E = E^i + E^s, \qquad H = H^i + H^s, \tag{5–1c}$$

where E^s, H^s represent the scattered field satisfying the Silver–Müller radiation condition

$$\lim_{r \to \infty} (H^s \times x - rE^s) = 0 \tag{5–1d}$$

uniformly in $\hat{x} = x/|x|$ (with $r = |x|$), and the incident field E^i, H^i is given by

$$E^i(x) = \frac{i}{k} \operatorname{curl} \operatorname{curl} p e^{ikx \cdot d} = ik(d \times p) \times d e^{ikx \cdot d}, \tag{5-1e}$$

$$H^i(x) = \operatorname{curl} p e^{ikx \cdot d} = ikd \times p e^{ikx \cdot d}, \tag{5-1f}$$

where the wave number k is positive, d is a unit vector giving the direction of propagation and p is the polarization vector. The existence and uniqueness of a solution to (5–1) is well known [14]. From the Stratton–Chu formula [14] it follows from (5–1) that E^s has the asymptotic behavior

$$E^s(x) = \frac{e^{ikr}}{r} E_\infty(\hat{x}, d, p) + O(r^{-2}), \tag{5-2}$$

where E_∞ is the *electric far field pattern* of the scattered electric field E^s. Note that since we are always assuming that k is fixed we have suppressed the dependence of E_∞ on k. It can easily be verified [14] that E_∞ is infinitely differentiable as a function of $\hat{x}$ and d, linear with respect to p and as a function of $\hat{x}$ is tangential to the unit sphere Ω.

We now turn our attention to the scattering of the electromagnetic plane wave (5–1e), (5–1f) by an inhomogeneous medium of compact support. In this case, under appropriate assumptions [14], our problem is to find $E, H \in C^1(R^3)$ satisfying

$$\begin{aligned} \operatorname{curl} E - ikH &= 0 \\ \operatorname{curl} H + ikn(x)E &= 0 \end{aligned} \quad \text{in } R^3, \tag{5-3a}$$

where n satisfies $C^{2,\alpha}(R^3)$ for some $0 < \alpha < 1$, $\operatorname{Re} n > 0$, $\operatorname{Im} n \geq 0$, $1 - n$ has compact support $\bar{D}$; and where

$$E = E^i + E^s, \qquad H = H^i + H^s \tag{5-3b}$$

such that E^i, H^i is given by (5–1e), (5–1f) and E^s, H^s again satisfies the Silver–Müller radiation condition

$$\lim_{r \to \infty} (H^s \times x - rE^s) = 0 \tag{5-3c}$$

uniformly in $\hat{x}$. The existence and uniqueness of a solution to (5–3) is again well known [14] and E^s can be shown to have the asymptotic behavior (5–2).

We are now in a position to define the electric far field operator and its connection to what are called electromagnetic Herglotz pairs. To this end, we define the Hilbert space $T^2(\Omega)$ by

$$T^2(\Omega) := \{a : \Omega \to \mathbb{C}^3 : a \in L^2(\Omega), \, a \cdot \hat{x} = 0 \text{ for } \hat{x} \in \Omega\}.$$

The *electric far field operator* $F : T^2(\Omega) \to T^2(\Omega)$ is then defined by

$$(Fg)(\hat{x}) := \int_\Omega E_\infty(\hat{x}, d, g(d)) \, ds(d), \quad \hat{x} \in \Omega \tag{5-4}$$

where $g \in T^2(\Omega)$. We note that F is a compact linear operator on $T^2(\Omega)$. An *electromagnetic Herglotz pair* is a pair of vector fields of the form

$$E(x) = \int_\Omega e^{ikx \cdot d} a(d) \, ds(d), \qquad H(x) = \frac{1}{ik} \operatorname{curl} E(x), \qquad (5\text{--}5)$$

for $x \in R^3$ where $a \in T^2(\Omega)$ is the *kernel* of E, H. In particular, Fg is the electric far field pattern for (5–1) or (5–3) respectively corresponding to the electromagnetic Herglotz pair with kernel ikg as incident field.

As in the case of the inverse scattering problem for acoustic waves, of basic importance is the fact that F is injective with dense range. The proof of the following two theorems and corollary follows along the same lines as previously discussed for the case of acoustic waves [14]. Recall that a Maxwell eigenfunction for D is a solution of Maxwell's equations (5–1a) in D satisfying (5–1b) on ∂D.

THEOREM 5.1. *The electric far field operator for (5–1) is injective with dense range if and only if there does not exist a Maxwell eigenfunction for D which is an electromagnetic Herglotz pair.*

THEOREM 5.2. *The electric far field operator for (5–3) is injective with dense range if and only if there does not exist $E_1, H_1 \in C^1(D) \cap C(\bar{D})$ and an electromagnetic Herglotz pair E_0, H_0 with kernel $a \neq 0$ such that E_0, H_0 and E_1, H_1 is a solution to the homogeneous electromagnetic interior transmission problem*

$$
\begin{array}{llll}
\operatorname{curl} E_0 - ikH_0 = 0, & \operatorname{curl} H_0 + ikE_0 = 0 & \text{in } D, & \\
\operatorname{curl} E_1 - ikH_1 = 0, & \operatorname{curl} H_1 + ikn(x)E_1 = 0 & \text{in } D, & (5\text{--}6a) \\
\nu \times (E_1 - E_0) = 0, & \nu \times (H_1 - H_0) = 0 & \text{on } \partial D. & (5\text{--}6b)
\end{array}
$$

COROLLARY 5.3. *If there exists an $x_0 \in D$ such that $\operatorname{Im} n(x_0) > 0$ then the electric far field operator for (5–3) is injective with dense range.*

Proceeding as in our discussion of acoustic waves, the next topic we consider is the uniqueness of the solution to the inverse scattering problem for (5–1) and (5–3) respectively. In particular, for (5–1) we want to determine whether or not a knowledge of E_∞ for fixed k uniquely determines D and in the case of (5–3) whether or not a knowledge of E_∞ for fixed k uniquely determines $n(x)$. To this end, we have the following theorems due to Colton and Kress [14] and Colton and Päivärinta [18] respectively. The proofs of both results are similar to the corresponding proofs in the acoustic case already discussed. However, for the inverse scattering problem associated with (5–3), serious technical problems arise due to the fact that we must now construct a solution E, H of (5–3a) such that E has the form

$$E(x) = e^{i\zeta \cdot x} \left(\eta + R_\zeta(x) \right)$$

where $\zeta, \eta \in \mathbb{C}^3$, $\eta \cdot \zeta = 0$ and $\zeta \cdot \zeta = k^2$ and, in contrast to the case of acoustic waves, it is no longer true that R_ζ decays to zero as $|\zeta|$ tends to infinity. For details of how this difficulty is resolved we refer the reader to [18] and [33]. We also note that the scattering problem (5–3) corresponds to the case when the magnetic permeability μ is constant and for uniqueness results in the case when μ is no longer constant see [59], [60] and [69].

THEOREM 5.4. *Assume that D_1 and D_2 are two domains such that the electric far field patterns corresponding to the scattering problem (5–1) coincide for all incident directions $d \in \Omega$ and all polarizations $p \in R^3$. Then $D_1 = D_2$.*

THEOREM 5.5. *The coefficient $n(x)$ in (5–3) is uniquely determined by a knowledge of the electric far field pattern for all incident directions $d \in \Omega$ and all polarizations $p \in R^3$.*

Having determined the uniqueness of a solution to the inverse scattering problem, the next step is to derive reconstruction algorithms which are numerically viable. It is here that the theory for electromagnetic waves lags well behind that for acoustic waves. In particular, although methods based on the weak scattering approximation have been used extensively, particularly in problems associated with synthetic aperature radar [3], [8], the nonlinear problem has only begun to be considered. Notable accomplishments in the case of nonlinear optimization techniques to solve the inverse scattering problem associated with (5–1) have been achieved by Haas, et.al [28] and Maponi, et.al [52] whereas the case of non-linear optimization techniques to solve the inverse scattering problem associated with (5–3) have recently been considered by Dorn, et.al [23]. Finally, based on Theorems 5.1 and 5.2 and the approximation properties of electromagnetic Herglotz pairs, Colton, Haddar and Monk [11] and Haddar and Monk [29] have used the linear sampling method to solve inverse scattering problems associated with (5–1) and (5–3) respectively. However, there is much to be done and we close this survey with a short list of open problems that await the input of new ideas for their solution:

(1) Extend the methods of Kirsch for acoustic waves discussed in Section 3 of this survey to the case of electromagnetic waves. For initial steps in this direction, see [49].

(2) Show that the set of transmission eigenvalues for the homogeneous electromagnetic interior transmission problem form a discrete set.

(3) Show that for the case of electromagnetic waves the support of an inhomogeneous anisotropic media in R^3 is uniquely determined by the corresponding electric far field pattern.

(4) Establish the mathematical basis of the linear sampling method for Maxwell's equations for an anisotropic medium. For partial results in a special case, see [62, Section 7.4].

Acknowledgement. The research was supported in part by a grant from the Air Force Office of Scientific Research.

References

[1] T. S. Angell, R. E. Kleinman and G. F. Roach, An inverse transmission problem for the Helmholtz equation, *Inverse Problems* **3** (1987), 149–180.

[2] L. T. Biegler, T. F. Coleman, A. R. Conn and F. Santosa, *Large-Scale Optimization with Applications, Part I: Optimization in Inverse Problems and Design*, IMA Volumes in Mathematics and its Applications, Vol. **92**, Springer-Verlag, New York, 1997.

[3] B. Borden, *Radar Imaging of Airborne Targets*, IOP Publishing, Bristol, 1999.

[4] F. Cakoni and D. Colton, On the mathematical basis of the linear sampling method, *Georgian Mathematical Journal*, to appear.

[5] F. Cakoni, D. Colton and H. Haddar, The linear sampling method for anisotropic media, *J. Comp. Appl. Math.*, **146** (2002), 285–299.

[6] F. Cakoni, D. Colton and P. Monk, The direct and inverse scattering problems for partially coated obstacles, *Inverse Problems*, **17** (2001), 1997–2015.

[7] F. Cakoni and H. Haddar, The linear sampling method for anisotropic media: Part II, MSRI Preprint No. 2001–026, 2001.

[8] M. Cheney, A mathematical tutorial on syntehtic aperature radar, *SIAM Review* **43** (2001), 301–312.

[9] D. Colton, J. Coyle and P. Monk, Recent developments in inverse acoustic scattering theory, *SIAM Review* **42** (2000), 369–414.

[10] D. Colton, K. Giebermann and P. Monk, A regularized sampling method for solving three dimensional inverse scattering problems, *SIAM J. Sci. Comput.* **21** (2000), 2316–2330.

[11] D. Colton, H. Haddar and P. Monk, The linear sampling method for solving the electromagnetic inverse scattering problem, *SIAM J. Sci. Comput.* **24** (2002), 719–731.

[12] D. Colton and A. Kirsch, A simple method for solving inverse scattering problems in the resonance region, *Inverse Problems* **12** (1996), 383–393.

[13] D. Colton, A. Kirsch and L. Päivärinta, Far field patterns for acoustic waves in an inhomogeneous medium, *SIAM J. Math. Anal.* **20** (1989), 1472–1483.

[14] D. Colton and R. Kress, *Inverse Acoustic and Electromagnetic Scattering Theory*, Second Edition., Springer–Verlag, New York, 1998.

[15] D. Colton and R. Kress, Eigenvalues of the far field operator for the Helmholtz equation in an absorbing medium, *SIAM J. Appl. Math.* **55** (1995), 1724–1735.

[16] D. Colton and R. Kress, On the denseness of Herglotz wave functions and electromagnetic Herglotz pairs in Sobolev spaces, *Math. Methods Applied Sciences* **24** (2001), 1289–1303.

[17] D. Colton, R. Kress and P. Monk, Inverse scattering from an orthotropic medium, *J. Comp. Applied Math.*, **81** (1997), 269–298.

[18] D. Colton and L. Päivärinta, The uniqueness of a solution to an inverse scattering problem for electromagnetic waves, *Arch. Rational Mech. Anal.* **119** (1992), 59–70.

[19] D. Colton, M. Piana and R. Potthast, A simple method using Morozov's discrepancy principle for solving inverse scattering problems, *Inverse Problems* **13** (1997), 1477–1493.

[20] D. Colton and B.D. Sleeman, Uniqueness theorems for the inverse problem of acoustic scattering, *IMA Jour. Applied Math.*, **31** (1983), 253–259.

[21] D. Colton and B. D. Sleeman, An approximation property of importance in inverse scattering theory, *Proc. Edinburgh Math. Soc.* **44** (2001), 449–454.

[22] A. Devaney, Acoustic tomography, in *Inverse Problems of Acoustic and Elastic Waves*, F. Santosa, et. al. editors, SIAM, Philadelphia, 1984, 250–273.

[23] O. Dorn, H. Bertete-Aguirre, J. G. Berryman and G. C. Papanicolaou, A nonlinear inversion method for 3D-electromagnetic imaging using adjoint fields, *Inverse Problems* **15** (1999), 1523–1558.

[24] H. W. Engl, M. Hanke and A. Neubauer, *Regularization of Inverse Problems*, Kluwer Academic Publisher, Dordrecht, 1996.

[25] G. Eskin, The inverse scattering problem in two dimensions at fixed energy, *Comm. Partial Diff. Equations* **26** (2001), 1055–1090.

[26] S. Gutman and M. Klibanov, Iterative method for multidimensional inverse scattering problems at fixed frequencies, *Inverse Problems* **10** (1994), 573–599.

[27] F. Gylys-Colwell, An inverse problem for the Helmholtz equation, *Inverse Problems* **12** (1996), 139–156.

[28] M. Haas, W. Rieger, W. Rucker and G. Lehner, Inverse 3D acoustic and electromagnetic obstacle scattering by iterative adaptation in *Inverse Problems of Wave Propagation and Diffraction*, G. Chavent and P. Sabatier, eds., Springer-Verlag, Heidelberg, 1997, 204–215.

[29] H. Haddar and P. Monk, The linear sampling method for solving the electromagnetic inverse medium problem, *Inverse Problems* **18** (2002), 891–906.

[30] P. Hähner, A periodic Faddeev-type soltuion operator, *J. Diff. Equations* **128** (1996), 300–308.

[31] P. Hähner, On the uniqueness of the shape of a penetrable, anisotropic obstacle, *J. Comp. Applied Math.* **116** (2000), 167–180.

[32] P. Hähner, Acoustic scattering, in *Scattering*, E. R. Pike and P. Sabatier, eds., Academic Press, London, 2001.

[33] P. Hähner, Electromagnetic waves scattering in *Scattering*, E. R. Pike and P. Sabatier, eds., Academic Press, London, 2001.

[34] F. Hettlich, Fréchet derivatives in inverse obstacle scattering, *Inverse Problems* **11** (1995), 371–382.

[35] M. Ikehata, Reconstruction of an obstacle from the scattering amplitude at a fixed frequency, *Inverse Problems* **14** (1998), 949–954.

[36] V. Isakov, *Inverse Problems for Partial Differential Equations*, Springer-Verlag, New York, 1998.

[37] V. Isakov, On uniqueness in the inverse transmission scattering problem, *Comm. Partial Diff. Equations* **15** (1990), 1565–1587.

[38] V. Isakov and A. Nachman, Global uniqueness in a two dimensional semilinear elliptic inverse problem, *Trans. Amer. Math. Soc.* **347** (1995), 3375–3390.

[39] V. Isakov and Z. Sun, The inverse scattering problem at fixed energies in two dimensions, *Indiana Univ. Math. Journal* **44** (1995), 883–396.

[40] A. Kirsch, *An Introduction to the Mathematical Theory of Inverse Problems*, Springer Verlag, Berlin, 1996.

[41] A. Kirsch, The domain derivative and two applicaitons in inverse scattering, *Inverse Problems* **9** (1993), 81–96.

[42] A. Kirsch, Characterization of the shape of the scattering obstacle by the spectral data of the far field operator, *Inverse Problems* **14**, 1489–1512 (1998).

[43] A. Kirsch, Factorization of the far-field operator for the inhomogeneous medium case and an application in inverse scattering theory, *Inverse Problems*, **15** (1999), 413–429.

[44] A. Kirsch, New characterizations of solutions in inverse scattering theory, *Applicable Analysis* **76** (2000), 319–350.

[45] A. Kirsch and R. Kress, An optimization method in inverse acoustic scattering, in *Boundary Elements IX*, C.A. Brebbia, et. al., editors, Springer-Verlag, Berlin, 1987, 3–18.

[46] A. Kirsch and R. Kress, Uniqueness in inverse obstacle scattering, *Inverse Problems*, **9** (1993), 285–299.

[47] R. Kress, *Linear Integral Equations*, Springer-Verlag, Berlin, Second Edition, 1999.

[48] R. Kress, Acoustic scattering in *Scattering*, E. R. Pike and P. Sabatier, eds. Academic Press, London, 2001.

[49] R. Kress, Electromagnetic waves scattering, in *Scattering* E. R. Pike and P. Sabatier, eds. Academic Press, London, 2001.

[50] R. Kress and W. Rundell, Inverse scattering for shape and impedance, *Inverse Problems* **17** (2001), 1075–1085.

[51] K. Langenberg, Applied inverse problems for acoustic, electromagnetic and elastic wave scattering, in *Basic Methods of Tomography and Inverse Problems*, P.C. Sabatier, ed., Adam Hilger, Bristol, 1987, 125–467.

[52] P. Maponi, M. Recchioni and F. Zirilli, The use of optimization in the reconstruction of obstacles from acoustic or electromagnetic scattering data, in *Large Scale Optimization with Applications, Part I: Optimization in Inverse Problems and Design*, L. Biegler, et. al. eds., IMA Volumes in Mathematics and its Applications, Vol. **92**, Springer-Verlag, New York, 1997, 81–100.

[53] A. Nachman, Reconstructions from boundary measurements, *Annals of Math.*, **128** (1988), 531–576.

[54] F. Natterer and F. Wübbeling, *Mathematical Methods in Image Reconstruction*, SIAM Publications, Phildelphia, 2001.

[55] F. Natterer and F. Wübbeling, A propagation–backpropagation method for ultrasound tomography, *Inverse Problems* **11** (1995), 1225–1232.

[56] A. N. Norris, A direct inverse scattering method for imaging obstacles with unknown surface conditions, *IMA Jour. Applied Math.* **61** (1998), 267–290.

[57] R. Novikov, Multidimensional inverse spectral problems for the equation $-\Delta\psi + (v(x) - Eu(x))\psi = 0$, *Trans. Functional Anal. Appl.*, **22** (1988), 263–272.

[58] R. Novikov, The inverse scattering problem on a fixed energy level for the two dimensional Schrödinger operator, *J. Funct. Anal.* **103** (1992), 409–463.

[59] P. Ola, L. Päivärinta, and E. Somersalo, An inverse boundary value problem in electrodynamics, *Duke Math. Jour.* **70** (1993) 617–653.

[60] P. Ola and E. Somersalo, Electromagnetic inverse problems and generalized Sommerfeld potentials, *SIAM J. Appl. Math* **56** (1996), 1129–1145.

[61] M. Piana, On uniqueness for anisotropic inhomogeneous inverse scattering problems, *Inverse Problems*, **14** (1998), 1565–1579.

[62] R. Potthast, *Point Sources and Multipoles in Inverse Scattering Theory*, Research Notes in Mathematics, Vol. **427**, Chapman and Hall/CRC, Boca Raton, Florida, 2001.

[63] R. Potthast, Fréchet differentiability of boundary integral operators in inverse acoustic scattering, *Inverse Problems*, **10** (1994), 431–447.

[64] R. Potthast, On a concept of uniqueness in inverse scattering for a finite number of incident waves, *SIAM J. Appl. Math.* **58** (1998), 666–682.

[65] R. Potthast, Stability estimates and reconstructions in inverse acoustic scattering using singular sources, *J. Comp. Applied Math.*, **114** (2000), 247–274.

[66] A. G. Ramm, Recovery of the potential from fixed-energy scattering data, *Inverse Problems*, **4** (1988), 877–886.

[67] B. P. Rynne and B.D. Sleeman, The interior transmission problem and inverse scattering from inhomogeneous media, *SIAM J. Math. Anal.*, **22** (1991), 1755–1762.

[68] Z. Sun and G. Uhlmann, Generic uniqueness for an inverse boundary value problem, *Duke Math. J.* **62** (1991), 131–155.

[69] Z. Sun and G. Uhlmann, An inverse boundary value problem for Maxwell's equations, *Arch. Rat. Mech. Anal.* **119** (1992), 71–93.

[70] Z. Sun and G. Uhlmann, Recovery of singularities for formally determined inverse problems, *Comm. Math. Physics* **153** (1993), 431–445.

[71] J. Sylvester and G. Uhlmann, A global uniqueness theorem for an inverse boundary value problem, *Annals of Math.*, **125** (1987), 153–169.

[72] P. M. Van Den Berg and R. E. Kleinman, Gradient methods in inverse acoustic and electromagnetic scattering, in *Large-Scale Optimization with Applications, Part I: Optimization in Inverse Problems and Design*, L. T. Biegler, et. al. eds. IMA Volumes in Mathematics and its Applications. Vol. **92**, Springer-Verlag, New York, 1997, 173–194.

[73] P. Werner, Beugungsprobleme der mathematischen Akustik, *Arch. Rat. Mech. Anal.*, **12** (1963), 155–184.

[74] Y. X. You, G. P. Miao and Y. Z. Liu, A fast method for acoustic imaging of multiple three dimensional objects, *J. Acoust. Soc. Am.* **108** (2000), 31–37.

DAVID COLTON
DEPARTMENT OF MATHEMATICAL SCIENCES
UNIVERSITY OF DELAWARE
NEWARK, DELAWARE 19716
UNITED STATES
   colton@math.udel.edu

Inverse Problems in Transport Theory

PLAMEN STEFANOV

ABSTRACT. We study an inverse problem for the transport equation in a bounded domain in $\mathbb{R}^n$. Given the incoming flux on the boundary, we measure the outgoing one. The inverse problem is to recover the absorption coefficient $\sigma_a(x)$ and the collision kernel $k(x, v', v)$ from this data. This paper is a survey of recent results about general k's without assuming that k depends on a reduced number of variables. We present uniqueness results in dimensions $n \geq 3$ for the time dependent and the stationary problem, and in the time dependent case we study the inverse scattering problem as well. The proofs are constructive and lead to direct procedures for recovering σ_a and k. For $n = 2$ the problem of recovering k is formally determined and we prove uniqueness for small k and a stability estimates.

1. Introduction

This paper is a review of the recent progress in the study of inverse problems for the transport equation in $\mathbb{R}^n$, $n \geq 2$ by the author and M. Choulli [CSt1], [CSt2], [CSt3], [CSt4] and the author and G. Uhlmann [StU]. We are focused here on the case when the collision kernel k introduced below depends on all of its variables x, v', v. There are a lot of works dealing with k's of the form $k(x, v' \cdot v)$ that is also physically important but we will not discuss those results here.

Define the transport operator T by

$$Tf = -v \cdot \nabla_x f(x, v) - \sigma_a(x, v) f(x, v) + \int_V k(x, v', v) f(x, v') \, dv', \qquad (1\text{--}1)$$

where $f = f(x, v)$ represents the density of a particle flow at the point $x \in X$ moving with velocity $v \in V$. Here $X \subset \mathbb{R}^n$ is a bounded domain with C^1-boundary, $V \subset \mathbb{R}^n$ is the velocity space. We assume that V is an open set in Sections 2 and 3, and that $V = S^{n-1}$ (and dv is replaced by dS_v) in Section 4. All of our results in Sections 2 and 3 hold in the case when $V = S^{n-1}$ with obvious modifications. In fact, the case $V = S^{n-1}$ leads to some simplifications,

The author was partly supported by NSF Grant DMS-0196440.

111

for example there is no need to work locally in the open set $V \setminus \{0\}$ in some cases and the measure $d\tilde{\xi}$ can be chosen to be $d\xi$ in Section 3. The coefficient $\sigma_a(x,v) \geq 0$ above measures the absorption of particles at the point (x,v) due to change of velocity or absorption by the surrounding media. The collision kernel $k(x,v',v) \geq 0$ is related to the number of particles that change their velocity from v' to v at the point x.

We study inverse problems for both the time-dependent transport equation

$$(\partial_t - T)f = 0 \tag{1-2}$$

and the stationary transport equation

$$Tf = 0. \tag{1-3}$$

Set $\Gamma_\pm = \{(x,v) \in \partial X \times V : \pm n(x) \cdot v > 0\}$, where $n(x)$ is the outer normal to ∂X at $x \in \partial X$. A typical boundary value problem for (1–2) or (1–3) is associated with boundary conditions

$$f|_{\Gamma_-} = f_-, \tag{1-4}$$

where f_- is a given function on Γ_-, that depends also on t in the first case. On the boundary we measure the outgoing flux f_+ generated by a given incoming flux f_-, i.e., we assume that we know that so called *albedo* operator $\mathcal{A}$ defined by

$$\mathcal{A} : f_- \longmapsto f_+ = f|_{\Gamma_+}. \tag{1-5}$$

The inverse problem that we study is this: Can we determine the functions σ_a and k from a knowledge of $\mathcal{A}$? In the time dependent case we study the inverse scattering problem as well — recovery of σ_a and k from the knowledge of the scattering operator S. In order to ensure uniqueness of recovery of σ_a, we assume that σ_a depends on x only, since otherwise there is clearly no uniqueness. Under this assumption and some natural assumptions in the stationary case that guarantee the solvability of the direct problem (1–2), we not only prove uniqueness but give an explicit solution for the inverse problem. Our approach is based on the study of the singularities of the Schwartz kernel of $\mathcal{A}$ and we show that all information about σ_a and k is contained in those singularities. For large σ_a, those singularities have very small amplitudes and are hard to measure in real applications, so then this approach is of less interest for the practical recovery of σ_a and k. In Section 4 we prove some stability estimates.

2. The Time Dependent Transport Equation

2.1. Main results In this section we present inverse problems results about the time dependent transport equation (1–2)

$$\frac{\partial}{\partial t}u(x,v,t) = -v \cdot \nabla_x u(x,v,t) - \sigma_a(x,v)u(x,v,t) + \int_V k(x,v',v)u(x,v',t)\,dv'. \tag{2-1}$$

We first introduce some terminology and notation. The production rate $\sigma_p(x, v')$ is defined by

$$\sigma_p(x, v') = \int_V k(x, v', v)\, dv.$$

Following [RS] we say that the pair (σ_a, k) is *admissible* if

(i) $0 \le \sigma_a \in L^\infty(\mathbb{R}^n \times V)$,
(ii) $0 \le k(x, v', \cdot) \in L^1(V)$ for a.e. $(x, v') \in \mathbb{R}^n \times V$ and $\sigma_p \in L^\infty(\mathbb{R}^n \times V)$.

Throughout this paper we assume that σ_a and k are extended as 0 for $x \notin X$.

Set $T_0 = -v \cdot \nabla_x$ with its maximal domain. It is well-known that T_0 is a generator of a strongly continuous group $U_0(t)f = f(x - tv, v)$ of isometries on $L^1(X \times V)$ preserving nonnegative functions. Following the notation in [RS], we introduce the operators

$$[A_1 f](x, v) = -\sigma_a(x, v)f(x, v), \qquad T_1 = T_0 + A_1, \qquad D(T_1) = D(T_0),$$
$$[A_2 f](x, v) = \int_V k(x, v', v)f(x, v')\, dv', \quad T = T_0 + A_1 + A_2 \quad D(T) = D(T_0),$$

and set $A = A_1 + A_2$. Operators A_1 and A_2 are bounded on $L^1(X \times V)$ and T_1, T are generators of strongly continuous groups $U_1(t) = e^{tT_1}$, $U(t) = e^{tT}$, respectively [RS]. For $U_1(t)$ we have an explicit formula

$$[U_1(t)f](x, v) = e^{-\int_0^t \sigma_a(x - sv, v)\, ds} f(x - tv, v), \tag{2-2}$$

while for $U(t)$ we have

$$\|U(t)\| \le e^{Ct}, \quad C = \|\sigma_p\|_{L^\infty}. \tag{2-3}$$

We work in the Banach space $L^1(X \times V)$, so here $\|U(t)\|$ is the operator norm of $U(t)$ in $\mathcal{L}(L^1(X \times V))$. It should be mentioned also that $U(t)$ preserves the cone of nonnegative functions for $t \ge 0$.

One can define the wave operators associated with T, T_0 by

$$W_- = \operatorname*{s-lim}_{t \to \infty} U(t)U_0(-t), \tag{2-4}$$

$$\tilde{W}_+ = \operatorname*{s-lim}_{t \to \infty} U_0(-t)U(t). \tag{2-5}$$

If W_- and $\tilde{W}_+$ exist, one can define the scattering operator

$$S = \tilde{W}_+ W_-$$

as a bounded operator in $L^1(X \times V)$. Scattering theory for (1-1) has been developed in [Hej], [Si], [V1] and we refer to these papers (see also [RS]) for sufficient conditions guaranteeing the existence of S. We would like to mention here also [P1], [U], [E], [St], [V2], [CMS]. An abstract approach based on the Limiting Absorption Principle has been proposed in [Mo]. We show below however that S can always be defined as an operator $S : L^1_c(\mathbb{R}^n \times V \backslash \{0\}) \to L^1_{\mathrm{loc}}(\mathbb{R}^n \times V \backslash \{0\})$.

The inverse scattering problem for (2–1) is the following: Does S determine uniquely σ_a, k? The answer is affirmative if σ_a is independent of v.

THEOREM 2.1 ([CSt1], [CSt2]). *Let (σ_a, k), $(\hat{\sigma}_a, \hat{k})$ be two admissible pairs such that σ_a, $\hat{\sigma}_a$ do not depend on v and denote by S, $\hat{S}$ the corresponding scattering operators. Then, if $S = \hat{S}$, we have $\sigma_a = \hat{\sigma}_a$, $k = \hat{k}$.*

The assumption that σ_a, $\hat{\sigma}_a$ depend on x only can be relaxed a little by assuming that they depend on x and $|v|$ only. In the general case however, there is no uniqueness. Assume, for example, that $k = 0$. Then

$$Sf = e^{-\int_{-\infty}^{\infty} \sigma_a(x-sv,v)\,ds}f, \quad (k=0) \tag{2-6}$$

and therefore S determines $\int_{-\infty}^{\infty} \sigma_a(x - sv, v)\,ds$ only for any $v \in V$. It is easy to see that those integrals do not determine σ_a uniquely. If σ_a is independent of $v/|v|$ however, then S determines the X-ray transform of σ_a and it therefore determines σ_a in this case.

Next we consider the albedo operator $\mathcal{A}$. Assume that X is convex and has C^1-smooth boundary ∂X. Consider the measure $d\xi = |n(x) \cdot v| d\mu(x)\, dv$ on $\Gamma_\pm$, where $d\mu(x)$ is the measure on ∂X. We will solve the problem

$$\begin{cases} (\partial_t - T)u = 0 & \text{in } \mathbb{R} \times X \times V, \\ u|_{\mathbb{R} \times \Gamma_-} = u_-, \\ u|_{t \ll 0} = 0 \end{cases} \tag{2-7}$$

for $u(t, x, v)$, where $u_- \in L_c^1(\mathbb{R}; L^1(\Gamma_-, d\xi))$. Problem (2–7) has a unique solution $u \in C(\mathbb{R}; L^1(X \times V))$ and one defines the albedo operator $\mathcal{A}$ by

$$\mathcal{A}g = u|_{\mathbb{R} \times \Gamma_+} \in L^1_{\text{loc}}(\mathbb{R}; L^1(\Gamma_+, d\xi)). \tag{2-8}$$

Therefore, $\mathcal{A} : L_c^1(\mathbb{R}; L^1(\Gamma_-, d\xi)) \to L^1_{\text{loc}}(\mathbb{R}; L^1(\Gamma_+, d\xi))$. It can be seen that $\mathcal{A}g$ can be defined more generally for $g \in L^1(\mathbb{R} \times \Gamma_-, dt\, d\xi)$ with $g = 0$ for $t \ll 0$. We show below that in fact $\mathcal{A}$ determines S uniquely and conversely, S determines $\mathcal{A}$ uniquely by means of explicit formulae in case when X is convex. This generalizes earlier results in [AE], [EP], [P2] showing that there is a relationship between S and $\mathcal{A}$. To this end, let us define the extension operators $J_\pm$ and the restriction (trace) operators $R_\pm$ as follows. Set

$$\Omega = \{(x, v) \in \mathbb{R}^n \times V \backslash \{0\} : x - tv \in X \text{ for some } t \in \mathbb{R}\}, \tag{2-9}$$

and define the functions

$$\tau_\pm(x, v) = \min\{t \in \mathbb{R} : (x \pm tv, v) \in \Gamma_\pm\}, \quad (x, v) \in \Omega. \tag{2-10}$$

Given $g \in L^1(\mathbb{R} \times \Gamma_\pm, dt\, d\xi)$, consider the following operators of extension:

$$(J_\pm g)(x, v) = \begin{cases} g(\pm\tau_\pm(x,v), x \pm \tau_\pm(x,v)v, v) & \text{if } (x, v) \in \Omega, \\ 0 & \text{otherwise.} \end{cases}$$

It is easy to check that $J_\pm : L^1(\mathbb{R} \times \Gamma_\pm, dt\, d\xi) \to L^1(X \times V)$ are isometric. Denote by $R_\pm$ the operator of restriction

$$R_\pm f = f|_{\Gamma_\pm}, \quad f \in C(\mathbb{R}^n \times V).$$

Although $R_\pm$ is not a bounded operator on $L^1(X \times V)$ (see [Ce1], [Ce2] and Theorem 3.2 below), we can show that $R_\pm U_0(t)f \in L^1(\mathbb{R} \times \Gamma_\pm, dt\, d\xi)$ is well defined for any $f \in L^1(X \times V)$. Denote by χ_Ω the characteristic function of Ω. We establish the following relationships between S and $\mathcal{A}$.

THEOREM 2.2 ([CSt1], [CSt2]). *Assume that X is convex. Then*

(a) $\mathcal{A}g = R_+ U_0(t)SJ_- g$ *for all* $g \in L_c^1(\mathbb{R} \times \Gamma_-, dt\, d\xi)$,
(b) $Sf = J_+ \mathcal{A}R_- U_0(t)f + (1 - \chi_\Omega)f$, $f \in L_c^1(\mathbb{R}^n \times V\backslash\{0\})$,
(c) $\mathcal{A}$ *extends to a bounded operator*

$$\mathcal{A} : L^1(\mathbb{R} \times \Gamma_-, dt\, d\xi) \to L^1(\mathbb{R} \times \Gamma_+, dt\, d\xi)$$

if and only if S extends to a bounded operator on $L^1(X \times V)$.

We decompose $L^1(\mathbb{R}^n \times V) = L^1(\Omega) \oplus L^1((\mathbb{R}^n \times V) \backslash \Omega)$. A similar decomposition holds for $L_c^1(\mathbb{R}^n \times V\backslash\{0\})$. Then S leaves invariant both spaces, moreover $S|_{L^1((\mathbb{R}^n \times V)\backslash\Omega)} = \mathrm{Id}$, so S can be decomposed as a direct sum $S = S_\Omega \oplus \mathrm{Id}$. Denote $\mathcal{R}_\pm = R_\pm U_0(\cdot) : L^1(\Omega) \to L^1(\mathbb{R} \times \Gamma_\pm, dt\, d\xi)$. Then $\mathcal{R}_\pm$ are isometric and invertible and $\mathcal{R}_\pm^{-1} = \mathcal{J}_\pm$ with $\mathcal{J}_\pm : L^1(\mathbb{R} \times \Gamma_\pm, dt\, d\xi) \to L^1(\Omega)$, $\mathcal{J}_\pm f := J_\pm f|_{L^1(\Omega)}$. Then we can rewrite Theorem 2.2 (a), (b) as follows:

$$\mathcal{A} = \mathcal{R}_+ S_\Omega \mathcal{J}_- \quad \text{on } L_c^1(\mathbb{R} \times \Gamma_-, dt\, d\xi),$$
$$S_\Omega = \mathcal{J}_+ \mathcal{A}\mathcal{R}_- \quad \text{on } L_c^1(\mathbb{R}^n \times V\backslash\{0\}).$$

Thus $\mathcal{A}$ can be obtained from S_Ω by a conjugation with invertible isometric operators and vice-versa.

An immediate consequence of Theorem 2.2 is that $\mathcal{A}$ determines σ_a and k uniquely for σ_a independent of v and X convex. In short, in this case the inverse boundary value problem is equivalent to the inverse scattering problem. However, we can prove uniqueness for the inverse boundary value problem for not necessarily convex domains as well independently of Theorems 2.1 and 2.2.

THEOREM 2.3 [CSt1], [CSt2]. *Let (σ_a, k), $(\hat{\sigma}_a, \hat{k})$ be two admissible pairs with σ_a, $\hat{\sigma}_a$ independent of v. Then, if the albedo operators $\mathcal{A}$, $\hat{\mathcal{A}}$ on ∂X coincide, we have $\sigma_a = \hat{\sigma}_a$, $k = \hat{k}$.*

2.2. Singular decomposition of the fundamental solution and the kernels of S and $\mathcal{A}$. Proof of the main results in section 2.1. The key to proving the uniqueness results above is to study the singularities of the Schwartz kernel of S and respectively $\mathcal{A}$. To this end we will study first the kernel of the

solution operator of the problem (2–7). Given $(x', v') \in \mathbb{R}^n \times V \setminus \{0\}$, consider the problem

$$\begin{cases} (\partial_t - T)u = 0 & \text{in } \mathbb{R} \times \mathbb{R}^n \times V, \\ u|_{t \ll 0} = \delta(x - x' - tv)\delta(v - v'), \end{cases} \tag{2–11}$$

δ being the Dirac delta function. Problem (2–1) has a unique solution

$$u^{\#}(t, x, v, x', v')$$

depending continuously on t with values in $\mathcal{D}'(\mathbb{R}_x^n \times V_v \times \mathbb{R}_{x'}^n \times V_{v'} \setminus \{0\})$. We have the following singular expansion of $u^{\#}$.

THEOREM 2.4 [CSt1], [CSt2]. *Problem (2–11) has the unique solution* $u^{\#} = u_0^{\#} + u_1^{\#} + u_2^{\#}$, *where*

$$u_0^{\#} = e^{-\int_0^{\infty} \sigma_a(x-sv, v)\, ds} \delta(x - x' - tv)\delta(v - v'),$$

$$u_1^{\#} = \int_0^{\infty} e^{-\int_0^s \sigma_a(x-\tau v, v)d\tau} e^{-\int_0^{\infty} \sigma_a(x-sv-\tau v', v')d\tau}$$

$$\times k(x-sv, v', v)\delta(x - sv - (t-s)v' - x')\, ds,$$

$$u_2^{\#} \in C\left(\mathbb{R}; L_{\text{loc}}^{\infty}(\mathbb{R}_{x'}^n \times V_{v'}; L^1(\mathbb{R}_x^n \times V_v))\right).$$

The proof of Theorem 2.4 is based on iterating twice Duhamel's formula

$$U(t-r) = U_1(t-r) + \int_r^t U(t-s)A_2 U_1(s-r)\, ds$$

and on estimating the remainder term.

To build the scattering theory for the transport equation, we first show that the wave operators W_-, $\tilde{W}_+$ (see (2–4), (2–5)) exist as operators between the spaces

$$W_- : L_c^1(\mathbb{R}^n \times V \setminus \{0\}) \longrightarrow L^1(X \times V),$$
$$\tilde{W}_+ : L^1(X \times V) \longrightarrow L_{\text{loc}}^1(\mathbb{R}^n \times V \setminus \{0\}).$$

Then we define the scattering operator

$$S = \tilde{W}_+ W_- : L_c^1(\mathbb{R}^n \times V \setminus \{0\}) \longrightarrow L_{\text{loc}}^1(\mathbb{R}^n \times V \setminus \{0\}). \tag{2–12}$$

It can be seen that S is well defined on the wider subset $\{f : \exists t_0 \text{ such that } U_0(t)f = 0 \text{ for } x \in X, t < -t_0\}$ (the incoming space).

The distribution $u^{\#}(t, x, v, x', v')$ is the Schwartz kernel of $U(t)W_-$. Let $S(x, v, x', v')$ be the Schwartz kernel of the scattering operator S. Then

$$S(x, v, x', v') = \lim_{t \to \infty} u^{\#}(t, x+tv, v, x', v').$$

This limit exists trivially, in fact for any $K \Subset \mathbb{R}^n \times V \setminus \{0\}$, the distribution $u^{\#}(t, x+tv, v, x', v')|_K$ is independent of t for $t \geq t_0(K)$. On the other hand, as mentioned in the Introduction, it is not trivial to show that under some condition, S is a kernel of a bounded operator in $L^1(\mathbb{R}^n \times V)$. One can also prove the following integral representation of the scattering kernel:

$$S(x, v, x', v') = e^{-\int_{-\infty}^{\infty} \sigma_a(x - \tau v, v) d\tau} \delta(x - x') \delta(v - v')$$

$$+ \int_{-\infty}^{\infty} e^{-\int_s^{\infty} \sigma_a(x + \tau v, v) d\tau} (A_2 u^{\#})(s, x + sv, v, x', v') \, ds. \quad (2\text{–}13)$$

This formula is an analogue of the representation of the scattering amplitude (in our setting, that would be the second term in the right-hand side above) for the Schrödinger equation.

Now, combining Theorem 2.4 and the representation above, we get the following express for the kernel S of the scattering operator S.

THEOREM 2.5 [CSt1], [CSt2]. *We have* $S = S_0 + S_1 + S_2$, *where the Schwartz kernels* $S_j(x, v, x', v')$ *of the operators* S_j, $j = 0, 1, 2$ *satisfy*

$$S_0 = e^{-\int_{-\infty}^{\infty} \sigma_a(x - \tau v, v) d\tau} \delta(x - x') \delta(v - v'),$$

$$S_1 = \int_{-\infty}^{\infty} e^{-\int_s^{\infty} \sigma_a(x + \tau v, v) d\tau} e^{-\int_0^{\infty} \sigma_a(x + sv - \tau v', v') d\tau}$$

$$k(x + sv, v', v) \delta(x - x' + s(v - v')) \, ds,$$

$$S_2 \in L^{\infty}_{\text{loc}}(\mathbb{R}^n_{x'} \times V_{v'} \backslash \{0\}; \ L^1_{\text{loc}}(\mathbb{R}^n_x \times V_v \backslash \{0\})).$$

We are ready now to complete the proof of Theorem 2.1. The idea of the proof is the following. Suppose we are given the scattering operator S corresponding to a unknown admissible pair (σ_a, k). Then we know the kernel $S = S_0 + S_1 + S_2$. It follows from Theorem 2.5 that S_0 is a singular distribution supported on the hyperplane $x = x'$, $v = v'$ of dimension $2n$, S_1 is supported on a $(3n + 1)$-dimensional surface (for $v \neq v'$), while S_2 is a function. Therefore, S_j, $j = 0, 1, 2$ have different degrees of singularities and given $S = S_0 + S_1 + S_2$, one can always recover S_0 and S_1. From S_0 one can recover the X-ray transform of σ_a and therefore σ_a itself, provided that σ_a is independent of v. Next, suppose for simplicity that σ_a and k are continuous. Then for fixed x, v, v' with $v \neq v'$, S_1 is a delta-function supported on the line $x' = x + s(v - v')$, $s \in \mathbb{R}$ with density a multiple of $k(x + sv, v', v)$. Therefore, one can recover that density for each s and in particular setting $s = 0$, we get $k(x, v', v)$. Moreover, based on this, we can write explicit formulae that extract σ_a and k from S by allowing S to act on a sequence of suitably chosen test functions that concentrate on one of the singular varieties described above, see [CSt2].

We will skip the proof of Theorem 2.2. We merely recall that it proves uniqueness for the inverse boundary value problem for convex X as a direct consequence of Theorem 2.5.

Now assume that X is not necessarily convex. We can still prove uniqueness for the inverse boundary value problem by showing that A determines uniquely $u^{\#}$ for x outside X by following arguments in [SyU], and then using (2–13) and Theorem 2.5. In order to give a constructive (in fact, explicit) reconstruction, we study next the Schwartz kernel of the operator $\mathcal{A}$ in the spirit of Theorem 2.5. A priori, this kernel is a distribution in $\mathcal{D}'(\mathbb{R} \times \Gamma_+ \times \mathbb{R} \times \Gamma_-)$. Denote by δ_1 the Dirac delta function on $\mathbb{R}^1$ and by $\delta_y(x)$ the Delta function on ∂X defined by $\int_{\partial X} \delta_y \varphi \, d\mu(x) = \varphi(y)$. Set

$$\theta(x, y) = \begin{cases} 1 & \text{if } px + (1-p)y \in X \text{ for each } p \in [0, 1], \\ 0 & \text{otherwise.} \end{cases}$$

THEOREM 2.6 [CSt1], [CSt2]. *The Schwartz kernel of $\mathcal{A}$ has the form*

$$\alpha(t - t', x, v, x', v'),$$

that is, formally,

$$(\mathcal{A}g)(t, x, v) = \int_{\mathbb{R} \times \Gamma_-} \alpha(t - t', x, v, x', v') g(t', x', v') \, d\xi(x', v')$$

with $\alpha = \alpha_0 + \alpha_1 + \alpha_2$, where the $\alpha_j(\tau, x, v, x', v')$ (with $(x, v) \in \Gamma_+$, $(x', v') \in \Gamma_-$) satisfy

$$\alpha_0 = |n(x') \cdot v'|^{-1} e^{-\int_0^{\tau_-(x,v)} \sigma_a(x-sv, v) \, ds} \delta_{\{x - \tau_-(x,v)v\}}(x') \delta(v - v') \delta_1(\tau - \tau_-(x, v)),$$

$$\alpha_1 = |n(x') \cdot v'|^{-1} \int e^{-\int_0^s \sigma_a(x-pv, v) \, dp} e^{-\int_0^{\tau_-(x-sv, v')} \sigma_a(x-sv-pv', v') \, dp} \times$$
$$\delta_1(\tau - s - \tau_-(x-sv, v')) k(x-sv, v', v) \delta_{\{x-sv-\tau_-(x-sv,v')v'\}}(x') \theta(x-sv, x) \, ds,$$

$$\alpha_2 \in L^\infty\left(\Gamma_-; L^1_{\text{loc}}(\mathbb{R}_\tau; L^1(\Gamma_+, d\xi))\right).$$

The proof of Theorem 2.3 now follows from the theorem above by analyzing the singularities of α as above. In this case, we can also wite explicit formulae for σ_a and k as certain limits of the distribution α acting on a sequence of test functions concentrating near the singularities of α_0 and α_1, respectively; see [CSt2].

3. The Stationary Transport Equation

We turn our attention now to the boundary value problem (1–3), (1–4) for the stationary transport equation

$$\begin{cases} -v \cdot \nabla_x f(x, v) - \sigma_a(x, v) f(x, v) + \int_V k(x, v', v) f(x, v') \, dv' = 0 & \text{in } X \times V, \\ \\ f|_{\Gamma_-} = f_-. \end{cases} \quad (3\text{–}1)$$

Here X does not need to be convex. We impose some conditions in order to ensure the unique solvability of the direct problem (3–1). Recall the definition (2–10) of $\tau_\pm$. Set $\tau = \tau_- + \tau_+$. We will consider two cases. First we assume that

$$\|\tau \sigma_a\|_{L^\infty} < \infty, \qquad \|\tau \sigma_p\|_{L^\infty} < 1. \qquad (3\text{–}2)$$

This condition in particular guarantees that the dynamics $U(t)$ corresponding to the time-dependent transport equation $(\partial_t - T)u = 0$ is subcritical [RS], that is, the L^1-norm of the solution is uniformly bounded for $t > 0$; compare with (2–3).

Note that (3–2) holds if in particular $\left\| |v|^{-1} \sigma_a \right\|_{L^\infty} < \infty$, $\operatorname{diam} X \left\| |v|^{-1} \sigma_p \right\|_{L^\infty} <$ 1. The second situation we will consider is when [DL]

$$0 \le \nu \le \sigma_a(x, v) - \sigma_p(x, v) \quad \text{for a.e. } (x, v) \in X \times V \qquad (3\text{–}3)$$

with some $\nu > 0$. This condition means that the absorption rate is greater than the production rate. This also implies that the corresponding dynamics is subcritical.

The main result in this section is this:

THEOREM 3.1 [CSt3], [CSt4]. *Assume that* (σ_a, k), $(\hat{\sigma}_a, \hat{k})$ *are two admissible pairs with* $\sigma_a = \sigma_a(x, |v|)$, $\hat{\sigma}_a = \hat{\sigma}_a(x, |v|)$ *and assume that they satisfy either* (3–2) *or* (3–3). *Assume that the corresponding albedo operators* A *and* $\hat{A}$ *coincide. Then*

(a) *if* $n \ge 3$, *then* $\sigma_a = \hat{\sigma}_a$, $k = \hat{k}$;
(b) *if* $n = 2$, *then* $\sigma_a = \hat{\sigma}_a$.

In the stationary case we have less data than in the time dependent one studied in Section 2 because the time variable is not present. The inverse boundary value problem is overdetermined in dimension $n \ge 3$ because the kernel of A depends on $4n - 2$ variables while k is a function of $3n$ variables, $\sigma_a(x, v/|v|)$ is a function of $2n - 1$ variables. In the 2D case however, the inverse problem for recovery of k is formally determined (but still overdetermined for the recovery of σ_a) and the theorem above does not provide uniqueness in this case. In Section 4 we formulate a uniqueness result in the 2D case for small k.

To prove Theorem 3.1, we again study the Schwartz kernel of the albedo operator A. Note that in the stationary case, A acts on functions on Γ_- and maps them to functions on Γ_+ (see next subsection for details). It turns out that, similarly to the time dependent case, the kernel of A has a singular decomposition $\alpha = \alpha_0 + \alpha_1 + \alpha_2$, where α_0 and α_1 are delta functions for $n \ge 3$ supported on varieties of different dimensions with densities that identify respectively σ_a and k. The third term α_2 is a locally L^1 function and can be distinguished form α_0 and α_1. If $n = 2$, α_0 is still a delta function, but α_1 is a locally L^1 function and cannot be distinguished from α_2 and this explains why this method does not work then. The 2D case is considered separately in next section.

3.1. Some estimates in the stationary case

Similarly to the measure defined in Section 2.1, we define a measure on $\Gamma_\pm$ by:

$$d\tilde{\xi} = \min\{\tau(x, v), \lambda\} \left| n(x) \cdot v \right| d\mu(x)\, dv,$$

where $\lambda > 0$ is an arbitrary constant. Using the trace Theorem 3.2 below, we can show that $A : L^1(\Gamma_-, d\xi) \to L^1(\Gamma_+, d\xi)$ is a bounded operator if (3–2) holds and $A : L^1(\Gamma_-, d\tilde{\xi}) \to L^1(\Gamma_+, d\tilde{\xi})$ is bounded if (3–3) holds; see also [DL]. Note that if a neighborhood of the origin is not included in V or in particular, if $V = S^{n-1}$, then we can choose $d\tilde{\xi} = \tau d\xi$.

THEOREM 3.2 (TRACE THEOREM). (a) [CSt3]

$$\|f|_{\Gamma_\pm}\|_{L^1(\Gamma_\pm, d\xi)} \le \|T_0 f\| + \|\tau^{-1} f\|.$$

(b) [Ce1], [Ce2]

$$\|f|_{\Gamma_\pm}\|_{L^1(\Gamma_\pm, d\tilde\xi)} \le \lambda \|T_0 f\| + \|f\|.$$

Note that (b) follows from (a) by setting $f = \min\{\lambda, \tau\} g$.

We introduce the spaces $\mathcal{W}$, $\tilde{\mathcal{W}}$ via the norms

$$\|f\|_{\mathcal{W}} = \|T_0 f\| + \|\tau^{-1} f\|, \qquad \|f\|_{\tilde{\mathcal{W}}} = \|T_0 f\| + \|f\|.$$

Then Theorem 3.2 says that taking the trace $f \mapsto f|_{\Gamma_\pm}$ is a continuous operator from $\mathcal{W}$ into $L^1(\Gamma_\pm, d\xi)$ and similarly from $\tilde{\mathcal{W}}$ into $L^1(\Gamma_\pm, d\tilde\xi)$.

Given $f_- \in L^1(\Gamma_-, d\xi)$, define Jf_- as the following prolongation of f_- inside $X \times V$:

$$Jf_- = e^{-\int_0^{\tau_-(x,v)} \sigma_a(x-sv,v)\, ds} f_-(x - \tau_-(x,v)v, v), \qquad (x,v) \in X \times V.$$

Jf_- is defined so that $T_1 Jf_- = 0$, $Jf_-|_{\Gamma_-} = f_-$, therefore J is the solution operator of the problem $T_1 f = 0$, $f|_{\Gamma_-} = f_-$.

PROPOSITION 3.1. (a) Assume that $\|\tau \sigma_a\|_{L^\infty} < \infty$. Then

$$\|Jf_-\|_{\mathcal{W}} \le C \|f_-\|_{L^1(\Gamma_-, d\xi)},$$

with $C = 1 + \|\tau\sigma_a\|_{L^\infty}$. If $\sigma_a = 0$, then we have equality above (and $C = 1$).

(b) Assume (3-3). For any $f_- \in L^1(\Gamma_-, d\tilde\xi)$ we have

$$\|Jf_-\|_{\tilde{\mathcal{W}}} \le C \|f_-\|_{L^1(\Gamma_-, d\tilde\xi)},$$

where $C = (1 + \|\sigma_a\|_{L^\infty}) \max\{1, (\nu\lambda)^{-1}\}$.

We reduce the boundary value problem (3-1) to an integral equation using standard arguments to get

$$(I + K)f = Jf_-, \tag{3-4}$$

where I stands for the identity and K is the integral operator

$$Kf = -\int_0^{\tau_-(x,v)} e^{-\int_0^t \sigma_a(x-sv,v)\, ds} (A_2 f)(x - tv, v)\, dt. \tag{3-5}$$

Formally, $K = T_1^{-1} A_2$, and for T_1^{-1} we have

$$T_1^{-1} f = -\int_0^{\tau_-(x,v)} e^{-\int_0^t \sigma_a(x-sv,v)\, ds} f(x - tv, v)\, dt.$$

PROPOSITION 3.2. Assume (3-2). Then:

(a) $\tau^{-1} T_1^{-1}$, $\tau^{-1} T^{-1}$ and $A_2 \tau$ are bounded operators in $L^1(X \times V)$ and therefore $K = T_1^{-1} A_2$ is a bounded operator in $L^1(X \times V; \tau^{-1}\, dx\, dv)$. Moreover, the operator norm of K is not greater than $\|\tau\sigma_p\|_{L^\infty} < 1$ and therefore $(I + K)^{-1}$ exists in this space.

(b) *The integral equation (3–4) and therefore the boundary value problem (3–1)
     are uniquely solvable for any $f_- \in L^1(\Gamma_-, d\xi)$ and then $f \in \mathcal{W}$.*
(c) *The albedo operator $\mathcal{A}$ is a bounded map $\mathcal{A} : L^1(\Gamma_-, d\xi) \to L^1(\Gamma_+, d\xi)$.*

PROPOSITION 3.3. *Assume (3–3). Then:*

(a) K, T_1^{-1} *and* T^{-1} *are bounded operators in $L^1(X \times V)$ and $K = T_1^{-1} A_2$.
     Further, $I + K$ is invertible and $(I + K)^{-1} = I - T^{-1} A_2$.*
(b) *The integral equation (3–4) and therefore the boundary value problem (3–1)
     are uniquely solvable for any $f_- \in L^1(\Gamma_-, d\tilde{\xi})$ and then $f \in \tilde{\mathcal{W}}$.*
(c) $\mathcal{A}$ *is a bounded map $\mathcal{A} : L^1(\Gamma_-, d\tilde{\xi}) \to L^1(\Gamma_+, d\tilde{\xi})$.*

3.2. The fundamental solution of the stationary transport equation.

We solve (3–1) with $f_- = \phi_-$, where

$$\phi_- = \frac{1}{|n(x') \cdot v'|} \delta_{\{x'\}}(x) \delta(v - v'),$$

where $(x', v') \in \Gamma_-$ are regarded as parameters, $n(x')$ is the outer normal, and
$\delta_{\{x'\}}$ is a distribution on ∂X defined by $(\delta_{\{x'\}}, \varphi) = \int \delta_{\{x'\}}(x) \varphi(x) d\mu(x) = \varphi(x')$.
On the other hand, we will denote by δ the ordinary Dirac delta function in $\mathbb{R}^n$.
The integral above is to be considered in distribution sense. Let us denote by
$\phi(x, v, x', v')$ the solution (in distribution sense) of

$$\begin{cases} T\phi = 0 & \text{in } X \times V, \\ \phi|_{\Gamma_-} = \phi_-. \end{cases} \tag{3-6}$$

To solve (3–6), we write

$$\varphi = J\varphi_- - KJ\varphi_- + (I + K)^{-1} K^2 J\varphi_-$$

and analyze each term. To treat the third term, we observe that

$$(I + K)^{-1} K^2 J\varphi_- = T^{-1} A_2 K J\varphi_-.$$

We are led to:

THEOREM 3.3. *Assume that (σ_a, k) is admissible and either (3–2) or (3–3) holds.
Then the solution $\phi(x, v, x', v')$ of (3–6) can be written as $\phi = \phi_0 + \phi_1 + \phi_2$, where*

$$\phi_0 = \int_0^{\tau_+(x',v')} e^{-\int_0^{\tau_-(x,v)} \sigma_a(x - pv, v)\, dp} \delta(x - x' - tv) \delta(v - v')\, dt,$$

$$\phi_1 = \int_0^{\tau_-(x,v)} \int_0^{\tau_+(x',v')} e^{-\int_0^s \sigma_a(x - pv, v)\, dp} e^{-\int_0^{\tau_-(x-sv,v')} \sigma_a(x - sv - pv', v')\, dp}$$
$$\times k(x - sv, v', v) \delta(x - x' - sv - tv')\, dt\, ds,$$

and

$$\phi_2 \in L^\infty(\Gamma_-; \mathcal{W}) \qquad \text{if (3–2) holds,}$$
$$\min\{\tau, \lambda\}^{-1} \phi_2 \in L^\infty(\Gamma_-; \tilde{\mathcal{W}}) \qquad \text{if (3–3) holds.}$$

The so constructed solution $\phi(x, v, x', v')$ is the distribution kernel of the solution operator $f_- \mapsto f$ of (3–1). In order to find the distribution kernel $\alpha(x, v, x', v')$ $((x, v) \in \Gamma_+, (x', v') \in \Gamma_-)$ of the albedo operator $\mathcal{A}$, it is enough to set

$$\alpha(x, v, x', v') := \phi(x, v, x', v')|_{(x,v)\in\Gamma_+}, \qquad (x', v') \in \Gamma_-.$$

Then, in the sense of distributions,

$$(\mathcal{A}f_-)(x, v) = \int_{\Gamma_-} \alpha(x, \theta, x', \theta') f_-(x', \theta')\, d\xi(x', \theta') \qquad \text{for all } f_- \in C_0^\infty(\Gamma_+).$$

Theorem 3.3 yields the following.

THEOREM 3.4. *Assume that (σ_a, k) is an admissible pair and that either (3–2) or (3–3) holds. Then the distribution kernel $\alpha(x, v, x', v')$ of $\mathcal{A}$ satisfies $\alpha = \alpha_0 + \alpha_1 + \alpha_2$ with*

$$\alpha_0 = \frac{1}{n(x) \cdot v} e^{-\int_0^{\tau_-(x,v)} \sigma_a(x-pv,v)\, dp} \delta_{\{x'+\tau_+(x',v')v'\}}(x)\delta(v - v'),$$

$$\alpha_1 = \frac{1}{n(x) \cdot v} \int_0^{\tau_+(x',v')} e^{-\int_0^{\tau_+(x'+tv',v)} \sigma_a(x-pv,v)\, dp} e^{-\int_0^t \sigma_a(x+pv',v')\, dp}$$
$$\times k(x + tv', v', v)\delta_{\{x'+tv'+\tau_+(x'+tv',v)\}}(x)\, dt,$$

and

$$\alpha_2 \in L^\infty(\Gamma_-; L^1(\Gamma_+, d\xi)) \qquad \text{if (3–2) holds,}$$

$$\min\{\tau(x', v'), \lambda\}^{-1}\alpha_2 \in L^\infty(\Gamma_-; L^1(\Gamma_+, d\tilde{\xi})) \qquad \text{if (3–3) holds.}$$

Theorem 3.4 implies the following way for proving Theorem 3.1. Assume that we are given the albedo operator $\mathcal{A}$, corresponding to some admissible pair (σ_a, k), satisfying either (3–2) or (3–3). Then we also know the distribution $\alpha(x, v, x', v')$. By Theorem 3.4, $\alpha = \alpha_0 + \alpha_1 + \alpha_2$. Here α_0 is a delta-type distribution supported on a $(2n - 1)$–dimensional variety in $\Gamma_+ \times \Gamma_-$. Next, α_1 is also a delta-type distribution (provided that $n \geq 3$) supported on a $3n$-dimensional variety in $\Gamma_+ \times \Gamma_-$, while α_2 is a (locally L^1) function on the $(4n-2)$-dimensional $\Gamma_+ \times \Gamma_-$. Notice that if $n = 2$, then α_1 is a function as well. Therefore, if $n \geq 3$, one can distinguish between $\alpha_0 + \alpha_1$ and α_2. Moreover, since α_0 and α_1 have different degrees of singularities, one can recover α_0 and α_1. Now, if $\sigma_a = \sigma_a(x, |v|)$, then α_0 determines the X-ray transform $\int \sigma_a(x + s\omega, |v|)\, ds$ of σ_a for all x, $|v|$ and ω in an open subset of S^{n-1} (for all $\omega \in S^{n-1}$ if V is spherically symmetric). This determines uniquely σ_a (see e.g. [H]). Next, once we know σ_a, from α_1 we can recover k. If $n = 2$, then we can recover α_0 and therefore σ_a, but we cannot (at least using those arguments) distinguish between α_1 and α_2 which are both functions and therefore this approach does not work for reconstructing k in two dimensions. Based on those arguments, we can write explicit formulas for recovering the X-ray transform of σ_a and for recovering k as limits of the action of α on certain sequences of test functions with supports shrinking to the singularities of α_0 and α_1, respectively; see [CSt4].

4. The Two-Dimensional Stationary Transport Equation

In this section we study the inverse problem for the stationary transport equation (1–3) in the 2D case. As explained in Section 2, the inverse problem of recovering k from the albedo operator $\mathcal{A}$ is formally determined in this case. We prove below that there exists unique solution provided that k is small enough in the L^∞ norm and we also derive a stability estimate. The results of this section are based on joint work by the author and G. Uhlmann [StU].

Let $X \subset \mathbb{R}^2$ be an open convex set with smooth boundary and let us write the 2D stationary transport equation (1–3) in the form

$$- v(\theta) \cdot \nabla_x f - \sigma_a(x, \theta) f + \int_{S^1} k(x, \theta', \theta) f(x, \theta') \, d\theta' = 0, \qquad (4\text{–}1)$$

where $x \in X$, $\theta, \theta' \in S^1$ and

$$v(\theta) = (\cos\theta, \sin\theta).$$

We assume that σ_a and k are in L^∞. For simplicity, we consider the case $V = S^1$ here.

In this paper we work with k sufficiently small and under this assumption the direct problem (3–1) is always uniquely solvable with $f_- \in L^\infty(X \times S^1)$ and $\mathcal{A} f_- \in L^\infty(\Gamma_+)$. For the purpose of the inverse problem however, it is enough to think of $\mathcal{A}$ as an operator mapping $C_0^\infty(\Gamma_-)$ to $L^\infty(\Gamma_+)$.

Uniqueness and stability in the 2D case are proved for small k in the case when $k = k(x, \cos(\theta - \theta'))$ in [R1], and in the case $k = k(\theta, \theta')$ uniqueness for small k is proved in [T]. Note that in those cases the inverse problem is still formally overdetermined. In this section we prove a uniqueness results for general (but still small) $k(x, \theta', \theta)$. Our first result addresses the uniqueness of this inverse problem.

THEOREM 4.1 [StU]. *Define the class*

$$\mathcal{U}_{\Sigma, \varepsilon} = \{(\sigma_a(x), k(x, \theta', \theta)) : \|\sigma_a\|_{L^\infty} \leq \Sigma, \ \|k\|_{L^\infty} \leq \varepsilon\}. \qquad (4\text{–}2)$$

For any $\Sigma > 0$ there exists $\varepsilon > 0$ such that a pair $(\sigma_a, k) \in \mathcal{U}_{\Sigma, \varepsilon}$ is uniquely determined by its albedo operator $\mathcal{A}$ in the class $\mathcal{U}_{\Sigma, \varepsilon}$.

To prove Theorem 4.1, we study again singularities of the distribution kernel $\alpha(x, \theta, x', \theta')$ of $\mathcal{A}$ as in Section 3. In the two dimensional case however $\alpha = \alpha_0 + \alpha_1 + \alpha_2$ with $\alpha_0, \alpha_1, \alpha_2$ as in Theorem 3.4, but α_1 is not a delta type distribution anymore, instead it is an L^1 function and cannot be distinguished from α_2 as in Section 3. We denote $b = \alpha_1 + \alpha_2$. The term α_1 does have a singularity (integrable) at $\theta = \pm\theta'$ and it is L^∞_{loc} outside this set. Similarly, α_2 has a weaker, logarithmic singularity, as shown below. We will prove below that $\sin(\theta - \theta') b \in L^\infty$. We can therefore write

$$\alpha = \frac{\delta(\theta - \theta') \delta_{\{x' + \tau_+(x', \theta') v(\theta')\}}(x)}{n(x) \cdot v(\theta)} e^{-a(x', \theta')} + b(x, \theta, x', \theta'), \qquad (4\text{–}3)$$

where

$$a(x', \theta') = \int_0^{\tau_+(x',\theta')} \sigma_a(x' + tv(\theta'), \theta')\, dt, \qquad \sin(\theta - \theta')b \in L^\infty. \qquad (4\text{--}4)$$

In particular, knowing $\mathcal{A}$, we can uniquely determine a and b. Let we have two pairs (σ_a, k) and $(\tilde{\sigma}_a, \tilde{k})$ with albedo operators $\mathcal{A}$ and $\tilde{\mathcal{A}}$, respectively. Set

$$\delta_1 = \|a - \tilde{a}\|_{H^1(\Gamma_-)}, \qquad \delta_2 = \|(b - \tilde{b})\sin(\theta - \theta')\|_{L^\infty(\Gamma_- \times \Gamma_+)}. \qquad (4\text{--}5)$$

Our second result is the following stability estimate.

THEOREM 4.2 [StU]. *Let*

$$\mathcal{V}_{\Sigma,\varepsilon}^s = \{(\sigma_a(x), k(x, \theta', \theta)) \in H^s(X) \times C(X \times S^1 \times S^1) :$$

$$\|\sigma_a\|_{H^s} \le \Sigma, \ \|k\|_{L^\infty} \le \varepsilon\}. \qquad (4\text{--}6)$$

Then, for any $s > 1$, $\Sigma > 0$, there exists $\varepsilon > 0$ such that for any $(\sigma_a, k) \in \mathcal{V}_{\Sigma,\varepsilon}^s$ and $(\tilde{\sigma}_a, \tilde{k}) \in \mathcal{V}_{\Sigma,\varepsilon}^s$ and $0 < \mu < 1 - 1/s$, there exists $C > 0$ such that

$$\|\sigma_a - \tilde{\sigma}_a\|_{L^\infty} \le C\delta_1^{1-1/s-\mu},$$

$$\|k - \tilde{k}\|_{L^\infty} \le C(\delta_1^{1-1/s-\mu} + \delta_2).$$

REMARK. It follows from (4–15) and (4–16) below that we can choose $\varepsilon = C(d)e^{-2d\Sigma}$, where $d = \operatorname{diam} D$.

IDEA OF THE PROOF OF THEOREM 4.1. Recall that σ_a and k are L^∞ in all variables. It is convenient to think later that σ_a and k are extended as 0 for $x \notin X$.

First we reduce the boundary value problem

$$\begin{cases} Tf = 0 & \text{in } X \times S^1, \\ f|_{\Gamma_-} = f_- \in C_0^\infty(\Gamma_+) \end{cases} \qquad (4\text{--}7)$$

to the integral equation (3–4). Then f is given by

$$f = (I + K)^{-1}Jf_-, \qquad (4\text{--}8)$$

provided that $I + K$ is invertible in a suitable space. The definition (3–5) of K implies immediately that

$$\|Kf\|_{L^\infty(X \times S^1)} \le C\|f\|_{L^\infty(X \times S^1)},$$

where $C = \operatorname{diam} X \|k\|_{L^\infty}$. Therefore, if $(\sigma_a, k) \in \mathcal{U}_{\Sigma,\varepsilon}$ and $\varepsilon < 1/\operatorname{diam} X$, then $I + K$ is invertible in $L^\infty(X \times S^1)$, and then the solution f to (4–7) is given by (4–8). By using Neumann series, it is not hard to see that the trace $(I+K)^{-1}f|_{\Gamma_+}$ is well defined in $L^\infty(\Gamma_+)$ for any $f \in L^\infty(X \times S^1)$. This proves in particular that $\mathcal{A}$ maps $C_0^\infty(\Gamma_-)$ into $L^\infty(\Gamma_+)$ under the smallness assumption on k above. The same arguments also show that $\mathcal{A}f_-$ can be defined for any $f_- \in L^\infty(\Gamma_-)$ as well but we will not need this since we work with the distribution kernel of $\mathcal{A}$.

Define the fundamental solution $\phi(x, \theta, x', \theta')$ of the boundary value problem (4–7) as in Section 3. For $(x', v') \in \Gamma_-$, let $\phi(x, v, x', v')$ solve

$$\begin{cases} T\phi = 0 & \text{in } X \times S^1, \\ \phi|_{\Gamma_-} = |n(x') \cdot v(\theta')|^{-1} \delta_{x'}(x) \delta(\theta - \theta'). \end{cases} \tag{4-9}$$

As before, the albedo operator $\mathcal{A}$ has distribution kernel

$$\alpha(x, \theta, x', \theta') = \phi(x, \theta, x', \theta')|_{(x,\theta) \in \Gamma_+},$$

with $(x', \theta') \in \Gamma_-$, $(x, \theta) \in \Gamma_+$.

As in Section 3, we construct a singular expansion $\phi = \phi_0 + \phi_1 + \phi_2$ as follows. Let

$$E(x, \theta, t) = e^{\mp \int_0^t \sigma_a(x + sv(\theta), \theta) \, ds}, \quad \pm t \geq 0$$

be the total absorption along the path $[x, x + tv(\theta)]$. Then

$$\phi_0 = J\phi_-, \qquad \phi_1 = KJ\phi_-, \qquad \phi_2 = (I + K)^{-1} K^2 J\phi_-, \tag{4-10}$$

and $\phi_- = |n(x') \cdot v(\theta')|^{-1} \delta_{x'}(x) \delta(\theta - \theta')$ as in (4-9). Next,

$$\phi_0 = E(x, \theta, -\infty) \delta(\theta - \theta') \int_0^{\tau_+(x', v(\theta'))} \delta(x - x' - tv(\theta')) \, dt$$

and

$$\phi_1 = \chi E(y, \theta', -\infty) \frac{k(y, \theta', \theta)}{|\sin(\theta - \theta')|} E(y, \theta, \infty), \tag{4-11}$$

where $y = y(x', \theta', x, v)$ is the point of intersection of the rays $(0, \infty) \ni s \mapsto x' + sv(\theta')$ and $(-\infty, 0) \ni t \mapsto x + tv(\theta)$ and $\chi = \chi(x, \theta, x', \theta')$ equals 1, if those two rays intersect in $\bar{X}$, otherwise $\chi = 0$. Recall that X is convex.

To estimate ϕ_2, we need a lemma.

LEMMA 4.1. *Let (σ_a, k) and $(\sigma_a, \tilde{k})$ be in L^∞. Let A_2, K and $\tilde{A}_2$, $\tilde{K}$ be related to k and $\tilde{k}$ (not necessarily nonnegative), respectively. Then there exists $C > 0$ depending on $\operatorname{diam} X$ only such that*

$$|(\tilde{K}K\phi_0)(x, \theta, x', \theta')| \leq C\|\tilde{k}\|_{L^\infty} \|k\|_{L^\infty} \left(1 + \log \frac{1}{\sin|\theta - \theta'|}\right)$$

almost everywhere on $X \times S^1 \times \Gamma_-$, and also almost everywhere on $\Gamma_+ \times \Gamma_-$.

The proof of this lemma is based on the estimate

$$|(\tilde{A}_2 K\phi_0)(x, \theta, x', \theta')| \leq \int_{y \in l(x', \theta')} \frac{|\tilde{k}(x, \arg(x-y), \theta) k(y, \theta', \arg(x-y))|}{|x - y|} \, dl(y),$$

where $l(x', \theta')$ is the line through x' parallel to θ' and dl is the Euclidean measure on it. Using the following elementary estimate

$$\int_{-A}^{A} \frac{ds}{\sqrt{\nu^2 + s^2}} \leq 2\left(1 + \log \frac{A}{\nu}\right), \qquad 0 < \nu \leq A,$$

we easily complete the proof of the lemma.

For ϕ_2 we therefore have $\phi_2 = (I + K)^{-1}\phi_2^\#$ with $\phi_2^\# = K^2\phi_0$ and by Lemma 4.1,

$$0 \le \phi_2^\#(x', \theta', x, \theta) \le C\|k\|_{L^\infty}^2 \left(1 + \log \frac{1}{|\sin(\theta - \theta')|}\right). \qquad (4\text{-}12)$$

This implies a similar estimate for ϕ_2, because $\phi_2 = \phi_2^\# + (I - K)^{-1}K\phi_2^\#$. We summarize those estimates:

PROPOSITION 4.1. *For $\varepsilon > 0$ small enough, the fundamental solution ϕ of (4–7) defined by (4–9) admits the representation $\phi = \phi_0 + \phi_1 + \phi_2$ with*

$$\phi_0 = E(x, \theta, -\infty)\delta(\theta - \theta') \int_0^{\tau_+(x', v(\theta'))} \delta(x - x' - tv(\theta')) \, dt,$$

$$\phi_1 = \chi E(y, \theta', -\infty) \frac{k(y, \theta', \theta)}{|\sin(\theta - \theta')|} E(y, \theta, \infty),$$

$$0 \le \phi_2 \le C\|k\|_{L^\infty}^2 \left(1 + \log \frac{1}{|\sin(\theta - \theta')|}\right),$$

where y, χ are as in (4–11) and $C = C(\operatorname{diam} X)$.

Note that ϕ_1 is not a delta function but it is still singular with singularity at $v(\theta) = v(\theta')$ (forward scattering) and $v(\theta) = -v(\theta')$ (back-scattering). This singularity is integrable however, in fact it is easy to see that $\int_{\Gamma_+} \phi_1 d\xi \le \int \sigma_p(x' + tv(\theta')) \, dt$. The term ϕ_2 is also singular at $v(\theta) = \pm v(\theta')$ with a weaker, logarithmic singularity. Therefore, we can still distinguish between the singularities of the three terms as in the case $n \ge 3$. This analysis however can give us information about k only near forward and backward directions. It is interesting to see whether by studying subsequent lower order terms we can recover all derivatives of $k(x, \theta', \theta)$ at $\theta = \theta'$ and $\theta = \theta' + \pi$. If so, this would allow us to recover collision kernels analytic in θ, θ' (actually, analytic in $\theta - \theta'$ would be enough) and to approximate k near $\theta = \theta'$ and $\theta = \theta' + \pi$ for smooth k. This would not require smallness assumptions on k but we still have to be sure that the direct problem is solvable.

We are ready now to sketch the proof of Theorem 4.1. Fix $\Sigma > 0$ and assume that we have two pairs (σ_a, k), $(\tilde{\sigma}_a, \tilde{k})$ in $\mathcal{U}_{\Sigma, \varepsilon}$ with the same albedo operator and $\sigma_a, \tilde{\sigma}_a$ depending on x only. Denote by ϕ_j and $\tilde{\phi}_j$, $j = 0, 1, 2$, the corresponding components of the fundamental solutions ϕ and $\tilde{\phi}$ as in Proposition 4.1. Then

$$\phi_0 + \phi_1 + \phi_2 = \tilde{\phi}_0 + \tilde{\phi}_1 + \tilde{\phi}_2 \qquad \text{for } (x, \theta) \in \Gamma_+. \qquad (4\text{-}13)$$

As in Section 3, the most singular terms must agree; therefore,

$$\phi_0(x', \theta', x, \theta) = \tilde{\phi}_0(x', \theta', x, \theta) \qquad \text{for } (x', \theta') \in \Gamma_-, \ (x, \theta) \in \Gamma_+. \qquad (4\text{-}14)$$

Thus the X-ray transform of $\sigma_a(x)$ and $\tilde{\sigma}_a(x)$ coincide. Therefore $\sigma_a(x) = \tilde{\sigma}_a(x)$. Next,

$$\chi E(y, \theta', -\infty)\frac{k(y, \theta', \theta) - \tilde{k}(y, \theta', \theta)}{|\sin(\theta - \theta')|}E(y, \theta, \infty) = \tilde{\phi}_2 - \phi_2 \qquad \text{on } \Gamma_- \times \Gamma_+,$$

where y, χ are as in (4–11). This, together with (4–13) and (4–14), leads to the inequality

$$\chi|k(y, \theta', \theta) - \tilde{k}(y, \theta', \theta)| \leq C|\sin(\theta - \theta')||\tilde{\phi}_1 - \phi_1|$$
$$= C|\sin(\theta - \theta')||\tilde{\phi}_2 - \phi_2| \qquad \text{on } \Gamma_- \times \Gamma_+, \qquad (4\text{–}15)$$

where $C = e^{2d\Sigma}$, $d = \operatorname{diam} D$. The rest of the proof is based on the estimate

$$\operatorname*{ess\,sup}_{\Gamma_- \times \Gamma_+}|\sin(\theta - \theta')||\tilde{\phi}_2 - \phi_2| \leq C\varepsilon\|k - \tilde{k}\|_{L^\infty(X \times S^1 \times S^1)} \qquad (4\text{–}16)$$

with $C > 0$ depending on $\operatorname{diam} D$ only. An essential role in its proof is played by Lemma 4.1. Observe that in particular, the factor $\sin(\theta - \theta')$ cancels the weaker logarithmic singularity of ϕ_2 and $\tilde{\phi}_2$. Then by (4–15), (4–16),

$$\|k - \tilde{k}\|_{L^\infty(X \times S^1 \times S^1)} \leq C\varepsilon\|k - \tilde{k}\|_{L^\infty(X \times S^1 \times S^1)},$$

and for $\varepsilon > 0$ small enough this implies $k = \tilde{k}$. $\qquad\qquad\qquad\qquad\square$

SKETCH OF THE PROOF OF THEOREM 4.2. According to Proposition 4.1, for the distribution kernel $\alpha(x, \theta, x', \theta')$ of $\mathcal{A}$ we have the representation (4–3), where a is as in (4–4), while $b = (\phi_1 + \phi_2)|_{(x,\theta) \in \Gamma_+}$. Then b is a function and an elementary calculation show that $b \in L^\infty(\Gamma_+, L^1(\Gamma_-))$. Proposition 4.1 shows that $b\sin(\theta - \theta') \in L^\infty$. Assume that we have two pairs of continuous functions (σ_a, k) and $(\tilde{\sigma}_a, \tilde{k})$ in $\mathcal{V}^s_{\Sigma, \varepsilon}$ with albedo operators $\mathcal{A}$ and $\tilde{\mathcal{A}}$, respectively. In what follows we will denote the quantities a, b, α, etc., related to the second pair by putting a tilde sign over it. Also, we will use the notation $\Delta\mathcal{A}$ to denote the difference $\Delta\mathcal{A} = \mathcal{A} - \tilde{\mathcal{A}}$, and similarly $\Delta a = a - \tilde{a}$, etc. In this paper, Δ never stands for the Laplacian.

Let δ_1 and δ_2 be as in (4–5). According to Proposition 4.1, $\delta_2 < \infty$. Our goal is to estimate $\Delta\sigma_a$ and Δk in terms of δ_1 and δ_2. Observe that, as in the uniqueness proof, Δe^{-a} and $\Delta(\alpha_1 + \alpha_2)$ can be recovered from $\Delta\alpha$ by separating the most singular part of $\Delta\alpha$ from the rest. Therefore, δ_1 measures the magnitude of the singular part of $\Delta\alpha$, while δ_2 measures the magnitude of the regular part.

We start by estimating $\Delta\sigma_a$. By a result of Mukhometov [Mu],

$$\|\Delta\sigma_a\|_{L^2} \leq C\delta_1. \qquad (4\text{–}17)$$

Next we estimate Δk in terms of δ_1 and δ_2. Set

$$E_1(y, \theta, \theta') = E(y, \theta', -\infty)E(y, \theta, \infty).$$

By (4–11), $\sin|\theta - \theta'|\alpha_1(x, \theta, x', \theta') = \chi(E_1 k)(y, \theta', \theta)$ with y, χ as in (4–11).

Our starting point is the relation $\Delta(E_1 k) = k\Delta E_1 + \tilde{E}_1 \Delta k$. Note first that $|\Delta E_1| \leq 2d|\Delta\sigma_a| \leq 2d\delta_1'$, where $\delta_1' = \|\Delta\sigma_a\|_{L^\infty}$. Hence, on supp χ,

$$
\begin{aligned}
\tilde{E}_1|\Delta k|(y,\theta,\theta') &\leq |\Delta(E_1 k)| + k|\Delta E_1| \\
&\leq |\Delta(\alpha_1 + \alpha_2)\sin|\theta - \theta'| - \Delta\alpha_2 \sin|\theta - \theta'|| + C\delta_1' \\
&\leq \delta_2 + |\Delta\alpha_2|\sin|\theta - \theta'| + C\delta_1'.
\end{aligned}
\tag{4-18}
$$

Therefore,

$$
\|\Delta k\|_{L^\infty} \leq C\big(\delta_1' + \delta_2 + \|\Delta\alpha_2 \sin|\theta - \theta'|\|_{L^\infty}\big)
\tag{4-19}
$$

The next step is to prove an estimate similar to (4–16) for the last term in the right-hand side above. Recall that in (4–16), we have $\sigma_a = \tilde{\sigma}_a$, while here we only know how to estimate the difference $\Delta\sigma_a = \sigma_a - \tilde{\sigma}_a$. Nevertheless, we can proceed along similar lines in order to get

$$
\||\Delta\alpha_2 \sin|\theta - \theta'|\|_{L^\infty} \leq C\varepsilon(\|\Delta k\|_{L^\infty} + \delta_1').
\tag{4-20}
$$

Therefore, for $\varepsilon > 0$ small enough, (4–19) implies the stability estimate

$$
\|\Delta k\|_{L^\infty} \leq C(\delta_1' + \delta_2).
\tag{4-21}
$$

Estimate (4–21) is the base of our stability estimate. Using an interpolation inequality, we conclude that, for any fixed $0 \leq \mu \leq 1 - 1/s$,

$$
\|\Delta\sigma_a\|_{1+s\mu} \leq C(\Sigma)\|\Delta\sigma_a\|_{L^2}^{1-1/s-\mu} \quad \text{for } \sigma_a, \tilde{\sigma}_a \in \mathcal{V}_{\Sigma,\varepsilon}^s.
$$

By a standard Sobolev embedding theorem and (4–17),

$$
\delta_1' = \|\Delta\sigma_a\|_{L^\infty} \leq C(\Sigma)\|\Delta\sigma_a\|_{L^2}^{1-1/s-\mu} \leq C'(\Sigma)\delta_1^{1-1/s-\mu}.
\tag{4-22}
$$

Therefore, (4–21) yields

$$
\|\Delta k\|_{L^\infty} \leq C\left(\delta_1^{1-1/s-\mu} + \delta_2\right).
\tag{4-23}
$$

Estimates (4–22) and (4–23) complete the proof of Theorem 4.2. $\qquad\square$

5. Open Problems

Our choice of open problems is subjective and mainly reflects personal taste.

Uniqueness for σ_a depending on both x and v. As we have demonstrated in the previous sections, if $k = 0$, then $\sigma_a(x,v)$ cannot be recovered from $\mathcal{A}$ (or from the scattering operator) because the line integrals $\int \sigma_a(x + tv, v)\, dt$ do not recover σ_a. Suppose however that $\sigma_p = \sigma_a$. Then the absorption is due only to the fact that particles may change velocity and each such event is interpreted as a particle instantly moving from the point (x, v') of the phase space (therefore absorption at (x, v')) to the point (x, v). Then the counter example above does not work. Can we recover $\sigma_a(x, v)$ in this case? If yes, then recovery of k goes along the same lines as above. More generally, one can assume that $\sigma_a(x, v) = \sigma_p(x, v) + a(x)$. Even more generally, the counter example above cannot be generalized in an obvious way if $k > 0$ in X. Is this condition alone enough for uniqueness?

Relaxing the smallness condition in the two-dimensional case. It would be interesting to prove uniqueness in the 2D case without smallness assumption on k. Some conditions on σ_a and k are needed even for the direct problem; see, for example, (3–1) and (3–3) — the first one does require k to be small (but with an explicit bound in general much larger than the one needed for the inverse problem) while the second one does not. As mentioned in Section 4, one can try to recover $k(x, v', v)$ near $v = \pm v'$ at infinite order by studying the singularities of the kernel of $\mathcal{A}$ which solves the problem for k analytic in v and v'. It would also be interesting to see whether one could recover the singularities of k from boundary measurements, at least if we assume that they are of jump type across some curve.

Stability for $n \geq 3$. Stability estimates in dimensions $n \geq 3$ have been proven by Romanov (see [R1], [R2], [R3], for example) and by Wang [W], under additional assumptions that k depend on fewer variables. In the general situation studied in Section 3, we know of no stability estimates, even for small k as in Theorem 4.2 (where $n = 2$). We believe that such an estimate should be possible to derive following the proof of Theorem 3.1. This is done in [W] under the additional assumption that $k = k(v', v)$.

Alternative recovery method for large σ_a and k. We do not impose smallness assumptions on the coefficients in dimensions $n \geq 3$, and our method gives in fact an explicit solution of the inverse problem which in particular implies a reconstruction method based on taking certain limits near the singularity of α. However, for large σ_a, the amplitude of the most singular part α_0 is exponentially small for large σ_a. For all practical purposes, measuring the leading singularity is hard or impossible in this case. Therefore, it would be important to develop a method for relatively large σ_a and k that does not rely on measuring the singularities of α. One possible way is to study the diffusion limit (replacing σ_a and k by $\lambda \sigma_a$ and λk and taking $\lambda \to \infty$) and an associated inverse problem.

References

[AB] Yu. E. Anikonov, B. A. Bubnov, *Inverse problems of transport theory*, Soviet Math. Doklady **37** (1988), 497–499.

[AE] P. Arianfar and H. Emamirad, *Relation between scattering and albedo operators in linear transport theory*, Transport Theory and Stat. Physics, **23**(4) (1994), 517–531.

[B] A. Bondarenko, *The structure of the fundamental solution of the time-independent transport equation*, J. Math. Anal. Appl. **221**:2 (1998), 430–451.

[Ce1] M. Cessenat, *Théorèmes de trace L^p pour des espaces de fonctions de la neutronique*, C. R. Acad. Sci. Paris, Série I **299** (1984), 831–834.

[Ce2] M. Cessenat, *Théorèmes de trace pour des espaces de fonctions de la neutronique*, C. R. Acad. Sci. Paris, Série I **300** (1985), 89–92.

[CMS] M. Chabi, M. Mokhtar-Kharroubi and P. Stefanov, *Scattering theory with two L^1 spaces: application to transport equations with obstacles*, Ann. Fac. Sci. Toulouse **6**(3) (1997), 511–523.

[CSt1] M. Choulli and P. Stefanov, *Scattering inverse pour l'équation du transport et relations entre les opérateurs de scattering et d'albédo*, C. R. Acad. Sci. Paris **320** (1995), 947–952.

[CSt2] M. Choulli and P. Stefanov, *Inverse scattering and inverse boundary value problems for the linear Boltzmann equation*, Comm. P.D.E. **21** (1996), 763–785.

[CSt3] M. Choulli and P. Stefanov, *Reconstruction of the coefficients of the stationary transport equation from boundary measurements*, Inverse Prob. **12** (1996), L19–L23.

[CSt4] M. Choulli and P. Stefanov, *An inverse boundary value problem for the stationary transport equation*, Osaka J. Math. **36**(1) (1999), 87–104.

[CZ] M. Choulli, A. Zeghal, *Laplace transform approach for an inverse problem*, Transport Theory and Stat. Phys., **24**(9) (1995), 1353–1367.

[DL] R. Dautray and J.-L. Lions, *Analyse Mathématique et Calcul Numérique pour les Sciences et les Techniques*, vol. 9, Masson, Paris, 1988.

[E] H. Emamirad, *On the Lax and Phillips scattering theory for transport equation*, J. Funct. Anal. **62** (1985), 276–303.

[EP] H. Emamirad and V. Protopopescu, *Relationship between the albedo and scattering operators for the Boltzmann equation with semi-transparent boundary conditions*, Math. Meth. Appl. Sci., to appear.

[Hej] J. Hejtmanek, *Scattering theory of the linear Boltzmann operator*, Commun. Math. Physics **43** (1975), 109–120.

[H] S. Helgason, *The Radon Transform*, Birkhäuser, Boston, Basel, 1980.

[L1] E. W. Larsen, *Solution of multidimensional inverse transport problems*, J. Math. Phys. **25**(1) (1984), 131–135.

[L2] E. Larsen, *Solution of three-dimensional inverse transport problems*, Transport Theory Statist. Phys. **17**:2–3 (1988), 147–167.

[MC] N. J. McCormick, *Recent developments in inverse scattering transport methods*, Trans. Theory and Stat. Phys. **13** (1984), 15–28.

[Mo] M. Mokhtar-Kharroubi, *Limiting absorption principle and wave operators on $L^1(\mu)$ spaces, Applications to Transport Theory*, J. Funct. Anal. **115** (1993), 119–145.

[Mu] R. Mukhometov, On the problem of integral geometry, *Math. Problems of Geophysics*, Akad. Nauk SSSR, Sibirs. Otdel., Vychisl. Tsentr, Novosibirsk, **6**:2 (1975), 212–242 (in Russian).

[PV] A. I. Prilepko, N. P. Volkov, *Inverse problems for determining the parameters of nonstationary kinetic transport equation from additional information on the traces of the unknown function*, Differentsialnye Uravneniya **24** (1988), 136–146.

[P1] V. Protopopescu, *On the scattering matrix for the linear Boltzmann equation*, Rev. Roum. Phys. **21** (1976), 991–994.

[P2] V. Protopopescu, *Relation entre les opérateurs d'albédo et de Scattering avec des conditions aux frontières non-transparentes*, C. R. Acad. Sci. Paris, Série I **318** (1994), 83–86.

[RS] M. Reed and B. Simon, *Methods of Modern Mathematical Physics*, Vol. 3, Academic Press, New York, 1979.

[R1] V.G. Romanov, *Estimation of stability in the problem of determining the attenuation coefficient and the scattering indicatrix for the transport equation*, Sibirsk. Mat. Zh. **37**:2 (1996), 361–377, iii; translation in Siberian Math. J. **37**:2 (1996), 308–324.

[R2] V.G. Romanov, *Stability estimates in the three-dimensional inverse problem for the transport equation*, J. Inverse Ill-Posed Probl. **5**:5 (1997), 463–475

[R3] V. Romanov, *A stability theorem in the problem of the joint determination of the attenuation coefficient and the scattering indicatrix for the stationary transport equation*, Mat. Tr. **1**:1 (1998), 78–115.

[Si] B. Simon, *Existence of the scattering matrix for linearized Boltzmann equation*, Commun. Math. Physics **41** (1975), 99–108.

[St] P. Stefanov, *Spectral and scattering theory for the linear Boltzmann equation*, Math. Nachr. **137** (1988), 63–77.

[StU] P. Stefanov and G. Uhlmann, *Optical Tomography in two dimensions*, to appear in Methods Appl. Anal.

[SyU] J. Sylvester and G. Uhlmann, *The Dirichlet to Neumann map and applications*, in: Inverse Problems in Partial Differential Equations, SIAM Proceedings Series List (1990), 101–139, ed. by D. Colton, R. Ewing and R. Rundell.

[T] A. Tamasan, *An inverse boundary value problem in two dimensional transport*, Inverse Problems **18** (2002), 209–219.

[U] T. Umeda, *Scattering and spectral theory for the linear Boltzmann operator*, J. Math. Kyoto Univ. **24** (1984), 208–218.

[Vi] I. Vidav, *Existence and uniqueness of non-negative eigenfunctions of the Boltzmann operator*, J. Math. Anal. Appl. **22** (1968), 144–155.

[V1] J. Voigt, *On the existence of the scattering operator for the linear Boltzmann equation*, J. Math. Anal. Appl. **58** (1977), 541–558.

[V2] J. Voigt, *Spectral properties of the neutron transport equation*, J. Math. Anal. Appl. **106** (1985), 140–153.

[W] J. Wang, *Stability estimates of an inverse problem for the stationary transport equation*, Ann. Inst. H. Poincaré Phys. Théor. **70**:5 (1999), 473–495.

PLAMEN STEFANOV
DEPARTMENT OF MATHEMATICS
PURDUE UNIVERSITY
WEST LAFAYETTE, IN 47907
UNITED STATES
stefanov@math.purdue.edu

Near-Field Tomography

P. SCOTT CARNEY AND JOHN C. SCHOTLAND

ABSTRACT. We consider the inverse scattering problem for wave fields containing evanescent components. Applications to near-field optics and tomographic imaging with subwavelength resolution are described.

CONTENTS

1. Introduction

This article is concerned with a class of inverse problems that arise in near-field optics. These problems may be considered to be special cases of the more general problem of inverse scattering with wave fields containing evanescent components. To place the work described herein in context, we begin with some background information on near-field optical microscopy.

133

Microscopy is both modern and ancient, beginning nearly four centuries ago with the advent of the Galilean microscope, around 1610. To the extent that Galileo's telescope changed our perspective on the place of humankind in the universe, the microscope changed our perception of the stuff of which we are made. For over 250 years microscopists saw lens design and manufacture as the limiting factors in resolving power. Abbe [1] and Rayleigh [2] separately determined that no matter the physical aparatus, the linear size of the smallest feature that may be resolved with monochromatic, or quasi-monochromatic, light is on the order of the central wavelength of the light. However, the analyses of Abbe and Rayleigh were predicated on the restriction that the imaging instrument is several wavelengths or more from the object. Near-field optical techniques surpass the Abbe–Rayleigh limits by doing away with this restriction.

Near-field optical microscopy has developed dramatically in recent years [3; 4; 5]. The ubiquity of the need for microscopic inspection techniques has brought the intellectual resources of several disparate disciplines to bear on the task of improving the basic methods and putting them to novel application. Conspicuous among those applications are the imaging of biological samples, the inspection and manipulation of nano-electronic components in semiconductor technology, and the inspection and activation of nano-optical devices.

The first proposal of a method to circumvent the Rayleigh resolution limit was put forward by Synge [6] in 1928. Synge proposed that a thin sample be illuminated through a subwavelength aperture. By recording the transmitted light as a function of aperture position, a subwavelength resolved image of the sample may be acquired. Today this method is known as near-field scanning optical microscopy (NSOM) [3; 4; 6; 7; 8; 9; 10; 11] or scanning near-field optical microscopy (SNOM); it is practiced in many variations including the reciprocal arrangement in which the sample is illuminated by a source in the far zone and light is collected through a small aperture. The role of the small aperture is now played by the tip of a tapered optical fiber, a technique not known to Synge. A number of other modalities that fall under the umbrella of near-field optics are subsequently discussed (Figure 1).

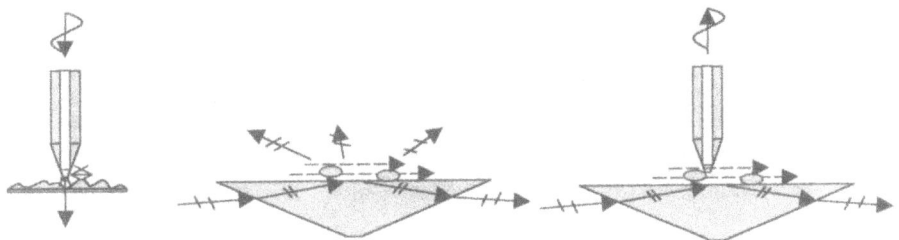

Figure 1. Basic experiments of near-field optics: left, near-field scanning optical microscopy (NSOM); middle, total internal reflection microscopy (TIRM); right, photon scanning tunneling microscopy (PSTM).

Near-field scanning optical microscopy. NSOM has attracted considerable attention as a technique to obtain images of surfaces with subwavelength resolution. This achievement is particularly important for imaging structures where spectroscopic concerns or sample handling requirements dictate the use of lower frequency fields and yet high spatial resolution is still required. Various experimental modalities are in practical use. Two prominent examples are collection mode NSOM and illumination mode NSOM. In illumination mode NSOM, a tapered fiber probe with a sub-wavelength size aperture serves as a source of illumination in the near-zone of the sample. The scattered field intensity is then measured and recorded as a function of the probe position while the probe is scanned over the sample. In collection mode NSOM, the fiber probe serves to detect the total field in the near-zone as the sample is illuminated by a source in the far zone.

There are certain limitations of NSOM as currently practiced. Despite the fact that the sample may present a complicated three-dimensional structure, NSOM produces only a two-dimensional image. Indeed, rather than being an imaging method, it is more accurate to say that NSOM maps the sub-wavelength structure of the optical near-field intensity above the sample. Under certain simplifying assumptions, such as homogeneity of the bulk optical properties of the sample [12; 13; 14; 15], the images produced in these experiments may be related to the sample structure. However, for the more general case in which the topography of the sample and the bulk optical properties both vary, the relationship between the near-field intensity and the sample structure has proven ambiguous [16]. To appreciate this point we consider simulations of the NSOM image of a collection of point scatterers as shown in Figure 2. As is well noted in the literature, the topmost layer of scatterers dominates the conventional (nontomographic) image. When the top layer is removed from the simulation, the conventional image, made with the tip now $\lambda/4$ from the nearest layer, manifests blurring similar to that observed in experiments in which the scanning tip is withdrawn from the sample [10]. See Figure 2.

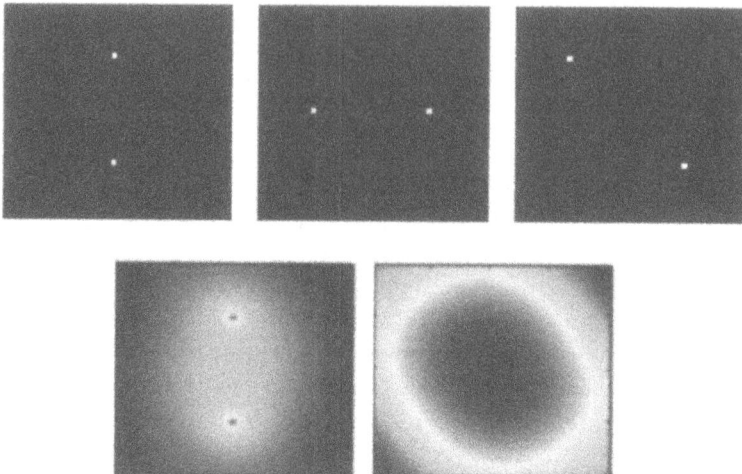

Figure 2. A simulated NSOM image. In each frame the field of view is $\lambda \times \lambda$. The model scatterer consists of six point scatterers distributed in three planes, $z = 0.05\lambda$, $z = 0.25\lambda$, and $z = 0.45\lambda$; the successively deeper planes are shown from left to right in the top row. The figure on the left in the bottom row shows the results of a scalar simulation of a collection mode NSOM image made by scanning in the $z = 0$ plane. Illumination is provided by a normally incident plane wave. The figure on the right in the bottom row is the same simulation with the scatterers in the $z = 0.05\lambda$ plane removed.

Total internal reflection microscopy. Essential to the near-field modality of NSOM is the presence of inhomogeneous, or evanescent, modes of the illumination field. Specifically, the illuminating field consists of a superposition of plane waves including the high spatial frequency evanescent plane waves. These waves are super-oscillatory parallel to some reference plane and are exponentially decaying away from the plane. The super-resolving capabilities of NSOM may be attributed to the high spatial frequency of the evanescent waves. Instead of generating these modes at the small aperture in NSOM, they may be generated at the interface of two media by total internal reflection as is done in total internal reflection microscopy (TIRM).

TIRM has been in practical use for decades. This technique has primarily been used as a means of surface inspection [17; 18], though the sensitivity of the field to distance along the decay axis has been used to advantage in applications such as the measurement of distance between two surfaces [19]. Until recently the opportunities for transverse superresolution made possible by the high spatial frequency content of the probe field have been largely over-looked. However, recently a direct imaging approach resulting from the marriage of standing-wave illumination techniques and TIRM has been described [20; 21], achieving transverse resolution of $\lambda/7$.

Photon scanning tunneling microscopy. At the intersection of NSOM and TIRM modalities is the method referred to as photon scanning tunneling microscopy (PSTM) [3; 22; 23]. In this technique the object is illuminated by an evanescent wave generated at the face of a prism or slide (as in TIRM), and the scattered field is detected via a tapered fiber probe in the near-zone of the sample (as in NSOM). Because PSTM is a dark field method, that is the signal is zero if the sample is absent, PSTM generally offers a better signal to noise ratio than NSOM.

Inverse scattering and near-field tomography. In all of the above mentioned modalities, the connection between the measured field and the sample properties has proven to be problematic. Variations in surface height may be indistinguishable from variations in the refractive index of the sample. To clarify the meaning of the measurements and to provide three-dimensional imaging capability, it is desirable to find a solution to the near-field inverse scattering problem.

There is an extensive body of literature on the far-field inverse scattering problem[24]. The inverse scattering problem for near field optics presents challenges and opportunities unlike those encountered in the far-field problem. When, as in the far-field, all waves are homogeneous, the scattering data may be related to a Fourier transform of the sample structure. In contrast, the near-field problem, due to the presence of evanescent fields, will generally involve data related to the object structure through a Fourier–Laplace transform. Because the inversion of the Laplace transform is ill-posed, inverse scattering in the near-field has been thought unfeasible. However, the inverse scattering problem is also over-determined so that it may be observed that there exist many copies of the Fourier–Laplace transform of the object structure encoded on the scattering data. This over-specification of the object structure effectively allows for the averaging of multiple unstable reconstructions to produce a stable reconstruction. The means to accomplish such averaging may be obtained from the analytic construction of the pseudoinverse of the forward scattering operator. Results in this direction have been reported for the TIRM, NSOM and PSTM modalities [25; 26; 27]. The use of inverse scattering methods to reconstruct tomographic images in near-field optics is called near-field tomography.

Near-field power extinction tomography. The preceding discussion focused on inverse scattering methods for extant near-field imaging modalities. It should be stressed that in order to carry-out the program of inverse scattering and object reconstruction described herein, the measurements must be phase sensitive. In certain modalities (NSOM, PSTM), the measurements are intrinsically holographic, producing clear interference between the illuminating field and the scattered field. For other modalities (the TIRM variants) the scattered field must be measured interferometrically with a separate reference field. Because these measurements must be made with multiple directions of illumination, the holographic approach may be somewhat challenging to implement. See Figure 3.

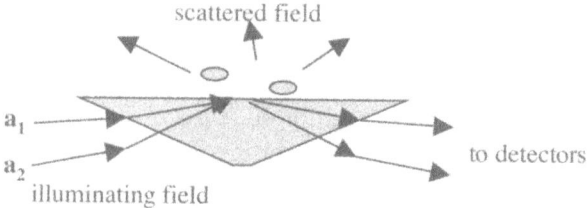

Figure 3. Illustrating the near-field power extinction tomography experiment. Two plane waves are totally internally reflected at the prism face generating evanescent wave. Frustration of the total internal reflection due to the presence of the scatterer generates a scattered field. The power content of the reflected beams is monitored.

To circumvent the phase problem it is desirable to effect reconstruction of the sample structure from measurements only of the power transmitted through the object as is done in X-ray computed tomography. For a scattering sample, a single illuminating wave will not provide sufficient information because the field fails to propagate on a line or even in a rectilinear manner as described by geometrical optics. It was recently shown that knowledge of the three dimensional structure of a scattering medium may be inferred from the power lost by the probe field to scattering and absorption [28; 29] when *two* coherent plane waves illuminate the sample. This result follows, in part, from a generalization of the optical cross-section theorem that applies when more than one wave is incident on the scatterer and allows for the incident waves to be evanescent. Making use of these results, a new modality has recently been proposed in which a sample is illuminated by two superposed evanescent plane waves at the surface of a prism or slab [30]. In the absence of a sample on the prism face, all of the power incident on the interface is reflected. In the presence of the sample, a certain amount of power is coupled into propagating modes and a certain amount of power is absorbed by the object. The total power lost from the incident field due to the presence of the sample is referred to as the extinguished power. The extinguished power carries information about the structure of the scatterer. In order to obtain a complete description of the object, the power content of the internally reflected beams is monitored as the phase between the illuminating waves is varied. Thus a phase measurement problem is traded for a phase control problem, a generally more tractable one. Varying the phase relationship between the two waves causes the interference pattern of the incident field to shift and probe other parts of the object. The higher the spatial frequency of the evanescent waves used to probe the sample, the more rapidly the field falls off as a function of depth into the sample. For this reason the resolution of tomographs generated from these measurements falls off with depth into the sample, as with the other near-field modalities. This method is known as near-field power extinction tomography, or when the probe field is light, near-field optical power extinction tomography.

The remainder of this article is organized as follows. In Section 2 those aspects of scattering theory, for both scalar and electromagnetic waves, that are necessary for the treatment of the forward scattering problem in near-field tomography are discussed. In Section 3 the forward problems for four separate experimental modalities are considered: scanning near-field tomography, total internal reflection tomography, photon scanning tunneling tomography, and near-field power extinction tomography. In each case the integral equations that relate the three-dimensional structure of the sample susceptibility to the scattered field are derived. In Section 4 a unified approach to the inverse problem is developed by construction of the singular value decompostion of the relevant forward scattering operators. Issues of sampling as well as effects due to incomplete and limited data are addressed and the results explored numerically. Finally, in Section 5 a summary of our results and a discussion of future research directions is presented.

2. Scattering Theory

In this section those aspects of scattering theory necessary for the formulation of the forward problem in near-field tomography are reviewed. Scalar waves are discussed first followed by a parallel treatment of the vector theory of electromagnetic scattering. It may be noted that the vector theory is essential for near-field optics since the scalar approximation to the scattering of electromagnetic waves is invalid when the dielectric susceptibility varies on subwavelength scales. Nevertheless, the scalar theory is of independent physical interest since it describes, for example, the scattering of acoustic waves in acoustic microscopy [31; 32; 33].

2A. Scalar Case

Basic equations. Consider an experiment in which an inhomogeneous sample is deposited on a homogeneous substrate. The substrate is assumed to be thick so that only one face need be considered, thus defining an interface between two half-spaces. The index of refraction in the lower half-space $z < 0$ (the substrate) has a constant value n. The index of refraction in the upper half-space $z \geq 0$ varies within the domain of the sample but otherwise has a value of unity. The sample is illuminated either from above (reflection geometry) or below (transmission geometry) by a monochromatic scalar wave of frequency $\omega = ck_0$. The scalar field $U(\boldsymbol{r})$ obeys the reduced wave equation

$$\nabla^2 U(\boldsymbol{r}) + k_0^2 \big(n^2(z) + 4\pi\eta(\boldsymbol{r})\big)U(\boldsymbol{r}) = -4\pi S(\boldsymbol{r}),$$

where $\eta(\boldsymbol{r})$ is the susceptibility of the sample, k_0 is the free-space wavenumber, $n(z)$ is the z-dependent index of refraction as described above, and $S(\boldsymbol{r})$ is the source density. The total field consists of two parts:

$$U(\boldsymbol{r}) = U_i(\boldsymbol{r}) + U_s(\boldsymbol{r}),$$

where $U_i(r)$ and $U_s(r)$ are the incident and scattered fields, respectively. The incident field may be identified as the field that would exist in the absence of the sample. U_i obeys the reduced wave equation

$$\nabla^2 U_i(r) + k_0^2 n^2(z) U_i(r) = -4\pi S(r). \tag{2-1}$$

It follows that the scattered field satisfies the equation

$$\nabla^2 U_s(r) + k_0^2 n^2(z) U_s(r) = -4\pi k_0^2 \eta(r) U(r).$$

Integral equations. The analysis will be facilitated by use of the Green's function, $G(r, r')$ which obeys the equation

$$\nabla^2 G(r, r') + n^2(z) k_0^2 G(r, r') = -4\pi \delta(r - r')$$

and satisfies the boundary conditions

$$G(r, r')\big|_{z=0^+} = G(r, r')\big|_{z=0^-},$$
$$\hat{z} \cdot \nabla G(r, r')\big|_{z=0^+} = \hat{z} \cdot \nabla G(r, r')\big|_{z=0^-}$$

on the $z = 0$ plane. Following standard procedures it may then be seen that the incident field is given by

$$U_i(r) = \int d^3 r' G(r, r') S(r').$$

The scattered field obeys the integral equation

$$U_s(r) = k_0^2 \int d^3 r' \, G(r, r') U(r') \eta(r'). \tag{2-2}$$

The Green's function $G(r, r')$ may be expressed in the plane-wave decomposition [34; 35]

$$G(r, r') = \int \frac{d^2 q}{(2\pi)^2} g(z, z'; q) \exp\big(iq \cdot (\rho - \rho')\big). \tag{2-3}$$

Explicit expressions for $g(z, z'; q)$ in the half-space geometry are given in the Appendix. In free space it can be shown that

$$g(z, z'; q) = \frac{2\pi i}{k_z(q)} \exp\big(ik_z(q)|z - z'|\big).$$

The notation should be understood to mean that $r = (\rho, z)$ and

$$k_z(q) = \sqrt{k_0^2 - q^2}.$$

The plane wave modes appearing in (2–3) are labeled by the transverse part of the wave vector q. The modes for which $|q| \leq k_0$ correspond to propagating waves while the modes with $|q| > k_0$ correspond to evanescent waves. For these modes $k_z(q)$ is pure imaginary. This leads to exponential decay of the field with propagation and a corresponding loss of high spatial frequency components.

The integral equation (2–2) has the form of the Lippmann–Schwinger equation of quantum scattering theory. It provides a complete description of the scattering of the incident wave within the sample and the substrate. If the scattered wave is much weaker than the incident field, then we may replace the field by the incident field in the right hand side of (2–2). Equation (2–2) thus becomes

$$U_s(\boldsymbol{r}) = \int d^3r' G(\boldsymbol{r},\boldsymbol{r}')U_i(\boldsymbol{r}')\eta(\boldsymbol{r}'). \qquad (2\text{–}4)$$

This result, which is referred to as the first Born approximation for the scattered field, is valid for small, weakly scattering objects.

The optical theorem. Energy conservation leads to a fundamental result in scattering theory known as the optical theorem. In its most general form [36], it expresses the total power P extinguished from the incident field as the integral

$$P = 4\pi k_0 \,\mathrm{Im} \int d^3r U_i^*(\boldsymbol{r})U(\boldsymbol{r})\eta(\boldsymbol{r}). \qquad (2\text{–}5)$$

The depletion of the power from the incident beam may be seen to arise from the interference between the total field and the incident field within the region of the scatterer.

The result is better known for a restricted case. If the incident field consists of a homogeneous, or propagating, plane wave with amplitude a and wave vector $\boldsymbol{k}$,

$$U_i(\boldsymbol{r}) = ae^{i\boldsymbol{k}\cdot\boldsymbol{r}},$$

then P is related to the scattering amplitude in the forward (incident) direction by the expression

$$P = |a|^2 \frac{4\pi}{k_0} \,\mathrm{Im}\, A(\boldsymbol{k},\boldsymbol{k}). \qquad (2\text{–}6)$$

Here $A(\boldsymbol{k},\boldsymbol{k}')$ is the scattering amplitude associated with the transition *via* scattering of a plane wave with wave vector $\boldsymbol{k}$ to a plane wave with wave vector $\boldsymbol{k}'$. It is defined by

$$A(\boldsymbol{k},\boldsymbol{k}') = k_0^2 a^{-1} \int d^3r e^{-i\boldsymbol{k}'\cdot\boldsymbol{r}} U(\boldsymbol{r})\eta(\boldsymbol{r}). \qquad (2\text{–}7)$$

Equation (2–6) is the classical form of the optical theorem known in quantum scattering theory [37; 38; 39].

2B. Vector Case. To adequately describe the physics of near-field optical microscopy, the vector theory of electromagnetic scattering must be invoked. The starting point for such an investigation is the set of Maxwell's equations. It will be assumed that the material of the sample is linear and nonmagnetic (a broad class of materials). As in the scalar case, the problem is set in a half-space with the upper ($z \geq 0$) half-space being the vacuum plus the sample and the lower ($z < 0$) half-space being filled with a material of refractive index n.

Basic equations. The fields are entirely descibed by specifying the electric field everywhere. The magnetic field thus may be ignored as redundant and it is sufficient to consider only the electric field E. For a monochromatic source, E satisfies the reduced wave equation

$$\nabla \times \nabla \times E(r) - k_0^2(n^2(z) + 4\pi\eta(r))E(r) = 4\pi k_0^2 P(r).$$

Here the dielectric suspectibilty $\eta(r)$ is related to the permittivity $\varepsilon(r)$ by $\varepsilon(r) = n^2(z) + 4\pi\eta(r)$, and $P(r)$ is the dielectric polarization (which acts as a source of the electric field). Again the field consists of two parts,

$$E(r) = E^i(r) + E^s(r). \tag{2-8}$$

The incident field $E^i(r)$ obeys the equation

$$\nabla \times \nabla \times E^i(r) - k_0^2 n^2(z)E^i(r) = 4\pi k_0^2 P(r).$$

Thus the scattered field $E^s(r)$ satisfies

$$\nabla \times \nabla \times E^s(r) - k_0^2 n^2(z)E^s(r) = 4\pi k_0^2 \eta(r)E(r).$$

Integral equations. The fields may be expressed in integral equation form by use of the Green's tensor $G(r, r')$ which satisfies

$$\nabla \times \nabla \times G(r, r') - k_0^2 n^2(z)G(r, r') = 4\pi\delta(r - r')I$$

where I is the unit tensor. The Green's tensor must also obey the boundary conditions

$$\hat{z} \times G(r, r')\big|_{z=0+} = \hat{z} \times G(r, r')\big|_{z=0-},$$
$$\hat{z} \times \nabla \times G(r, r')\big|_{z=0+} = \hat{z} \times \nabla \times G(r, r')\big|_{z=0-}$$

on the $z = 0$ plane. For later reference we note the plane-wave decomposition of $G_{\alpha\beta}(r, r')$ [40] :

$$G_{\alpha\beta}(r, r') = \int \frac{d^2q}{(2\pi)^2} g_{\alpha\beta}(q, z) \exp(iq \cdot (\rho - \rho')). \tag{2-9}$$

Explicit expressions for $g_{\alpha\beta}(q, z)$ in the half-space geometry are given in the Appendix. In free space it can be shown that

$$g_{\alpha\beta}(z, z'; q) = \frac{2\pi i}{k_z(q)} \left(\delta_{\alpha\beta} - k_0^{-2} k_\alpha(q)k_\beta(q)\right) \exp(ik_z(q)|z - z'|),$$

where $k(q) = (q, k_z(q))$. Using these results, it may be seen that the incident field is given by

$$E_\alpha^i(r) = k_0^2 \int d^3r' G_{\alpha\beta}(r, r')P_\beta(r'), \tag{2-10}$$

where the summation convention over repeated indices applies. The scattered field obeys the integral equation

$$E_\alpha^s(r) = k_0^2 \int d^3r' G_{\alpha\beta}(r, r')E_\beta(r')\eta(r').$$

Within the accuracy of the first Born approximation, the electric field may be replaced by the incident field in the right hand side of the above equation thus obtaining

$$E_\alpha^s(r) = k_0^2 \int d^3r' G_{\alpha\beta}(r,r') E_\beta^i(r') \eta(r'). \qquad (2\text{--}11)$$

The optical theorem. Energy conservation for Maxwell's equations leads to the optical theorem for electromagnetic waves. The extinguished power may be shown to be given by the expression [36]

$$P = \frac{1}{2} c k_0 \, \mathrm{Im} \int d^3r E^{i*}(r) \cdot E(r) \eta(r). \qquad (2\text{--}12)$$

If the incident field is a propagating plane wave with amplitude a and wave vector k

$$E^i(r) = a e^{ik \cdot r},$$

then P is related to the scattering amplitude by

$$P = |a|^2 \frac{4\pi}{k_0} \, \mathrm{Im}\big(A(k,k) \cdot \hat{e}^*\big),$$

where $\hat{e}$ is a unit vector in the a direction. Here $A(k,k')$ is the vector scattering amplitude defined by

$$A(k,k') = \frac{k_0^2 c}{8\pi |a|} \int d^3r e^{-ik' \cdot r} E(r) \eta(r).$$

3. Forward Problem

The forward problem in near-field tomography is the problem of computing the scattered field from the susceptibility. Analyses are carried out here separately for each experimental modality, making use of the scalar and vector scattering theory developed in Section 2. A common form relating the susceptibility to the scattering data will emerge.

3A. Scanning Near-Field Tomography. Scanning near-field tomography is based on the scanning modalities of near-field optics, namely NSOM. The two principal genres of NSOM are illumination mode and collection mode. The nomenclature reflects the role of the probe in each modality. In collection mode NSOM, a sharp, tapered optical fiber tip is coated with metal and a small aperture is exposed at the end of the fiber probe. The sample is illuminated by a source located in the far-zone of the sample and the total field is collected in the near-zone of the sample through the small aperture in the probe as the probe is scanned over the sample. In illumination mode NSOM, a probe, like that used in collection mode, is scanned over the sample in the near-zone. However, the source of illumination is light transmitted through the fiber and the small aperture at the tip. The field scattered by the sample is then collected and measured in the far-zone and recorded as a function of probe position.

To effect tomographic reconstruction of the sample it is necessary to perform all measurements with phase sensitivity. In the case of collection mode NSOM, the scattered field is naturally in superposition with the incident field and so the measurements are intrinsically holographic [42; 5]—a situation analogous to the Gabor hologram[43; 44; 45]. Illumination mode, however, is not intrinsically holographic and so the phase of the scattered field must be ascertained by some other means, possibly by interference with some reference field coherent with the illuminating field [46] as is done in the far field problem to produce the Leith–Upatnieks hologram [43; 47; 48]. In this article, it will be assumed that the scattered field is measured with phase sensitivity and we will not dwell on the experimental particulars, though it should be noted that the problem is nontrivial.

3A1. Scalar case

Illumination mode. An illustration of illumination mode NSOM is shown in Figure 4. It is assumed that the sample lies on a substrate which is defined by the plane $z = 0$ and is illuminated with a point source which lies in the plane $z = z_s$. The sample lies in the region $0 \leq z \leq z_s$ and is described by a susceptibility $\eta(r)$. If the point source has unit amplitude and is located at the position $r_1 = (\rho_1, z_s)$ then, according to (2–1) and (2–2), $S(r) = \delta(r - r_1)$ and hence $U_i(r) = G(r, r_1)$. The scattered field is then given by

$$U_s(r) = k_0^2 \int d^3r' G(r, r') G(r', r_1) \eta(r').$$

Suppose that the observation point r is in the far field of the sample and that the field is measured in the $z > 0$ half-space. This situation is referred to as the reflection geometry; the transmission geometry, in which the field is measured in the $z < 0$ half-space, will not be considered here but is amenable to a similar treatment. It may be seen that for $|r| \gg |r'|$ the leading term in the asymptotic expansion of the Green's function is given by

$$G(r, r') \sim \frac{e^{ik_0 r}}{r} e^{-ik(q) \cdot r'} \left(1 + R(q) e^{2ik_z(q)z'}\right), \qquad (3\text{–}1)$$

where $k(q)$ lies in the direction of r, $|q| \leq k_0$ and $R(q)$ is the reflection coefficient defined in the Appendix. The scattered field behaves as an outgoing spherical wave. It may be expressed in terms of the scattering amplitude $A(\rho_1, q)$ which depends on the position of the source and the propagation vector $k(q)$ in the direction of observation:

$$U_s(r) \sim \frac{e^{ik_0 r}}{r} A(\rho_1, q),$$

where

$$A(\rho_1, q) = k_0^2 \int d^3r\, e^{-ik(q) \cdot r} \left(1 + R(q) e^{2ik_z(q)z}\right) G(r, r_1) \eta(r). \qquad (3\text{–}2)$$

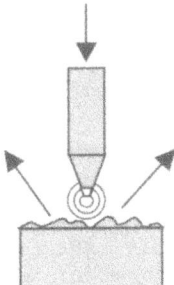

Figure 4. Illumination mode geometry. The sample is illuminated by light from the probe tip in the near-zone. The scattered field is measured in the far zone and recorded as a function of tip position.

We assume that the source is scanned over a square lattice with lattice spacing h, thus sampling the scattering amplitude. It will prove useful to define a data function $\Phi(\boldsymbol{q}_1, \boldsymbol{q}_2)$ by the lattice Fourier transform

$$\Phi(\boldsymbol{q}_1, \boldsymbol{q}_2) = \sum_{\boldsymbol{\rho}_1} e^{i\boldsymbol{q}_1 \cdot \boldsymbol{\rho}_1} A(\boldsymbol{\rho}_1, \boldsymbol{q}_2),$$

where the sum over $\boldsymbol{\rho}_1$ is carried out over all lattice vectors and $\boldsymbol{q}_1$ belongs to the first Brillouin zone (FBZ) of the lattice. In this case FBZ=$[-\pi/h, \pi/h] \times [-\pi/h, \pi/h]$. It may be observed that if $\boldsymbol{q}_1$ is not limited to the FBZ, the data outside the FBZ are redundant. Making use of (2–3), (3–2) and the identity

$$\sum_{\boldsymbol{\rho}} e^{i\boldsymbol{q} \cdot \boldsymbol{\rho}} = \left(\frac{2\pi}{h}\right)^2 \sum_{\boldsymbol{q}'} \delta(\boldsymbol{q} - \boldsymbol{q}'),$$

where $\boldsymbol{q}'$ denotes a reciprocal lattice vector [1], we find that

$$\Phi(\boldsymbol{q}_1, \boldsymbol{q}_2) = \left(\frac{k_0}{h}\right)^2 \int d^3r \sum_{\boldsymbol{q}} \exp\big(i(\boldsymbol{q}_1 - \boldsymbol{q}_2 - \boldsymbol{q}) \cdot \boldsymbol{\rho}\big)\big(1 + R(\boldsymbol{q}_2)e^{2ik_z(\boldsymbol{q}_2)z}\big)$$

$$\times e^{ik_z(\boldsymbol{q}_2)z} g(z, z_s; \boldsymbol{q}_1 - \boldsymbol{q})\eta(\boldsymbol{r}). \quad (3\text{–}3)$$

It is natural to reconstruct $\eta(\boldsymbol{r})$ on the same lattice that field is sampled. In that case the inverse problem will prove more tractable if $\eta(\boldsymbol{r})$ is assumed to be band limited so as to be consistent with the lattice on which $A(\boldsymbol{\rho}_1, \boldsymbol{q}_2)$ is sampled. Then the sum over $\boldsymbol{q}$ may be truncated and only the $\boldsymbol{q} = 0$ term contributes to $\Phi(\boldsymbol{q}_1, \boldsymbol{q}_2)$. Thus (3–3) may be written in the form of the integral equation

$$\Phi(\boldsymbol{q}_1, \boldsymbol{q}_2) = \int d^3r K(\boldsymbol{q}_1, \boldsymbol{q}_2; \boldsymbol{r})\eta(\boldsymbol{r}). \quad (3\text{–}4)$$

Here the kernel

$$K(\boldsymbol{q}_1, \boldsymbol{q}_2; \boldsymbol{r}) = e^{i(\boldsymbol{q}_1 - \boldsymbol{q}_2) \cdot \boldsymbol{\rho}} \kappa(\boldsymbol{q}_1, \boldsymbol{q}_2; z),$$

---

[1]The reciprocal lattice consists of all points in the plane of the form $(2n\pi/h, 2m\pi/h)$ with n, m being integers.

where

$$\kappa(\boldsymbol{q}_1, \boldsymbol{q}_2; z) = \left(\frac{k_0}{h}\right)^2 \left(1 + R(\boldsymbol{q}_2)e^{2ik_z(\boldsymbol{q}_2)z}\right)e^{ik_z(\boldsymbol{q}_2)z}g(z, z_s; \boldsymbol{q}_1).$$

Equation (3–4) expresses the forward problem in scalar illumination mode scanning near-field tomography.

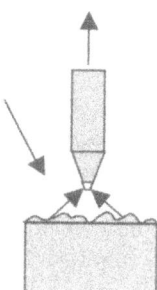

Figure 5. Collection mode geometry. The sample is illuminated from the far zone. The field is collected at the probe tip and recorded as a function of tip position.

Collection mode. An illustration of collection mode NSOM in the reflection geometry is shown in Figure 5. In this situation the sample is illuminated from the far zone by an incident plane wave and the scattered field is detected in the near-zone by means of an idealized point detector. Evidently, the incident wave may be reflected from the boundary and is of the form

$$U_i(\boldsymbol{r}) = \left(1 + R(\boldsymbol{q})e^{2ik_z(\boldsymbol{q})z}\right)e^{i\boldsymbol{k}_i(\boldsymbol{q})\cdot\boldsymbol{r}},$$

where $\boldsymbol{k}_i(\boldsymbol{q}) = (\boldsymbol{q}, -k_z(\boldsymbol{q}))$, $R(\boldsymbol{q})$ is the reflection coefficient (defined in the Appendix), and it has been assumed that the incident wave is of unit amplitude. The scattered field, measured in the $z = z_d$ plane at a point with coordinate $\boldsymbol{r} = (\boldsymbol{\rho}, z_d)$, is given by the expession

$$U_s(\boldsymbol{r}) = k_0^2 \int d^3r' \left(1 + R(\boldsymbol{q})e^{2ik_z(\boldsymbol{q})z'}\right)e^{i\boldsymbol{k}_i(\boldsymbol{q})\cdot\boldsymbol{r}'}G(\boldsymbol{r}, \boldsymbol{r}')\eta(\boldsymbol{r}'),$$

which follows from (2–4).

The scattered field is sampled on a square lattice with lattice spacing h and, as in the case of illumination mode, the data function is defined by the lattice Fourier transform

$$\Phi(\boldsymbol{q}_1, \boldsymbol{q}_2) = \sum_{\boldsymbol{\rho}} e^{-i\boldsymbol{q}_2\cdot\boldsymbol{\rho}}U_s(\boldsymbol{\rho}, z_d; \boldsymbol{q}_1),$$

where $\boldsymbol{q}_2 \in$ FBZ and the dependence of the scattered field on the transverse component $\boldsymbol{q}_1$ of the incident wave vector has been made explicit. Using the plane wave decomposition of the Green's function (2–3) and carying out the Fourier transform, we find that

$$\Phi(\boldsymbol{q}_1, \boldsymbol{q}_2) = \left(\frac{k_0}{h}\right)^2 \int d^3r \sum_{\boldsymbol{q}} \exp(i(\boldsymbol{q}_1 - \boldsymbol{q}_2 - \boldsymbol{q}) \cdot \boldsymbol{\rho})$$

$$\times \left(1 + R(\boldsymbol{q}_1)e^{2ik_z(\boldsymbol{q}_1)z}\right)e^{-ik_z(\boldsymbol{q}_1)z}g(z_d, z; \boldsymbol{q}_2 - \boldsymbol{q})\eta(\boldsymbol{r}). \quad (3\text{–}5)$$

As discussed earlier, $\eta(\boldsymbol{r})$ is assumed to be transversely band limited with a band limit commensurate with the lattice structure. Then (3–5) may be written in the form of the integral equation (3–4) with

$$\kappa(\boldsymbol{q}_1, \boldsymbol{q}_2; z) = \left(\frac{k_0}{h}\right)^2 \left(1 + R(\boldsymbol{q}_1)e^{2ik_z(\boldsymbol{q}_1)z}\right)e^{-ik_z(\boldsymbol{q}_1)z}g(z_d, z; \boldsymbol{q}_2).$$

3A2. Vector case. The mathematical treatment of the vector forward problem for scanning near-field tomography follows closely the scalar case.

Illumination mode. In illumination mode the sample is illuminated by a point source with position $\boldsymbol{r}_1 = (\boldsymbol{\rho}_1, z_s)$ which is scanned in the $z = z_s$ plane. The incident field is obtained from (2–10) with the dielectric polarization $\boldsymbol{P}(\boldsymbol{r}) = \boldsymbol{p}\delta(\boldsymbol{r} - \boldsymbol{r}_1)$, $\boldsymbol{p}$ being the dipole moment of the source of the field. Thus

$$E_\alpha^i(\boldsymbol{r}) = k_0^2 G_{\alpha\beta}(\boldsymbol{r}, \boldsymbol{r}_1)p_\beta.$$

Using (2–11) the scattered field is seen to be given by

$$E_\alpha^s(\boldsymbol{r}) = k_0^4 \int d^3r' G_{\alpha\beta}(\boldsymbol{r}, \boldsymbol{r}')G_{\beta\gamma}(\boldsymbol{r}', \boldsymbol{r}_1)p_\gamma \eta(\boldsymbol{r}').$$

In the upper half space, for $|\boldsymbol{r}| \gg |\boldsymbol{r}'|$, the Green's tensor assumes the asymptotic form

$$G_{\alpha\beta}(\boldsymbol{r}, \boldsymbol{r}') \sim S_{\alpha\gamma}^{-1}(\boldsymbol{q})\tilde{g}_{\gamma\delta}(\boldsymbol{q}, z)S_{\delta\beta}(\boldsymbol{q})\frac{\exp(ik_0 r)}{r}\exp(-i\boldsymbol{k}(\boldsymbol{q}) \cdot \boldsymbol{r}'), \qquad (3\text{–}6)$$

where $\boldsymbol{k}(\boldsymbol{q})$ lies in the direction of $\boldsymbol{r}$ and the prefactors are defined in the Appendix. Thus the scattered field in the far-zone in the upper half space takes the form

$$E_\alpha^s(\boldsymbol{r}) \sim \frac{e^{ik_0 r}}{r}A_\alpha(\boldsymbol{\rho}_1, \boldsymbol{q}).$$

Here the scattering amplitude is given by

$$A_\alpha(\boldsymbol{\rho}_1, \boldsymbol{q}) = k_0^2 \int d^3r e^{-i\boldsymbol{k}(\boldsymbol{q}) \cdot \boldsymbol{r}}w_{\alpha\beta}(\boldsymbol{q}, z)G_{\beta\gamma}(\boldsymbol{r}, \boldsymbol{r}_1)p_\gamma \eta(\boldsymbol{r}),$$

where

$$w_{\alpha\beta}(\boldsymbol{q}, z) = k_0^2 S_{\alpha\gamma}^{-1}(\boldsymbol{q})\tilde{g}_{\gamma\delta}(\boldsymbol{q}, z)S_{\delta\beta}(\boldsymbol{q}). \qquad (3\text{–}7)$$

As before, a data function is defined through a lattice Fourier transform of the sampled scattering amplitude

$$\Phi_\alpha(\boldsymbol{q}_1, \boldsymbol{q}_2) = \sum_{\boldsymbol{\rho}_1} e^{i\boldsymbol{q}_1 \cdot \boldsymbol{\rho}_1} A_\alpha(\boldsymbol{\rho}_1, \boldsymbol{q}_2),$$

where $q_1 \in$ FBZ. Making use of the plane wave decomposition (2–9), it is found that

$$\Phi_\alpha(q_1, q_2) = \left(\frac{k_0}{h}\right)^2 \int d^3 r \sum_q \exp\big(i(q_1 - q_2 - q) \cdot \rho\big)$$
$$\times e^{ik_z(q_2)z} w_{\alpha\beta}(q_2, z) g_{\beta\gamma}(z, z_s; q_1 - q) p_\gamma \eta(r). \quad (3\text{–}8)$$

If $\eta(r)$ is transversely band limited, as in the previous cases, with a band limit commensurate with the lattice structure then we find that (3–8) may be written in the form of the integral equation

$$\Phi_\alpha(q_1, q_2) = \int d^3 r K_\alpha(q_1, q_2; r) \eta(r). \quad (3\text{–}9)$$

Here the kernel

$$K_\alpha(q_1, q_2; r) = \exp\big(i(q_1 - q_2) \cdot \rho\big) \kappa_\alpha(q_1, q_2; z),$$

where

$$\kappa_\alpha(q_1, q_2; z) = \left(\frac{k_0}{h}\right)^2 e^{ik_z(q_2)z} w_{\alpha\beta}(q_2, z) g_{\beta\gamma}(z, z_s; q_1) p_\gamma.$$

Equation (3–9) expresses the forward problem for vector illumination mode scanning near-field tomography.

Collection mode. As in the scalar case, collection mode NSOM in the reflection geometry involves illumination of the sample by a source in the far-zone and collection of the near-zone scattered field in the upper half-space. The incident field will be taken to linearly polarized,

$$E^i_\alpha(r) = E^{(0)}_\alpha \big(1 + R(q)e^{2ik_z(q)z}\big) e^{ik_i(q) \cdot r},$$

where $E^{(0)}_\alpha$ is the polarization of the incident field, $R(q)$ is the appropriate Fresnel reflection coefficient for the electric field and $k_i(q) = (q, -k_z(q))$. For simplicity, we will consider the incident field to have TE polarization, that is the polarization vector of the incident fields is parallel to the boundary of the half-space. More general states of polarization may also be considered and it is important to note that the signal in NSOM has a strong polarization dependence [16]. Using (2–11), we find that the scattered field, measured in the plane $z = z_d$ is given by the expression

$$E^s_\alpha(r) = k_0^2 \int d^3 r' \big(1 + R(q)e^{2ik_z(q)z'}\big) e^{ik_i(q) \cdot r'} G_{\alpha\beta}(r, r') E^{(0)}_\beta \eta(r').$$

The data function is defined as the lattice Fourier transform

$$\Phi_\alpha(q_1, q_2) = \sum_\rho e^{-iq_2 \cdot \rho} E^s_\alpha(\rho, z_d; q_1), \quad (3\text{–}10)$$

where $q_2 \in$ FBZ. The plane wave decomposition (2–9) may be utilized to obtain

$$\Phi_\alpha(\boldsymbol{q}_1, \boldsymbol{q}_2) = \left(\frac{k_0}{h}\right)^2 \int d^3r \sum_q \exp\bigl(i(\boldsymbol{q}_1 - \boldsymbol{q}_2 - \boldsymbol{q}) \cdot \boldsymbol{\rho}\bigr)\bigl(1 + R(\boldsymbol{q}_1)e^{2ik_z(\boldsymbol{q}_1)z}\bigr)$$
$$\times\, e^{-ik_z(\boldsymbol{q}_1)z} g_{\alpha\beta}(z_d, z; \boldsymbol{q}_2 - \boldsymbol{q}) E_\beta^{(0)} \eta(\boldsymbol{r}). \quad (3\text{–}11)$$

If $\eta(\boldsymbol{r})$ is transversely band limited with a band limit consistent with the lattice structure, then we find that (3–11) may be written in the form of the integral equation (3–9) with

$$\kappa_\alpha(\boldsymbol{q}_1, \boldsymbol{q}_2; z) = \left(\frac{k_0}{h}\right)^2 \bigl(1 + R(\boldsymbol{q}_1)e^{2ik_z(\boldsymbol{q}_1)z}\bigr)e^{-ik_z(\boldsymbol{q}_1)z} g_{\alpha\beta}(z_d, z; \boldsymbol{q}_2) E_\beta^{(0)}.$$

3B. Total Internal Reflection Tomography. In total internal reflection tomography, the sample is illuminated by an evanescent wave that is generated by total internal reflection. The scattered field is then measured in the far zone of the scatterer as the direction of the incident wave is varied. See Figure 1. It should be noted that the scattered field must be measured with phase sensitivity.

3B1. Scalar case. The sample resides in vacuum in the geometry described earlier. The sample is illluminated by an evanescent plane wave which is generated by total internal reflection in the half-space with $n > 1$. The field incident on the sample is of the form

$$U_i(\boldsymbol{r}) = e^{i\boldsymbol{k}_1(\boldsymbol{q}_1)\cdot\boldsymbol{r}}, \quad (3\text{–}12)$$

where $\boldsymbol{k}_1(\boldsymbol{q}_1) = (\boldsymbol{q}_1, k_z(\boldsymbol{q}_1))$ is the incident wave vector. The transverse wave vector $\boldsymbol{q}$ satisfies $k_0 \leq |\boldsymbol{q}_1| \leq nk_0$, where n the index of refraction of the lower half space. Note that k_z is imaginary with the choice of sign dictated by the physical requirement that the field decay exponentially with increasing values of z. In the far zone, the scattered field behaves as an outgoing spherical wave determined by the wave vector $\boldsymbol{k}_2(\boldsymbol{q}_2) = (\boldsymbol{q}_2, k_z(\boldsymbol{q}_2))$ with $|\boldsymbol{q}_2| \leq k_0$. Making use of the integral equation (2–4) and the asymptotic form of the outgoing Green's function (3–1), it may be seen that the scattered field in the far zone is given by

$$U_s \sim \frac{e^{ik_0r}}{r} A(\boldsymbol{q}_1, \boldsymbol{q}_2).$$

Here $A(\boldsymbol{q}_1, \boldsymbol{q}_2)$, which is the scattering amplitude associated with the scattering of evanescent plane waves with transverse wave vector $\boldsymbol{q}_1$ into homogeneous plane waves with transverse wave vector $\boldsymbol{q}_2$, is related to the susceptibility of the scattering object by the expression

$$A(\boldsymbol{q}_1, \boldsymbol{q}_2) = k_0^2 \int d^3r \bigl(1 + R(\boldsymbol{q}_2)e^{2ik_z(\boldsymbol{q}_2)z}\bigr)\exp\bigl(i(\boldsymbol{k}_1(\boldsymbol{q}_1) - \boldsymbol{k}_2(\boldsymbol{q}_2)) \cdot \boldsymbol{r}\bigr)\eta(\boldsymbol{r}).$$

Note that this result may be rewritten in the form of the integral equation (3–4) where $A(\boldsymbol{q}_1, \boldsymbol{q}_2)$ is identified with the data function $\Phi(\boldsymbol{q}_1, \boldsymbol{q}_2)$ and

$$\kappa(\boldsymbol{q}_1, \boldsymbol{q}_2; \boldsymbol{r}) = k_0^2\bigl(1 + R(\boldsymbol{q}_2)e^{2ik_z(\boldsymbol{q}_2)z}\bigr)\exp\bigl(i(k_z(\boldsymbol{q}_1) - k_z(\boldsymbol{q}_2))z\bigr).$$

3B2. Vector case. As in the scalar case, the incident field is taken to be an evanescent plane wave with polarization $E_\alpha^{(0)}$

$$E_\alpha^i(\boldsymbol{r}) = E_\alpha^{(0)} e^{i\boldsymbol{k}_1(\boldsymbol{q}_1)\cdot\boldsymbol{r}}, \tag{3-13}$$

where $\boldsymbol{k}_1(\boldsymbol{q}_1) = (\boldsymbol{q}_1, k_z(\boldsymbol{q}_1))$ is the incident wave vector and $k_0 \leq |\boldsymbol{q}_1| \leq nk_0$. In the far zone the scattered field is characterized by the wave vector $\boldsymbol{k}_2(\boldsymbol{q}_2) = (\boldsymbol{q}_2, k_z(\boldsymbol{q}_2))$ with $|\boldsymbol{q}_2| \leq k_0$. Using the integral equation (2–11) for the scattered field and (3–6), it may be seen that asymptotic form of the scattered field in the far zone is given by

$$E_\alpha^s \sim \frac{e^{ik_0 r}}{r} A_\alpha(\boldsymbol{q}_1, \boldsymbol{q}_2).$$

Here $A_\alpha(\boldsymbol{q}_1, \boldsymbol{q}_2)$ denotes the vector scattering amplitude which is related to the susceptibility by

$$A_\alpha(\boldsymbol{q}_1, \boldsymbol{q}_2) = \int d^3 r \, w_{\alpha\beta}(\boldsymbol{q}_2, z) E_\beta^{(0)} \exp\big(i \left(\boldsymbol{k}(\boldsymbol{q}_1) - \boldsymbol{k}(\boldsymbol{q}_2)\right) \cdot \boldsymbol{r}\big) \eta(\boldsymbol{r}).$$

This relation may be written in the form of the vector integral equation (3–9) where the scattering amplitude $A_\alpha(\boldsymbol{q}_1, \boldsymbol{q}_2)$ is identified with the data function $\Phi_\alpha(\boldsymbol{q}_1, \boldsymbol{q}_2)$ and

$$\kappa_\alpha(\boldsymbol{q}_1, \boldsymbol{q}_2; \boldsymbol{r}) = w_{\alpha\beta}(\boldsymbol{q}_2, z) E_\beta^{(0)} \exp\big(i \left(k_z(\boldsymbol{q}_1) - k_z(\boldsymbol{q}_2)\right) z\big),$$

with $w_{\alpha\beta}$ given by (3–7).

3C. Photon Scanning Tunneling Tomography. In photon scanning tunneling tomography, the sample is illuminated by an evanescent wave and the scattered field is detected in the near zone. See Figure 1. Photon scanning tunneling tomography is a hybrid of total internal reflection tomography and scanning near-field tomography. The analyses follows those in Section 3A1 for the scalar case and in Section 3A2 for the vector case.

3C1. Scalar case. The sample is illuminated by an evanescent plane wave of the form (3–12). The scattered field is then given by (2–4). If the scattered field is sampled on the plane $z = z_d$ and the data function $\Phi(\boldsymbol{q}_1, \boldsymbol{q}_2)$ is defined by (3–5), it can be seen that

$$\Phi(\boldsymbol{q}_1, \boldsymbol{q}_2) = \left(\frac{k_0}{h}\right)^2 \int d^3 r \sum_{\boldsymbol{q}} \exp\big(i(\boldsymbol{q}_1 - \boldsymbol{q}_2 - \boldsymbol{q}) \cdot \boldsymbol{\rho}\big) e^{ik_z(\boldsymbol{q}_1)z} g(z_d, z; \boldsymbol{q}_2 - \boldsymbol{q}) \eta(\boldsymbol{r}).$$

$$\tag{3-14}$$

Note that if $\eta(\boldsymbol{r})$ is transversely band-limited then (3–14) may be put in the form of the scalar integral equation (3–4) with

$$\kappa(\boldsymbol{q}_1, \boldsymbol{q}_2; z) = \left(\frac{k_0}{h}\right)^2 e^{ik_z(\boldsymbol{q}_1)z} g(z_d, z; \boldsymbol{q}_2).$$

3C2. Vector case. The sample is illuminated with an evanescent plane wave of the form (3–13) and the scattered field is given by (2–11). The data function, defined by (3–10), may be seen to be given by

$$\Phi_\alpha(\boldsymbol{q}_1, \boldsymbol{q}_2) = \left(\frac{k_0}{h}\right)^2 \int d^3r \sum_{\boldsymbol{p}} \exp\left(i(\boldsymbol{q}_1 - \boldsymbol{q}_2 - \boldsymbol{p}) \cdot \boldsymbol{\rho}\right)$$
$$\times\, e^{ik_z(\boldsymbol{q}_1)z} g_{\alpha\beta}(z_d, z; \boldsymbol{q}_2 - \boldsymbol{p}) E_\beta^{(0)} \eta(\boldsymbol{r}). \quad (3\text{--}15)$$

If $\eta(\boldsymbol{r})$ is transversely band-limited, then (3–15) has the form of the vector integral equation (3–9) with

$$\kappa_\alpha(\boldsymbol{q}_1, \boldsymbol{q}_2; z) = \left(\frac{k_0}{h}\right)^2 e^{ik_z(\boldsymbol{q}_1)z} g_{\alpha\beta}(z_d, z; \boldsymbol{q}_2) E_\beta^{(0)}. \quad (3\text{--}16)$$

3D. Near-Field Power Extinction Tomography. In near-field power extinction tomography, the sample is illuminated by a coherent beam consisting of a superposition of two evanescent plane waves as shown in Figure 3. The intensity of the incident field is then structured due to the interference between the plane waves as may be seen in Figure 6. Some of the power carried by the

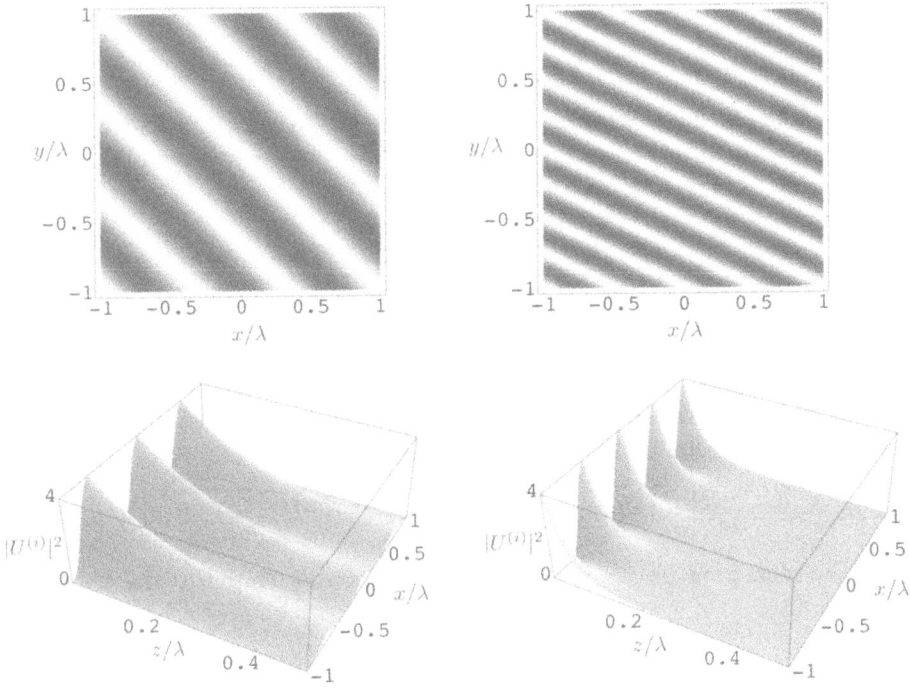

Figure 6. Two sets of two evanescent plane waves. The top row displays the intensity in the $z = 0$ plane resulting from the addition of the evanescent waves. The bottom figures display the intensity of the resultant field as a function of depth, z, and one of the transverse coordinates, x, in the $y = 0$ plane.

incident field is absorbed by the sample and some is scattered into propagating modes of the scattered field in both the upper and lower half-spaces. The total power lost from the incident field due to the presence of the sample is called the extinguished power. For the scattering of single homogeneous plane waves, the extinguished power is usually normalized by the incident power per unit area and the resulting quantity is called the total cross-section. The power extinguished from the beam is monitored at the output of the totally reflected component plane waves. To obtain the structure of the sample, the extinguished power is recorded as both the relative phase and the the orientation of the constituent evanescent waves are varied. Thus the measurements need not be phase sensitive.

3D1. Scalar case. Consider a scattering experiment with an incident field consisting of a superposition of two evanescent waves of the form

$$U_i(\mathbf{r}) = a_1 e^{i\mathbf{k}_1 \cdot \mathbf{r}} + a_2 e^{i\mathbf{k}_2 \cdot \mathbf{r}},$$

where a_1, a_2 denote the amplitudes of the waves and $\mathbf{k}_1, \mathbf{k}_2$ their wave vectors. According to the generalized optical theorem (2–5), the extinguished power is given by

$$P(a_1, a_2) = \frac{4\pi}{k_0}$$
$$\times \text{Im}\left(|a_1|^2 A(\mathbf{k}_1, \mathbf{k}_1^*) + a_1^* a_2 A(\mathbf{k}_2, \mathbf{k}_1^*) + a_2^* a_1 A(\mathbf{k}_1, \mathbf{k}_2^*) + |a_2|^2 A(\mathbf{k}_2, \mathbf{k}_2^*)\right). \quad (3\text{–}17)$$

Here $A(\mathbf{k}_1, \mathbf{k}_2)$ is the scattering amplitude associated with the scattering of a plane wave with wave vector $\mathbf{k}_1$ into a plane wave with wave vector $\mathbf{k}_2$. Using (2–7), it may be seen that to lowest order in $\eta(\mathbf{r})$

$$A(\mathbf{k}_1, \mathbf{k}_2) = k_0^2 \int d^3 r\, e^{i(\mathbf{k}_1 - \mathbf{k}_2)\cdot \mathbf{r}} \eta(\mathbf{r}). \quad (3\text{–}18)$$

It will prove useful to extract the cross-terms from (3–17), that is to gain information about the scattering amplitude for nonzero momentum transfer. This can be accomplished for any set of $\mathbf{k}_1$ and $\mathbf{k}_2$ through four measurements of the extinguished power where the relative phases are varied between measurements. To this end it is useful define the data function

$$\Phi(\mathbf{k}_1, \mathbf{k}_2) = \frac{k_0}{8\pi a_1^* a_2}\left(P(a_1, ia_2) - P(a_1, -ia_2) + i\left(P(a_1, a_2) - P(a_1, -a_2)\right)\right).$$

It may be seen from (3–17) that the data function is related to the scattering amplitude by

$$\Phi(\mathbf{k}_1, \mathbf{k}_2) = A(\mathbf{k}_1, \mathbf{k}_2^*) - A^*(\mathbf{k}_2, \mathbf{k}_1^*). \quad (3\text{–}19)$$

Let $\alpha(\mathbf{r}) \equiv \text{Im}\,\eta(\mathbf{r})$ denote the absorptive part of the susceptibility $\eta(\mathbf{r})$. Then (3–18) and (3–19) yield

$$\Phi(\mathbf{k}_1, \mathbf{k}_2) = \int d^3 r\, e^{i(\mathbf{k}_1 - \mathbf{k}_2^*)\cdot \mathbf{r}} \alpha(\mathbf{r}).$$

Note that this result may be rewritten in the form of the integral equation (3–4) where

$$\kappa(\boldsymbol{q}_1, \boldsymbol{q}_2; \boldsymbol{r}) = 2ik_0^2 \exp\big(i\,(k_z(\boldsymbol{q}_1) - k_z^*(\boldsymbol{q}_2))\,z\big),$$

and the dependence of $\boldsymbol{k}_1, \boldsymbol{k}_2$ on their transverse parts $\boldsymbol{q}_1, \boldsymbol{q}_2$ has been made explicit.

3D2. Vector case. As in the scalar case the experiment consists of a sample in vacuum in the upper half-space and a set of totally internally reflected plane waves in the lower half-space where $n > 1$. The incident field on the vacuum side then consists of a coherent superposition of two evanescent waves of the form

$$\boldsymbol{E}^i = \boldsymbol{a}_1 e^{i\boldsymbol{k}_1 \cdot \boldsymbol{r}} + \boldsymbol{a}_2 e^{i\boldsymbol{k}_2 \cdot \boldsymbol{r}},$$

where $\boldsymbol{a}_1, \boldsymbol{a}_2$ denote the vector amplitudes of the waves and $\boldsymbol{k}_1, \boldsymbol{k}_2$ their wave vectors. The extinguished power is given by

$$\begin{aligned}
P(\boldsymbol{a}_1, \boldsymbol{a}_2) = \frac{4\pi}{k_0} \,\mathrm{Im}\,\big(&|a_1|^2 \boldsymbol{A}(\boldsymbol{k}_1, \boldsymbol{k}_1^*) \cdot \hat{\boldsymbol{e}}_1^* + a_1^* a_2 \boldsymbol{A}(\boldsymbol{k}_2, \boldsymbol{k}_1^*) \cdot \hat{\boldsymbol{e}}_1^* \\
&+ a_2^* a_1 \boldsymbol{A}(\boldsymbol{k}_1, \boldsymbol{k}_2^*) \cdot \hat{\boldsymbol{e}}_2^* + |a_2|^2 \boldsymbol{A}(\boldsymbol{k}_2, \boldsymbol{k}_2^*) \cdot \hat{\boldsymbol{e}}_2^*\big), \quad (3\text{–}20)
\end{aligned}$$

where $\boldsymbol{a}_1 = a_1 \hat{\boldsymbol{e}}_1$ and $\boldsymbol{a}_2 = a_2 \hat{\boldsymbol{e}}_2$, $\hat{\boldsymbol{e}}_1, \hat{\boldsymbol{e}}_2$ denote unit vectors in the directions of $\boldsymbol{a}_1, \boldsymbol{a}_2$, and we have used the generalized optical theorem (2–12). Here, to lowest order in $\eta(\boldsymbol{r})$, the vector scattering amplitude is given by

$$\boldsymbol{A}(\boldsymbol{k}_1, \boldsymbol{k}_2) = \frac{k_0^2 c}{8\pi}\, \hat{\boldsymbol{e}} \int d^3 r\, e^{i(\boldsymbol{k}_1 - \boldsymbol{k}_2) \cdot \boldsymbol{r}} \eta(\boldsymbol{r}).$$

Note that within the first Born approximation $\boldsymbol{A}(\boldsymbol{k}_1, \boldsymbol{k}_2) \cdot \hat{\boldsymbol{e}} = c/8\pi A(\boldsymbol{k}_1, \boldsymbol{k}_2)$, where $A(\boldsymbol{k}_1, \boldsymbol{k}_2)$ is the scalar scattering amplitude defined in (3–18).

The data function $\Phi(\boldsymbol{k}_1, \boldsymbol{k}_2)$ is defined by

$$\begin{aligned}
\Phi(\boldsymbol{k}_1, \boldsymbol{k}_2) = \;&\frac{k_0}{c\,\boldsymbol{a}_1^* \cdot \boldsymbol{a}_2} \\
&\times \big(P(\boldsymbol{a}_1, i\boldsymbol{a}_2) - P(\boldsymbol{a}_1, -i\boldsymbol{a}_2) + i\big(P(\boldsymbol{a}_1, \boldsymbol{a}_2) - P(\boldsymbol{a}_1, -\boldsymbol{a}_2)\big)\big). \quad (3\text{–}21)
\end{aligned}$$

Then it is readily seen, within the accuracy of the first Born approximation, that $\Phi(\boldsymbol{k}_1, \boldsymbol{k}_2)$ is related to the scalar scattering amplitude by

$$\Phi(\boldsymbol{k}_1, \boldsymbol{k}_2) = A(\boldsymbol{k}_1, \boldsymbol{k}_2^*) - A^*(\boldsymbol{k}_2, \boldsymbol{k}_1^*).$$

This result is identical to (3–19). As a consequence, with suitable modifications we may apply the scalar theory to vector near-field power extinction tomography.

4. Inverse Problem

The inverse problem consists of recovering the susceptibility $\eta(r)$ from the data function $\Phi(q_1, q_2)$. To this end, the pseudoinverse solution to the integral equations (3–4) and (3–9) will be systematically constructed. First, a brief review the singular value decomposition (SVD) of linear operators on Hilbert spaces[49] is given.

4A. Singular Value Decomposition. Let K denote a linear operator with kernel $K(x, y)$ which maps the Hilbert space $\mathcal{H}_1$ into the Hilbert space $\mathcal{H}_2$. The SVD of K is a representation of the form

$$K(x, y) = \sum_n \sigma_n g_n(x) f_n^*(y),$$

where σ_n is the singular value associated with the singular functions f_n and g_n. The $\{f_n\}$ and $\{g_n\}$ are orthonormal bases of $\mathcal{H}_1$ and $\mathcal{H}_2$, respectively and are eigenfunctions with eigenvalues σ_n^2 of the positive self-adjoint operators K^*K and KK^*:

$$K^*K f_n = \sigma_n^2 f_n, \qquad KK^* g_n = \sigma_n^2 g_n.$$

In addition, the f_n and g_n are related by

$$K f_n = \sigma_n g_n, \qquad K^* g_n = \sigma_n f_n.$$

The pseudoinverse solution to the equation $Kf = g$ is defined to be the minimizer of $\|Kf - g\|$ with smallest norm. This well-defined element $f^+ \in N(K)^\perp$ is unique and may be shown[49] to be of the form $f^+ = K^+ g$, where the pseudoinverse operator K^+ is given by $K^+ = K^*(KK^*)^{-1}$ and $N(K)^\perp$ is the orthogonal complement of the null space of K. The SVD of K may be used to express K^+ as

$$K^+(x, y) = \sum_n \frac{1}{\sigma_n} f_n(x) g_n^*(y). \tag{4–1}$$

4B. Scalar Case. Consider the scalar integral equation

$$\Phi(q_1, q_2) = \int d^3 r K(q_1, q_2; r) \eta(r),$$

where

$$K(q_1, q_2; r) = \exp(i(q_1 - q_2) \cdot \rho) \kappa(q_1, q_2; z).$$

This equation describes the scalar forward problem for each of the experimental modalities we have considered. In each case, only the functional form of $\kappa(q_1, q_2; z)$ must be altered. In addition, the wave vectors q_1, q_2 take values in a set which depends upon the choice of experiment and the available data. It is convenient to introduce a data set Ω that specifies the available wave vectors and a function $\chi(q_1, q_2)$ that is unity if $(q_1, q_2) \in \Omega$ and is zero otherwise. The function $\kappa(q_1, q_2; z)$ is then modified so that $\kappa(q_1, q_2; z) \to \kappa(q_1, q_2; z) \chi(q_1, q_2)$.

To obtain the SVD of $K(\boldsymbol{q}_1, \boldsymbol{q}_2; \boldsymbol{r})$ it will prove useful to introduce the following identity:

$$K(\boldsymbol{q}_1, \boldsymbol{q}_2; \boldsymbol{r}) = \int d^2 Q \exp(i\boldsymbol{Q} \cdot \boldsymbol{\rho}) \delta(\boldsymbol{Q} + \boldsymbol{q}_2 - \boldsymbol{q}_1) \kappa(\boldsymbol{Q} + \boldsymbol{q}_2, \boldsymbol{q}_2; z).$$

Using this result, the matrix elements of the operator KK^* are seen to be given by

$$KK^*(\boldsymbol{q}_1, \boldsymbol{q}_2; \boldsymbol{q}_1', \boldsymbol{q}_2') = \int d^2 Q M(\boldsymbol{q}_2, \boldsymbol{q}_2'; \boldsymbol{Q}) \delta(\boldsymbol{Q} + \boldsymbol{q}_2 - \boldsymbol{q}_1) \delta(\boldsymbol{Q} + \boldsymbol{q}_2' - \boldsymbol{q}_1'), \quad (4\text{--}2)$$

where

$$M(\boldsymbol{q}_2, \boldsymbol{q}_2'; \boldsymbol{Q}) = \int_0^L dz \kappa(\boldsymbol{Q} + \boldsymbol{q}_2, \boldsymbol{q}_2; z) \kappa^*(\boldsymbol{Q} + \boldsymbol{q}_2', \boldsymbol{q}_2'; z),$$

with L the range of $\eta(\boldsymbol{r})$ in the $\hat{\boldsymbol{z}}$ direction. The singular vectors $g_{\boldsymbol{QQ}'}$ of K satisfy

$$KK^* g_{\boldsymbol{QQ}'} = \sigma_{\boldsymbol{QQ}'}^2 g_{\boldsymbol{QQ}'},$$

and may be constructed by making the *ansatz* that

$$g_{\boldsymbol{QQ}'}(\boldsymbol{q}_1, \boldsymbol{q}_2) = C_{\boldsymbol{Q}'}(\boldsymbol{q}_2; \boldsymbol{Q}) \delta(\boldsymbol{Q} + \boldsymbol{q}_2 - \boldsymbol{q}_1), \quad (4\text{--}3)$$

for some $C_{\boldsymbol{Q}'}(\boldsymbol{q}_2; \boldsymbol{Q})$. Equation (4–2) now implies that

$$\int d^2 q' M(\boldsymbol{q}, \boldsymbol{q}'; \boldsymbol{Q}) C_{\boldsymbol{Q}'}(\boldsymbol{q}'; \boldsymbol{Q}) = \sigma_{\boldsymbol{QQ}'}^2 C_{\boldsymbol{Q}'}(\boldsymbol{q}; \boldsymbol{Q}).$$

Thus $C_{\boldsymbol{Q}'}(\boldsymbol{q}_2; \boldsymbol{Q})$ is an eigenvector of $M(\boldsymbol{Q})$ labeled by $\boldsymbol{Q}'$ with eigenvalue $\sigma_{\boldsymbol{QQ}'}^2$. Since $M(\boldsymbol{Q})$ is self-adjoint, the $C_{\boldsymbol{Q}'}(\boldsymbol{q}_2; \boldsymbol{Q})$ may be taken to orthonormal. Next, the $f_{\boldsymbol{QQ}'}$ may be found from $K^* g_{\boldsymbol{QQ}'} = \sigma_{\boldsymbol{QQ}'} f_{\boldsymbol{QQ}'}$ and are given by

$$f_{\boldsymbol{QQ}'}(\boldsymbol{r}) = \frac{1}{\sigma_{\boldsymbol{QQ}'}} \int d^2 q \exp(-i\boldsymbol{Q} \cdot \boldsymbol{\rho}) \kappa^*(\boldsymbol{Q} + \boldsymbol{q}, \boldsymbol{q}; z) C_{\boldsymbol{Q}'}^*(\boldsymbol{q}; \boldsymbol{Q}). \quad (4\text{--}4)$$

It follows that the SVD of $K(\boldsymbol{q}_1, \boldsymbol{q}_2; \boldsymbol{r})$ is given by the expression

$$K(\boldsymbol{q}_1, \boldsymbol{q}_2; \boldsymbol{r}) = \int d^2 Q d^2 Q' \sigma_{\boldsymbol{QQ}'} f_{\boldsymbol{QQ}'}^*(\boldsymbol{r}) g_{\boldsymbol{QQ}'}(\boldsymbol{q}_1, \boldsymbol{q}_2). \quad (4\text{--}5)$$

The SVD (4–5) may now be used to obtain the pseudoinverse solution to the integral equation (3–4):

$$\eta^+(\boldsymbol{r}) = \int d^2 q_1 d^2 q_2 K^+(\boldsymbol{r}; \boldsymbol{q}_1, \boldsymbol{q}_2) \Phi(\boldsymbol{q}_1, \boldsymbol{q}_2),$$

where $K^+(\boldsymbol{r}; \boldsymbol{q}_1, \boldsymbol{q}_2)$ is the pseudoinverse of $K(\boldsymbol{q}_1, \boldsymbol{q}_2; \boldsymbol{r})$. Using the result (4–1), the pseudoinverse K^+ may be seen to be given by

$$K^+(\boldsymbol{r}; \boldsymbol{q}_1, \boldsymbol{q}_2) = \int d^2 Q d^2 Q' \frac{1}{\sigma_{\boldsymbol{QQ}'}} f_{\boldsymbol{QQ}'}(\boldsymbol{r}) g_{\boldsymbol{QQ}'}^*(\boldsymbol{q}_1, \boldsymbol{q}_2). \quad (4\text{--}6)$$

Substituting (4–3) and (4–4) into (4–6) and using the spectral decomposition

$$\int d^2Q' \frac{1}{\sigma^2_{QQ'}} C_{Q'}(q;Q)C^*_{Q'}(q';Q) = M^{-1}(q,q';Q),$$

where $M^{-1}(q,q';Q)$ is the qq' matrix element of $M^{-1}(Q)$ we obtain

$$\eta^+(r) = \int d^2q_1 d^2q_2 d^2q'_2 \int d^2Q \exp(-iQ \cdot \rho)\,\delta(Q + q_2 - q_1)$$
$$\times\, M^{-1}(q_2,q'_2;Q)\kappa^*(Q + q'_2,q'_2;z)\Phi(q_1,q_2), \quad (4\text{–}7)$$

which is the inversion formula for scalar near-field tomography.

4C. Vector Case. Consider the vector integral equation

$$\Phi_\alpha(q_1,q_2) = \int d^3r K_\alpha(q_1,q_2;r)\eta(r),$$

where

$$K_\alpha(q_1,q_2;r) = \exp\big(i(q_1-q_2)\cdot\rho\big)\kappa_\alpha(q_1,q_2;z).$$

The functional form of $\kappa_\alpha(q_1,q_2;r)$ is determined by the experimental modality which is under consideration. As in the scalar case, it is assumed that $\Phi_\alpha(q_1,q_2)$ is specified for (q_1,q_2) in some data set and an appropriate blocking function $\chi(q_1,q_2)$ is introduced. The vector integral equation (3–9) differs from its scalar counterpart (3–4) only by a factor associated with the polarization. Evidently, by measuring a fixed component of the scattered field we see that the scalar inversion formula (4–7) may be used to reconstruct $\eta(r)$.

The SVD for the general vector case may be obtained by an analysis similar to the scalar case. Following the previous development it may be seen that the SVD of $K_\alpha(q_1,q_2;r)$ takes the form

$$K_\alpha(q_1,q_2;r) = \int d^2Q d^2Q' \sigma_{QQ'} f^*_{QQ'}(r)g^\alpha_{QQ'}(q_1,q_2).$$

Here the singular functions are given by

$$g^\alpha_{QQ'}(q_1,q_2) = C^\alpha_{Q'}(q_2;Q)\delta(Q + q_2 - q_1),$$
$$f_{QQ'}(r) = \frac{1}{\sigma_{QQ'}} \int d^2q \exp(-iQ \cdot \rho)\kappa^*_\alpha(Q + q,q;z)C^{\alpha*}_{Q'}(q;Q).$$

The $C^\alpha_{Q'}(q_2;Q)$ are eigenfuntions of $M_{\alpha\beta}(q_2,q'_2;Q)$ with eigenvalues $\sigma^2_{QQ'}$

$$\int d^2q' M_{\alpha\beta}(q,q';Q)C^\beta_{Q'}(q';Q) = \sigma^2_{QQ'}C^\alpha_{Q'}(q;Q),$$

where

$$M_{\alpha\beta}(q_2,q'_2;Q) = \int_0^L dz\kappa_\alpha(Q + q_2,q_2;z)\kappa^*_\beta(Q + q'_2,q'_2;z).$$

The pseudoinverse solution to the integral equation (3–9) is given by

$$\eta^+(\boldsymbol{r}) = \int d^2q_1 d^2q_2 K_\alpha^+(\boldsymbol{r}; \boldsymbol{q}_1, \boldsymbol{q}_2) \Phi_\alpha(\boldsymbol{q}_1, \boldsymbol{q}_2),$$

where

$$K_\alpha^+(\boldsymbol{r}; \boldsymbol{q}_1, \boldsymbol{q}_2) = \int d^2Q d^2Q' \frac{1}{\sigma_{\boldsymbol{Q}\boldsymbol{Q}'}} f_{\boldsymbol{Q}\boldsymbol{Q}'}(\boldsymbol{r}) g_{\boldsymbol{Q}\boldsymbol{Q}'}^{\alpha*}(\boldsymbol{q}_1, \boldsymbol{q}_2).$$

More explicitly,

$$\eta^+(\boldsymbol{r}) = \int d^2q_1 d^2q_2 d^2q_2' \int d^2Q \exp(-i\boldsymbol{Q} \cdot \boldsymbol{\rho}) \delta(\boldsymbol{Q} + \boldsymbol{q}_2 - \boldsymbol{q}_1)$$
$$\times \left(M^{-1}(\boldsymbol{Q})\right)_{\alpha\beta}(\boldsymbol{q}_2, \boldsymbol{q}_2') \kappa_\alpha^*(\boldsymbol{Q} + \boldsymbol{q}_2', \boldsymbol{q}_2'; z) \Phi_\beta(\boldsymbol{q}_1, \boldsymbol{q}_2), \quad (4\text{--}8)$$

which is the inversion formula for vector near-field tomography.

4D. Regularization and Resolution. In order to avoid numerical instabilty and set the resolution of the reconstructed image to be comensurate with the available data, the SVD inversion formulas must be regularized. In particular, $1/\sigma$ is replaced in the inversion formulas (4–7) and (4–8) by $R(\sigma)$ where $R(\sigma)$ is a suitable regularizer The role of regularization is to limit the contribution of small singular values to the reconstruction. This has the effect of replacing an ill-posed problem with a well-posed one that closely approximates the original. A simple choice for $R(\sigma)$ consists of truncation, that is,

$$R(\sigma_{\boldsymbol{Q}\boldsymbol{Q}'}) = \begin{cases} \sigma_{\boldsymbol{Q}\boldsymbol{Q}'}^{-1} & \text{if } \sigma_{\boldsymbol{Q}\boldsymbol{Q}'} \geq \sigma_{\min}, \\ 0 & \text{if } \sigma_{\boldsymbol{Q}\boldsymbol{Q}'} < \sigma_{\min}. \end{cases}$$

for some $\sigma_{\min}$. If Tikhonov regularization is used

$$R(\sigma_{\boldsymbol{Q}\boldsymbol{Q}'}) = \frac{\sigma_{\boldsymbol{Q}\boldsymbol{Q}'}}{\lambda + \sigma_{\boldsymbol{Q}\boldsymbol{Q}'}^2},$$

where λ is the regularization parameter. This choice leads to smoothing of $\eta^+(\boldsymbol{r})$ by penalizing functions with large L^2 norm. Other regularization schemes may be appropriate depending on the noise model and experimental particulars.

Regularization effectively filters the reconstructed image. In the near-field inverse scattering problem, the evanescent waves that decay most rapidly tend to be filtered out. Since the most rapidly decaying evanescent waves are also the waves on which the high spatial frequency information is encoded, regularization limits the resolution achievable.

4E. Numerical Simulations. The preceding results may be better understood with the aid of numerical simulations of reconstructions in several different forms of near-field tomography. Collection mode scalar near-field scanning tomography is considered first [26]. The model system consists of a three-dimensional distribution of six point scatterers, two on the horizontal, vertical or diagonal axis of each of three planes as shown in Figure 2. For simplicity the indices of refraction of both half-spaces are chosen to have the vacuum value of unity. The field was

Figure 7. Reconstructed tomographs of the scattering object shown in Figure 2.

sampled on a lattice with spacing $\lambda/20$ and the sample was illuminated with 21 different plane waves. Figure 7 shows the reconstructions obtained. Observe that the simulated NSOM image obtained with the tip withdrawn a distance $\lambda/4$ from the nearest plane of scatterers, as shown in Figure 2, is blurred, whereas the scatterers in the reconstructions are clearly identifiable. Because the high frequency components of the field fall off exponentially with distance from the scatterer, the resolution of the image is dependent on the depth of the slice, the deeper layers being less well resolved.

Figure 8 explores the robustness of the inversion procedure in the presence of noise [26]. Two point scatterers are located a distance 0.51λ from the scan plane with noise added to the signal at various levels as indicated. The noise was taken to be Gaussian and of zero mean, with a variance proportional to the square of the signal at each pixel on the measurement plane.

Next, total internal reflection tomography with scalar waves is considered. The reconstruction of $\eta(\boldsymbol{r})$ for a collection of spherical scatterers was performed. The forward data was calculated by considering the scattering of evanescent waves from a homogeneous sphere including multiple scattering by means of a partial wave expansion. For a sphere of radius a centered at $\boldsymbol{r}_0$ with refractive index n, it may be found that

$$A(\boldsymbol{k}_1, \boldsymbol{k}_2) = \exp(i(\boldsymbol{k}_1 - \boldsymbol{k}_2) \cdot \boldsymbol{r}_0) \sum_{\ell=0}^{\infty} (2\ell + 1) A_\ell P_\ell(\hat{\boldsymbol{k}}_1 \cdot \hat{\boldsymbol{k}}_2), \qquad (4\text{--}9)$$

where A_ℓ are the usual partial wave expansion coefficients(see, for instance [50]), P_ℓ are the Legendre polynomials and the caret has the meaning $\hat{\boldsymbol{k}} = \boldsymbol{k}/\sqrt{\boldsymbol{k} \cdot \boldsymbol{k}^*}$. To treat the scattering of evanescent waves, the argument of the Legendre polynomials in (4–9) must exceed unity. The series may nonetheless be shown to be convergent due to the rapid decay of the A_ℓ with increasing ℓ.

The forward data was obtained for a collection of four spheres of radius $\lambda/20$. All scatterers are present simultaneously in the forward simulation with inter-sphere scattering neglected. The spheres were arranged in two layers, one equatorial plane coincident with the $z = \lambda/20$ plane, the other with the $z = \lambda/4$ plane. In each layer, one sphere was taken to have index $n = 1.2$ and one sphere

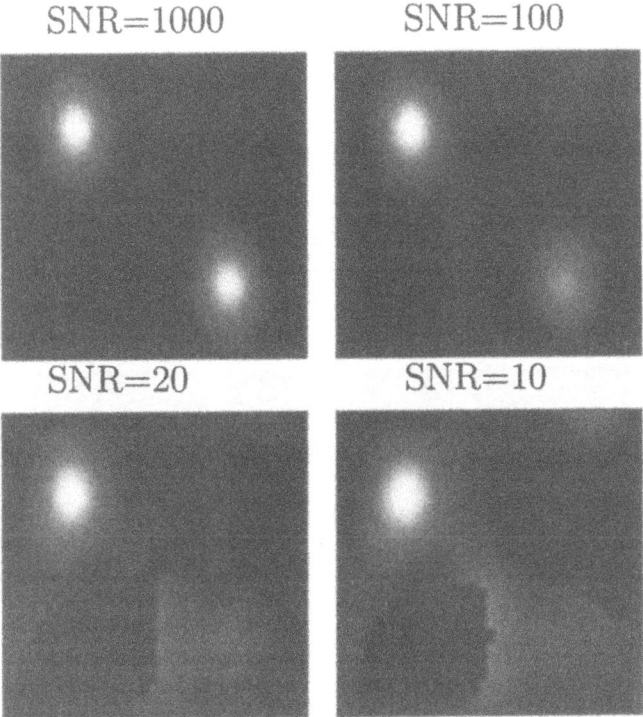

SNR=1000   SNR=100

SNR=20   SNR=10

Figure 8. Demonstrating reconstruction from noisy data sets. The SNR is the ratio of the magnitude of the signal to the standard deviation of the noise at each data point. The noise in the data set for each image in identical except for a scale factor.

was taken to have index $n = 1.2 + 0.2i$. In each of the simulations complex Gaussian noise of zero mean was added to the signal at various levels as indicated. Simulations were performed for two different prisms, one (Figure 9) with an index of $n = 10$, as might be encountered in the infrared, and another (Figure 10) with an index of $n = 4$. Further details are available in [25].

It may be seen from the reconstructions that the real and imaginary parts of the susceptibility may be found separately and that the reconstructions are subwavelength resolved. The resolution depends both on the size of the regulariztion parameter that indirectly sets the number of singular functions used in the reconstruction, and on the depth, a consequence of the fact that the probe fields decay exponentially into the sample resulting in the loss of high frequency Fourier components of the susceptibility. The tomographs at the $z = \lambda/20$ layer are more highly resolved for the higher index prism than the lower index prism, but there is little difference at the $z = \lambda/4$ layer.

Next considered is photon scanning tunneling tomography for vector waves. Two point scatterers are located on the prism face separated by 0.3λ. The field and intensity were computed in the measurement plane for three scan heights,

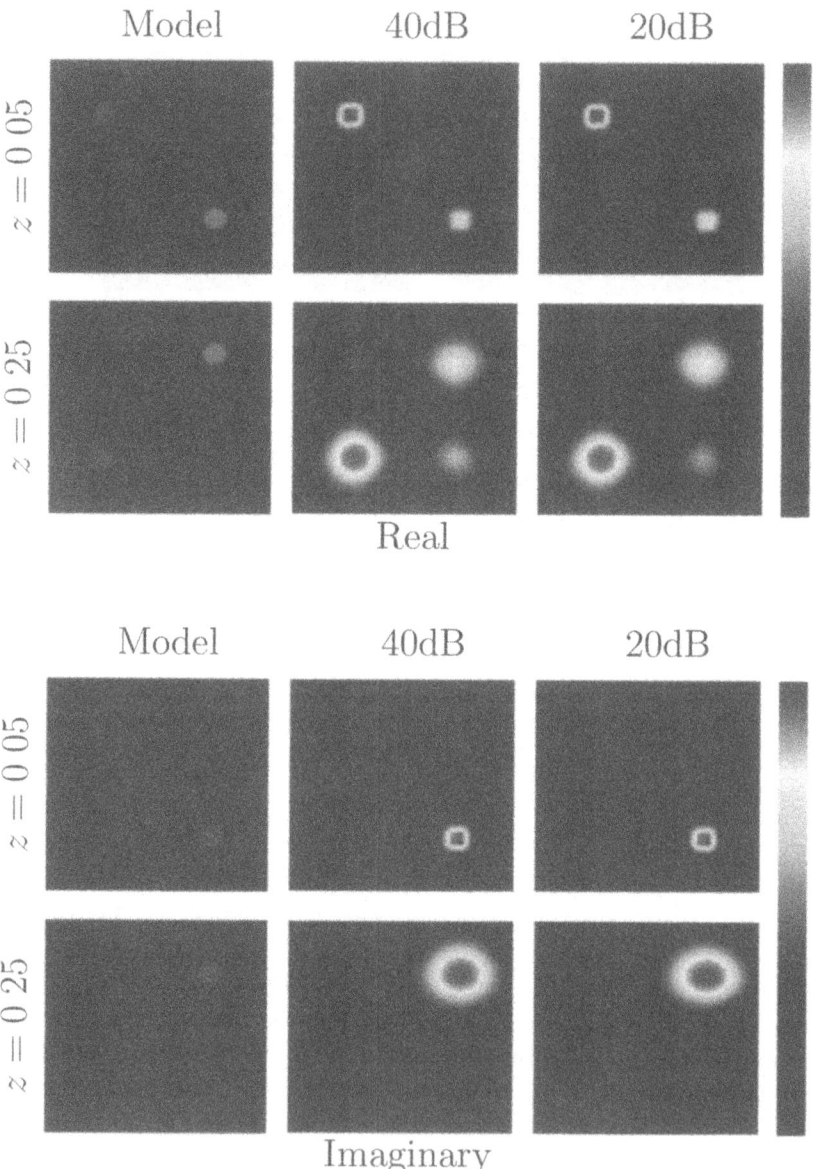

Figure 9. The reconstructed tomographs of the real and imaginary parts of the susceptibility using a prism of refractive index $n = 10$. The signal to noise ratio is given in dB above each column. The images were plotted using the linear color scale indicated to the right. The field of view in each image is $\lambda \times \lambda$.

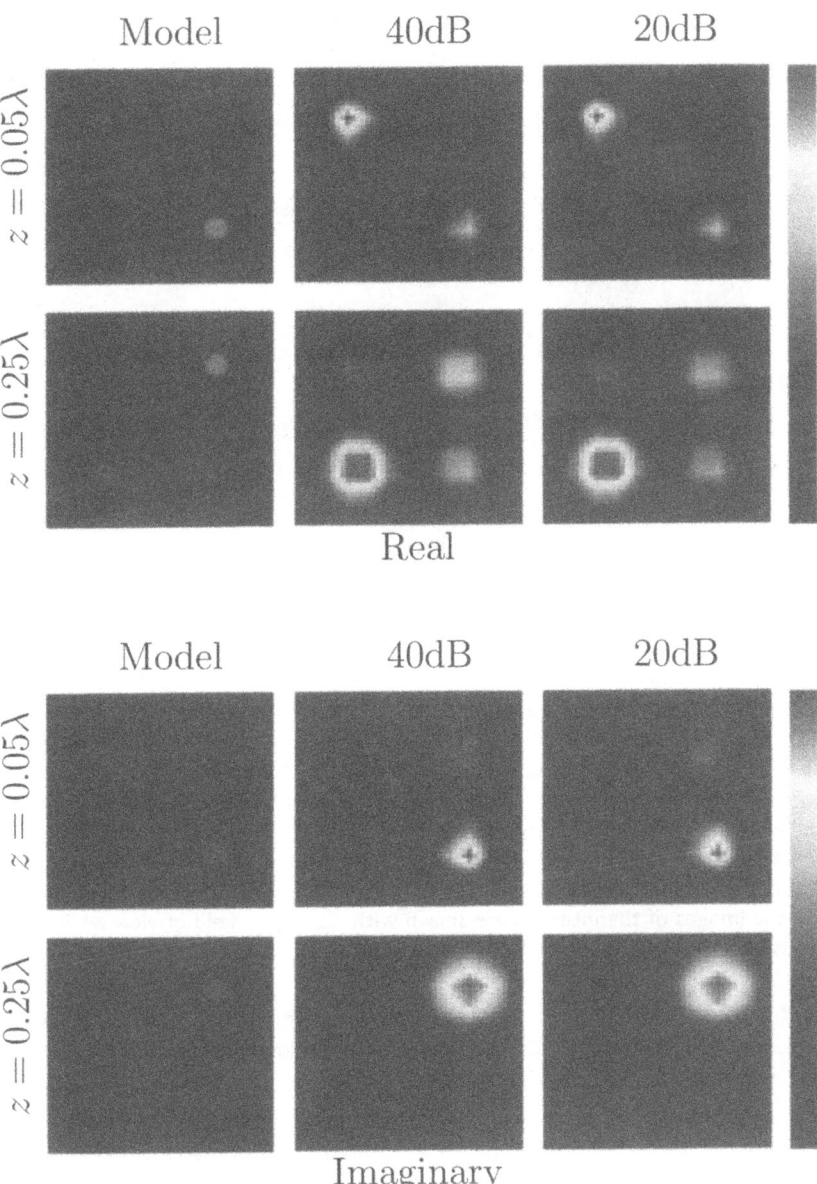

Figure 10. The reconstructed tomographs of the real and imaginary parts of the susceptibility using a prism of refractive index $n = 4$. All other parameters are as indicated in Figure 9.

$q_x = 2.5k_0$ $q_x = 0$ reconstructed

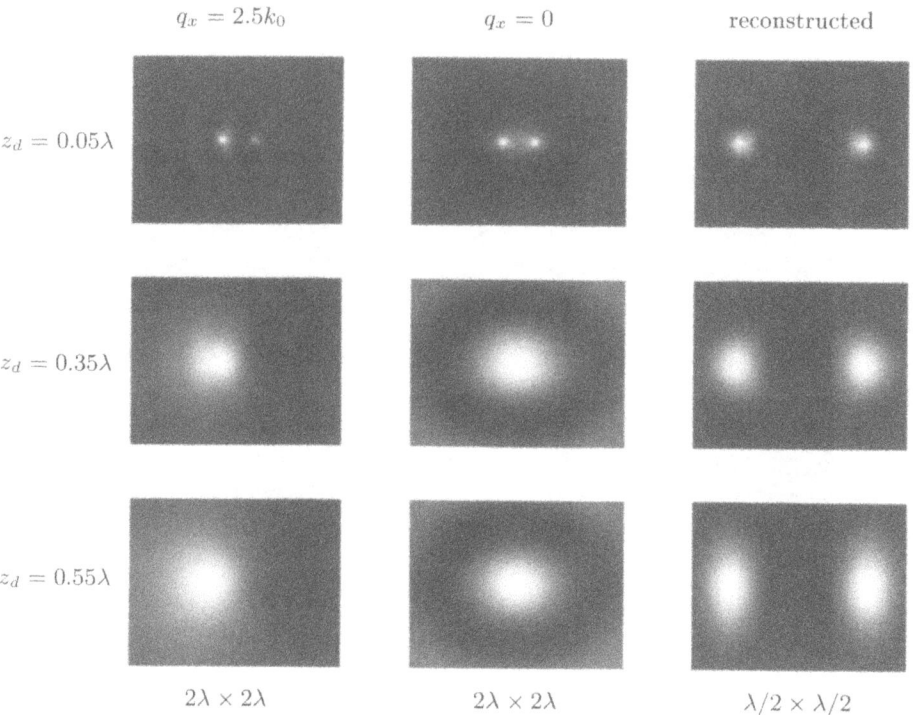

$2\lambda \times 2\lambda$ $2\lambda \times 2\lambda$ $\lambda/2 \times \lambda/2$

Figure 11. Demonstrating the observable intensities and reconstructed images for two point scatterers separated by 0.3λ with the measurement plane at various distances from the prism face. The left column shows the simulated intensity in the measurement plane $z = z_d$ for an illuminating field with transverse wave vector, q_x set to the maximum value attainable in a prism with index $n = 2.5$. The middle column shows the simulated intensity with the illuminating wave incident normal to the plane of observation. The right column displays the image reconstructed from multiple views obtained with different illuminating fields. Note that the images of the intensity are shown with a $2\lambda \times 2\lambda$ field of view while the reconstructed image is shown with a $\lambda/2 \times \lambda/2$ field of view.

z_d, of the probe. Only the TE polarization, *i.e.* the polarization vector parallel to the prism face, is used. The scattered field is computed on a $4\lambda \times 4\lambda$ window and sampled on a cartesian grid at a spacing of $\lambda/10$ with a total of 41 illuminating plane waves all with $q_y = 0$ and q_x on equally spaced points in the range $[-2.5k_0, 2.5k_0]$, corresponding to a range attainable with a prism of index $n = 2.5$. The computed fields are shown in Figure 11 in a $2\lambda \times 2\lambda$ field of view, while the reconstructed scatterer is shown in a $\lambda/2 \times \lambda/2$ field of view. The scan made furthest from the prism face with an illuminating evanescent wave may be seen to be dominated by the scattered wave due to the exponential decay of the illuminating field. It may be observed that the object structure, which is unclear in the direct measurements made farthest from the sample, is still clearly evident in the reconstructions. The increased spread of the points may

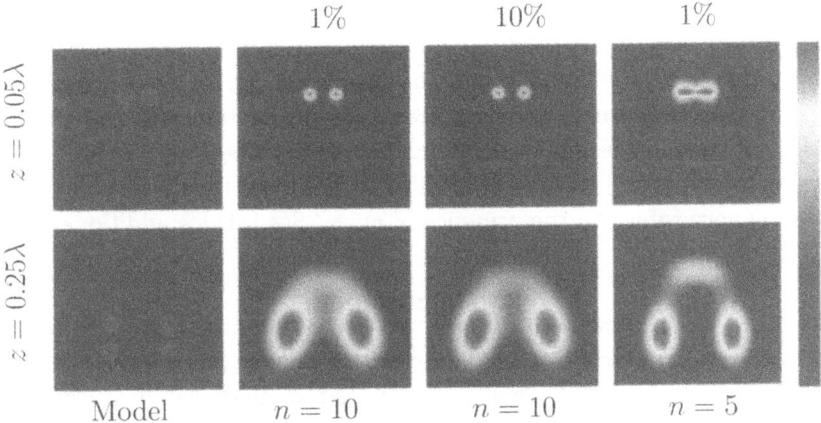

Figure 12. The simulated tomographs in power extinction tomography. The field of view is $\lambda \times \lambda$ in each image. The scatterers used in the forward simulation are shown in the column labeled Model. The numbers across the top indicate the level of noise relative to the signal. The indices listed across the bottom indicate the index of refraction of the prism used in the simulations. Each reconstruction was normalized by its maximum value and imaged using the linear color scale shown to the right.

be attributed to the loss of high spatial frequency components of the scattered field to exponential decay of evanescent waves. See [27] for further information.

Finally, reconstructions for near-field power extinction tomography with scalar waves are simulated. To demonstrate the feasibility of the inversion, the reconstruction of $\alpha(r)$ for a collection of spherical scatterers have been numerically simulated. The forward data was calculated from the partial wave expansion (4–9). The forward data was obtained for a collection of six spheres of radius $\lambda/20$ and index of refraction $n = 1.1 + 0.2i$, distributed on three planes as shown in Figure 12. All scatterers are present simultaneously in the forward simulation with inter-sphere scattering neglected. Simulations of experiments done with two different prisms, one with index of refraction $n = 5$ the other with $n = 10$ are presented. The reconstructions obtained at depths of $.05\lambda$, and $.25\lambda$ which correspond to the two separate equatorial planes of the original distribution of scatterers are displayed. Complex Gaussian noise of zero mean was added to the data function at various levels as indicated.

The resolution of the reconstruction is seen to be controlled by several factors including the index of the prism, the depth of the slice, and choice of regularization parameters. These effects may be understood by observing that the resolution is governed by the low pass filtering that is inherent in the transverse Fourier integral in (4–7) and additionally by the exponential decay of high-frequency components of the scattered field with increasing degree of evanescence. In general, a useful rule of thumb is that for a prism of index n the transverse resolution will be on the order of $\lambda/2n$ at a depth of $\lambda/2n$ after which it falls off linearly.

This is seen in the $n = 10$ simulations where the spheres whose edges are separated by $\lambda/20$ may be resolved in the slice at a depth of $\lambda/20$. However, the spheres in the next layer at $\lambda/4$ with the same spacing are not resolvable, but the groups of spheres which are spaced at $\lambda/4$ may be resolved. For the $n = 5$ case the scatterers in the top layer are not as well resolved, but the scatterers in the deeper layer are well resolved. It may be observed that the reconstruction algorithm is very robust in the presence of noise. See [51] for additional details.

5. Discussion

The mathematical structure of the near-field inverse scattering problem has been reviewed. It has been demonstrated that in the weak scattering limit, where the forward scattering problem may be linearized, an analytic solution to the inverse problem may be obtained. The data required to implement this method may be obtained from a variety of near-field optical experiments.

Several directions for further research are apparent. The solution to the inverse problem discussed here is based on a linearization of the forward probem by means of the first Born approximation. A solution applicable when multiple scattering becomes significant is desirable. An interesting novel method for inverse scattering in diffusion tomography beyond the linear model has recently been developed [52]. Diffusion tomography shares a great deal of formal mathematical structure with the near-field problem and so it may be possible to apply that method to near-field tomography. The treatment of sampling given here is adequate and outlines the basic approach, but more sophisticated sampling schemes, possibly even adaptive sampling methods, will need to be explored. Prior constraints on the sample may be used to great advantage in the inverse problem. In the work presented here, the finite thickness of the sample is always incorporated in the solution to the inverse problem. Such prior knowledge lends stability to the solution to the inverse problem and greatly improves the imaging of those parts of the sample farthest from the scan plane or the prism face. Methods to include other prior information may be expected to be similarly useful.

Near-field tomography offers improved imaging tools for a wide range of disciplines including the rapidly developing areas of research in nanotechnology. Applications may also be found in nonimaging optics as well. For instance, near-field tomography may provide a means to read out three dimensional optical data storage devices with data encoded on sub-wavelength scales. Beyond these applications, the work presented here provides new insight into the physics of highly localized wave fields and the propagation of light on very small scales.

Acknowledgments

Much of the work described in this review was carried out while we were members of the Department of Electrical Engineering at Washington University in St. Louis. We would like to express our gratitude to Professors J. A. O'Sullivan, D. L. Snyder, and B. E. Spielman for their support. It is also a pleasure to thank Professor V. A. Markel for stimulating discussions. PSC wishes to acknowledge support from the US National Aeronautics and Space Administration under Grant NAG3-2764.

Appendix

The scalar Green's function in the half-space geometry is given by

$$G(r, r') = \frac{i}{2\pi} \int d^2q \, k_z^{-1}(q)\{1 + R(q)\exp(2ik_z(q)z')\}\exp(ik(q) \cdot (r - r')).$$

Here $R(q)$ is the reflection coefficient

$$R(q) = \frac{k_z(q) - k_z'(q)}{k_z(q) + k_z'(q)},$$

with $k_z(q) = \sqrt{k_0^2 - q^2}$, $k_z'(q) = \sqrt{n^2 k_0^2 - q^2}$, and $k(q) = (q, k_z(q))$.

The Green's tensor in the half-space geometry is given by

$$G_{\alpha\beta}(r, r') = \frac{i}{2\pi} \int \frac{d^2q}{k_z(q)} S_{\alpha\gamma}^{-1}(q)\tilde{g}_{\gamma\delta}(q)S_{\delta\beta}(q)\exp(ik(q) \cdot (r - r')),$$

where $S(q)$ is the matrix that rotates $k(q)$ into the xz plane, or more explicitly

$$S(q) = |q|^{-1} \begin{pmatrix} q_x & q_y & 0 \\ -q_y & q_x & 0 \\ 0 & 0 & |q| \end{pmatrix}$$

and

$$\tilde{g}_{xx} = \left(\frac{k_z(q)}{k_0}\right)^2 (1 + R'(q)\exp(2ik_z(q)z')),$$

$$\tilde{g}_{yy} = 1 + R(q)\exp(2ik_z(q)z'),$$

$$\tilde{g}_{zz} = \left(\frac{|q|}{k_0}\right)^2 (1 - R'(q)\exp(2ik_z(q)z')),$$

$$\tilde{g}_{zx} = -\frac{|q|k_z(q)}{k_0^2}(1 + R'(q)\exp(2ik_z(q)z')),$$

$$\tilde{g}_{xz} = -\frac{|q|k_z(q)}{k_0^2}(1 - R'(q)\exp(2ik_z(q)z')),$$

all other elements of g being zero. In addition,

$$R'(q) = \frac{k_z'(q) - nk_z(q)}{k_z'(q) + nk_z(q)}.$$

Bibliography

[1] E. Abbe. *Archiv f. Mikroscopische Anat.*, 9:413, 1873.

[2] Lord Rayleigh. *Phil. Mag.*, 8:261, 1879.

[3] D. Courjon, K. Sarayeddine, and M. Spajer. Scanning tunneling optical microscopy. *Opt. Comm.*, 71:23–28, 1989.

[4] C. Girard and A. Dereux. Near-field optics theories. *Rep. Prog. Phys.*, 59:657–699, 1996.

[5] J-J. Greffet and R. Carminati. Image formation in near-field optics. *Prog. Surf. Sci.*, 56:133–237, 1997.

[6] E. Synge. A suggested method for extending microscopic resolution into the ultra-microscopic region. *Phil. Mag.*, 6:356–362, 1928.

[7] E. Ash and G. Nicholls. Super-resolution aperture scanning microscope. *Nature*, 237:510–512, 1972.

[8] A. Lewis, M. Isaacson, A. Harootunian, and A. Muray. Development of a 500 Å spatial resolution light microscope. I. Light is efficiently transmitted through $\lambda/16$ diameter apertures. *Ultramicroscopy*, 13:227–231, 1984.

[9] D. W. Pohl, W. Denk, and M. Lanz. Optical stethoscopy: Image recording with resolution $\lambda/20$. *Appl. Phys. Lett.*, 44:651–653, 1984.

[10] E. Betzig and J. K. Trautman. Near-field optics: microscopy, spectroscopy, and surface modification beyond the diffraction limit. *Science*, 257:189–195, 1992.

[11] R. Dickson, D. Norris, Y-L. Tzeng, and W. Moerner. Three-dimensional imaging of single molecules solvated in pores of poly(acrylamide) gels. *Science*, 274:966–969, 1996.

[12] N. Garcia and M. Nieto-Vesperinas. Near-field optics inverse-scattering reconstruction of reflective surfaces. *Opt. Lett.*, 18:2090–2092, 1993.

[13] N. Garcia and M. Nieto-Vesperinas. Direct solution to the inverse scattering problem for surfaces from near-field intensities without phase retrieval. *Opt. Lett.*, 20:949–951, 1995.

[14] R. Carminati and J-J. Greffet. Reconstruction of the dielctric contrast profile from near-field data. *Ultramicroscopy*, 61:11–16, 1995.

[15] R. Carminati, J-J. Greffet, N. Garcia, and M. Nieto-Vesperinas. Direct reconstruction of surfaces from near-field intensity under spatially incoherent illumination. *Opt. Lett.*, 21:501–503, 1996.

[16] J-J. Greffet, A. Sentenac, and R. Carminati. Surface profile reconstruction using near-field data. *Opt. Commun.*, 116:20–24, 1995.

[17] C. W. McCutchen. Optical systems for observing surface topography by frustrated total internal reflection and interference. *Rev. Sci. Instr.*, 35:1340–1345, 1964.

[18] P. A. Temple. Total internal reflection microscopy: a surface inspection technique. *Appl. Opt.*, 20:2656–2664, 1981.

[19] P. J. Sides and J. Lo. Measurement of linear nanometric distances between smooth plane parallel bodies by total internal reflection. *Appl. Phys. Lett.*, 69:141–142, 1996.

[20] G. E. Cragg and P. T. C. So. Standing wave total internal reflection microscopy - breaking the diffraction resolution limit. *Biophys. J.*, 78:248a, 2000.

[21] P. T. C. So, H-S. Kwon, and C. Dong. Resolution enhancement in standing-wave total internal reflection microscopy: a point-spread-function engineering approach. *Journ. Opt. Soc. Am. A*, 18:2833–2845, 2001.

[22] R. C. Reddick, R. J. Warmack, and T. L. Ferrell. New form of scanning optical microscopy. *Phys. Rev.*, B39:767–770, 1989.

[23] J. Guerra and W. Plummer. U.s. patenet no. 4681451 (1987).

[24] E. Wolf. Principles and development of diffraction tomography. In Anna Consortini, editor, *Trends in Optics*, pages 83–110. Academic Press, San Diego, 1996.

[25] P. S. Carney and J. C. Schotland. Three-dimensional total internal reflection microscopy. *Opt. Lett.*, 26:1072–1074, 2001.

[26] P. S. Carney and J. C. Schotland. Inverse scattering for near-field microscopy. *Appl. Phys. Lett.*, 77:2798–2800, 2000.

[27] P. S. Carney and J. C. Schotland. Determination of three-dimensional structure in photon scanning tunneling microscopy. *Journ. Opt. A: Pure Appl. Opt.*, 4:S140–S144, 2002.

[28] P. S. Carney, E. Wolf, and G. S. Agarwal. Statistical generalizations of the optical theorem with applications to inverse scattering. *Journ. Opt. Soc. Am. A*, 14:3366–3371, 1997.

[29] P. S. Carney, E. Wolf, and G. S. Agarwal. Diffraction tomography using power extinction measurements. *Journ. Opt. Soc. Am. A*, 16:2643–2648, 1999.

[30] P. S. Carney, V. A. Markel, and J. C. Schotland. Near-field tomography without phase retrieval. *Phys. Rev. Lett.*, 86:5874–5876, 2001.

[31] Z. Yu and S. Boseck. Scanning acoustic microscopy and its applications to material characterization. *Rev. Mod. Phys.*, 67:863–891, 1995.

[32] G. A. D. Briggs. *Acoustic Microscopy*. Clarendon, Oxford, 1992.

[33] R. A. Lemons and C. F. Quate. In W. P. Mason and R. N. Thurston, editors, *Physical Acoustics Vol. 14*, page 1. Academic, London, 1979.

[34] P. C. Clemmow. *The Plane Wave Spectrum Representation of Electromagnetic Fields*. Pergamon Press, Oxford, 1996.

[35] H. Weyl. Ausbreitung elektromagnetischer wellen über einem ebenen leiter. *Ann. Phys. (Leipzig)*, 60:481–500, 1919.

[36] P. S. Carney, J. C. Schotland, and E. Wolf. Reflection, transmission and extinction of optical power: the generalized optical theorem. *in preparation*, 2002.

[37] M. Born and E. Wolf. *Principles of Optics*. Cambridge University Press, 7th edition, 1999.

[38] E. Feenberg. The scattering of slow electrons by neutral atoms. *Phys. Rev.*, 40:40–54, 1932.

[39] H. C van de Hulst. On the attenuation of plane waves by obstacles of arbitrary size and form. *Physica*, pages 740–746, 1949.

[40] A. A. Maradudin and D. L. Mills. Scattering and absorption of electromagnetic radiation by a semi-infinite medium in the presence of surface roughness. *Phys. Rev. B*, 11:1392–1415, 1975.

[41] P. S. Carney. *Optical Theorems in Statistical Wavefields with Applications.* PhD thesis, University of Rochester, Rochester,NY, June 1999.

[42] S. I. Bozhevolnyi and B. Vohnsen. Near-field optical holography. *Phys. Rev. Lett.*, 71:3351–3355, 1996.

[43] P. Hariharan. *Optical Holography.* Cambridge University Press, 1996.

[44] D. Gabor. A new microscopic principle. *Nature*, 161:777–778, 1948.

[45] D. Gabor. Microscopy by reconstructed wavefronts. *Proc. Roy. Soc. A*, 197:454–487, 1949.

[46] P. L. Phillips, J. C. Knight, J. M. Pottage, G. Kakarantzas, and P. St J. Russel. Direct measurement of optical phase in the near field. *Appl. Phys. Lett.*, 76:541–543, 2000.

[47] E. N. Leith and J. Upatnieks. Reconstructed wavefronts and communication theory. *J. Opt. Soc. Am*, 52:1123–1130, 1962.

[48] E. N. Leith and J. Upatnieks. Wavefront reconstruction with continuous-tone objects. *J. Opt. Soc. Am*, 53:1377–1381, 1963.

[49] F. Natterer. *The Mathematics of Computerized Tomography.* Wiley, New York, 1986.

[50] C. Cohen-Tannoudji, B. Diu, and F. Laloe. *Quantum Mechanics.* Hermann, Paris, 1977.

[51] P. S. Carney, V. A. Markel, and J. C. Schotland. Near-field tomography without phase retrieval. *Phys. Rev. Lett.*, 86:5874–5876, 2001.

[52] V. A. Markel and J. C. Schotland. Inverse problem in optical diffusion tomography IV: Nonlinear inversion formulas. *Journ. Opt. Soc. Am. A*, 20: 903–912, 2003.

P. SCOTT CARNEY
UNIVERSITY OF ILLINOIS AT URBANA–CHAMPAIGN
URBANA, IL 61801
UNITED STATES
carney@uiuc.edu

JOHN C. SCHOTLAND
UNIVERSITY OF PENNSYLVANIA
PHILADELPHIA, PA 19104
UNITED STATES
schotland@seas.upenn.edu

Inverse Problems for Time Harmonic Electrodynamics

PETRI OLA, LASSI PÄIVÄRINTA, AND ERKKI SOMERSALO

ABSTRACT. We study the inverse boundary value and inverse scattering problems for time-harmonic Maxwell's equations. The goal is to recover electromagnetic material parameters (permittivity, conductivity and permeability) in an unaccessible region of space from field measurements outside this region. We review the known results concerning the isotropic material parameters. Maxwell's equations are formulated here using differential forms. This representation is found particularly useful when anisotropies are allowed.

Introduction

In his famous article *A Dynamical Theory of Electromagnetic Field* of 1864 James Clerk Maxwell wrote down differential equations that describe the laws of electromagnetism in full generality. The four equations of Maxwell,

$$\nabla \cdot \boldsymbol{D}(x,t) = \rho(x,t), \tag{0-1}$$

$$\nabla \cdot \boldsymbol{B}(x,t) = 0, \tag{0-2}$$

$$\frac{\partial \boldsymbol{B}(x,t)}{\partial t} + \nabla \times \boldsymbol{E}(x,t) = 0, \tag{0-3}$$

$$-\frac{\partial \boldsymbol{D}(x,t)}{\partial t} + \nabla \times \boldsymbol{H}(x,t) = \boldsymbol{J}(x,t), \tag{0-4}$$

describe the dynamics of the five vector fields $\boldsymbol{E}$, $\boldsymbol{D}$, $\boldsymbol{B}$, $\boldsymbol{H}$ and $\boldsymbol{J}$. Here $\boldsymbol{E}(x,t)$ is the electric field, $\boldsymbol{D}(x,t)$ the *electric displacement*, $\boldsymbol{B}$ the *magnetic induction* or *magnetic flux density*, $\boldsymbol{H}(x,t)$ is the *magnetic field* and, finally, $\boldsymbol{J}(x,t)$ is the *electric current density*. Since modern vector calculus was unknown to Maxwell, he formulated these equations as twenty scalar equations. The present form of these equations originates from Oliver Heaviside from the 1880's.

Equation (0–1) is Gauss' law and it says that infinitesimally the total flux of the electric displacement is equal to the density of free charges. The scalar field ρ here is the *free charge density*. Equation (0–2) is the magnetic analogue of Gauss' law saying that there are no free magnetic charges. Equation (0–3), called Faraday's law, explains how a changing magnetic flux creates an electric current in a conductive loop, a law that is based on a series of experiments that Faraday performed during 1831 and 1832. At that time the phenomenon dual to Faraday's law of induction was known as Ampère's law. It explains how an electric current in a loop creates a magnetic field and in our notation reads

$$\nabla \times \boldsymbol{H}(x,t) = \boldsymbol{J}(x,t). \tag{0–5}$$

The asymmetry in the equations (0–3) and (0–5) worried Maxwell and he started to think about Faraday's idea of polarization. Under the influence of an electric field a medium starts to polarize. This results in a small change in the position of charges and hence an electric current. He added a new current term $\frac{\partial}{\partial t}\boldsymbol{D}(x,t)$ to Ampère's law and as a result of purely theoretical reasoning discovered, among other things, electromagnetic waves. The existence of these waves was later verified by the experiments of Herz.

We call equations (0–1) to (0–4) macroscopic, because they deal directly with observable physical quantities and explain how they are related to each other. In particular, the structure of the medium is of no consequence. Also, without any additional assumptions, these equations are not enough to determine the fields uniquely, as a moment's reflection reveals. In addition to his four differential equations, Maxwell described four so-called *structural* or *constitutive equations* that relate $\boldsymbol{E}$ with $\boldsymbol{D}$, $\boldsymbol{B}$ with $\boldsymbol{H}$ and $\boldsymbol{J}$ with $\boldsymbol{E}$:

$$\boldsymbol{D}(x,t) = \varepsilon(x)\boldsymbol{E}(x,t),$$
$$\boldsymbol{B}(x,t) = \mu(x)\boldsymbol{H}(x,t),$$
$$\boldsymbol{J}(x,t) = \boldsymbol{J}_0(x,t) + \sigma(x)\boldsymbol{E}(x,t).$$

Here $\varepsilon(x)$ is the *electric permittivity* or *dielectricity*, $\mu(x)$ the *magnetic permeability* and $\sigma(x)$ the *electric conductivity*. The current density is divided in two parts, $\boldsymbol{J}_0$ being the forced current density, while the second term is the ohmic (or volume) current density driven by the electric field. Roughly speaking, ε expresses the tendency of the material to form electric dipoles under the influence of an external electric field, while the conductivity is related to the mobility of free charges in the material. The permeability μ is analogous to ε, expressing the magnitude in which the material is forming magnetic dipoles in an external magnetic field.

The goal in electromagnetic inverse problems is to determine these parameters in an inaccessible region in a noninvasive way from field measurements outside this region. The application areas include geophysical prospecting, nondestructive testing and medical imaging. As an example, we mention here the problem of detecting leukemia by using electromagnetic waves. This is made possible by the fact that leukemia causes a change of electric permittivity in the bone marrow by a factor of up to two. For more details, we refer to [3] and Chapter 2 in [1].

In this article we review the uniqueness results and reconstruction algorithms for time-harmonic fixed frequency inverse problems. This means that the time dependence of all fields is assumed to be $e^{-i\omega t}$, the frequency $\omega > 0$ being fixed. Instead of describing the electromagnetic fields as vector fields in a Euclidean space, we have chosen to define them as differential forms on a Riemannian manifold. This not only gives additional generality but also clarifies the nature of different physical fields. As an example, the electric displacement and magnetic induction have a physically well defined flux through a surface, hence they are integrable over two dimensional surfaces and consequently they correspond naturally to 2-forms. The formulation using forms makes obvious the invariance properties of Maxwell's equations. At the same time some formulas, like the radiation condition, are considerably simplified.

The structure of this article is as follows. In the first two sections we describe the problems to be considered starting from Maxwell's equations, and set up the mathematical framework that is going to be used. We also offer references to aspects of the problem that are not covered in detail in these notes. In Section 3 we rescale Maxwell's equations and then complete them into a Dirac type elliptic system. For a similar time domain formulation for Maxwell's equations in a more general setting, see [9]. In Section 4 we introduce the exponentially growing fundamental solution, and use this to find a large enough family of solutions that we can test the media with. In Section 5 we introduce an integration by parts formula that connects the parameters in the interior to our boundary measurement. To keep the reconstruction algrorithm constructive (at least mathematically), the next step is to show that our boundary data makes it possible to determine the Cauchy data of these special solutions, and this is done in Section 6. In the final section we explain how the unique determination of the parameters is proved.

1. Time-Harmonic Maxwell Equations

We now assume that the time dependence of all fields in (0–1) to (0–4) is harmonic with frequency $\omega > 0$, i.e., all time dependent fields above are of the form $f(x,t) = e^{-i\omega t} f(x)$. Cancelling out the oscillatory exponential we end up with the system

$$i\omega \, \boldsymbol{D}(x) + \nabla \times \boldsymbol{H}(x) = \boldsymbol{J}(x), \qquad (1\text{--}1)$$

$$\nabla \cdot \boldsymbol{D}(x) = \rho(x), \qquad (1\text{--}2)$$

$$-i\omega \, \boldsymbol{B}(x) + \nabla \times \boldsymbol{E}(x) = 0, \qquad (1\text{--}3)$$

$$\nabla \cdot \boldsymbol{B}(x) = 0. \qquad (1\text{--}4)$$

We intepret the electric field and magnetic field as 1-forms by identifying a vector field $\boldsymbol{F} = F_1\boldsymbol{i} + F_2\boldsymbol{j} + F_3\boldsymbol{k}$ with the 1-form (we are using the Einstein summation convention whenever convenient) $F = F_i dx^i$. To intepret the equations above in terms of forms we also have to identify vector fields with 2-forms as follows: The vector field $\boldsymbol{F} = F_1\boldsymbol{i} + F_2\boldsymbol{j} + F_3\boldsymbol{k}$ is identified with the 2-form $F_1 \left(dx^2 \wedge dx^3 \right) + F_2 \left(dx^3 \wedge dx^1 \right) + F_3 (dx^1 \wedge dx^2)$. In terms of the Euclidean Hodge-star operator $*_e$, this 2-form can be expressed as as $*_e(F_i dx^i)$. Equations (1–1) to (1–4) now take the form

$$i\omega D(x) + d\,H(x) = J(x), \tag{1–5}$$

$$d\,D(x) = \tilde{\rho}(x), \tag{1–6}$$

$$-i\omega\,B(x) + d\,E(x) = 0, \tag{1–7}$$

$$d\,B(x) = 0. \tag{1–8}$$

where $D = *_e(D_i dx^i)$, $B = *_e(B_i dx^i)$, and $\tilde{\rho} = \rho\,dV_e$, with $dV_e = dx^1 \wedge dx^2 \wedge dx^3$ is the Euclidean volume element.

For the moment we consider the system above on an arbitrary smooth differentiable and orientable three-manifold M, with or without a boundary. The microscopic structure of the medium in the domain is modelled by introducing a Riemannian metric g on M, and postulating that the magnetic induction and the electric displacement (which are 2-forms) are related to the magnetic and electric fields via

$$D = \gamma(x) * E, \qquad B = \mu(x) * H. \tag{1–9}$$

Here $*$ is the Hodge-star operator with recpect to the metric g, and μ and γ are smooth scalar functions,

$$\gamma(x) = \varepsilon(x) + i\,\frac{\sigma(x)}{\omega}.$$

Furthermore, we assume that ε and μ are equal to constants ε_0 and μ_0, respectively, outside a compact set, both are bounded and strictly positive, and σ is a nonnegative compactly supported function. In this formulation, the ohmic part of the current density J is merged with the electric displacement D, and J in (1–5) represents the forced current density J_0. Note that Maxwell's equations (1–5) to (1–8) are purely topological, i.e., there is no reference to the underlying metric. The metric properties appear, as expected. in the constitutive equations (1–9). As Kepler wrote in his 1602 thesis, albeit in a different context, *Ubi materia, ibi geometria*: Where there is matter, there is geometry.

In the sequel, we shall assume throughout that on the manifold M,

$$\tilde{\rho} = 0 \quad \text{and} \quad J_0 = 0.$$

We arrive at Maxwell's equations for the so called *perfect media*,

$$d\,H(x) + i\omega\gamma(x) * E(x) = 0, \tag{1–10}$$

$$d\gamma * E(x) = 0, \tag{1–11}$$

$$d\,E(x) - i\omega\mu(x) * H(x) = 0, \qquad\qquad (1\text{-}12)$$

$$d\,\mu * H(x) = 0. \qquad\qquad (1\text{-}13)$$

We remark that in reality not all media obey the constitutive relations (1–9) used here. First of all, in some applications the functions $\mu(x)$ and $\gamma(x)$ also depend on the frequency via so called *dispersion relations*. In the time domain, the frequency dependency corresponds to the *memory* of the matter, i.e., the responses of the material, such as the polarization, are not instantaneous but depend on the past values of the fields. Mathematically, this means that in the time domain, the constitutive relations become causal time convolutions. Secondly, not all media are *isotropic*. The medium is isotropic if one can choose $\gamma(x)$ and $\mu(x)$ to be scalar functions. For example muscle tissue is anisotropic and these functions have to be allowed to be more general tensors. We refer to [9] for a discussion of Maxwell's equations for forms in anisotropic media. Finally, the constitutive relations might be more complicated. For example, both D and B can depend on a linear combination of E and H, which leads to *chiral media*. The dependence can also be nonlinear. Such materials are in abundance in nature. For example, several crystals are chiral and metals in strong magnetic fields behave in a nonlinear fashion. In this work we limit ourselves to perfect media.

2. Inverse Problems

In this section we formulate the inverse boundary value problem as well as the inverse scattering problems for Maxwell's equations.

We start by fixing certain notations. Assume first that M is a smooth compact oriented 3-manifold with $\partial M \neq \varnothing$. We denote by $\Omega^k M$, $0 \leq k \leq 3$ the vector bundle of smooth k-forms on M. Let $i : \partial M \to M$ denote the canonical imbedding. We define the *tangential trace* of k-forms as

$$t : \Omega^k M \to \Omega^k \partial M, \ t\omega = i^*\omega \quad \text{for } \omega \in \Omega^k M, \ 0 \leq k \leq 2.$$

where i^* is the pull-back of i. Similarly, we define the *normal trace* as

$$n : \Omega^k M \to \Omega^{3-k} \partial M, \ n\omega = i^*(*\omega) \quad \text{for } \omega \in \Omega^k M, \ 1 \leq k \leq 3.$$

Observe that for 1-forms, the tangential component corresponds to the tangential component of the vector field while for 2-forms, it corresponds to the transversal flux through the boundary. For the normal trace, the roles are interchanged. For more precise discussion, see e.g. [20]. (In fact, the definition of the normal trace here differs from that given in the cited reference.)

Stokes' formula can be written now as follows: Let $\delta : \Omega^k M \to \Omega^{k-1} M$ denote the codifferential for k-forms,

$$\delta = (-1)^{n(k+1)+1} * d* = (-1)^k * d * \quad \text{for dimension } n = 3.$$

We denote the inner product of k-forms over M as

$$(\omega, \eta) = \int_M \omega \wedge *\bar{\eta},$$

while at the boundary we denote

$$\langle \omega, \eta \rangle = \int_{\partial M} \omega \wedge \bar{\eta} \quad \text{for } \omega \in \Omega^k \partial M, \ \eta \in \Omega^{2-k} \partial M.$$

We have the identity

$$(d\omega, \eta) - (\omega, \delta\eta) = \langle t\omega, n\eta \rangle \quad \text{for } \omega \in \Omega^k M, \ \eta \in \Omega^{k+1} M. \tag{2–1}$$

With these notations, we define the *admittance map* for Maxwell's equations at the boundary: Assume for simplicity that $\gamma - \varepsilon_0, \mu - \mu_0 \in C_0^\infty(\text{int}(M))$, i.e., the material parameters near the boundary ∂M are constants $\varepsilon_0 > 0$ and $\mu_0 > 0$, respectively. We define

$$\Lambda : t(\varepsilon_0^{1/2} E) \mapsto t(\mu_0^{1/2} H).$$

The inverse boundary value problem (IBP) we consider here can be stated as follows:

IBP. *From the knowledge of the admittance map Λ at the boundary, determine the material parameters γ and μ in M.*

One can also consider this problem for chiral media; see [11]. For anisotropic media, if γ and μ are conformally related to each other, then the linarization suggests that the nonuniqueness arises solely from boundary preserving diffeomorphisms of M to itself, see [22]. This result is proved in the time domain in [9].

Equally natural is the inverse scattering problem. For simplicity, we assume here that $M = \mathbb{R}^3$ endowed with the Euclidean metric $g = g_e$, and furthermore, $\varepsilon(x) = \varepsilon_0$ and $\mu(x) = \mu_0$ for $x \notin D$. Consider the following plane-wave solution of Maxwell's equations in vacuum,

$$E_i(x) = e^{i\langle x, k \rangle} p, \qquad H_i(x) = e^{i\langle x, k \rangle} q,$$

where k satisfies $|k|^2 = \varepsilon_0 \mu_0 \omega^2$. To satisfy equations (1–10) and (1–12), we require that the polarization 1-forms p and q satisfy

$$k \wedge p = \omega \mu_0 * q, \qquad k \wedge q = -\omega \varepsilon_0 * p,$$

where we have identified the vector k with a 1-form through $k(v) = \langle k, v \rangle$. It follows then that

$$\mu_0 \|q\|^2 = \mu_0 q \wedge *q = \varepsilon_0 \|p\|^2, \qquad p \wedge *q = 0,$$

and furthermore, the equations (1–11) and (1–13) require that

$$k \wedge *p = 0, \qquad k \wedge *q = 0.$$

The total field is written as a sum of the incoming field above plus the scattered field,

$$E = E_i + E_{sc}, \qquad H = H_i + H_{sc}.$$

The scattered field needs to satisfy a radiation condition at infinity. To understand better the the radiation condition for differential forms, let us go back for a while to the physical time domain picture. In the discussion below, we write concisely $E_{sc} = E_{sc}(x, t)$ and $H_{sc} = H_{sc}(x, t)$ for the physical (real valued) scattered time domain fields. Consider a ball B_R of radius $R > 0$ containing the inhomogeneity D. The total energy of the scattered field in $B_R \setminus \overline{D}$ expressed as

$$\mathcal{E} = \tfrac{1}{2}\varepsilon_0 \|E_{sc}\|_R^2 + \tfrac{1}{2}\mu_0 \|H_{sc}\|_R^2 = \mathcal{E}_E + \mathcal{E}_H,$$

where we write

$$\|E_{sc}\|_R^2 = \int_{B_R \setminus \overline{D}} E_{sc} \wedge *E_{sc} = (E_{sc}, E_{sc})_R,$$

and similarly for H_{sc}. Consider the electric part of the energy. The time derivative of it gives us

$$\frac{\mathcal{E}_E}{\partial t} = \varepsilon_0 \left(\frac{\partial E_{sc}}{\partial t}, E_{sc} \right)_R = \frac{1}{\mu_0}(\delta B_{sc}, E_{sc})_R,$$

and further, by applying Stokes' law,

$$\frac{\mathcal{E}_E}{\partial t} = \frac{1}{\mu_0}(B_{sc}, dE_{sc})_R + \frac{1}{\mu_0}\left(\langle nB_{sc}, tE_{sc} \rangle_{\partial B_R} + \langle nB_{sc}, tE_{sc} \rangle_{\partial D} \right). \qquad (2\text{--}2)$$

Again, from Maxwell's equations, we obtain

$$\frac{1}{\mu_0}(B_{sc}, dE_{sc})_R = -\frac{1}{\mu_0}\left(B_{sc}, \frac{\partial B_{sc}}{\partial t} \right) = -\frac{\partial \mathcal{E}_H}{\partial t}.$$

By substituting this identity into equation (2–2), we find that the total change of energy equals the flux through the boundaries ∂B_R and ∂D, i.e.,

$$\frac{\partial \mathcal{E}}{\partial t} = \frac{\partial \mathcal{E}_E}{\partial t} + \frac{\partial \mathcal{E}_H}{\partial t} = \frac{1}{\mu_0}\left(\langle nB_{sc}, tE_{sc} \rangle_{\partial B_R} + \langle nB_{sc}, tE_{sc} \rangle_{\partial D} \right)$$

$$= \langle tH_{sc}, tE_{sc} \rangle_{\partial B_R} + \langle tH_{sc}, tE_{sc} \rangle_{\partial D}.$$

The radiation conditions are now defined in such a way that for large R, the energy flux through ∂B_R either becomes negative (outgoing waves) or positive (incoming wave). For the outgoing wave, we write

$$\langle tH_{sc}, tE_{sc} \rangle_{\partial B_R} = -\left(\frac{\varepsilon_0}{\mu_0} \right)^{1/2} \langle tE_{sc}, tE_{sc} \rangle_{\partial B_R} + \left\langle tH_{sc} + \left(\frac{\varepsilon_0}{\mu_0} \right)^{1/2} tE_{sc}, tE_{sc} \right\rangle_{\partial B_R}.$$

Hence, to assure that the last term has an asymptotically vanishing effect, we set the radiation condition

$$t(\varepsilon_0^{1/2} E_{sc} + \mu_0^{1/2} H_{sc}) = o\left(\frac{1}{|x|} \right). \qquad (2\text{--}3)$$

This is the outgoing *radiation condition* for differential forms that we impose for the scattered field in the frequency domain. Compared with the *Silver–Müller radiation condition* in the vector formalism, this appears strikingly simple.

By using the representations of the scattered fields in terms of Green's functions, it is possible to derive an asymptotic representation of the fields,

$$tE_{\mathrm{sc}}(x) = E_\infty(\hat{x}; k; p)\frac{e^{i|k||x|}}{|x|} + o(|x|^{-1}),$$

$$tH_{\mathrm{sc}}(x) = H_\infty(\hat{x}; k; p)\frac{e^{i|k||x|}}{|x|} + o(|x|^{-1}),$$

where $\hat{x} = x/|x|$, and the mutually orthogonal 1-forms (defined on the unit sphere) E_∞ and H_∞ are the electric and magnetic far-field patterns, respectively, corresponding to the polarization p and incidence direction k with $|k|^2 = \varepsilon_0\mu_0\omega^2$. Note that one only needs to specify one of these, the other one can then be immediately obtained from the radiation conditions. The inverse scattering problem (ISP) can now be formulated as follows.

ISP. *From the knowledge of $E_\infty(\hat{x}; k; p)$ for all $\hat{x} \in S^2$, $k \in \mathbb{R}^3$ with $|k|$ fixed and for three linearly independent polarizations p determine the material parameters γ and μ.*

If one knows the admittance map on a smooth surface Γ enclosing the inhomogeneity, then the boundary value on Γ of the rescaled scattered field $e_{\mathrm{sc}} = \gamma^{1/2}E_{\mathrm{sc}}$ corresponding to the incoming plane wave with electric component $e_{\mathrm{i}} = \varepsilon_0^{1/2}pe^{i\langle x,k\rangle}$ can be solved for from the boundary integral equation

$$\tfrac{1}{2}te_{\mathrm{sc}} = te_{\mathrm{i}} + D_k\Lambda te_{\mathrm{sc}} - K_k te_{\mathrm{sc}}$$

on Γ. (The argument is similar the one used to derive Equation (6–2) below.) Here, the operators D_k and K_k are defined analogously to D and K introduced in Section 6 but using the standard outgoing fundamental solution $-e^{ik|x|}/4\pi|x|$ instead of Faddeev's Green's function. The mapping properties are unchanged in this replacement, and the unique solvability follows in a standard manner assuming that ω is not a resonance frequency. Hence we know the tangential boundary values of e and h on Γ, and thus also the far-fields are determined by the impedance map.

To be able to reduce the ISP to IBP, we need to go to the opposite direction, and this is not as simple. The difficulty (and also its resolution) is similar to the acoustic case, so we are rather brief in describing it. If one knows the far-fields of all waves scattered by plane waves, one also knows, using Rellich's argument, the scattered fields in the complement of the union of the supports of $\gamma - \varepsilon_0$ and $\mu - \mu_0$, and thus one also knows their boundary values on Γ, i.e., one knows the restriction of the impedance map to all total fields corresponding to incoming plane waves. The problem is to show that this data determines the impedance map. This was shown to be true for Maxwell's equations in [19] (Section 6.4.) and hence the scattering problem is reduced to IBP. The argument is a modification of the idea of Nachman [13] and Ramm [18]. In fact, the argument in [19] deals with the acoustic and electromagnetic cases simultaneously. Using this argument, however, one has to assume that the interface Γ is chosen so that ω is not a magnetic resonance frequency. Of course one is free to choose the interface so that this is avoided, but in practice it is not easy to determine when one is close to a resonant frequency.

One can also deal with the inverse scattering problem directly, without reducing it to the IBP: For Maxwell's equations this was done in [4] assuming that $\mu = \mu_0$, and stability results were obtained in [5]. The crucial part is again the construction of the exponentially increasing solutions, and in these proofs one does not need any assumptions on ω.

3. The Scaled System

In this section we follow the idea of [17] and rescale the electromagnetic forms in such a way that we only have to deal with one metric and complete this system to an elliptic system of Dirac type. In [17] this was only done for the case of Euclidean background metric, but the principle remains the same in a more general setting, see [9]. The basic idea is of course old and well-known: Even though the divergence conditions (1–6) and (1–8) are implied by the two other equations (1–5) and (1–7), they make it possible to reduce the system to an elliptic system that in a homogenous medium coincides with the Helmholtz equation for E and H, respectively. Also, the divergence conditions are crucial when analyzing the low-frequency limit, since they single out the right limit value (remember that Maxwell's equations have an infinite dimensional kernel when $\omega = 0$). The approach we follow is a modification of the argument originally due to R. Picard (see [15]). The idea is to get a first order elliptic and symmetric system (at least in the principal part) that under some conditions reduces to Maxwell's equations. The ellipticity is achieved by including the divergence conditions to the system, but to make it symmetric one needs to modify it further. We start with the system (1–10) to (1–13) and introduce 3-forms Φ and Ψ by

$$i\omega\Phi = d(\gamma * E), \qquad i\omega\Psi = d(\mu * H). \tag{3-1}$$

Of course, if E and H satisfy Maxwell's equations, these forms vanish. Now we modify (1–10) and (1–12) to

$$dH - \frac{1}{\mu} * d * \left(\frac{1}{\gamma}\Phi\right) + i\omega\gamma * E = 0 \tag{3-2}$$

$$dE + \frac{1}{\gamma} * d * \left(\frac{1}{\mu}\Psi\right) - i\omega\mu * H = 0. \tag{3-3}$$

The principal part of this system still depends on γ and μ, and in order to make it more simple we scale the unknown fields: Let

$$e = \gamma^{1/2} E, \quad h = \mu^{1/2} H, \quad \phi = \frac{1}{\gamma\mu^{1/2}} \Phi, \quad \psi = \frac{1}{\gamma^{1/2}\mu} \Psi. \tag{3-4}$$

This makes the physical dimensions of the unknown 1- and 3-forms equal, and it also makes the principal part of the system depend only on the background metric g as we shall see. To get a full graded algebra, let us further define the 0- and 2-forms φ and b as

$$\varphi = *\phi, \qquad b = *h.$$

We introduce the notation

$$\Omega M = \Omega^0 M \times \Omega^1 M \times \Omega^2 M \times \Omega^3 M,$$

for the full Grassmannian bundle and endow ΩM with the obvious inner product: If $u = (u^0, u^1, u^2, u^3)$, $v = (v^0, v^1, v^2, v^3) \in \Omega M$, we set

$$(u, v) = \sum_{j=0}^{3} \int_M u^j \wedge *\overline{v^j}.$$

Define a graded form

$$X = (\varphi, e, b, \psi) \in \Omega M.$$

A straightforward insertion into the augmented equations (3–1) to (3–3) along with the identities $\delta = (-1)^k *d*$ and $** = (-1)^{k(n-k)} = 1$ for forms of degree k in $\mathbb{R}^n$ with $n = 3$ give that X satisfies the system

$$(P - i\kappa)X + VX = 0, \tag{3-5}$$

where the principal part is

$$P = \begin{pmatrix} 0-\delta & 0 & 0 & 0 \\ d & 0-\delta & 0 & 0 \\ 0 & d & 0-\delta & 0 \\ 0 & 0 & d & 0 \end{pmatrix},$$

the scalar $\kappa = \kappa(x)$ is the nonconstant wave number,

$$\kappa = \omega(\gamma\mu)^{1/2},$$

and V is a local potential given by

$$V = \begin{pmatrix} 0 & *(d\alpha \wedge * \cdot) & 0 & 0 \\ d\beta \wedge \cdot & 0 & *(d\beta \wedge * \cdot) & 0 \\ 0 & -d\alpha \wedge \cdot & 0 & *(d\alpha \wedge * \cdot) \\ 0 & 0 & d\beta \wedge \cdot & 0 \end{pmatrix}.$$

Here

$$\alpha = \tfrac{1}{2}\ln\gamma, \qquad \beta = \tfrac{1}{2}\ln\mu.$$

The first order operator has several important properties. First, Stokes' formula (2–1) implies that

$$(Pu, v) + (u, Pv) = \langle tu, nv \rangle + \overline{\langle tv, nu \rangle}, \tag{3-6}$$

where we introduced the shorthand notation tv, $nv \in \Omega \partial M = \Omega^0 \partial M \times \Omega^1 \partial M \times \Omega^2 \partial M$,

$$tv = (tv^0, tv^1, tv^2), \qquad nv = (nv^3, nv^2, nv^1),$$

and

$$\langle tu, nv \rangle = \langle tu^0, nv^1 \rangle + \langle tu^1, nv^2 \rangle + \langle tu^2, nv^3 \rangle.$$

Observe that this expression does not define an inner product on ∂M.

Second, we observe immediately that

$$P^2 = -\mathbf{\Delta} = -\mathrm{diag}(\Delta^0, \Delta^1, \Delta^2, \Delta^3),$$

where $\Delta^k = \delta d + d\delta$ is the Laplace–Beltrami operator for k-forms. But more is true. Namely, splitting the Grassmann algebra into its even and odd degree parts,

$$\Omega M = \Omega^+ M \oplus \Omega^- M,$$

where $\Omega^+ M = \Omega^0 M \times \Omega^2 M$ and $\Omega^- M = \Omega^1 M \times \Omega^3 M$, we can write the potential as an off-diagonal block matrix

$$V = \begin{pmatrix} 0 & V_- \\ V_+ & 0 \end{pmatrix} = \begin{pmatrix} 0 & 0 & *(d\alpha \wedge * \cdot) & 0 \\ 0 & 0 & -d\alpha \wedge \cdot & *(d\alpha \wedge * \cdot) \\ d\beta \wedge \cdot *(d\beta \wedge * \cdot) & 0 & 0 \\ 0 & d\beta \wedge \cdot & 0 & 0 \end{pmatrix},$$

with $V_\pm : \Omega^\pm M \to \Omega^\mp M$. In this representation, the operator P becomes just the Dirac type operator $\mathbf{D} = d - \delta : \Omega^\pm M \to \Omega^\mp M$. Hence we can write (3–5) equivalently as

$$(\mathbf{D} - i\kappa + V)X = 0.$$

Now let $\tilde{V}$ be the adjoint of V:

$$\tilde{V} = \begin{pmatrix} 0 & \tilde{V}_- \\ \tilde{V}_+ & 0 \end{pmatrix} = \begin{pmatrix} 0 & 0 & *(d\beta \wedge * \cdot) & 0 \\ 0 & 0 & -d\beta \wedge \cdot & *(d\beta \wedge * \cdot) \\ d\alpha \wedge \cdot *(d\alpha \wedge * \cdot) & 0 & 0 \\ 0 & d\alpha \wedge \cdot & 0 & 0 \end{pmatrix}.$$

LEMMA 3.1. *The first order terms of the product*

$$(\mathbf{D} - i\kappa(x) + V(x))(\mathbf{D} + i\kappa(x) - \tilde{V}(x)) \tag{3-7}$$

vanish, i.e., the product is of the form $-\mathbf{\Delta} + k^2 + Q(x)$, *where Q is a zeroth order pointwise multiplier.*

PROOF. The nontrivial part of the lemma is of course the vanishing of the first order derivatives of the commutator-like term $VD - D\tilde{V}$. Since both D and V change the parity of the degree, we may consider only the even degree part. The odd degree case is handled similarly. By a direct computation we get for $u^+ = (u^0, u^2) \in \Omega^+ M$ that

$$V_-(d - \delta)u^+ = \begin{pmatrix} *(d\alpha \wedge *(du^0 - \delta u^2)) \\ -d\alpha \wedge (du^0 - \delta u^2) + *(d\alpha \wedge *d\omega^2) \end{pmatrix}, \qquad (3\text{--}8)$$

and similarly,

$$(d - \delta)\tilde{V}_+ u^+ = \begin{pmatrix} -\delta(d\alpha \wedge u^0 + *(d\alpha \wedge *u^2)) \\ d(d\alpha \wedge u^0 + *(d\alpha \wedge *u^2)) - \delta(d\alpha \wedge u^2) \end{pmatrix}. \qquad (3\text{--}9)$$

By using a local orthonormal coframe, a straightforward computation shows that the differences are of order zero in u^+. For later use, we demonstrate this explicitly for the 0-form component. The first term in the upper component of (3–9) is

$$-\delta(d\alpha \wedge u^0) = *d * (d\alpha \wedge u^0) = *(d\alpha \wedge *du^0) - u^0 \Delta^0 \alpha,$$

and the second term gives

$$-\delta * (d\alpha \wedge *u^2) = *d(d\alpha \wedge *u^2) = -*(d\alpha \wedge d * u^2) = -*(d\alpha \wedge *\delta u^2)$$

By comparing to the dirst component of (3–8), we observe that

$$(V_-(d - \delta)u^+ - (d - \delta)\tilde{V}_+ u^+)^0 = u^0 \Delta^0 \alpha. \qquad (3\text{--}10)$$

The calculation of the second component is slightly more tedious but straightforward.

Observe also that the potential $V\tilde{V}$ is diagonal. Indeed, we have

$$V\tilde{V} = \mathrm{diag}(|d\alpha|^2, |d\alpha|^2, |d\beta|^2, |d\beta|^2),$$

where $|d\alpha|^2 = *(d\alpha \wedge *d\alpha)$. This result is used later. $\square$

4. Green's Function

In this section we derive exponentially growing (or Faddeev's) Green's function for the complete Maxwell system (3–5) treated in the previous section. Although the discussion of the previous section can be carried out in more general metric, here we have to confine ourselves to the Euclidian case. Thus, we shall assume that $g = g_e$ is the Euclidian metric, and the Euclidian normal coordiantes are denoted by (x^1, x^2, x^3).

We start by recaling the definition of scalar Faddeev's Green's function: For any $\zeta \in \mathbb{C}^3$, set

$$G(x) = G_\zeta(x) = e^{i\langle x, \zeta\rangle} g_\zeta(x), \qquad g_\zeta(x) = \left(\frac{1}{2\pi}\right)^3 \int_{\mathbb{R}^3} \frac{e^{i\langle x, \xi\rangle}}{|\xi|^2 + 2\langle \xi, \zeta\rangle} \, d\xi,$$

where the inner products are the real inner products, i.e., no complex conjugation is included. When ζ is chosen in such a way that

$$\langle \zeta, \zeta \rangle = k^2, \tag{4-1}$$

the function G is indeed Green's function for the Helmholtz operator,

$$(\Delta - k^2)G(x) = \delta(x).$$

Note that our Δ is now the geometer's Laplacian $d\delta + \delta d$, which is a positive operator.

This scalar Green's function has the following important asymptotic property as $|\zeta| \to \infty$ along the variety $\{\zeta \in \mathbb{C}^3; \langle \zeta, \zeta \rangle = k^2\}$. Letting L_δ^2 be the weighted L^2-space with norm

$$\|f\|_\delta^2 = \int_{\mathbb{R}^3} |f|^2 (1 + |x|^2)^\delta \, dx \tag{4-2}$$

we have the following estimate due to Sylvester and Uhlmann ([23]):

PROPOSITION 4.1. *For $|\zeta|$ large, we have*

$$\|g_\zeta * f\|_\delta \leq \frac{C}{|\zeta|} \|f\|_{\delta+1},$$

where $-1 < \delta < 0$.

By using this scalar Green's function, we define an exponentially growing Green's tensor for $\Delta - k^2$ by setting

$$\boldsymbol{G}(x - y) = G(x-y)\left(1, \sum_{j=1}^{3} dx^j \otimes dy^j, \sum_{j=1}^{3} \theta_j \otimes \nu_j, dV_x \otimes dV_y\right)$$

$$= G(x-y)\boldsymbol{I},$$

where $\theta_j = \frac{1}{2}\varepsilon_{jk\ell}dx^k \wedge dx^\ell$ and $\nu_j = \frac{1}{2}\varepsilon_{jk\ell}dy^k \wedge dy^\ell$ and $dV_x = dx^1 \wedge dx^2 \wedge dx^3$, $dV_y = dy^1 \wedge dy^2 \wedge dy^3$. Oberve that

$$\boldsymbol{I} \wedge *(\lambda_j dy^j) = \lambda_j dx^j.$$

For later reference, note that $\boldsymbol{I}$ can be written componentwise as

$$\boldsymbol{I} = \sum_{j=1}^{8} \omega_x^j \otimes \omega_y^j, \tag{4-3}$$

where $\omega_x^1 = \omega_y^1 = (1, 0, 0, 0)$, $\omega_x^2 = (0, dx^1, 0, 0)$, $\omega_y^2 = (0, dy^1, 0, 0)$ and so on.

With the help of this Green's tensor, we define now a graded form that could be called a *generalized Sommerfeld potential*. Let $Y_0 \in \Omega M$ be any graded form satisfying

$$(-\boldsymbol{\Delta} + k^2)Y_0 = 0 \tag{4-4}$$

in $\mathbb{R}^3$. We seek to solve for the potential $Y \in \Omega M$ from the Lippmann–Schwinger type equation

$$Y(x) = Y_0(x) - \int_M \boldsymbol{G}(x - y) \wedge *(Q(y)Y(y)). \tag{4-5}$$

The existence of such a solution for large $|\zeta|$ is guaranteed by Proposition 4.1. Also, we observe that Y satisfies the Schrödinger equation

$$(-\boldsymbol{\Delta} + k^2 + Q(x))Y(x) = 0. \tag{4-6}$$

From Proposition 4.1, we obtain also the important information of the asymptotic behaviour of Y for large $|\zeta|$.

THEOREM 4.2. *For $|\zeta|$ large enough, $-1 < \delta < 0$, and for any constant coefficient form y_0 which is bounded in ζ, the equation (4–5) has a unique solution $Y_\zeta = e^{i\langle x, \zeta\rangle}(y_0 + w_\zeta)$, where $\|w_\zeta\|_\delta < C/|\zeta|$.*

For later use, we fix already here the form Y_0 and require that it is of the form

$$Y_0(x) = e^{i\langle x, \zeta\rangle} y_0,$$

where we assume that $\zeta \in \mathbb{C}^3$ satisfies the condition (4–1), guaranteeing equation (4–4) to be valid. Furthermore, the constant graded form $y_0 = (y^0, y^1, y^2, y^3)$ is required to satisfy

$$ky^0 = -*(\zeta \wedge *y^1), \qquad ky^3 = -\zeta \wedge y^2, \tag{4-7}$$

where we identified ζ with a complex 1-form by $\zeta(u) = \langle \zeta, u\rangle$. The conditions above imply that

$$((P + ik)Y_0)^0 = 0, \qquad ((P + ik)Y_0)^3 = 0, \tag{4-8}$$

i.e., the 0-form and 3-form components of $(P + ik)Y_0(x)$ vanish.

Now we use the decomposition property of Lemma 3.1. By setting

$$X_\zeta(x) = (P + i\kappa(x) - \tilde{V}(x))Y_\zeta(x),$$

we find that X_ζ satisfies the complete Maxwell's system

$$(P - i\kappa(x) + V(x))X_\zeta(x) = 0. \tag{4-9}$$

We call this solution the *exponentially growing solution* of the complete Maxwell's system.

From the foregoing definition, it is not obvious that $X_\zeta = (X_\zeta^0, X_\zeta^1, X_\zeta^2, X_\zeta^3)$ is indeed a solution to the original Maxwell's system, i.e., that $X_\zeta^0 = 0$ and $X_\zeta^3 = 0$ as they should in order that the pair (X_ζ^1, X_ζ^2) would represent scaled electric and magnetic fields. However, one can prove the following result.

LEMMA 4.3. *Assume that Y_0 is chosen so that the conditions (4–8) are satisfied. Then, for large $|\zeta|$, we have $X_\zeta^0 = 0$ and $X_\zeta^3 = 0$.*

PROOF. In view of what was said, it only remains to check that for $|\zeta|$ large enough the 0- and 3-form components of X_ζ vanish. We show this for the first component, the last component being handled similarly.

Since X_ζ satisfies the complete Maxwell's system (4–9), we have

$$(P + i\kappa - \tilde{V})(P - i\kappa + V)X_\zeta = 0.$$

A calculation similar to the one in the proof of Lemma 3.1 shows now that X_ζ^0 satisfies

$$(-\Delta + k^2)X_\zeta^0 + qX_\zeta^0 = 0 \qquad (4\text{--}10)$$

where the potential $q(x)$ is given by

$$q = \Delta^0\beta - |d\beta|^2 + (\kappa^2 - k^2). \qquad (4\text{--}11)$$

On the other hand, by using the particular form of the solution Y_ζ defined in Theorem 4.2, we can decompose X_ζ by a straightworward substitution of Y_ζ into the definition X_ζ as

$$X_\zeta = (P + ik)Y_0 + e^{i\langle x,\zeta\rangle}w_\zeta,$$

where $w_\zeta \in L^2_{-\delta}$. Furthermore, from the equation (4–10), it follows further that w_ζ^0 must satisfy the integral equation

$$w_\zeta^0 = w_0^0 - g_\zeta * (qw_\zeta^0),$$

with

$$w_0^0 = e^{-i\langle x,\zeta\rangle}((P + ik)Y_0)^0 = 0$$

by the assumption of Y_0. It follows from Proposition 4.1 that for large $|\zeta|$, $w_\zeta^0 = 0$. For details, see [17]. $\qquad \square$

5. From Inside to Boundary

In this section, we derive a formula that relates the material parameters inside M to the boundary values of the exponentially growing solution. Here, M is bounded and Euclidean with a smooth boundary, and $\gamma = \varepsilon_0$, $\mu = \mu_0$ near the boundary. The formula is related to the energy integral appearing in electrical impedance tomography, but due to the complexity of the complete Maxwell's system it is more involved.

To begin with, let $Y_0^* \in \Omega M$ be any solution of the homogenous space problem,

$$(P - ik)Y_0^* = 0. \qquad (5\text{--}1)$$

By using equation (4–6) and the decomposition of Δ, we have

$$(QY_\zeta, Y_0^*) = -((-\Delta + k^2)Y_\zeta, Y_0^*) = -((P - ik)(P + ik)Y_\zeta, Y_0^*)$$
$$= -((P - ik)\tilde{X}_\zeta, Y_0^*), \qquad (5\text{--}2)$$

where we denoted $\tilde{X}_\zeta = (P + ik)Y_\zeta$. Observe that when $\gamma = \varepsilon_0$ and $\mu = \mu_0$, we have $\tilde{X}_\zeta = X_\zeta$ by the definition of X_ζ. By using Stokes' formula (3–6) for P and equation (5–1), we find that

$$(QY_\zeta, Y_0^*) = -\langle tX_\zeta, nY_0^* \rangle - \overline{\langle tY_0^*, nX_\zeta \rangle}.$$

Here, we used the fact that at ∂M, $X_\zeta = \tilde{X}_\zeta$. Hence, if we know the boundary data $\{tX_\zeta, nX_\zeta\}$, we obtain an integral involving the potential Q over M.

To understand the significance of this relation better, we look at the linearization of the left hand side of (5–2) with a particular choice of the form Y_0^*. The linearization means the approximation

$$(QY_\zeta, Y_0^*) \approx (QY_0, Y_0^*). \tag{5–3}$$

In view of Theorem 4.2, this approximation is asymptotically valid as $|\zeta| \to \infty$. Following the original ideas of Calderón, we choose $Y_0(x)$ as in the previous section. Similarly, we set

$$Y_0^*(x) = e^{i\langle x, \zeta^* \rangle} y_0^*.$$

where

$$\zeta - \overline{\zeta^*} = \xi,$$

$\xi \in \mathbb{R}^3$ being a fixed vector. We require further that

$$\langle \zeta, \zeta \rangle = \langle \zeta^*, \zeta^* \rangle = k^2.$$

As we shall see in Section 7, in $\mathbb{C}^3$ there is enough space to make such a choice. The constant graded form y_0^* must be chosen again in such a way that equation (5–1) holds. It is easy to see that such a choice is obtained if we set, e.g.,

$$y_0^* = \frac{1}{|\zeta|}(P(i\zeta^*) + ik)z,$$

where $z = (z^0, z^1, z^2, z^3)$ is an arbitrary constant coefficient graded form and $P(i\zeta^*)$ is the symbol of the operator P, i.e.,

$$P(i\zeta^*) = e^{-i\langle x, \zeta^* \rangle} P e^{i\langle x, \zeta^* \rangle}.$$

With these choices, we obtain

$$(QY_0, Y_0^*) = \int_M e^{i\langle x, \xi \rangle}(Q(x)y_0, y_0^*). \tag{5–4}$$

Hence, we see that within the linearization, the boundary values of X determine the Fourier transform of $(Q(x)y_0, y_0^*)$ and thus the function itself. In Section 7, we show how the material parameters $\mu(x)$ and $\gamma(x)$ can be recovered from this data.

6. From Λ to Boundary Values of X

In the previous section, we showed how the boundary values of X determine the integral (5–4). In this section, we show that the knowledge of the admittance map determines the boundary values of X.

The idea is to derive a version of the Stratton–Chu representation formula for the field X. To this end, we start with the Lippmann–Schwinger type equation for Y, and by writing

$$\tilde{X}(y) = (P + ik)Y(y),$$

we have

$$Y(x) = Y_0(x) - \int_M G(x - y) \wedge *Q(y)Y(y)$$

$$= Y_0(x) + \int_M G(x - y) \wedge *(-\Delta + k^2)Y(y)$$

$$= Y_0(x) + \int_M G(x - y) \wedge *(P - ik)\tilde{X}(y).$$

By writing $G(x - y)$ in terms of the components of I as in (4–3), we obtain through integration by parts the equation

$$Y(x) = Y_0(x) + \sum \omega_x^j \int_M G(x-y)\omega_y^j \wedge *(P - ik)\tilde{X}(y)$$

$$= Y_0(x) + \sum \omega_x^j \int_M (-P_y - ik)G(x-y)\omega_y^j \wedge *\tilde{X}(y)$$

$$+ \sum \omega_x^j \left(\int_{\partial M} tG(x-y)\omega_y^j \wedge nX(y) + \int_{\partial M} tX(y) \wedge nG(x-y)\omega_y^j \right).$$

Here we used the fact that at the boundary, $\tilde{X}(y) = X(y)$. We substitute the integral representation of $Y(x)$ in this formula and use the fact that for $x \neq y$, we have

$$(P_x + ik)(-P_y - ik)G(x-y)I = 0,$$

and we arrive at the identity

$$X(x) = X_0(x) + (P + ik) \sum \omega_x^j \left(\int_{\partial M} tG(x-y)\omega_y^j \wedge nX(y) \right.$$

$$\left. + \int_{\partial M} tX(y) \wedge nG(x-y)\omega_y^j \right),$$

where

$$X_0(x) = (P + ik)Y_0(x).$$

Now assume that $|\zeta|$ is large. Then, by Lemma 4.3, we have $X = (0, e, h, 0)$, and the boundary integral above takes the form

$$X(x) = X_0(x) + (P+ik)\bigg(\int_{\partial M} G(x-y)\boldsymbol{n}e, \ \sum_{j=1}^{3} dx^j \int_{\partial M} G(x-y)\boldsymbol{t}dy^j \wedge \boldsymbol{n}b,$$

$$\sum_{j=1}^{3}\theta_j \int_{\partial M} G(x-y)\boldsymbol{t}e \wedge \boldsymbol{n}\nu_j, \ dV \int_{\partial M} G(x-y)\boldsymbol{t}h\bigg). \quad (6\text{-}1)$$

Letting the point x approach the boundary ∂M from the exterior domain $\mathbb{R}^3 \setminus M$ we obtain an integral equation for the boundary values of X. However, assuming that the impedance map Λ is known, it suffices to solve the tangential component of the electric field. Indeed, we have

$$\boldsymbol{n}b = \Lambda \boldsymbol{t}e,$$

and assuming that $|\zeta|$ is large, from Maxwell's equations

$$-\delta b + ike = 0, \qquad de + ikb = 0,$$

we find that

$$\boldsymbol{n}e = \frac{1}{ik}\boldsymbol{n}\delta b = \frac{1}{ik}\boldsymbol{t}d * b.$$

Since the exterior derivative and the tangential trace commute, we have further

$$\boldsymbol{n}e = \frac{1}{ik}d_\partial \boldsymbol{n}b = \frac{1}{ik}d_\partial \Lambda \boldsymbol{t}e.$$

Here, d_∂ denotes the exterior derivative on ∂M. Similarly, we have

$$\boldsymbol{t}b = -\frac{1}{ik}\boldsymbol{t}de = -\frac{1}{ik}d_\partial \boldsymbol{t}e.$$

Summarizing,

$$\boldsymbol{t}X = (0, \ \boldsymbol{t}e, \ \boldsymbol{t}b) = \bigg(0, \boldsymbol{t}e, -\frac{1}{ik}d_\partial \boldsymbol{t}e\bigg),$$

$$\boldsymbol{n}X = (0, \boldsymbol{n}b, \boldsymbol{n}e) = \bigg(0, \Lambda \boldsymbol{t}e, \frac{1}{ik}d_\partial \Lambda \boldsymbol{t}e\bigg).$$

Thus, we shall consider only the 1-form component of the system (6–1) and solve it for $\boldsymbol{t}e$. Denoting by e_0 the 1-form component of X_0 we have, for $x \in \mathbb{R}^3 \setminus M$,

$$e = e_0 + d\int_{\partial M} G\boldsymbol{n}e - \delta \sum_{j=1}^{3}\theta_j \int_{\partial M} G\boldsymbol{t}e \wedge \boldsymbol{n}\nu_j + ik\sum_{j=1}^{3}dx^j \int_{\partial M} G\boldsymbol{t}dy^j \wedge \boldsymbol{n}b$$

$$= e_0 + \frac{1}{ik}d\int_{\partial M} Gd\Lambda \boldsymbol{t}e - \delta \sum_{j=1}^{3}\theta_j \int_{\partial M} G\boldsymbol{t}e \wedge \boldsymbol{n}\nu_j + ik\sum_{j=1}^{3}dx^j \int_{\partial M} G\boldsymbol{t}dy^j \wedge \Lambda \boldsymbol{t}e.$$

Here the arguments of the functions are suppressed for brevity. Now we need to apply the tangential boundary trace from the exterior domain to both sides of this equation. To get a boundary integral equation, we need to take into account the jump relations of the layer potentials. Consider first the second integral on the right. By using the identites

$$n\nu_j = t dy^j, \qquad \delta\theta_j f(x) = *df(x) \wedge dx^j,$$

we obtain

$$\delta \sum_{j=1}^{3} \theta_j \int_{\partial M} Gte \wedge n\nu_j = \sum_{j=1}^{3} \left(\int_{\partial M} \frac{\partial G}{\partial x^k} te \wedge dy^j \right) * (dx^k \wedge dx^j).$$

For simplicity, assume for a while that we use tangent-normal coordinates such that $M = \{x^3 \geq 0\}$. Then $te \wedge dy^3 = 0$, while $t(*(dx^k \wedge dx^j)) = 0$ for $j = 1, 2$ and $k \neq 3$, so finally

$$t \sum_{j=1}^{3} \left(\int_{\partial M} \frac{\partial G}{\partial x^k} te \wedge dy^j \right) *(dx^k \wedge dx^j)$$

$$= -\left(\int_{\partial M} \frac{\partial G}{\partial x^3} e_1 dy^1 \wedge dy^2 \right)\Big|_{\partial M}^{+} dx^1 - \left(\int_{\partial M} \frac{\partial G}{\partial x^3} e_2 dy^1 \wedge dy^2 \right)\Big|_{\partial M}^{+} dx^2$$

$$= \tfrac{1}{2} te - \sum_{j=1}^{3} n \left(\int_{\partial M} d_x Gte \wedge dy^j \right) \wedge dx^j,$$

the normal trace of the singular integral being understood in the sense of the principal value.

In a similar fashion we treat the first integral. Here we observe that since the tangential trace and the exterior derivative commute, the integral kernel has no derivatives of Green's function in the normal direction and hence the jump relations produce no extra terms besides the principal value integral.

By combining the terms, we reach the identity

$$\tfrac{1}{2} te = te_0 + D\Lambda te - Kte, \tag{6-2}$$

where the operators D and K are given as

$$D\omega(x) = \frac{1}{ik} \left(d_\partial t \int_{\partial M} G(x-y) d\omega(y) + k^2 \sum_{j=1}^{3} \int_{\partial M} G(x-y) t dy_j \wedge \omega(y) \right),$$

$$K\omega(x) = \sum_{j=1}^{3} n \left(\int_{\partial M} d_x G(x-y)\omega(y) \wedge dy^j \right) \wedge dx^j,$$

and where $x \in \partial M$, $\omega \in \Omega^1 \partial M$ and the singular integrals are understood in the sense of principal values. Introduce the spaces

$$H(d, \Omega^k M) = \{ f \in L^2(\Omega^k M) : df \in L^2(\Omega^{k+1} M)\},$$
$$H(\delta, \Omega^k M) = \{ f \in L^2(\Omega^k M) : \delta f \in L^2(\Omega^{k-1} M)\}$$

and on the boundary

$$H^{-1/2}(d, \Omega^k \partial M) = \{g \in H^{-1/2}(\Omega^k \partial M); d_{\partial} g \in H^{-1/2}(\Omega^{k+1} \partial M)\}.$$

Then we have the bounded trace maps

$$\boldsymbol{t} : H(d, \Omega^k M) \to H^{-1/2}(d, \Omega^k \partial M),$$

$$\boldsymbol{n} : H(\delta, \Omega^k M) \to H^{-1/2}(d, \Omega^{3-k} \partial M),$$

and these maps are onto. Also, K maps $H^{-1/2}(d, \Omega^k \partial M)$ compactly to itself, and D just boundedly. For more details on this the reader is referred to [14]. Also there is large literature on layer potential techniques on (subdomains) of Riemannian manifolds, even with Lipschitz boundaries, see [12] and references therein. Of course, it is not known what is the analogue of the exponentially increasing Green's function in the general metric case.

It turns out that the equation (6–2) is of Fredholm type and has a unique solution exactly when ω is not an eigenfrequency for the interior Maxwell problem with vanishing tangential electric field. We shall not go into details here but refer to the (vector version) of this equation in the references [16] and [17].

7. From (Y_0^*, QY_0) to γ and μ

It turns out that for the reconstruction of γ and μ in the interior one does not need to recover the whole matrix Q. Indeed, since we do not know the relevant boundary data for the second order system, this cannot be done starting from the impedance map. However, as remarked in Section 5, one can still hope to extract information on Q. We start by making some explicit choices for the constant form Y_0. Fix $\xi \in \mathbb{R}^3$, and choose coordinates so that $\xi = (|\xi|, 0, 0)$. Then for $R > 0$ let

$$\zeta = \zeta(R) = (|\xi|/2, i(|\xi|^2/4 + R^2)^{1/2}, (R^2 + k^2)^{1/2}),$$

$$\zeta^* = \bar{\zeta} - \xi.$$

Now $\langle \zeta, \zeta \rangle = \langle \zeta^*, \zeta^* \rangle = k^2$, and for some constant 1- and 2-forms y^1 and y^2 to be chosen later, let

$$y_0 = \frac{1}{|\zeta|} \left(-*(\zeta \wedge *y^1), ky^1, ky^2, -\zeta \wedge y^2 \right).$$

This choice guarantees the conditions (4–7). Further, let

$$y_0^* = \frac{1}{|\zeta|} (P(i\zeta^*) + ik)z.$$

Now, as $|\zeta| \to \infty$ we have the limits

$$\lim_{|\zeta| \to \infty} y_0 = -\left(*(\hat{\zeta} \wedge *y^1), 0, 0, \hat{\zeta} \wedge y^2 \right),$$

$$\lim_{|\zeta| \to \infty} y_0^* = P(i\hat{\zeta})z,$$

where $\hat{\zeta} = \lim \zeta/|\zeta| = 1/\sqrt{2}(0, i, 1)$. By choosing y^1 so that $-*(\hat{\zeta} \wedge *y^1) = 1$ and y^2, we have

$$\lim_{|\zeta| \to \infty} y_0 = (1, 0, 0, 0).$$

On the other hand, by choosing $z = (0, z^1, 0, 0)$, we have

$$\lim_{|\zeta| \to \infty} y_0^* = (-i * (\hat{\zeta} \wedge *z^1), 0, i\hat{\zeta} \wedge z^1, 0).$$

If z^1 satisfies $-i *(\hat{\zeta} \wedge *z^1) = 1$, we use the equality $Q_{0,j} = 0$ for $j \neq 0$, where Q is the potential of Lemma 3.1, to obtain

$$\lim_{|\zeta| \to \infty} (QY_\zeta, Y_0^*) = \hat{Q}_{0,0}(\xi),$$

where $Q_{0,0}$ denotes the component of the potential that maps 0-forms to 0-forms. Similarly, we may choose the forms to yield

$$\lim_{|\zeta| \to \infty} (QY_\zeta, Y_0^*) = \hat{Q}_{3,3}(\xi).$$

From Lemma 3.1, we find that

$$Q_{0,0} = \Delta^0 \alpha - |d\alpha|^2 + (\kappa^2 - k^2),$$

and similarly

$$Q_{3,3} = \Delta^0 \beta - |d\beta|^2 + (\kappa^2 - k^2).$$

By denoting $u = \gamma^{1/2}$, $v = \mu^{1/2}$ so that $\alpha = \log u$, $\beta = \log v$ and $\kappa = kuv$, the equations above simplify further as

$$Q_{0,0} = \frac{1}{u}(\Delta^0 u - k^2 u(uv - 1)),$$

$$Q_{0,0} = \frac{1}{v}(\Delta^0 v - k^2 v(uv - 1)).$$

We assumed that in a neighbourhood of ∂M, u and v are known constants. An application of the unique continuation principle for elliptic equations then shows that u and v, i.e., μ and γ are uniquely determined by the admittance map.

References

[1] K. Chadan, D. Colton, L. Päivärinta and W. Rundell: *An introduction to inverse scattering and inverse spectral problems*, SIAM Monographs on Mathematical Modelling and Computation, Philadelphia 1997.

[2] D. Colton and R. Kress: *Inverse acoustic and electromagnetic scattering theory*, Springer–Verlag, Berlin 1992.

[3] D. Colton and P. Monk: The detection and monitoring of leukemia using electromagnetic waves, *Inverse Problems* **10** (1994), 1235-1251; 11 (1995) 329-342.

[4] D. Colton and L. Päivärinta: The uniqueness of a solution to an inverse scattering problem for electromagnetic waves, *Arch. Rational Mech. Anal.* **119** (1992) 59–70.

[5] P. Hähner: *On acoustic, electromagnetic, and elastic scattering problems in inhomogeneous media*, Habilitationschrift, University of Göttingen 1998.

[6] V. Isakov: Completeness of produscts of solutions and some inverse problems for PDE, *J. Diff. Eq.* **92** (1991) 305–316.

[7] M. Joshi and S. McDowall: Total determination of material parameters from electromagnetic boundary information, *Pacific J. Math.* **193** (2000) 107–129.

[8] A. Katchalov, Y. Kurylev and M. Lassas: *Inverse boundary spectral problems*, Monographs and surveys in pure and applied mathematics 123, Chapman and Hall/CRC, 2001.

[9] Y. Kurylev, M. Lassas and E. Somersalo: Direct and inverse problems for Maxwell's equations with scalar impedance. Helsinki University of Technology Research Reports **A455** (2003) pp 55. Submitted for publication.

[10] M. Lassas: Non–selfadjoint inverse spectral problems and their applications to random bodies, *Ann. Acad. Sci. Fenn. Math. Diss.* **103** (1995).

[11] S. McDowall: An electrodynamic inverse problem in chiral media, *Trans. Amer. Math. Soc.* **352** (2000) 2993–3013.

[12] D. Mitrea and M. Mitrea: Finite energy solutions of Maxwell's equations and constructive Hodge–decompositions on nonsmooth Riemannian manifolds, *J. Funct. Anal.* **190** (2002) 339–417.

[13] A. I. Nachman: Recostructions from boudary measurements, *Ann. of Math* **128** (1988) 531–576.

[14] L. Paquet: Problèmes mixtes pour le système de Maxwell. *Ann. Fac. Sci. Toulouse Math* **(5)4** (1982) no.2, 103–141.

[15] R. Picard: On the low frequency asymptotics in electromagnetic theory, *J. Reine Angew. Math.* **394** (1984) 50–73.

[16] P. Ola, L. Päivärinta and E. Somersalo: An inverse boundary value problem in electrodynamics, *Duke Math. J.* **70** (1993) 617–653.

[17] P. Ola and E. Somersalo: Electromagnetic inverse problems and generalized Sommerfeld's potentials, *SIAM J. Appl.Math* **56** (1996) 1129–1145.

[18] A. Ramm: *Multidimensional inverse scattering problems*, Pitman Monographs and Surveys in Pure and Applied Mathematics, Longman Scientific and Technical 1992.

[19] E. Sarkola: A unified approach to direct and inverse scattering for acoustic and electromagnetic waves, *Ann. Acad. Sci. Fenn. Math. Diss.* **101**, 1995.

[20] G. Schwarz: *Hodge Decomposition – A Method for Solving Boundary Value Problems*. Lecture Notes in Mathematics 1607, Springer-Verlag, Berlin (1995)

[21] Z. Sun and G. Uhlmann: An inverse boundary value problem for Maxwell's equations, *Arch. Rational Mech. Anal.* **119** (1992) 71–93.

[22] J. Sylvester: Linearizations of anisotropic inverse problems, *Inverse problems in Mathematical Physics*, Proc. Saariselkä, Finland 1992, ed. L. Päivärinta and E. Somersalo, Springer Lecture Notes in Physics 422, Springer– Verlag 1993, 231–241.

[23] J. Sylvester and G. Uhlmann: A global uniqueness theorem for an inverse boundary value problem, *Ann. of Math.* **125** (1987) 153–169.

PETRI OLA
UNIVERSITY OF OULU
DEPARTMENT OF MATHEMATICAL SCIENCES
UNIVERSITY OF OULU
PO BOX 3000
FIN–90014
FINLAND
   Petri.Ola@oulu.fi

LASSI PÄIVÄRINTA
UNIVERSITY OF OULU
DEPARTMENT OF MATHEMATICAL SCIENCES
UNIVERSITY OF OULU
PO BOX 3000
FIN–90014
FINLAND
   lassi@tols16.oulu.fi

ERKKI SOMERSALO
HELSINKI UNIVERSITY OF TECHNOLOGY
INSTITUTE OF MATHEMATICS
PO BOX 1100
FIN–02015
FINLAND
   Erkki.Somersalo@hut.fi

Microlocal Analysis of the X-Ray Transform
with Sources on a Curve

DAVID FINCH, IH-REN LAN, AND GUNTHER UHLMANN

ABSTRACT. We survey several settings where distributions associated to paired Lagrangians appear in inverse problems. We make a closer study of a particular case: the microlocal analysis of the X-ray transform with sources on a curve.

1. Introduction

As mentioned in this book's preface, microlocal analysis (MA) is very useful in inverse problems in determining singularities of the medium parameters. In this chapter we survey several such applications, including an elaboration of some applications of MA to tomography that were already mentioned in Section 5 of Faridani's chapter (pages 11–14). We recall below the general setting.

While in two-dimensional tomography it is often possible to irradiate an unknown object from all directions, in three dimensions it is usually not practical to obtain this many data. Moreover, since the manifold of lines in $\mathbb{R}^3$ is four dimensional, while the object under investigation is a function of three variables, it should suffice to restrict the measurements to a three-dimensional submanifold of lines. Of course, in practice one has only finitely many measurements, but the considerations of the continuous case can be used to guide the design of algorithms and sampling geometries.

Reconstruction from line integral data is never local. That is, to reconstruct a function f at a point $x \in \mathbb{R}^n$ requires more than the data of the line integrals of f over all lines passing through a neighborhood of x. However, if the full line integral transform is composed with its adjoint, the resulting operator is an elliptic pseudodifferential operator, which preserves singular supports. Moreover, to compute this composition at a point x only requires line integrals for lines which pass through a neighborhood of x, and so useful information can be determined about the unknown function from just local data. In the planar setting, these observations have been elaborated upon and built into a useful tool for

Uhlmann's work was partly supported by the NSF and a John Simon Guggenheim Fellowship.

microtomography (see the discussion and references in section 6 of Faridani's chapter). In three dimensions, the observation is less useful since it still involves X-rays from all directions, and it is natural to wonder what might be done for local reconstruction of the singular support, or more refined information about the singularities of the object under consideration, when only those lines in a three-dimensional family which also pass through an arbitrarily small neighborhood of the point x are used. That is, what may be reconstructed when the data is local, in the sense of this paragraph, and restricted in the sense of the preceding paragraph? The first tools to handle the restricted X-ray transform were developed in [GrU1], where the setting was the more general geodesic transform on a Riemannian manifold. This will be further discussed below. The theory of paired Lagrangian distributions plays an important role in the analysis.

In this chapter we also describe other situations in inverse problems where the theory of paired Lagrangian distributions is important. These distributions, whose wave front set consists of two cleanly intersecting conic Lagrangian manifolds, were initially defined in [MU], and further developed in the papers [GuU], [AU], [GrU], [GrU2]. For a summary of some of the results see Section 2.

The forward fundamental solution for the wave equation $\Box^{-1}$ is an operator whose wave front set consists of two cleanly intersecting conic Lagrangian manifolds One is the diagonal and the other is the forward flowout by the Hamiltonian vector field H_p of the characteristic variety, $\{p = 0\}$, of the wave equation. In fact, as it is shown in [MU], this is valid for any operator of real-principal type (see Section 3). The diagonal part preserves singularities while the flowout moves them.

This fact was used in [GrU] to show that from the singularities of the backscattering data one can determine the singularities of a conormal potential with some restrictions on the type of singularity but including bounded potentials. A conormal potential has wave front set contained in the normal bundle of a submanifold. A fundamental step in the proof of this result is the construction of geometrical optics solutions for the wave equation plus the conormal potential q with data a plane wave in the far past,

$$\begin{cases} (\Box + q(x))u(x,t,\omega) = 0 & \text{on} \quad \mathbb{R}^{n-1} \times S^{n-1}, \\ u(x,t,\omega) = \delta(t - x \cdot \omega), & t \ll 0, \end{cases}$$

where $\Box = \partial^2/\partial t^2 - \Delta_{\mathbb{R}^n}$ is the wave operator on $\mathbb{R}^{n+1}$ acting independently of ω.

The solution is constructed in the form

$$u = \delta(t - x \cdot \omega) + u_1 + u_2$$

where

$$u_1 = \Box^{-1}(q\delta(t - x \cdot \omega))$$

and u_2 is a smoother distribution.

In [GrU] a detailed study of the singularities of u_1 was made, using the fact that we know exactly how $\Box^{-1}$ propagates singularities. This is reviewed in Section 4.

Another inverse problem where the theory of paired Lagrangian distributions appears naturally is Calderón's problem [C]. The question is whether one can determine a conductivity $\gamma > 0$ by making voltage and current measurements at the boundary. This information is encoded in the so-called Dirichlet-to-Neumann map.

An important technique in the study of this inverse problem has been the construction of complex geometrical optics solutions for the conductivity equation div $\gamma \nabla u = 0$. Let $\rho \in \mathbb{C}^n \backslash 0$ satisfy $\rho \cdot \rho = 0$. For $|\rho|$ large these solutions have the form

$$u_\rho(x) = e^{x \cdot \rho} \gamma^{-1/2}(1 + \psi_1(x, \rho) + \psi_2(x, \rho))$$

Furthermore ψ_1 decays in $|\rho|$ for $|\rho|$ large uniformly in compact sets and ψ_2 decays in $|\rho|$ faster than ψ_1. See [U] for a recent survey and further developments and references.

The term ψ_1 is constructed by solving

$$\Delta_\rho \psi_1 = q$$

where $q = (\Delta \sqrt{\gamma})/\sqrt{\gamma}$ and

$$\Delta_\rho u = (\Delta + 2\rho \nabla)u.$$

As we show in Section 5, by taking the inverse Fourier transform in ρ, the operator Δ_ρ can be viewed as an operator of complex principal type and its inverse is in the class of operators associated to two intersecting Lagrangian manifolds, one being the diagonal and the other a flowout of a codimension two characteristic variety. The use of this fact to study Calderón's problem for conormal bounded conductivities is investigated in [GrLU1]. The case of conormal potentials has been considered in [GrLU].

In Section 6 we review the microlocal approach of [GuS], [G] to invert a class of generalized Radon transforms $\mathcal{R}$. In particular under the Bolker condition $\mathcal{R}^t \circ \mathcal{R}$ is an elliptic pseudodifferential operator and preserves singularities. Here $\mathcal{R}^t$ denotes the transpose of $\mathcal{R}$. For an application of this to seismic imaging see the chapter by de Hoop in this volume. Guillemin and Sternberg showed that the range of a generalized Radon transform satisfying the Bolker condition can be characterized as the solution set of a system of pseudodifferential equations. A left parametrix for this system is another example of a paired Lagrangian distribution.

In Section 7 we describe in more detail the setting of restricted Radon transforms studied in [GrU1] that were already mentioned above. If $\mathcal{C}$ is a geodesic complex satisfying certain geometric conditions, and $\mathcal{R}_\mathcal{C}$ is the restricted Radon (geodesic) transform for $\mathcal{C}$, then $\mathcal{R}_\mathcal{C}$ is a Fourier integral operator and $\mathcal{R}_\mathcal{C}^t \circ \mathcal{R}_\mathcal{C}$, microlocalized away from certain bad points, falls in the class of operators associated to the diagonal and a flowout Lagrangian. The symbol of this operator is calculated on the diagonal and a relative left parametrix is constructed.

In Section 8, we specialize the discussion of Section 7 to the complex of lines meeting a space curve in $\mathbb{R}^3$: this is motivated by the tomographic scanner design wherein an X-ray source moves on a trajectory in space and for each source point measurements are made on a two-dimensional detector. One of the results of the analysis of the operators $\mathcal{R}_C$ and $\mathcal{R}_C^t$ in [GrU1] is the relation of the wave front set of a distribution μ and those of $\mathcal{R}_C\mu$ and $\mathcal{R}_C^t\circ\mathcal{R}_C\mu$. (This was used by Quinto [Q1], who gave a more elementary presentation of the relation of the first two. His work is better known in the tomography community.) At about the same time as [Q1], Louis and Maass proposed $\mathcal{R}_C^t \circ \Delta \circ \mathcal{R}_C$ as a local tomography operator and made some experimental reconstructions, [LoM]. (The operator Δ was the Laplacian on the sphere. Their weighting in the adjoint is different, but that is immaterial to the analysis.) They wrote down an integral which gave the symbol for their operator. Subsequently, A. Katsevich [Ka] studied the mapping properties of $\mathcal{R}_C^t \circ \mathcal{R}_C$ (with the adjoint weighting of Louis and Maass) on wave front sets, found an expression for the principal symbol, and computed some asymptotic expansions of $\mathcal{R}_C^t\circ\mathcal{R}_C\mu$ near the additional geometric singularities, in the case where μ is a piecewise smooth function. At the same time, and independently, the second author studied $\mathcal{R}_C^t\circ\mathcal{R}_C$ in his thesis, [La]. He computed the symbol of $\mathcal{R}_C^t\circ\mathcal{R}_C$ on the diagonal and on the flowout Lagrangian, and the symbol for $\mathcal{R}_C^t\circ\mathcal{R}_C\mu$, when μ is a conormal distribution (satisfying certain geometric hypotheses on its wave front set). These results, somewhat reworked, are presented here for the first time, along with a few subsequent developments. A microlocal analysis of the restricted Doppler transform has been done recently in [R]. Other applications of the calculus of paired Lagrangian distributions to reflection seismology are considered in [Ha] and [N]. Applications to synthetic aperture radar (SAR) are given in [NC].

Acknowledgement. We thank Karthik Ramaseshan for his very useful comments on an earlier version of the paper.

2. Spaces of Paired Lagrangian Distributions

In this section we recall the spaces of conormal distributions and distributions associated with either a single Lagrangian manifold or two cleanly intersecting Lagrangian manifolds.

Let X be an n-dimensional smooth manifold, and $\Lambda \subset T^*X\backslash 0$ a conic Lagrangian manifold. The Hörmander space $I^m(\Lambda)$ of Lagrangian distributions on X associated with Λ consists [H] of all locally finite sums of distributions of the form

$$u(x) = \int_{\mathbb{R}^N} e^{i\phi(x,\theta)}a(x,\theta)d\theta,$$

where $\phi(x, \theta)$ is a nondegenerate phase function parametrizing Λ and

$$a \in S^{m+(n/4)-(N/2)}\left(X \times (\mathbb{R}^N \backslash 0)\right) = \{a \in C^\infty\left(X \times (\mathbb{R}^N \backslash 0)\right):$$
$$|\partial_x^\alpha \partial_\theta^\beta a(x, \theta)| \le C_{\alpha\beta K} \langle\theta\rangle^{m+(n/4)-(N/2)-|\alpha|} \quad \forall \alpha \in \mathbb{Z}_+^N, \ \beta \in \mathbb{Z}_+^n, \ x \in K \Subset X\}.$$

(Here we use the standard notation $\langle\theta\rangle = (1 + |\theta|^2)^{1/2}$.) For $u \in I^m(\Lambda)$, the wave front set $WF(u) \subset \Lambda$.

Let X and Y be smooth manifolds. The operators $F : C_0^\infty(X) \to \mathcal{D}'(Y)$ whose Schwartz kernel $K_F \in \mathcal{D}'(X \times Y)$ is a Lagrangian distribution associated to a conic Lagrangian manifold Γ (also called canonical relation) with respect to the twisted symplectic form $\omega_{T^*(X \times Y)} = \omega_{T^*(X)} - \omega_{T^*(Y)}$ are called Fourier integral operators. Here $\omega_{T^*(X)}, \omega_{T^*(Y)}$ denote the symplectic forms on $T^*(X), T^*(Y)$ respectively. We have the twisted wave front set $WF'(K_F) \subset \Lambda$ where

$$WF'(K_F) = \{(x, y, \xi, \eta) \in T^*(X \times Y)\backslash 0 : (x, y, \xi, -\eta) \in WF(K_F)\}.$$

Now let $S \subset X$ be a smooth submanifold of codimension k. Then the conormal bundle of S,

$$N^*S = \{(x, \xi) \in T^*X \backslash 0 : x \in S, \xi \perp T_x S\},$$

is a Lagrangian submanifold of $T^*X \backslash 0$; the space of distributions on X conormal to S is by definition

$$I^\mu(S) = I^{\mu+(k/2)-(n/4)}(N^*S).$$

If $h \in C^\infty(X, \mathbb{R}^k)$ is a defining function for S, with rank $(dh) = k$ at S, then $u(x) \in I^\mu(S) \implies$

$$u(x) = \int_{\mathbb{R}^k} e^{ih(x)\cdot\theta} a(x, \theta) \, d\theta, \quad a \in S^\mu\left(X \times (\mathbb{R}^k \backslash 0)\right).$$

For example, if δ_S is a smooth density on S, then $\delta_S \in I^0(S)$, while a distribution on $X \backslash S$ having a Heaviside-type singularity at S belongs to $I^{-k}(S)$. One easily sees that

$$I^\mu(S) \subset L^p_{\text{loc}}(X) \quad \text{if } \mu < -k(1 - 1/p).$$

Now, let Λ_0, $\Lambda_1 \subset T^*X \backslash 0$ be a cleanly intersecting pair of conic Lagrangians in the sense of [MU]. Thus, $\Sigma = \Lambda_0 \cap \Lambda_1$ is smooth and

$$T_{\lambda_0}\Sigma = T_{\lambda_0}\Lambda_0 \cap T_{\lambda_0}\Lambda_1 \quad \text{for all } \lambda_0 \in \Sigma.$$

Associated to the pair (Λ_0, Λ_1) is a class of Lagrangian distributions, $I^{p,\ell}(\Lambda_0, \Lambda_1)$, indexed by $p, \ell \in \mathbb{R}$, which satisfy $WF(u) \subset \Lambda_0 \cup \Lambda_1$ [MU],[GuU]. Microlocally, away from Σ,

$$I^{p,\ell}(\Lambda_0, \Lambda_1) \subset I^{p+\ell}(\Lambda_0 \backslash \Lambda_1) \text{ and } I^{p,\ell}(\Lambda_0, \Lambda_1) \subset I^p(\Lambda_1). \tag{2-1}$$

We have

$$\bigcap_\ell I^{p,\ell}(\Lambda_0, \Lambda_1) = I^p(X, \Lambda_1), \quad \bigcap_p I^{p,\ell}(\Lambda_0, \Lambda_1) = C^\infty(X).$$

The principal symbol of a paired Lagrangian distribution $I^{p,\ell}(\Lambda_0, \Lambda_1)$ consists of the pair of symbols $(\sigma^0_{p+\ell}, \sigma^1_\ell)$ of the Lagrangian distributions $I^{p+\ell}(\Lambda_0 \backslash \Lambda_1)$ and $I^p(\Lambda_1 \backslash \Lambda_0)$ away from the intersection Σ. For the definition of the symbol of a Lagrangian distribution see [H] and section 8 in this paper for more details. The symbols σ^0 and σ^1 each have a conormal singularity as they approach the intersection and the singularities satisfy a compatibility condition at the intersection (see [GuU] for more details).

The symbol calculus of [GuU] implies

THEOREM 2.1. Let $u \in I^{p,\ell}(\Lambda_0, \Lambda_1)$. If $\sigma_{p+\ell}(u) = 0$ on $\Lambda_0 \backslash \Sigma$ then $u \in I^{p,\ell-1} + I^{p-1,\ell}(\Lambda_0, \Lambda_1)$.

If $Y_2 \subset Y_1 \subset X$ are smooth submanifolds with

$$\text{codim}_X(Y_1) = d_1 \quad \text{and} \quad \text{codim}_X(Y_2) = d_1 + d_2,$$

then $N^* Y_1$ and $N^* Y_2$ intersect cleanly in codimension d_2. The space of distributions on X conormal to the pair (Y_1, Y_2) of orders μ, μ' is

$$I^{\mu,\mu'}(Y_1, Y_2) = I^{\mu+\mu'+(d_1+d_2)/2-n/4,\, -d_2/2-\mu'}(N^* Y_1, N^* Y_2)$$

$$= I^{\mu+(d_1/2)-(n/4),\, \mu'+(d_2/2)}(N^* Y_2, N^* Y_1).$$

If one introduces local coordinates $(x_1, \ldots, x_n)$ on X such that

$$Y_1 = \{x_1 = \cdots = x_{d_1} = 0\} = \{x' = 0\},$$

$$Y_2 = \{x_1 = \cdots = x_{d_1+d_2} = 0\} = \{x' = 0, x'' = 0\},$$

then $u(x)$ belongs to $I^{\mu,\mu'}(Y_1, Y_2)$ if and only if it can be written locally as

$$u(x) = \int_{\mathbb{R}^{d_1+d_2}} e^{i(x' \cdot \xi' + x'' \cdot \xi'')} a(x; \xi'; \xi'')\, d\xi'\, d\xi''$$

with $a(x; \xi'; \xi'')$ belonging to the product-type symbol class

$$S^{\mu,\mu'}(X \times (\mathbb{R}^{d_1} \backslash 0) \times \mathbb{R}^{d_2})$$

$$= \{a \in C^\infty : |\partial_x^\gamma \partial_{\xi''}^\beta \partial_\xi^\alpha a(x, \xi)| \le C_{\alpha\beta\gamma K} \langle \xi', \xi'' \rangle^{\mu-|\alpha|} \langle \xi'' \rangle^{\mu'-|\beta|}\}.$$

Let X be a smooth manifold of dimension n. We denote the diagonal

$$D = \{(x, \xi, x, \xi) : (x, \xi) \in T^*(X) \backslash 0\}.$$

The class of operators $F : C_0^\infty(X) \to \mathcal{D}'(X)$ whose twisted wave front set consists of two intersecting conic Lagrangian manifolds, one being the diagonal, is called the class of pseudodifferential operators with singular symbols. An important class of pseudodifferential operators with singular symbols are those whose other Lagrangian manifold, Λ_Σ, is a flowout. Let $\Sigma \subset T^* X \backslash 0$ be a smooth, codimension k conic submanifold, $1 \le k < n$ which is involutive with respect to the symplectic form $\omega_{T^*(X)}$ (that is, the ideal of functions vanishing on Σ is closed under the Poisson bracket). Thus $T_{(x,\xi)} \Sigma^\omega \subset T_{(x,\xi)} \Sigma$ is a k plane for all

$(x, \xi) \in \Sigma$, where $T_{(x,\xi)}\Sigma^\omega$ denotes the orthogonal complement of $T_{(x,\xi)}\Sigma$ with respect to the symplectic form. The distribution $\{T_{(x,\xi)}\Sigma^\omega\}$ is integral with integral submanifolds $\Xi_{(x,\xi)}$ called the bicharacteristic leaves of Σ. The flowout of Σ is the canonical relation $\Lambda_\Sigma \subset (T^*(X)\backslash 0) \times T^*(Y)\backslash 0$ given by

$$\Lambda_\Sigma = \{(x, \xi, y, \eta) \in \Sigma \times \Sigma : (y, \eta) \in \Xi_{(x,\xi)}\}.$$

In [AU] a composition calculus was developed for pseudodifferential operators with singular symbols when the other Lagrangian Λ_Σ is a flow out. Notice that $D \circ D = D$, $D \circ \Lambda_\Sigma = \Lambda_\Sigma$, $\Lambda_\Sigma \circ D = \Lambda_\Sigma$ and $\Lambda_\Sigma \circ \Lambda_\Sigma = \Lambda_\Sigma$. Here $C_1 \circ C_2$ denotes the composition of the relations C_1 and C_2. Thus one can expect that the composition of pseudodifferential operators with singular symbols for which the second Lagrangian is a flowout is again in the same class. A theorem of [AU] shows that this indeed the case. More precisely we have

THEOREM 2.2. *Let* $A_i \in I^{p_i, \ell_i}(D, \Lambda_\Sigma), i = 1, 2$, *with* Λ_Σ *a flowout as above.* *Then* $A_1 A_2 \in I^{p_1+p_2+k/2, \, l_1+l_2-k/2}(D, \Lambda_\Sigma)$. *The principal symbol of* $A_1 A_2$ *on* D, *away from the intersection, is given by*

$$\sigma(A_1 A_2)|_{D\backslash(D\cap\Lambda_\Sigma)} = (\sigma(A_1)\sigma(A_2))|_{D\backslash(D\cap\Lambda_\Sigma)}.$$

In [AU] the symbol of $A_1 A_2$ is also computed on the flowout Lagrangian away from the intersection with the diagonal.

Using this calculus one can prove the following estimate:

THEOREM 2.3. *Let* $A \in I^{p,\ell}(X, D, \Lambda_\Sigma)$ *with* Λ_Σ *a flowout as above.* *Then*

$$A : H^s_{\text{comp}}(X) \to H^{s+s_0}_{\text{loc}}(X) \quad \text{for all } s \in \mathbb{R}$$

if

$$\max(p + (k/2), \, p + l) \leq -s_0.$$

3. Parametrices for Principal Type Operators

The simplest example of an operator of principal type on $\mathbb{R}^n$, for $n \geq 2$, is the operator $D_{x_1} = (1/i)\partial/\partial x_1$. The forward fundamental solution is given by

$$E_+ f(x) = i \int_{-\infty}^{x_1} f(s, x') \, ds,$$

where we are using coordinates $x = (x_1, x')$. Let

$$\Lambda_+ = \{(x_1, x', \xi_1, \xi', y_1, x', \xi_1, \xi') \in T^*(\mathbb{R}^n) \times T^*(\mathbb{R}^n) : y_1 \geq x_1\}.$$

It is readily seen that Λ_+ is the forward flowout from $D \cap \{\xi_1 = 0\}$ by H_{ξ_1} where H_p denotes the Hamiltonian vector field of p. We have

$$WF'E_+ = D \cup \Lambda_+,$$

and in fact $E_+ \in I^{-1/2, -1/2}(D, \Lambda_+)$.

Another example of the class of operators whose Schwartz kernel has wave front set in two conic Lagrangian manifolds which intersect cleanly is the forward fundamental solution of the wave operator $\Box = \partial_t^2 - \sum_{i=1}^n \partial_{x_i}^2$. The forward fundamental solution is given by

$$\Box^{-1} f(t,x) = \int_0^t \int \frac{((t-s)^2 - |x-y|^2)_+^{-(n-1)/2}}{\Gamma\left(\frac{-n+3}{2}\right)} f(s,y)\,dy\,ds.$$

(The distribution $x_+^s/\Gamma(s+1)$ is defined by analytic continuation; see [H1, Section 3.2]). Notice that if $n \geq 3$ is odd, then $x_+^{-(n-1)/2}/\Gamma\left(\frac{-n+3}{2}\right) = \delta^{(n-3)/2}$. We have $\Box^{-1} \in I^{-3/2,-1/2}(D, \Lambda)$, where Λ is the forward flowout from $\Delta \cap \{p = 0\}$ by the Hamiltonian vector field H_p. Here p denotes the principal symbol of the wave operator: $p(t,x,\tau,\xi,t',x',\tau',\xi') = \tau^2 - |\xi|^2$.

The paper [MU] contains a symbolic construction of the forward parametrices ($PE = I + R$, with R smoothing) for pseudodifferential operators of real principal type. These parametrices were first studied in [DH]. We recall,

DEFINITION 3.1. Let $P(x,D)$ be an mth order classical pseudodifferential operator, with real homogeneous principal symbol $p_m(x,\xi)$. We say that P is of real principal type if (a) $dp_m \neq 0$ at char $P = \{(x,\xi) \in T^*X \backslash 0 : p_m(x,\xi) = 0\}$ so that char P is smooth, and (b) char P has no characteristics trapped over a compact set of X.

For $(x,\xi) \in$ char P, let $\Xi_{(x,\xi)}$ be the bicharacteristic of $P(x,D)$ (i.e., integral curve of H_{p_m}) through (x,ξ). Then the flowout canonical relation generated by char P,

$$\Lambda_P = \{(x,\xi;y,\eta) : (x,\xi) \in \text{char } P, \ (y,\eta) \in \Xi_{(x,\xi)}\}, \qquad (3\text{--}1)$$

intersects the diagonal D cleanly in codimension 1. In [MU], it was shown that $P(x,D)$ has a parametrix $Q \in I^{(1/2)-m,-1/2}(D, \Lambda_P)$.

We now review the mapping properties of a parametrix for a pseudodifferential operator of real principal type, acting on the spaces of distributions associated with one and two Lagrangians described in Section 2 (see [GrU]).

PROPOSITION 3.1. *Suppose $\Lambda_0 \subset T^*X \backslash 0$ is a conic Lagrangian intersecting char P transversally and such that each bicharacteristic of P intersects Λ_0 a finite number of times. Then, if $T \in I^{p,\ell}(D, \Lambda_P)$*

$$T : I^r(\Lambda_0) \to I^{r+p,\ell}(\Lambda_0, \Lambda_1),$$

where $\Lambda_1 = \Lambda_P \circ \Lambda_0$ is the flowout from Λ_0 on char P. Furthermore, for $(x,\xi) \in \Lambda_1 \backslash \Lambda_0$,

$$\sigma(Tu)(x,\xi) = \sum_j \sigma(T)(x,\xi;y_j,\eta_j)\sigma(u)(y_j,\eta_j),$$

where $\{(y_j,\eta_j)\} = \Lambda_0 \cap \Xi_{(x,\xi)}$.

The action of $I^{p,\ell}(D, \Lambda_p)$ on the class $I^{p',\ell'}(\Lambda_0, \Lambda_1)$ is described in the next proposition.

PROPOSITION 3.2. *Under the assumptions of Proposition* 3.1,

$$T : I^{p',\ell'}(\Lambda_0, \Lambda_1) \to I^{p+p'+1/2,\,\ell+\ell'-1/2}(\Lambda_0, \Lambda_1).$$

Thus, if Q is a parametrix for $P(x, D)$,

$$Q : I^{p',\ell'}(\Lambda_0, \Lambda_1) \to I^{p'+1-m,\,\ell'-1}(\Lambda_0, \Lambda_1).$$

The following result is also useful.

PROPOSITION 3.3. *Suppose $\Lambda_1 \subset T^*X\backslash 0$ is a conic Lagrangian which is characteristic for $P : \Lambda_1 \subset \operatorname{char} P$. Then, if $T \in I^{p,\ell}(D, \Lambda_P)$,*

$$T : I^r(\Lambda_1) \to I^{r+p+1/2}(\Lambda_1)$$

and thus

$$Q : I^r(\Lambda_1) \to I^{r+1-m}(\Lambda_1).$$

4. The Inverse Backscattering Problem for a Conormal Potential

In the wave equation approach to the inverse backscattering problem in the framework of the Lax–Phillips theory of scattering, the continuation problem of solving the wave equation plus a potential with data a plane wave in the far past is fundamental [GrU]:

$$\begin{cases} (\Box + q(x))u(x, t, \omega) = 0 & \text{on } \mathbb{R}^{n-1} \times S^{n-1} \\ u(x, t, \omega) = \delta(t - x \cdot \omega), & t \ll 0, \end{cases} \tag{4-1}$$

where $\Box = \partial^2/\partial t^2 - \Delta_{\mathbb{R}^n}$ is the wave operator on $\mathbb{R}^{n+1}$ acting independently of ω. For the inverse scattering problem one needs to understand the behavior of the solution $u(x, t, \omega)$ for t large.

In the case that q is a compactly supported smooth function we can write a solution of (4–1) in the form

$$u = \delta(t - x \cdot \omega) + a(t, x, \omega)H(t - x \cdot \omega).$$

where $H(x)$ denotes the Heaviside function and a is a smooth function of all variables. We have then that the wave front set of the solution satisfies

$$WFu \subset N^*\{t = x \cdot \omega)\} =: \Lambda_+.$$

Thus singularities propagate forward as time increases.

We now sketch the construction of the solution of (4–1) under the assumption that the potential $q(x)$ is conormal to a smooth codimension k submanifold. Let S be given by a defining function,

$$S = \{x \in \mathbb{R}^n : h(x) = 0\},$$

where $h \in C^\infty(\mathbb{R}^n, \mathbb{R}^k)$ satisfies $\mathrm{rank}(dh(x)) = k$ for $x \in S$; in addition we assume S has compact closure. Let

$$q(x) \in I^\mu(S) \quad \text{with } \mu < \begin{cases} -\max((1 - 2/n)k, \, k-1) & \text{if } n \geq 5, \\ -\max(k/2, \, k-1) & \text{if } n = 3 \text{ or } 4 \end{cases}$$

be compactly supported and real-valued. We have $q \in L^p(\mathbb{R}^n)$ for $p = n/2$ when $n \geq 5$ and $p > 2$ when $n = 3$ or 4.

Now define

$$S_1 = \{(x, t, \omega) \in \mathbb{R}^{n-1} \times S^{n-1} : x \in S\};$$

regarding $q(x)$ as a distribution on $\mathbb{R}^{n-1} \times S^{n-1}$ independent of t and ω, one has

$$q \in I^\mu(S_1).$$

We wish to find an approximate solution to (4–1). We look for an approximation

$$u \sim u_0 + u_1 + \cdots + u_j + \cdots$$

where $u_0(x, t, \omega) = \delta(t - x \cdot \omega)$ and such that the series on the right is (formally) telescoping when $\square + q$ is applied. The terms in the series are increasingly smooth. This type of solution is called a *geometrical optics* solution. Thus, $u_{j+1} = -\square^{-1}(q(x)u_j(x, t, \omega))$, where $\square^{-1}$ is the forward fundamental solution of $\square$. We only consider the first two more singular terms, the other terms are smoother as shown in [GrU]. We have

$$u_0 + u_1 = \delta(t - x \cdot \omega) - \square^{-1}(q(x)\delta(t - x \cdot \omega)). \tag{4-2}$$

Now, the most singular term in the expansion is

$$u_0(x, t, \omega) = \delta(t - x \cdot \omega) \in I^0(S_+),$$

where

$$S_+ = \{(x, t, \omega) \in \mathbb{R}^{n+1} \times S^{n-1} : t - x \cdot \omega = 0\}.$$

The submanifolds S_+ and S_1 intersect transversally; let $S_2 = S_+ \cap S_1$ be the resulting codimension $k + 1$ submanifold of $\mathbb{R}^{n+1} \times S^{n-1}$. Let $\Lambda_1 = N^*S_1, \Lambda_+ = N^*S_+$ and $\Lambda_2 = N^*S_2$ be the respective conormal bundles, which are conic Lagrangian submanifolds of $T^*(\mathbb{R}^{n+1} \times S^{n-1})\backslash 0$. The geometry of how these submanifolds intersect is summarized thus:

PROPOSITION 4.1. (1) $WF(q) \subset \Lambda_1$ and $WF(u_0) \subset \Lambda_+$.

(2) Λ_1 and Λ_+ are disjoint.

(3) Λ_2 intersects Λ_1 and Λ_+ cleanly in codimensions 1 and k, respectively, so that (Λ_1, Λ_2) and (Λ_+, Λ_2) are intersecting pairs.

The second term in (4–2) is

$$u_1 = -\square^{-1}\big(q(x,t)\delta(t - x \cdot \omega)\big),$$

where $\square^{-1}$ acts only in the (x, t) variables. We have

$$q(x,t) \cdot \delta(t - x \cdot \omega) \in I^{0,\mu}(S_+, S_2),$$

so that

$$WF(q \cdot \delta) \subset \Lambda_+ \cup \Lambda_2.$$

To obtain $WF(u_1)$, recall that

$$WF(\square^{-1}v) \subset (D \cup \Lambda_\square) \circ WF(v) \quad \text{for all } v \in \mathcal{E}'(\mathbb{R}^{n+1} \times S^{n-1}), \qquad (4\text{–}3)$$

where D is the diagonal of $T^*(\mathbb{R}^{n+1} \times S^{n-1})\backslash 0$ and $\Lambda_\square$ is the flowout of the characteristic variety

$$\operatorname{char}\square = \{(x, t, \omega; \xi, \tau, \Omega) : |\tau|^2 = |\xi|^2\}$$

of $\square$ (acting on $\mathbb{R}^{n+1} \times S^{n-1}$). In (4–3), $D \cup \Lambda_\square$ acts as a relation between subsets of $T^*(\mathbb{R}^{n+1} \times S^{n-1})\backslash 0$; of course, D acts as the identity. Also, $\Lambda_\square \circ \Lambda_+ = \Lambda_+$ since Λ_+ is characteristic for $\square$. Thus

$$WF(u_1) \subset \Lambda_+ \cup \Lambda_2 \cup \Lambda_\square \circ \Lambda_2.$$

Compared with the case of a smooth potential, u_1 has the additional singularity $\Lambda_2 \cup \Lambda_\square \circ \Lambda_2$.

An analysis of this contribution (see [GrU]) gives

PROPOSITION 4.2. $u_1 \in I^{-(n+1)/2}(\Lambda_+\backslash L) + I^{\mu+(k-2-n)/2}(\Lambda_-\backslash L), \quad t \gg 0.$

We note that when q is smooth, $u_1 \in I^{-(n+1)/2}(\Lambda_+)$. We describe below what Λ_- and L are. We have

$$L = \Lambda_\square \circ \Theta$$

where Θ is a conic neighborhood of $\Sigma_3 = \Lambda_2|_{S_3}$ with S_3 the set of points where the incoming plane wave and the surface S are tangent. We denote by $\Sigma = \Lambda_2 \cap \operatorname{char}\square$. Now

$$\Sigma = \Sigma_+ \cup \Sigma_-$$

with $\Sigma_+ = \Lambda_+ \cap \Lambda_2$. The "new" Lagrangian Λ_- is the flowout of $\Sigma_-\backslash\Sigma_3$ by H_p.

Using this additional singularity it is shown in [GrU] that we can recover the symbol of q from the singularities of the backscattering kernel; that is, the location and strength of the singularities of the q is determined by the singularities of the backscattering kernel. The crucial element in the proof, of importance in its own right, is the construction of geometrical optics solutions of (4–1) for q conormal as above.

For the case that q is a general potential, the operator

$$U : \mathcal{E}'(\mathbb{R}^n) \longrightarrow \mathcal{D}'(\mathbb{R}^n \times \mathbb{R} \times S^{n-1})$$

defined by

$$Uq(x, t, \omega) = \Box^{-1}(q(x) \cdot \delta(t - x \cdot \omega))$$

was studied in [GrU3], where the two following results are proved.

THEOREM 4.1.

$$U \in I^{-(n+4)/4, -1/2}(\Lambda_1, \Lambda_2),$$

where

$$\Lambda_1 = N^*\{(x, t, \omega, y;)x = y, t = x \cdot \omega\}, \quad \Lambda_2 = N^*\{|t - y \cdot \omega|^2 = |x - y|^2\}.$$

THEOREM 4.2.

$$U : H^s_{\text{comp}}(\mathbb{R}^n) \longrightarrow H^{s+1}_{\text{loc}}(\mathbb{R}^n \times \mathbb{R} \times S^{n-1}) \quad \text{for all } s < -\tfrac{1}{2},$$

with the endpoint result

$$U : H^{-1/2}_{\text{comp}}(\mathbb{R}^n) \longrightarrow B^{1/2}_{2,\infty,\text{comp}}(\mathbb{R}^n \times \mathbb{R} \times S^{n-1}).$$

Here $B^s_{p,\infty}$ denotes the standard Besov spaces of distributions with s derivatives having Littlewood–Paley components associated with large frequencies uniformly in L^p.

5. Operators of Complex Principal Type and Calderón's Problem

We first recall the inverse conductivity problem, also known as Calderón's problem. Let $\gamma \in C^2(\overline{\Omega})$ be a strictly positive function on $\overline{\Omega}$. The equation for the potential in the interior, with conductivity γ under the assumption of no sinks or sources of current in Ω, is

$$\text{div}(\gamma \nabla u) = 0 \text{ in } \Omega, \quad u|_{\partial\Omega} = f.$$

The Dirichlet-to-Neumann map is defined in this case as follows:

$$\Lambda_\gamma(f) = \left(\gamma \frac{\partial u}{\partial \nu} \right) \Big|_{\partial\Omega}.$$

Let $\rho \in \mathbb{C}^n \backslash 0$ satisfy $\rho \cdot \rho = 0$. A key step in the proof of unique determination of the conductivity γ from Λ_γ, and in the study of several other inverse problems, is the construction of *complex geometrical optics* solutions to the conductivity equation found in [SyU, SyU1]. (See also [U] and the references there for the applications of complex geometrical optics solutions to Calderón's problem and to other inverse problems.) For sufficiently large $|\rho|$ one can construct solutions to $\text{div}(\gamma \nabla u) = 0$ in $\mathbb{R}^n$ (extending γ to be 1 outside a large ball) of the form

$$u_\rho(x) = e^{x \cdot \rho} \gamma^{-1/2}(1 + \psi(x, \rho))$$

with

$$\|\psi(x, \rho)\|_{L^2(K)} \leq \frac{C}{|\rho|}$$

for every compact set K.

The term ψ is constructed by solving the equation

$$\Delta_\rho \psi = q(1 + \psi)$$

where $q = (\Delta\sqrt{\gamma})/\sqrt{\gamma}$ and

$$\Delta_\rho u = (\Delta + 2\rho\nabla)u.$$

The solution ψ is written in the form

$$\psi = \sum_{j=0}^{\infty} \psi_j$$

where

$$\psi_{j+1} = \Delta_\rho^{-1}(q\psi_j), \quad \psi_0 = 1.$$

The fundamental property of Δ_ρ^{-1} is that satisfies the estimate

$$\|\Delta_\rho^{-1} f\|_{L^2_\delta(\mathbb{R}^n)} \leq C \frac{\|f\|_{L^2_{\delta+1}(\mathbb{R}^n)}}{|\rho|} \quad \text{for } 0 < \delta < -1,$$

(see [SyU], [SyU1]), where $\|f\|^2_{L^2_\alpha(\mathbb{R}^n)} = \int |f(x)|^2 (1 + |x|^2)^\alpha \, dx$. Here we show that Δ_ρ can be viewed as an operator of complex principal type in the sense of Duistermaat–Hörmander [DH] and Δ_ρ^{-1} is a pseudodifferential operator with a singular symbol.

We take the Fourier transform of u in the $|\rho|$ variable.

$$v(x, r, \omega) = \int_{\mathbb{R}} e^{-ir\lambda} u(x, \lambda\omega) \, d\lambda.$$

Then the operator Δ_ρ is transformed into the operator

$$\Box_* = \Delta_x + 2i(\omega \cdot \nabla_x) \cdot \frac{\partial}{\partial r}$$

If $\rho \cdot \rho = 0$, we can write $\rho = (|\rho|/\sqrt{2})(\omega_R + i\omega_I)$ with $\omega_r, \omega_I \in S^{n-1}$ and $\omega_R \cdot \omega_I = 0$. Let

$$\mathcal{V} = \{\omega_R + i\omega_I \in S^{n-1} + iS^{n-1} : \omega_R \cdot \omega_I = 0\}.$$

Considered as an operator acting only in the (x, r) variables $\Box_*$ is of complex principal type in the sense of Duistermaat and Hörmander: $\Box_*$ has symbol

$$\sigma(\Box_*)(\xi, \tau; \omega) = -(|\xi|^2 - 2i(\omega \cdot \xi)\tau)$$
$$= -((|\xi|^2 - 2(\omega_I \cdot \xi)\tau) + i(\omega_R \cdot \xi)\tau) =: -(p_R + ip_I).$$

The functions p_R and p_I have linearly independent gradients and Poisson commute ($\{p_R, p_I\} = 0$) and so the characteristic variety Σ is codimension two and involutive. Moreover, the two-dimensional bicharacteristics are not trapped over a compact set, so $\Box_*$ is locally solvable. Now, $\Box_*$ actually acts on $\mathcal{D}'(\mathbb{R}^{n+1}_{x,r} \times \mathcal{V})$, with coefficients that depend on ω but without differentiation in the ω directions, and the above facts remain true as long as we work away from $0_{T^*\mathbb{R}^{n+1}} \times T^*\mathcal{V}$, which will always be the case below. Away from there, $\square_*$ possesses a parametrix $\square_*^{-1} \in I^{-2,0}(D, C_\Sigma)$ as we describe below.

We can write the (complex) Hamiltonian vector field of $-\frac{1}{2}\sigma(\square_*)$ as $H_R + iH_I$, where

$$H_R = (\xi - \tau\omega_I) \cdot \frac{\partial}{\partial x} - (\omega_I \cdot \xi)\frac{\partial}{\partial r} + \tau i_{\omega_I}^*(\xi) \cdot \frac{\partial}{\partial \Omega_I},$$

$$H_I = \tau\omega_R \cdot \frac{\partial}{\partial x} + (\omega_R \cdot \xi)\frac{\partial}{\partial r} - \tau i_{\omega_R}^*(\xi) \cdot \frac{\partial}{\partial \Omega_R}.$$

Here $\Omega = (\Omega_R, \Omega_I) \in T_\omega^*\mathcal{V}$ and $i_{\omega_A} : T_{\omega_A}S^{n-1} \hookrightarrow T_{\omega_A}\mathbb{R}^n$ is the natural inclusion for $A = R, I$. H_R and H_I span the annihilator $T\Sigma^\perp$ of $T\Sigma$ with respect to the canonical symplectic form on $T^*(\mathbb{R}^{n+1} \times \mathcal{V})$, and Σ is nonradial since the radial vector field

$$\xi \cdot \frac{\partial}{\partial \xi} + \tau\frac{\partial}{\partial \tau} + \Omega_R \cdot \frac{\partial}{\partial \Omega_R} + \omega_I \cdot \frac{\partial}{\partial \Omega_I} \notin T\Sigma^\perp.$$

(Recall — from [DH, 7.2.4], for example — that Σ *nonradial* means that the (two-dimensional) annihilator of $T\Sigma$ with respect to the symplectic form σ on T^*X does not contain the radial vector field $\sum_{i=1}^N \xi_i \frac{\partial}{\partial \xi_i}$ at any point.) The family of two dimensional subspaces $T\Sigma^\perp$ forms an integrable distribution in the sense of Frobenius, and its integral surfaces are the bicharacteristic leaves of Σ. It is easy to verify that no bicharacteristic leaf is trapped over a compact set and that the bicharacteristic foliation is regular (see [DH,§ 7]). The flowout of Σ is then the canonical relation $C_\Sigma \subset (T^*(\mathbb{R}^{n+1} \times \mathcal{V})\backslash 0) \times (T^*(\mathbb{R}^{n+1} \times \mathcal{V})\backslash 0)$ defined by

$$C_\Sigma = \{(x, r, \omega, \xi, \tau, \Omega; x', t', \omega', \xi', \tau', \Omega') : (x, r, \omega, \xi, \tau, \Omega) \in \Sigma,$$
$$(x', r', \omega', \xi', \tau', \Omega') = \exp(sH_R + tH_I)(x, r, \omega, \xi, \tau, \Omega) \text{ for some } (s, t) \in \mathbb{R}^2\}.$$

By the results of [DH], $\square_*$ is locally solvable, and admits a right-parametrix which we will denote by $\square_*^{-1}$, so that $\square_*\square_*^{-1} = I + E$ with E a smoothing operator. Although not stated in this way, since [DH] predates [MU] and [GuU], the parametrix of [DH] has a Schwartz kernel belonging to $I^{-2,0}(D', C'_\Sigma)$. Here, D is the diagonal as before and the prime denotes the twisting

$$(x, r, \omega, \xi, \tau, \Omega; \tilde{x}, \tilde{r}, \tilde{\omega}, \tilde{\xi}, \tilde{\tau}, \tilde{\Omega}) \to (x, r, \omega, \xi, \tau, \Omega; \tilde{x}, \tilde{r}, \tilde{\omega}, -\tilde{\xi}, -\tilde{\tau}, -\tilde{\Omega}).$$

Greenleaf, Lassas and Uhlmann [GLU] are using this microlocal approach to consider Calderón's problem when the conductivity γ has conormal singularities.

6. Microlocal Characterization of the Range of Radon Transforms

By a well known theorem of Fritz John, the range of the X-ray transform in $\mathbb{R}^3$ is characterized as a solution of an ultrahyperbolic equation. Guillemin and Sternberg in [GuS1] characterized microlocally the range of a very general class of Radon transforms. It was shown in [GuU] that the projection onto the range is an operator in the class of intersecting Lagrangians. In order to state these results we first describe the microlocal approach to the double fibration of Gelfand and Helgason.

Let X and Y be smooth manifolds with $\dim X = n$ and $\dim X \leq \dim Y$. Let Z be an embedded submanifold of $X \times Y$ of codimension $k < n$. We consider the double fibration diagram

$$
\begin{array}{ccc}
 & Z & \\
\rho \swarrow & & \searrow \pi \\
Y & & X
\end{array}
\qquad (6\text{--}1)
$$

where π and ρ are the natural projections onto X and Y, respectively. We also assume that π is proper.

We denote by G_x the fibers of the projection $\pi : Z \to X$, considered as submanifolds of Y and H_y the fibers of $\rho : Z \to Y$, considered as submanifolds of X. If μ is a smooth, nonvanishing measure on Z, then μ induces measures $d\mu_x$ on G_x and $d\mu_y$ on H_y. This gives rise to the generalized Radon transform, defined for $f \in C_0^\infty(X)$ by

$$
\mathcal{R}f(y) = \int_{H_y} f(x) \, d\mu_y(x), \ y \in Y. \qquad (6\text{--}2)
$$

The formal adjoint of $\mathcal{R}$ is given by

$$
\mathcal{R}^t g(x) = \int_{G_x} g(y) \, d\bar{\mu}_x(y), \ x \in X. \qquad (6\text{--}3)
$$

By standard duality arguments, $\mathcal{R}$ and $\mathcal{R}^t$ extend to act on distributions, $\mathcal{R} : \mathscr{E}'(X) \to \mathscr{D}'(Y)$ and $\mathcal{R}^t : \mathscr{D}'(Y) \to \mathscr{D}'(X)$.

It follows immediately from (6–2) that the Schwartz kernel of $\mathcal{R}$ is δ_Z, the delta function supported on Z defined by μ. Guillemin and Sternberg (see [Gu], [GuS]) first introduced microlocal techniques to the study of generalized Radon transforms noting that δ_Z is a Fourier integral distribution, and then studying the microlocal analogue of the double fibration (6–1). It follows from Hörmander's theory [H] that $\mathcal{R}$ is a Fourier integral operator of order $(\dim Y - \dim Z)/2$ associated with the canonical relation $\Gamma = N^* Z'$. Similarly, $\mathcal{R}^t$ is a Fourier integral operator associated with the canonical relation $\Gamma^t \subset T^*X \times T^*Y$, which is simply Γ with (x, ξ) and (y, η) interchanged.

Now consider the microlocal diagram

$$
\begin{array}{ccc}
 & \Gamma & \\
\rho \swarrow & & \searrow \pi \\
T^*Y & & T^*X
\end{array}
\qquad (6\text{--}4)
$$

where π and ρ again denote the natural projections, this time onto T^*X and T^*Y, respectively. We analyze the normal operator $\mathcal{R}^t \circ \mathcal{R}$. Concerning the wave front sets we have, by a theorem of Hörmander and Sato (see [H]), that for $f \in \mathcal{E}'(X)$,

$$WF((\mathcal{R}^t \circ \mathcal{R})f) \subset (\Gamma^t \circ \Gamma)(WF(f)).$$

In general $\Gamma^t \circ \Gamma$ can be a quite complicated object, but under certain assumptions one can prove that it is a canonical relation, in fact the diagonal D.

For example, if Γ is a canonical graph (i.e. the graph of a canonical transformation $\chi : T^*X \to T^*Y$), then this is the case and on the operator level, Hörmander's composition calculus applies to yield that $\mathcal{R}^t \circ \mathcal{R}$ is a pseudodifferential operator on X. This happens if π and ρ are local diffeomorphisms and ρ is 1-1. This is the case for the generalized Radon transforms considered in [B] and [Q].

If $\dim X < \dim Y$, however, Γ cannot be a canonical graph. This is the case for the X-ray transforms and geodesic X-ray transforms in dimensions ≥ 3. Guillemin [G], motivated by work of Bolker on the discrete Radon transform, introduced a condition that guarantees that $\Gamma^t \circ \Gamma$ is still the diagonal and allows the "clean intersection" composition calculus of Duistermaat and Guillemin [DG] to be applied. The Bolker condition is that

the map ρ in (6–4) is an embedding.

Guillemin then proved the following result:

THEOREM 6.1. *If the Bolker condition is satisfied, then $\mathcal{R}^t \circ \mathcal{R}$ is an elliptic pseudodifferential operator on X of order $\dim Y - \dim Z$. Hence, $\mathscr{R}$ is locally invertible. Moreover,*

$$\mathcal{R} : H^s_{\mathrm{comp}}(X) \to H^{s+(\dim Z - \dim Y)/2}_{\mathrm{loc}}(Y).$$

Other examples of generalized Radon transforms to which Theorem 6.3 applies include X-ray transforms and, more generally a class of geodesic X-ray transforms (see [GrU1], section 2 for the precise class of geodesic X-ray transforms). A consequence of the fact that the X-ray transform P is a Fourier integral operator is a precise description of the wave front set of Pf in terms of the wave front set of f, and also microlocal description of the Sobolev singularities. An elementary account is given in Theorem 3.1 of [Q1].

If the Bolker condition is satisfied then $\rho(\Gamma) =: \Sigma$ is a co-isotropic submanifold of $T^*(Y)\backslash 0$ of codimension $k = \dim Y - \dim X$. Locally the submanifold Σ is defined by $p_1(y, \eta) = \cdots = p_k(y, \eta) = 0$ such that the Poisson brackets of all the $p_i's$ vanish, i.e. $\{p_i, p_j\} = 0, i, j = 1, \ldots, k$ on Σ. The problem of showing that the Radon transform has for its range the solution set of a system of pseudodifferential equations is reduced in [GS1] to the construction of left parametrices for pseudodifferential equations of the form

$$P = P_1^2 + \cdots + P_k^2 + \sum_{i=1}^{n} A_i P_i + B,$$

where the principal symbol of P_i is p_i, $A'_i s$ and B are pseudodifferential operators of order zero.

It was shown in [GuU] that the left parametrix E lies in $I^{p,\ell}(D,\Lambda)$, where Λ denotes the joint flow out from $D \cap \Sigma$ by the H_{p_j}, $j = 1, \ldots, n$. The sets D and Λ intersect cleanly of codimension k on Σ.

We remark that the principal symbol of E on $D \setminus \Sigma$ is $(p_1^2 + \cdots + p_k^2)^{-1}$.

7. Restricted X-ray Transforms

If $W \subset Y$ is a submanifold, the restricted generalized Radon transform $\mathcal{R}_W f = \mathcal{R}f|_W$ will typically not satisfy the Bolker condition even if $\mathcal{R}$ does. It is then of interest to study what injectivity properties and estimates $\mathcal{R}_W$ satisfies (as compared with $\mathcal{R}$) and the operator theory associated with $\mathcal{R}_W$. This was done for the geodesic X-ray transform in [GrU1]. We denote by $X = (M, g)$ a complete, n dimensional simply connected Riemannian manifold. We assume, as in §2 of [GrU1], that the space of geodesics $Y =: \mathcal{M}$ is a smooth manifold of dimension $2n - 2$.

We now describe the structure of the microlocal diagram (6–4) for $\mathcal{C} \subset \mathcal{M}$, a geodesic complex satisfying an analogue of Gelfand's cone condition for the case of an admissible line complex [GGr]. Let

$$\mathcal{C}_x = \bigcup \{\gamma \in \mathcal{C} : x \in \gamma\}$$

which generates a cone with vertex at x

$$\Sigma_x = \bigcup \{\dot{\gamma} : \gamma \in \mathcal{C}_x\}.$$

Let $\gamma \in \mathcal{C}_x$ and $y \in \gamma$. The cone condition states that the tangent planes of Σ_x and Σ_y along $\gamma \in \mathcal{C}$ are parallel translates of each other.

We now describe the projections $\pi : \Gamma \to T^*M\setminus 0$ and $\rho : \Gamma \to T^*\mathcal{C}\setminus 0$ in the language of singularity theory. (Note that (6–4) is a diagram of smooth maps between manifolds, all of dimension $2n$.) First, one makes (see[GrU1,p.215) a curvature assumption on the cones Σ_x which guarantees that π has a Whitney fold (see[GoGu]), at least away from a codimension 3 submanifold of Γ (automatically empty if $n = 3$); furthermore, one microlocalizes away from the critical points of the complex. We denote by L the fold hypersurface of π, so that $\pi(L) \subset T^*M\setminus 0$ is an immersed hypersurface. Microlocally, the image $\pi(\Gamma)$ is a half-space in $T^*M\setminus 0$ with boundary $\pi(L)$. In fact, $\pi(\Gamma)$ is the support of the Crofton symbol $Cr_{\mathcal{C}}(x,\xi)$ of $\mathcal{C}$, defined by Gelfand and Gindikin [GGi]:

$$Cr_{\mathcal{C}}(x,\xi) = \#\{\gamma \in \mathcal{C}_x : \dot{\gamma} \perp \xi\}$$

if finite and 0 otherwise. $Cr_{\mathcal{C}}$ is piecewise constant and jumps by 2 across $\pi(L)$. So far, we have only used the curvature assumption, not the cone condition. The projection $\rho : \Gamma \to T^*\mathcal{C}\setminus 0$ is necessarily singular at L, since π is (this is a general fact about canonical relations), but for an arbitrary geodesic complex, little can be said about the structure of ρ. However, using Jacobi fields, one can show [GrU1,p.225] that, assuming that the curvature operator can be smoothly diagonalized, the cone condition forces ρ to be a *blow $-$ down* at L; that is, ρ has the singularity type of polar coordinates in $\mathbb{R}^2$ at the origin (crossed with a diffeomorphism in the remaining $2n-2$ variables). Thus, ρ is 1-1 away from L, $\rho|_L$ has 1-dimensional fibers, and $\rho(L)$, which is thus of codimension 2, is symplectic (noninvolutive) in the sense that $\omega_{T^*\mathscr{C}}|_{\rho(L)}$ is nondegenerate. Furthermore, the fibers of ρ are the lifts by π of the bicharacteristic curves of the hypersurface $\pi(L) \subset T^*M \backslash 0$.

Some canonical relations having the singular structure described above were independently considered by Guillemin [Gu1], for reasons arising in Lorentzian integral geometry.

We denote by $\mathcal{R}_\mathcal{C}$, $\mathcal{R}_\mathcal{C}^t$ the geodesic transform restricted to $\mathscr{C}$ and its transpose. In [GrU1] it is proven that

THEOREM 7.1. *Denoting by $\Lambda_{\Pi(L)}$ the flowout of $\Pi(L)$, we have*

$$\mathcal{R}_\mathcal{C}^t \circ \mathcal{R}_\mathcal{C} \in I^{-1,0}(D, \Lambda_{\Pi(L)}).$$

The symbol on D away from the intersection is computed in [GrU1]. By using Theorem 1.1 and the functional calculus of [AU] a relative left parametrix is constructed for $\mathcal{R}_\mathcal{C}$.

We remark that this result has as corollary Theorem 5.3 in Faridani's chapter (page 13), which was explicitly stated in Theorem 4.1 in [Q1].

More details are given in the next sections on the computation of the symbols in both Lagrangians for the case that the complex of curves are straight lines going through a curve satisfying some additional conditions.

8. The Complex of Lines Through a Curve in $\mathbb{R}^3$

In this section, we will study in more detail a specific case of a restricted X-ray transform, that of the complex of line passing through a curve in $\mathbb{R}^3$. Our goal is to compute the principal symbol on the diagonal, and the symbol on the flowout Lagrangian. In view of applications to limited data problems in computed tomography, we suppose that the restricted transform acts on functions (distributions) with support contained in a given set, and that the curve lies outside this set. Specifically, we suppose that Ω is a bounded open set in $\mathbb{R}^3$, that the curve C_v lies outside the closed convex hull of Ω, and that the tangent to C_v never points into Ω. Taking $a(t)$ to be an arc length parametrization, we parametrize the family of lines passing though C_v by $C_v \times S^2$, where the pair (a, θ) is associated to the line $a + \mathbb{R}\theta$ through a in direction θ. Notice that there is some redundancy, since the same line is also associated to $(a, -\theta)$, and any line which meets the curve C_v more than once is counted multiple times. Rewriting

(6–2) and (6–3) for this specific restricted transform we have

$$\mathcal{R}_C f(a, \theta) = \int_{\mathbb{R}} f(a + s\theta)\, ds \qquad (8\text{--}1)$$

where s is an arc length parameter on $\mathbb{R}$. The formal adjoint $\mathcal{R}_C^t$, which here maps $C^\infty(C_v \times S^2)$ to $C^\infty(\Omega)$, is defined for $x \in \Omega$ by

$$\mathcal{R}_C^t g(x) = \int_{C_v} g\left(a(t), \frac{x - a(t)}{|x - a(t)|}\right) \frac{1}{|x - a(t)|^2}\, dt. \qquad (8\text{--}2)$$

The operators $\mathcal{R}_C$ and $\mathcal{R}_C^t$ satisfy

$$\int_{C_v \times S^2} \mathcal{R}_C f(a, \theta) g(a, \theta)\, dt\, d\theta = \int_{\Omega} f(x) \mathcal{R}_C^t g(x)\, dx \qquad (8\text{--}3)$$

for smooth f and g, with f compactly supported in Ω. This relation is used to extend $\mathcal{R}_C$, by duality, to compactly supported distributions in Ω.

In the situation of the complex C described above, it is possible to study the geometry of $\Gamma^t \circ \Gamma$ directly to find the intersecting Lagrangians. (The definition of Γ is slightly modified, since it is only an immersed submanifold.) It is found that that $\Gamma^t \circ \Gamma$ is the union of a subset of the diagonal relation consisting of all $(x, \xi, x, \xi) \in T^*(\Omega)\backslash 0 \times T^*(\Omega)\backslash 0$ such that the plane through x with normal ξ intersects the curve C_v, and another set consisting of all $(x, \xi, y, \eta) \in T^*(\Omega)\backslash 0 \times T^*(\Omega)\backslash 0$ subject to the condition that x and y lie in a line through C_v, ξ and η are normal to the line and to the tangent to the curve at the point of intersection, and $s_2 \xi = s_1 \eta$, where s_1 (resp. s_2) is the distance from x (resp. y) to the point of intersection. This is parametrized by $(t, s_1, s_2, \theta, u) \to (a(t) + s_1\theta, us_2\beta, a(t) + s_2\theta, us_1\beta)$ where β is a conormal vector at $a(t)$ annihilating both the tangent vector to the curve and the tangent vector to the line in direction θ. Computations in local coordinates show that this map is an immersion when $s_1 \neq s_2$ and also when $s_1 = s_2$ provided that $a''(t)$, $a'(t)$ and θ are linearly independent. Moreover, in the second case, this is found to be precisely the condition for clean intersection between the image and the diagonal relation (see also the discussion prior to (3.20) in [GrU2]). We let $\overline{\Lambda}'$ be the full set, and Λ' be the image of the relatively open subset where $a''(t)$, $a'(t)$, and θ are linearly independent. For (x_0, ξ_0) such that (x_0, ξ_0, x_0, ξ_0) lies in Λ', choose one (if there be more than one) t_0 such that $x_0 = a(t_0) + s\theta$ and ξ_0 is normal to θ and $a'(t_0)$, and then $a'(t_0) \cdot \xi_0 = 0$ while $a''(t_0) \cdot \xi_0 \neq 0$. By the implicit function theorem, there is a conic neighborhood of ξ_0 and a smooth function $t(\xi)$, homogeneous of degree 0, such that $a'(t(\xi)) \cdot \xi = 0$. Defining $p(x, \xi) = (x - a(t(\xi))) \cdot \xi$ it is then the case that the sheet of Λ parametrized using $t(\xi)$ is the H_p flowout of $p = 0$.

THEOREM 8.1. *The symbol of $\mathcal{R}_C^t \circ \mathcal{R}_C$ on $D\backslash\Lambda'$ is given by*

$$\sigma^0(x, \xi) = \sum_{\{t : a(t) \in (x + \xi^\perp) \cap C_v\}} \frac{2\pi}{|\xi \cdot a'(t)||x - a(t)|^{n-2}} \sigma_{Id}^0(x, \xi), \qquad (8\text{--}4)$$

where σ^0_{Id} is the symbol of the identity operator considered as a reference section of $L \otimes \Omega^{1/2}$, and where it is also assumed that the sum is finite. The symbol of $\mathcal{R}^t_C \circ \mathcal{R}_C$ on $\overline{\Lambda}' \backslash D$ is given by

$$\sigma^1(x, \xi, y, \eta) = c \frac{1}{\sqrt{|y-a|-|x-a|}} |d\nu|^{1/2},$$

where $|d\nu|^{1/2}$ is the half-density on Λ' induced from the parametrization above, a is the point where the line through x and y meets the curve, and c incorporates some powers of 2π and of i.

The hypothesis that there are only finitely many intersections between any plane $x + \xi^{\perp}$ and C_v is true generically. Lan has proved also

THEOREM 8.2. *Let C be a compact smooth space curve. If C has non-vanishing torsion, then the set of intersection numbers of C with planes is bounded above.*

We will outline the proof of the symbol result, but the calculations on the flowout are too lengthy to be presented in detail here. Most of them can be found in [La], and will also be reported in another work in preparation. The principal symbol (i.e. on the diagonal) can be calculated by several methods. The easiest is to find a Fourier representation of $\mathcal{R}^t_C \circ \mathcal{R}_C f$ by carrying through the calculations of formula (3.6) in [GrU1]. As this can be done expeditiously, we include it here. Moreover, for this part of the calculation, the dimension n may be greater than three as well.

Since the line integral of f through $a(t)$ in the direction $\xi = \frac{x-a}{|x-a|}$ is equal to the line integral of f through x in the same direction, we have

$$\begin{aligned}
\mathcal{R}^t_C \circ \mathcal{R}_C f(x) &= \int_{C_v} \int_R f(a(t) + s\xi)|x - a(t)|^{1-n} \, ds \, dt \\
&= \int_{C_v} \int_R f(x + s\xi)|x - a(t)|^{1-n} \, ds \, dt \\
&= (2\pi)^{1-n} \int_{C_v} \int_{\xi^{\perp}} e^{ix \cdot \eta} \hat{f}(\eta) \, dv_{\xi^{\perp}}(\eta)|x - a(t)|^{1-n} \, dt \\
&= (2\pi)^{-n} \int_{R^n} e^{ix \cdot \eta} b(x, \eta) \hat{f}(\eta) \, d\eta,
\end{aligned}$$

where $b(x, \eta)$ is the pushforward of $2\pi |x - a(t)|^{1-n} \, dv_{(x-a)^{\perp}}(\eta) \, dt$ under the map $(\eta, t) \to \eta$, with $dv_{(x-a)^{\perp}}(\eta)$ the Lebesgue measure on $(x-a)^{\perp}$ and dt the arc length measure on C_v. In writing this, we have presumed that the pushforward measure has a density with respect to Lebesgue measure on $\mathbb{R}^n$. This will hold provided the set of critical points has measure zero [GuS, p. 304]. Let $\{u_1(t), \dots, u_{n-1}(t)\}$ be a smoothly varying orthonormal basis for $(x - a(t))^{\perp}$ and let $\zeta = (\zeta_1, \dots, \zeta_{n-1})$ be the coordinates of $\eta \in (x - a(t))^{\perp}$ with respect to this basis. Then the map $(\eta, t) \to \eta$ is given by $G(\zeta, t) = \sum \zeta_i u_i(t)$. The differential is given by

$$dG = [u_1, \ldots, u_{n-1}, \sum \zeta_i u'_i],$$

and since the $\{u_i\}_{i=1}^{n-1}$ span $(x - a)^{\perp}$, the absolute value of the determinant is just $\left| \sum \zeta_i u'_i \cdot \frac{x-a}{|x-a|} \right|$. Now since $u_i \cdot \frac{x-a}{|x-a|} = 0$, we have $u'_i \cdot \frac{x-a}{|x-a|} = -u_i \cdot \frac{d}{dt} \frac{x-a(t)}{|x-a(t)|}$ and thus

$$|\det(dG)| = \left| \sum \zeta_i u'_i \cdot \frac{x-a}{|x-a|} \right| = \left| \eta \cdot \frac{d}{dt} \frac{x-a}{|x-a|} \right|.$$

Simplifying, we obtain,

$$|\det(dG)| = |\eta \cdot a'| |x - a|^{-1}.$$

From this is evident that the set of critical points has measure zero. Moreover, the density of the pushforward at a regular value is given by the sum over the preimages of (x, η) of the value of density $|x - a|^{1-n}$ times the reciprocal of the Jacobian, and thus

$$b(x, \eta) = 2\pi \sum_{\{t : a(t) \in (x+\eta^{\perp}) \cap C_v\}} \frac{1}{|\eta \cdot a'(t)| |x - a(t)|^{n-2}}.$$

We note that for a given x, η is a regular value if and only if the plane $x + \eta^{\perp}$ has only transversal intersections with the curve C_v which holds precisely when $(x, \eta, x, \eta) \in D \backslash \Lambda'$.

The symbol of a Fourier integral operator is usually expressed in terms of an amplitude and phase function, when the Schwartz kernel of the operator is given explicitly as an oscillatory integral. We do not have an explicit phase function parametrizing the Lagrangian Λ, so we must approach the problem differently. Here we use an intrinsic characterization based on the asymptotics of testing the Schwartz kernel against localized oscillatory functions. A development can be found in [H, Sections 3.2 and 3.3], and, in the specific form in which we use it, in [D, Section 4.1].

DEFINITION 8.1. The principal symbol of order m of a Fourier integral distribution K of order m associated to the conic Lagrangian manifold Λ is the element in

$$S^{m+(n/4)}(\Lambda, \Omega_{1/2} \otimes L) / S^{m+(n/4)-1}(\Lambda, \Omega_{1/2} \otimes L) \qquad (8\text{–}5)$$

given by

$$\alpha \longrightarrow e^{i\psi(\pi(\alpha), \alpha)} \langle u e^{-i\psi(x, \alpha)}, K \rangle. \qquad (8\text{–}6)$$

Here $S^{\mu}(\Lambda, \Omega_{1/2} \otimes L)$ denotes the symbol space of sections of the complex line bundle $\Omega_{1/2} \otimes L$ over Λ, of growth order μ; moreover $u \in C_0^{\infty}(X, \Omega_{1/2})$, $\psi(x, \alpha) \in C^{\infty}(X \times \Lambda)$ is homogeneous of degree 1 in α, and the graph of $x \mapsto d_x \psi(x, \alpha)$ intersects Λ transversally at α.

To apply this, we need to find candidate functions ψ for which the graph of the differential is transverse to the Lagrangian where we wish to evaluate the symbol. If we can find such a ψ, which moreover has the form $\psi(x, y, w) = \psi_1(x, w) + \psi_2(y, w)$, where w is the point in the Lagrangian where the symbol is to be calculated, then we can evaluate the pairing when K is the Schwartz kernel of $\mathcal{R}_{\mathcal{C}}^t \circ \mathcal{R}_{\mathcal{C}}$ by

$$\langle K, e^{i(\psi_1+\psi_2)}\rho_1(x) \otimes \rho_2(y)\rangle = \langle \mathcal{R}_{\mathcal{C}}^t \circ \mathcal{R}_{\mathcal{C}} e^{i\psi_2}\rho_2, e^{i\psi_1}\rho_1\rangle$$
$$= \langle \mathcal{R}_{\mathcal{C}} e^{i\psi_2}\rho_2, \mathcal{R}_{\mathcal{C}} e^{i\psi_1}\rho_1\rangle.$$

This last pairing is an ordinary five dimensional integral over the product of $\mathcal{C}$ with two copies of $\mathbb{R}$ (for the line integrals). We will evaluate its asymptotics using the method of stationary phase.

Initially, we will assume that a point $w = (x_0, y_0, \xi_0, -\eta_0) \in \Lambda \backslash \Sigma$ is given, and that it lies in the flowout of the clean intersection subset of Σ, and define

$$\psi(x, y, w) = \langle x-x_0, \xi_0\rangle + \langle y-y_0, -\eta_0\rangle$$
$$+ \tfrac{1}{2}\langle y-y_0, y-y_0\rangle k(\xi_0, -\eta_0) + \tfrac{1}{2}\langle x-x_0, x-x_0\rangle h(\xi_0, -\eta_0), \quad (8\text{--}7)$$

where k and h are homogeneous of degree one in ξ_0 and η_0. When h is non-vanishing and k is identically zero, or the reverse, the graph of $d\psi$ is transverse to Λ at w. This is proved by showing that the 12×12 matrix whose first six columns represent the differential of the parametrization of Λ and whose last six columns represent the differential of the graph mapping has full rank. (The hypotheses of vanishing and non-vanishing of k and h are only to simplify the rank calculation.)

Next we substitute ψ for $w = (x_0, \tilde{\xi}, y_0, \tilde{\eta}) \in \Lambda \backslash \Sigma$ into the pairing (8) to obtain

$$\langle K, e^{i\psi}\rho_2(y) \otimes \rho_1(x)\rangle = \int \rho_1\left(a(t) + s_1\theta\right)\rho_2\left(a(t) + s_2\theta\right) e^{-i\tilde{\psi}} \, ds_1 \, ds_2 \, d\theta \, dt,$$

where $\tilde{\psi}$ is ψ evaluated at $(a(t) + s_1\theta, a(t) + s_2\theta)$. It is checked that $\tilde{\psi}$ has only the critical point corresponding to w if ρ_1 and ρ_2 have small enough support, then an application of stationary phase as $\tau \to \infty$ for $\tilde{\xi} = \tau\xi_0$, $\tilde{\eta} = \tau\eta_0$ gives the asymptotic expansion of the pairing. It is found, when exactly one of h, k is nonzero, that the leading term of the asymptotic expansion is given by

$$\frac{(2\pi)^{5/2}e^{\pi i\sigma/4}\rho_1\rho_2(x_0, y_0)}{k(\tilde{\xi}, \tilde{\eta})|\tilde{\xi}|\,\big||y_0 - a| - |x_0 - a|\big|^{1/2}|a(t(\tilde{\xi})) - y_0|^{1/2}|a'' \cdot \tilde{\xi}|^{1/2}} \quad \text{if } h = 0, \, k \neq 0,$$

$$\frac{(2\pi)^{5/2}e^{\pi i\sigma/4}\rho_1\rho_2(x_0, y_0)}{h(\tilde{\xi}, \tilde{\eta})|\tilde{\eta}|\,\big||y_0 - a| - |x_0 - a|\big|^{1/2}|a(t(\tilde{\xi})) - x_0|^{1/2}|a'' \cdot \tilde{\eta}|^{1/2}} \quad \text{if } k = 0, \, h \neq 0,$$

where the signature factor σ is given by

$$\sigma = 2 + \operatorname{sgn}((a''(t) \cdot \tilde{\xi})(|\tilde{\xi}| - |\tilde{\eta}|)).$$

We note that since we have assumed that w lies in the flowout of the clean intersection subset we have $a''(t) \cdot \tilde{\xi} \neq 0$. Now we must divide by the value of $\rho_1 \rho_2$ and account for the dependence of the asymptotics on the transverse Lagrangian, graph $d\psi$. Following the analysis in [H] or [D], it can be seen that the invariant expression of the symbol will be obtained by multiplying this asymptotic expression by $|P_L^* \omega|^{1/2}$, where P_L is the linear projection of the tangent space to the Lagrangian at w onto the tangent space to the fiber of $T^*(\Omega \times \Omega)$ along the tangent space to the graph of $d\psi$ at w, and ω is the volume induced in the fiber as the quotient of the volume from the symplectic form and the volume on the base. (The projection is non-singular by the hypothesis of transversal intersection.) These may be evaluated when $h = 0, k \neq 0$ and $h \neq 0, k = 0$ using the same coordinates as were used in the preceding calculations. Multiplying the leading terms above by these half-density factors, it is found that both expressions produce

$$\frac{(2\pi)^{5/2} e^{i\pi\sigma/4}}{\left||y_0 - a| - |x_0 - a|\right|^{1/2}}. \tag{8–8}$$

Taking account that the signature factor changes by $\pm i$ along any line in the flowout when passing through the diagonal, we may incorporate this in the denominator, to obtain the square root of the difference of $|y_0 - a|$ and $|x_0 - a|$. This analysis breaks down when w corresponds to a point in $\overline{\Lambda} \backslash \Lambda$. However, one can also approach the analysis of the Schwartz kernel of $\mathcal{R}_C^t \circ \mathcal{R}_C$ by another method. It can also be expressed as the pushforward under the natural projection from $C_v \times \Omega \times \Omega$ to $\Omega \times \Omega$ of the pullback by a submersion of a conormal distribution on the product of two two-spheres. One can then check that the tranversality condition of [GuS] is satisfied above $\overline{\Lambda}$ away from the diagonal, so that the Schwartz kernel is a Lagrangian distribution (in fact, conormal) on the flowout of the nonclean intersection subset as well. Since the symbol must be smooth, we can extend by continuity the formula obtained above.

The method used above for computing the symbol σ^1 using (8) can also be used to obtain the specific form of the principal symbol σ^0; the details are worked out in [La]. This was the version used by Ramaseshan in [R] where he needed to compute the principal symbol of the Doppler transform restricted to the complex of straight line through a space curve.

Finally we would like to point out that Propositions 3.1 and 3.3 have some interesting consequences in tomography. It is sometimes taken as a useful approximation to represent the object to be reconstructed as a superposition of products of a smooth function with the characteristic function of a set with smooth boundary, which places it in the category of conormal distributions considered in Section 2. The tomographer is interested in reconstructing the discontinuities (singularities) of the object, but Propositions 3.1 and 3.3 say that a local method (applying a differential or pseudodifferential operator to $\mathcal{R}_C^t \circ \mathcal{R}_C$) will always produce artifacts due to the flowout Lagrangian. More specifically, suppose μ is a conormal distribution associated to a surface S and that $(x_0, \xi_0, x_0, \xi_0) \in \Sigma$. Let $p(x, \xi) = (x - a(t(\xi))) \cdot \xi$, for ξ in a conic neighborhood of ξ_0 be as described prior to Theorem 8.1, let $P(x, D)$ be a pseudodifferential operator with symbol $p(x, \xi)$, so that microlocally $\Lambda' = \Lambda_P$ with Λ_P as in (3–1). One can then prove that char P intersects N^*S transversally at (x_0, ξ_0) provided that $x_0 - a(t(\xi_0))$ is not an asymptotic vector to the surface at x_0. (Of course, this holds automatically if the surface has positive Gaussian curvature at x_0.) Using $\Lambda_0 = N^*S$ and $\Lambda_1 = \Lambda_P \circ \Lambda_0$ as in Proposition 3.1, we have from Theorem 7.1, Proposition 3.1, and Theorem 8.1 that if $\mu \in I^r(N^*S)$ has non-zero symbol at (x_0, ξ_0) then $\mathcal{R}_C^t \circ \mathcal{R}_C \mu \in I^{r-1}(\Lambda_0 \backslash \Lambda_1)$ and $\mathcal{R}_C^t \circ \mathcal{R}_C \mu \in I^{r-1}(\Lambda_1 \backslash \Lambda_0)$, and the symbol of the latter is non-zero. This means that the propagated singularities have the same strength as the singularities which were to be recovered. (A similar observation, in a specific case, was also made by Katsevich in [Ka].) However, the structure of Λ_1 in this case, being the conormal bundle of a ruled surface, may provide evidence that it could be an artifact. Furthermore, applying $P(x, D)$ to $\mathcal{R}_C^t \circ \mathcal{R}_C \mu$ would decrease the order of the singularities in the flowout, though at the expense of changing the symbol on Λ_0 as well.

References

[AU] J. L. Antoniano and G. A. Uhlmann, *A functional calculus for a class of pseudodifferential operators with singular symbols*, Pseudodifferential Operators and Applications, Proceedings of Symposia in Pure Mathematics, vol. 43, AMS, 1985, pp. 5–16.

[B] G. Beylkin, *The inversion problem and applications of the generalized Radon transform*, Comm. Pure Appl. Math. **37**(1984), 579–599.

[BoQ] J. Boman and E. T. Quinto, *Support theorems for real-analytic Radon transforms*, Duke Math. J. **55** (1987), no. 4, 943–948.

[C] A. P. Calderón, *On an inverse boundary value problem*, Seminar on Numerical Analysis and its Applications to Continuum Physics, Soc. Brasileira de Matemática, Río de Janeiro, 65-73, 1980.

[D] J. J. Duistermaat, *Fourier integral operators*, Birkhäuser, Boston, 1996, corrected reprint of 1973 Courant notes.

[DH] J. J. Duistermaat and L. Hörmander, *Fourier integral operators II*, Acta. Math., **128** (1972), 183–269.

[GGi] I. M. Gelfand and S. Gindikin, *Nonlocal inversion formulas in real integral geometry*, Functional Analysis and its Applications. **11** (1977), 173-179.

[GGr] I. M. Gelfand and M. I. Graev, *Line complexes in the space $\mathbb{C}^n$*, Functional Analysis and its Applications **1** (1967), 14-27.

[GGvS] I. M. Gelfand, M. I. Graev, and Z. Ya. Shapiro, *Differential forms and integral geometry*, Functional Analysis and its Applications **3** (1969), 24–40.

[GoGu] M. Golubitsky and V. Gullemin, *Stable mappings and their singularities*, Graduate Texts in Mathematics **14**, Springer-Verlag, New York, 1973.

[GrU] A. Greenleaf and G. Uhlmann, *Recovering singularities of a potential from singularities of a potential from singularities of backscattering data*, Comm. Math. Phys., **157**(1993), 549-572.

[GrU1] A. Greenleaf and G. Uhlmann, *Nonlocal inversion formulas for the X-ray transform*, Duke Math. J. **58** (1989), no. 1, 205–240.

[GrU2] A. Greenleaf and G. Uhlmann, *Estimates for singular Radon transforms and pseudodifferential operators with singular symbols*, J. Func. Anal., **89** (1990), 202–232.

[GrU3] A. Greenleaf and G. Uhlmann, *Characteristic space-time estimates for the wave equation*, Math. Zeitschrift, **236**, (2001), 113-131.

[GrLU] A. Greenleaf, M. Lassas and G. Uhlmann, *The Calderón problem for conormal potentials I: Global uniqueness and reconstruction*, Comm. Pure Appl. Math., **56** (2003), 328-352.

[GrLU1] A. Greenleaf, M. Lassas and G. Uhlmann, manuscript in preparation.

[Gu] V. Guillemin, *On some results of Gel'fand in integral geometry*, in Pseudodifferential operators and applications (Notre Dame, Ind., 1984), pp. 149–155. Amer. Math. Soc., Providence, RI, 1985.

[Gu1] V. Guillemin, *Cosmology in (2+1)-dimensions, cyclic models, and deformations of $M_{2,1}$*, Annals of Math. Studies 121, Princeton U. Press, Princeton, N. J., 1989.

[GuS] V. Guillemin and S. Sternberg, *Geometric asymptotics*, Mathematical Surveys and Monographs, vol. 14, American Mathematical Society, Providence, Rhode Island, 1977.

[GuS1] V. Guillemin and S. Sternberg, *Some problems in integral geometry and some related problems in micro-local analysis*, Am. J. Math., **101**, (1979), 915–955.

[GuU] V. Guillemin and G. Uhlmann, *Oscillatory integrals with singular symbols*, Duke Math. J **48** (1981), no. 1, 251–267.

[H] L. Hörmander, *Fourier integral operators I*, Acta Math. **127** (1971), 79–183.

[H1] L. Hörmander, *The analysis of linear partial differential operators I*, Springer-Verlag, Berlin, 1983.

[Ha] S. Hansen, *Solution of an inverse problem by linearization*, Comm. PDE **16** (1991), 291–309.

[Ka] A. Katsevich, *Cone beam local tomography*, SIAM J. Appl. Math. **59** (1999), no. 6, 2224–2246.

[La] I.-R. Lan, *On an operator associated to a restricted X-ray transform*, Ph.D. thesis, Oregon State University, July 1999.

[LoMa] A. K. Louis and P. Maass, *Contour reconstruction in 3-D X-Ray CT*, IEEE Trans. Med. Imag. **12** (1993), 764–769.

[MU] R. Melrose and G. Uhlmann, *Lagrangian intersection and the Cauchy problem*, Comm. Pure Appl. Math **32** (1979), 483–519.

[N] C. Nolan, *Scattering near a fold caustic*, SIAM J. Appl. Math. **61** (2000), 659–673.

[NC] C. Nolan and M. Cheney, *Microlocal analysis of synthetic aperture radar imaging*, to appear in J. Fourier Analysis and Applications.

[Q] E. T. Quinto, *The dependence of the generalized Radon transform on defining measures*, Trans. Amer. Math. Soc. **257** (1980), no. 2, 331–346.

[Q1] E. T. Quinto, *Singularities of the X-ray transform and limited data tomography in $\mathbf{R}^2$ and $\mathbf{R}^3$*, SIAM J. Math. Anal. **24** (1993), no. 5, 1215–1225.

[R] K. Ramaseshan, *Microlocal analysis of the restricted Doppler transform*, to appear J. Fourier Analysis and Applications.

[SyU] J. Sylvester and G. Uhlmann, *A global uniqueness theorem for an inverse boundary problem*, Ann. of Math., **125** (1987), 153–169.

[SyU1] J. Sylvester and G. Uhlmann, *A uniqueness theorem for an inverse boundary value problem arising in electrical prospection*, Comm. Pure Appl. Math., **39**(1986), 91-112.

[U] G. Uhlmann, *Developments in inverse problems since Calderón's foundational paper*, Essays on Harmonic Analysis and Partial Differential Equations, pp. 295–345, Chicago Lectures in Math., Univ. Chicago Press, Chicago, IL, 1999, edited by M. Christ, C. Kenig and C. Sadosky.

DAVID FINCH
DEPARTMENT OF MATHEMATICS
OREGON STATE UNIVERSITY
CORVALLIS, OR, 97331
UNITED STATES
    finch@math.orst.edu

IH-REN LAN
VERITAS DGC INC.
10300 TOWN PARK DRIVE
HOUSTON, TX 77072
UNITED STATES
    ih-ren_lan@veritasdgc.com

GUNTHER UHLMANN
DEPARTMENT OF MATHEMATICS
UNIVERSITY OF WASHINGTON
SEATTLE, WA, 98195
UNITED STATES
    gunther@math.washington.edu

Microlocal Analysis of Seismic Inverse Scattering

MAARTEN V. DE HOOP

ABSTRACT. We review applications of microlocal analysis (MA) to reflection seismology. In this inverse method one attempts to estimate the index of refraction of waves in the earth from seismic data measured at the Earth's surface. Seismic imaging creates images of the Earth's upper crust using seismic waves generated by artificial sources and recorded into extensive arrays of sensors (geophones or hydrophones). The technology is based on a complex, and rapidly evolving, mathematical theory that employs advanced solutions to a wave equation as tools to solve approximately the general seismic inverse problem, with complications introduced by the heterogeneity and anisotropy of the Earth's crust. We describe several important developments using MA to generate these wave-solutions by manipulating the wavefields directly on their phase space. We also consider some recent applications of MA to global seismology.

1. Introduction

Microlocal analysis plays an increasingly important role in seismology, particularly in the imaging and inversion of seismic data. Here we consider imaging and inversion via the generalized Radon transform (GRT), concentrating on advances since the work of Beylkin [9], applying the work of Guillemin [51] and Taylor [98]. It is the aim of this exposition to connect microlocal analysis with seismology in the context of inverse scattering. The analysis of a related problem, the X-ray transform (see Greenleaf and Uhlmann [48; 49]) also contributes to the further understanding of the GRT in seismology. Microlocal analysis and the general theory of Fourier integral operators are described in the books by Hörmander [60; 61; 62], Duistermaat [43], and Treves [101; 102].

The author thanks The Mathematical Sciences Research Institute for partial support, through NSF grant DMS-9810361. He also thanks the members of the Mathematical Sciences Research Institute, and in particular Gunther Uhlmann, for providing a very stimulating environment during the Inverse Problems program in Fall 2001. The author thanks John Stockwell for his many suggestions to improve this survey.

Exploration seismology versus global seismology. Exploration seismology concerns the investigation of sedimentary structures in the upper crust of the earth, whereas global seismology concerns the investigation of the entire earth. In exploration reflection seismology, the following processes are amenable to the application of microlocal analysis: dip and azimuth moveout and seismic wavefield continuation, (map) migration, imaging, amplitude versus scattering angles analysis, inversion and resolution analysis, and migration velocity analysis. (Some of these processes also find application in acoustic emission and sonic borehole imaging.) In global seismology, the primary phases amenable to the application of microlocal analysis are earthquake generated body waves, both those that interact with the main transitions in the deep earth (such as the core mantle boundary) and those used in transmission wave-equation tomography. Also, spectral asymptotics applies to the study of free oscillations of the earth; this subject is beyond the scope of this survey. Here we discuss primarily exploration seismology, and we provide a brief outlook on global seismology.

Inverse scattering in seismology, in principle, yields an estimate of a distribution representing the elastic stiffness tensor in the earth. This tensor appears as coefficients in the elastic wave equation. From a geoscientist's perspective, however, stiffness is a manifestion of geodynamical processes such as mantle convection, magneto-hydrodynamics of the outer core, deformation and subduction of the continental crust, and sedimentary processes, with their own underlying mathematical models. Thus the inverse scattering problem becomes one of seismic waves coupled to one of these dynamical processes.

Caustics. The importance of microlocal techniques becomes apparent if caustics are formed in propagating wavefields in the earth. Caustics arise due to the heterogeneity and anisotropy of the elastic properties of the subsurface. It may be a matter of scale, though, whether or not some of the anisotropy originates from heterogeneity. Caustics form progressively in heterogeneous media, but may be intrinsic to the presence of anisotropy. Caustics due to heterogeneity are ubiquitous. For example, in models with small, smooth, random fluctuations in wave speed, which vary on a length scale large compared to a wave length but small compared with the propagation distance, caustic formation will occur with probability one; see White *et al.* [110].

Historical perspective. Some of the notions developed in microlocal analysis appear to have been independently discovered in seismology. Most notably, Hagedoorn [53] invented a purely graphical method for seismic imaging that is recognizable as a Fourier integral operator and its canonical relation. Rieber [85] and later Riabinkin [84] determined and exploited the "slopes" in addition to arrival times of the seismic events, which relates directly to the wavefront set of the data, to unravel complexities in the wavefield. Stolt [93] carefully used the notion of *migration dip* in imaging, which aids in the reconstruction of the wavefront set of the subsurface's stiffness. In *map migration* [108; 52; 109] (for the state of the art, see [66]) an injectivity condition is assumed that appears in Guillemin's [51] Bolker condition in the treatment of the generalized Radon transform.

1A. Exploration seismology. In a seismic experiment one generates elastic waves in the earth using active sources at the earth's surface. The waves that return to the surface of the earth are observed; see Figure 1 (in fact, sources and receivers are not always on the surface of the earth; this case is also considered). The problem is to reconstruct the elastic properties of the subsurface from the data thus obtained.

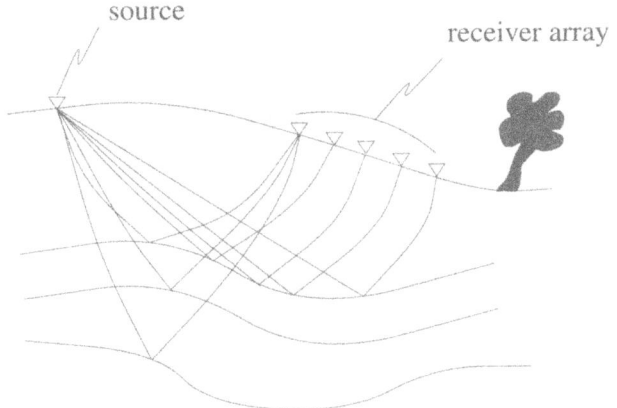

Figure 1. Schematic depth section showing reflection ray paths; a seismic experiment.

Hagedoorn's early approach to imaging of seismic reflection data can be summarized and illustrated as follows. If we restrict the seismic experiment to the acquisition of coinciding sources and receivers (*zero offset*) then the inverse scattering problem is formally determined. Figure 2(b) shows an earth model with a single reflecting surface; Figure 2(a) is a display of reflection data (seismic "traces") that would be produced at 81 locations at zero depth above the reflector. The source was a regularized delta function generating a pulse. For a pulse to travel down and up 1 km in upper medium with a speed of 2 km/s takes one second. The specular reflections occur at those points where the characteristic (ray) from the source/receiver location is normal to the reflector. The reflection can be mapped into the reflector as follows. On Figure 2(a) choose a source/receiver location (at the surface) and draw an isochrone curve (a circle for constant velocity) through any event on the corresponding seismic trace. This is illustrated in Figure 2 (bottom) for all seismic traces in Figure 2(a). The envelope of the isochrones delineates the reflector surface.

The Hagedoorn-derived methods being based totally on geometrical optics did not explicitly consider amplitudes. Amplitudes were considered by the inverse scattering based methods that followed. The approach presented in this survey originates with the work of Beylkin [9; 10; 11] and other authors (see the references below), applying microlocal analysis to the seismic inverse problem. Beylkin [10] considered the seismic inverse scattering problem in acoustic media with constant density. He modeled the data using the Born approximation, wherein the scattering is linearized in the medium coefficients. The medium perturbation $\delta c(x)$ acts as a distribution of scatterers superimposed on a smooth background medium $c(x)$. Given the background medium $c(x)$ an operator was given to reconstruct $\delta c(x)$ microlocally from an n-dimensional subset of the data (from data that constitute a function of n variables). Beylkin-derived methods excluded caustics, assumed scalar wave propagation, and isotropy.

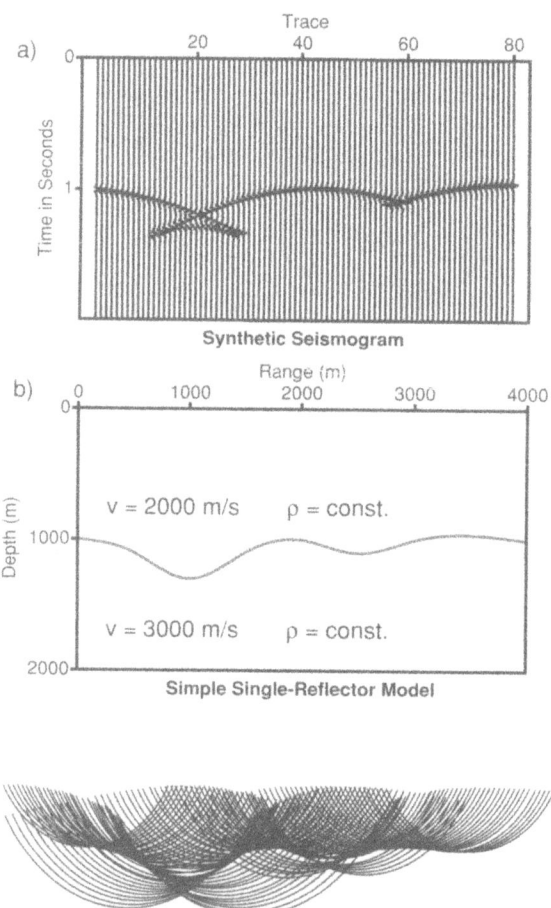

Figure 2. Top: zero-offset (coinciding source/receiver) reflection. Middle: reflector, where v = acoustic wave speed, ρ = density. Bottom: compass construction. After [18].

Present perspective on modeling and inversion. Data that are redundant in the sense that they are a function of more than n variables (*multiple offset* data) can be seen as a family of n-dimensional datasets, where each n-dimensional subset in the family has a fixed value of some coordinate, which we refer to as e. The result of the inversion, some manifestation of a reflectivity $r(x, e)$, should not depend on e. This is the criterion that must be used to estimate the background medium from the data; see for instance Symes [95]. Under the assumptions made by Beylkin [10], there exists microlocally an invertible map, transforming seismic data to a reflectivity function $r(x, e)$, of which the singular part should not depend on e. We consider such a transformation in a general framework that allows the presence of caustics and in anisotropic elastic rather than isotropic acoustic media. The treatment of elastic waves is based upon the decoupling of the hyperbolic system into n scalar equations (see Taylor [98], Ivrii [67], Dencker [39]) after which Fourier integral operator techniques are invoked. Each scalar equation governs the propagation of a particular mode, such as qP and $S_{1,\ldots,n-1}$.

It is common to distinguish two ways of modeling reflection data. In the first way, we assume the Born approximation. This approximation is essentially a linearization, wherein the medium parameters are written as the sum of a background medium and a perturbation that is assumed to be small and localized. It is assumed that the background is smooth and that the perturbation contains the singularities of the medium. In the second, it is assumed that the medium consists of different regions separated by smooth interfaces. The medium parameters are assumed to be smooth on each region, and smoothly extendible across each interface, but they vary discontinuously at an interface. Such interfaces are the *seismic reflectors*. We discuss how to model the high-frequency part of the data using Fourier integral operators, following the approach of Taylor [98], obtaining a generalization of the Kirchhoff approximation.

Subject to certain geometrical assumptions, including the Bolker condition [51], the multi-modal data can be written as an invertible Fourier integral operator, H_{MN} say, acting on a *reflectivity* distribution, $r_{MN}(x, e)$, that is a function of subsurface position x and the additional variable e, essentially parametrizing the scattering angle and scattering azimuths between an incoming and outgoing characteristic. The position of the singularities of $r_{MN}(x, e)$ does not depend on e. In the Kirchhoff approximation for elastic media, the function $r_{MN}(x, e)$ equals to highest order $R_{MN}(x, e) \|\partial z_n / \partial x\| \delta(z_n(x))$, where $R_{MN}(x, e)$ is the appropriately normalized reflection coefficient for the pair of elastic modes (M, N), and $\|\partial z_n / \partial x\| \delta(z_n(x))$ is the singular function of the interface given as a level set. For the Born approximation $r_{MN}(x, e)$ is given by pseudodifferential operators that take into account the radiation patterns acting on the medium perturbation. The coordinate e is *a priori* defined only on the coisotropic submanifold $\mathcal{L} \subset T^*Y \setminus 0$ — where Y represents the acquisition manifold to which the scattered wavefield (δG) is restricted via an operator $\mathcal{R}$ — that contains the wavefront set of the data. To construct an invertible Fourier integral operator from data to the function $r_{MN}(x, e)$, the coordinate e has to be defined on an open part of $T^*Y \setminus 0$; in Stolk and De Hoop [91] an extension is constructed of the coordinate function e from $\mathcal{L}$ to an open neighborhood of $\mathcal{L}$ in $T^*Y \setminus 0$ (which is not unique).

For the Born approximation, an inverse is also obtained in the least-squares sense via the parametrix of a normal operator, N_{MN}. The normal operator and its regularized inverse, $\langle N_{MN}^{-1} \rangle$ render a coupled spatial-parameter resolution analysis of the seismic experiment. The normal operator also provides a means to analyze nonmicrolocal contributions occurring if the Bolker condition is violated and replaced by a weaker condition leading to a characterization of artifacts.

When the data are redundant there is in addition a criterion to determine if the medium above the interface (the background medium in the Born approximation) is correctly chosen. The position of the singularities of the function $r_{MN}(x, e)$, obtained by acting with H_{MN}^{-1} on the data, should not depend on e. There exist pseudodifferential operators, W_{MN}, that, if the medium above the interface is correctly chosen, annihilate the data. This allows one to carry out an extension of differential semblance optimization in elastic media with caustics.

As mentioned, the wavefront set of the data is contained, under certain conditions, in a coisotropic submanifold of the acquisition cotangent bundle. It reveals a structure of characteristic strips. Restricting in the imaging operator the seismic data to a common coordinate value on these strips, yields a generalized Radon transform [10; 34; 38] that maps the reflection data into a seismic image. (Under certain conditions this generalized Radon transform is a Fourier integral operator.) Collecting these seismic images from the points on the characteristic strips corresponding to available data results in the set of so-called common-image-point gathers. In the presence of caustics, a filter needs to be designed and applied prior to extracting a trace from each of the common-image-point gathers in the set, to form a model image of the singular component of the medium.

From this image, we model seismic data that correspond to a different coordinate value on the characteristic strips. The result of this procedure is a composition of Fourier integral operators yielding seismic wavefield continuation, be it in the single scattering approximation. Relevant examples of seismic wavefield continuation are the *transformation to zero offset* [1] and the *transformation to common (prescribed) azimuth* [13]. The distribution kernel of transformation to zero offset is called dip moveout; the distribution kernel of transformation to common azimuth is called azimuth moveout.

Table 1 summarizes the operators that will be introduced in this survey.

Synopsis of publications. Many publications exist about high-frequency methods to invert seismic data in acoustic media. These methods date back to Hagedoorn [53]; from a seismic perspective, it has taken twenty years and more to develop the basic analysis of them [87; 31; 94; 75; 86]. From a mathematical perspective, the analysis started with the reconstruction of the singular compo-

Section		
[2,3]	modeling (FIO)	$[\delta G]$
[3]	acquisition (FIO)	$[F = \mathcal{R}\delta G]$
	imaging (FIO)	$[F^*]$
[4,5]	normal (Ψdo + nonmicrolocal operator)	$[N = F^*F]$
	resolution (Ψdo)	$[\langle N^{-1}\rangle N]$
	inversion (Least Squares, FIO)	$[\langle N^{-1}\rangle F^*]$
[6,7]	extended modeling (invertible FIO)	$[H, H^{-1}]$
[8]	annihilator (Ψdo)	$[W]$
[9]	generalized Radon transform (FIO)	$[L]$
[10]	"continuation" (FIO)	$[\mathcal{R}^c F \langle N^{-1}\rangle L]$

Table 1. Operators we will discuss. FIO stands for Fourier integral operator and Ψdo for pseudodifferential operator.

nent of the medium coefficients in the Born approximation, without caustics, by Beylkin [10]. Bleistein [17] discussed the case of a smooth jump using Beylkin's results. Rakesh [82] showed that the modeling operator in the Born approximation is a Fourier integral operator in the presence of caustics. Hansen [56] analyzed the inversion in an acoustic medium with multipathing for both the Born approximation and the case of a smooth jump. Ten Kroode *et al.* [99] extended the work of Hansen. Guillemin [51] discussed the Bolker condition in the context of generalized Radon transforms, that ensures invertibility of the modeling operator in the least-squares sense (see, e.g., De Hoop and Brandsberg–Dahl [33] and Stolk and De Hoop [91]). Stolk [89] simplified the analysis considering a case when the Bolker condition is violated. Nolan and Symes [80] discussed the imaging and inversion of seismic data with different (restricted) acquisition geometries.

The mathematical treatment of systems of equations, such as the elastic equations, in the high-frequency approximation has been given by Taylor [98]. This fundamental paper also discusses the interface problem. Beylkin and Burridge [12] discussed the imaging of seismic data in the Born approximation in isotropic elastic media, under a no-caustics assumption. De Hoop and Bleistein [32] discussed the imaging and inversion in general anisotropic elastic media, using a Kirchhoff-type approximation. In the presence of caustics, the foundations of this approximation were given by Stolk and De Hoop [91]. The generalized Radon transform in elastic media was developed in De Hoop *et al.* [34; 38]. The Born approximation for seismic data with maximal acquisition geometry in anisotropic elastic media allowing for multipathing was discussed and analyzed by De Hoop and Brandsberg–Dahl [33] and Stolk and De Hoop [91].

We mention two alternative (finite-frequency) but related approaches to inverse scattering of seismic data: The optimization approach (e.g., Tarantola [96; 97], De Hoop and De Hoop [35]), which falls into the category of *reverse time migration*, and the wavefield decomposition/double-square-root equation approach

(Claerbout [30], De Hoop *et al.* [36] and Stolk and De Hoop [92]), which falls into the category of *downward continuation migration.*

1B. Sedimentary environment. The medium parameters, stiffness c_{ijkl} and density ρ appear as the coefficients in the hyperbolic system of partial differential equations. Their properties, such as symmetry, are constrained by the types of rocks occurring in the subsurface and their microstructure as well as the ambient state of stress. Sedimentary rocks of interest include shales, sandstones, and carbonates (limestone, chalk, marlstone and dolomite).

Microscopically shales have anisotropic properties owing to the orientation of mineral grains; see Figure 3. Macroscopically, shales exhibit anisotropy due to the orientation of laminations owing to bedding or crossbedding. Their characteristic properties have been measured ultrasonically in the laboratory [68]. Their typical symmetry is hexagonal, with the restriction that triplication of the shear wave does not occur on the symmetry axis. Starting from their microstructure, shales were modeled mathematically by Hornby [63; 64] using contact theory [106; 107] on the one hand, and a combination of self consistent [58; 24] and differential effective medium approximations on the other. In view of their strongly anisotropic permeability, shales can be effective as seals over hydrocarbon reservoirs.

Reservoir rocks must be porous and permeable. These often consist of sandstones or carbonates. The porosity and permeability may result from intergrain voids or from fractures or from a combination of the two; see Figure 4. A sandstone is porous, and presumably filled with a fluid-gas mixture. Though sandstone by itself often may be assumed to be isotropic, fracturing breaks this symmetry typically to orthotropic. A commonly applied mathematical model for crack-induced anisotropy can be found in Hudson [65]. Even though the scattering theory presented here is valid in the "high-frequency" regime, the length

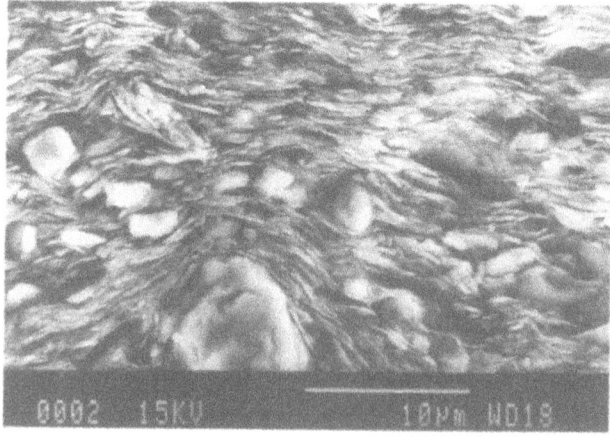

Figure 3. SEM picture of a typical shale (from Hornby *et al.* [64]).

scale of the microstructure of rocks is vastly smaller than the dominant seismic wavelengths. Effectively, the fluid-filled poro-elastic medium behaves like an (an)elastic solid. Some of the most established linear mathematical models for poro-elasticity, in particular those concerning a porous rock saturated with a fluid, were developed by Gassmann [47], Biot [14; 15; 16], Brown and Korringa [23], and Berryman and Milton [8]. Biot's equations were derived from microstructure using a homogenization approach by Burridge and Keller [26]. A theory that replaces the fluid in the pores by a gas-fluid mixture was developed by Batzle and Wang [7].

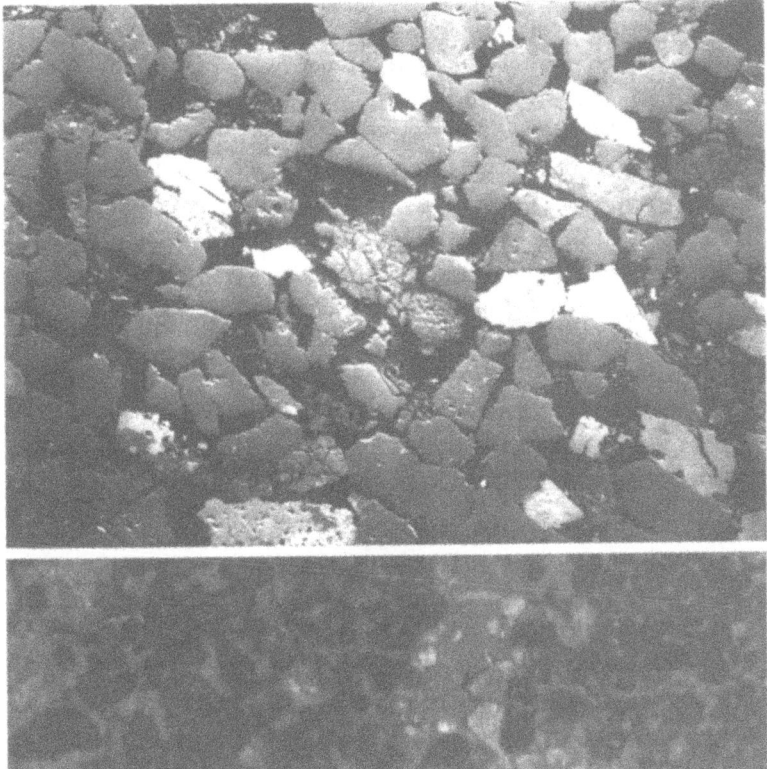

Figure 4. SEM picture of a sandstone (top) with a crack (light against darker background, running across the bottom image).

Seismic waves scatter at singularities in the elastic properties of the subsurface. These singularities are typically attributable to geological transitions (interfaces), unconformities, faults, as well as the interior structure of a formation such as one consisting of sand channels.

1C. Notation. Propagation of seismic waves occurs in the Earth's interior. The aim of seismic inverse scattering is to obtain information about selected target regions within the interior. The target is contained within an open set

$X \subset \mathbb{R}^n$. In practice $n = 2$ or 3, but we leave it unspecified. In exploration geophysics the subsurface refers to the shallow interior of the Earth. Subsurface position is denoted by x. Sources and receivers will be contained in open subsets O_s, O_r respectively, of the boundary ∂X of X. Their position is denoted by $\tilde{x}, \hat{x}$, respectively. Measurement of data takes place during a time interval $(0, T)$. The set of $(\hat{x}, \tilde{x}, t)$ for which data are taken is called the acquisition manifold Y; we assume that coordinates y' on Y are given. We assume that the particle displacement of the waves is measured for point sources at $\tilde{x}, t = 0$ with all its components, both at the source and at the receiver. Thus we assume that (after preprocessing) the data match the Green's function $G_{il}(\hat{x}, \tilde{x}, t)$, for $(\hat{x}, \tilde{x}, t) \in Y$.

We refer to the codimension of the set of $Y \subset \partial X \times \partial X \times (0, T)$ as the codimension of the acquisition manifold, and we denote it by c. Owing to the practicalities of data acquisition limits exist on the dimension of the set Y, which is expressed by c. For example, in marine data acquisition the receivers may lie along a line behind the source, in which case we have $n = 3, c = 1, \partial X = \{x \in \mathbb{R}^n : x_3 = 0\}$, $Y = \{(\hat{x}, \tilde{x}, t) \in \mathbb{R}^3 \times \mathbb{R}^3 \times (0, T) \mid \hat{x}_3 = \tilde{x}_3 = \hat{x}_2 - \tilde{x}_2 = 0\}$. We call such acquisition geometries common azimuth. Thus the data are a function of $2n - 1 - c$ variables. However, from such data we aim to determine a function of n variables; hence the data have redundancy with dimension $n - 1 - c$. The inverse problem is thus formally overdetermined.

The material presented in this survey has been published in the following papers: De Hoop et al. [34], De Hoop and Bleistein [32], Burridge et al. [25], De Hoop et al. [38], De Hoop and Brandsberg–Dahl [33], Stolk [89], Stolk and De Hoop [91], and Stolk [90].

2. Propagation of Elastic Waves in Smoothly Varying Media

Seismic wave amplitudes are sufficiently small such that the linearized theory of infinitesimal deformation applies. When combined with the equation of motion, this yields the elastic wave equation

$$\left(\rho \, \delta_{il} \frac{\partial^2}{\partial t^2} - \frac{\partial}{\partial x_j} c_{ijkl} \frac{\partial}{\partial x_k} \right) (\text{displacement})_l = (\text{volume force density})_i. \quad (2\text{--}1)$$

Here $\rho(x)$ is the volume density of mass and $c_{ijkl}(x)$ is the elastic stiffness tensor, with $i, j, k, l = 1, \ldots, n$.

2A. Decoupling the modes. In general, the elastic wave equation supports different wave types (modes). Seismologists easily identify the individual modes on seismograms. It is advantageous in the formulation of inverse scattering to trace the individual modes. Decoupling of the modes is accomplished by the diagonalization of system (2–1). To diagonalize this system, it is convenient to remove the x-dependent coefficient ρ multiplying the time derivative. Thus we introduce the equivalent system

$$P_{il}u_l = f_i, \qquad (2\text{-}2)$$

where

$$u_l = \sqrt{\rho}(\text{displacement})_l, \qquad f_i = \frac{1}{\sqrt{\rho}}(\text{volume force density})_i, \qquad (2\text{-}3)$$

and

$$P_{il} = \delta_{il}\frac{\partial^2}{\partial t^2} - \frac{\partial}{\partial x_j}\frac{c_{ijkl}}{\rho}\frac{\partial}{\partial x_k} + \text{l.o.t.} \qquad (2\text{-}4)$$

is the partial differential operator. Here we use the assumption that ρ is smooth and bounded away from zero. Both systems (2-1) and (2-2) are real, time reversal invariant, and their solutions satisfy reciprocity.

We describe how the system (2-2) can be decoupled by transforming it with appropriate pseudodifferential operators; see Taylor [98], Ivrii [67] and Dencker [39]. The goal is to transform the operator P_{il} by conjugation with a matrix-valued pseudodifferential operator $Q(x, D)_{iM}$, $D = D_x = -i\frac{\partial}{\partial x}$, to an operator that is of diagonal form, modulo a regularizing part,

$$Q(x, D)_{Mi}^{-1} P_{il}(x, D, D_t) Q(x, D)_{lN} = \text{diag}(P_M(x, D, D_t)\,;\, M = 1, \dots, n)_{MN}, \qquad (2\text{-}5)$$

where $D_t = -i\frac{\partial}{\partial t}$. The indices M, N denote the mode of propagation, and refer to qP and $S_{1,\dots,n-1}$ wave propagation. In fact, for the construction of Fourier integral operator solutions developed in the scalar wave case, it is sufficient to transform the partial differential operator to block-diagonal form, where each of the blocks $P_M(x, D, D_t)$ has scalar principal part (proportional to the identity matrix). In this case we will use the indices M, N to denote the block, and we will omit indices for the components within each block. Let

$$u_M = Q(x, D)_{Mi}^{-1}u_i, \qquad f_M = Q(x, D)_{Mi}^{-1}f_i. \qquad (2\text{-}6)$$

The system (2-2) is then equivalent to the uncoupled equations

$$P_M(x, D, D_t)u_M = f_M. \qquad (2\text{-}7)$$

Since the time derivative in P_{il} is already in diagonal form, it remains only to diagonalize its spatial part,

$$A_{il}(x, D) = -\frac{\partial}{\partial x_j}\frac{c_{ijkl}}{\rho}\frac{\partial}{\partial x_k} + \text{l.o.t.}$$

The goal becomes finding Q_{iM} and A_M such that (2-5) is valid with P_{il}, P_M replaced by A_{il}, A_M. The operator P_M is now

$$P_M(x, D, D_t) = \frac{\partial^2}{\partial t^2} + A_M(x, D).$$

Because of the properties of stiffness related to (i) the conservation of angular momentum, (ii) the properties of the strain-energy function, and (iii) the positivity of strain energy, subject to the adiabatic and isothermal conditions, the principal symbol $A_{il}^{\mathrm{prin}}(x, \xi)$ of $A_{il}(x, D)$ is a positive symmetric matrix. Hence, it can be diagonalized by an orthogonal matrix. On the level of principal symbols, composition of pseudodifferential operators reduces to multiplication. Therefore, we let $Q_{iM}^{\mathrm{prin}}(x, \xi)$ be this orthogonal matrix, and we let $A_M^{\mathrm{prin}}(x, \xi)$ be the eigenvalues of $A_{il}^{\mathrm{prin}}(x, \xi)$, so that

$$Q_{Mi}^{\mathrm{prin}}(x, \xi)^{-1} A_{il}^{\mathrm{prin}}(x, \xi) Q_{lN}^{\mathrm{prin}}(x, \xi) = \mathrm{diag}(A_M^{\mathrm{prin}}(x, \xi))_{MN}. \qquad (2\text{--}8)$$

(The eigenvalue-eigenvector system is sometimes referred to as the system of Christoffel equations.) The principal symbol $Q_{iM}^{\mathrm{prin}}(x, \xi)$ is the matrix that has as its columns the orthonormalized polarization vectors associated with the modes of propagation.

If the multiplicities of the eigenvalues $A_M^{\mathrm{prin}}(x, \xi)$ are constant, then the principal symbol $Q_{iM}^{\mathrm{prin}}(x, \xi)$ depends smoothly on (x, ξ) and microlocally equation (2–8) carries over to an operator equation. Taylor [98] has shown that if this condition is satisfied, then decoupling can be accomplished to all orders, where each block corresponds to a different eigenvalue. In fact, he proved the following slightly more general result.

LEMMA 2.1 (TAYLOR). *Suppose the pseudodifferential operator $Q_{iM}(x, D)$ of order 0 is such that*

$$Q(x, D)_{Mi}^{-1} A(x, D)_{il} Q(x, D)_{lN} = \begin{pmatrix} A_{(1)}(x, D) & 0 \\ 0 & A_{(2)}(x, D) \end{pmatrix}_{MN} + a(x, D)_{MN},$$

where the symbols $A_{(1)}(x, \xi)$ and $A_{(2)}(x, \xi)$ are homogeneous of order two and $a(x, \xi)_{MN}$ is polyhomogeneous of order one. Suppose the spectra of $A_{(1)}(x, \xi)$ and $A_{(2)}(x, \xi)$ are disjoint on a conic neighborhood of some $(x_0, \xi_0) \in T^*X \setminus 0$. Then by modifying Q with lower order terms the system can be transformed such that

$$a(x, D)_{MN} = \begin{pmatrix} a_{(1)}(x, D) & 0 \\ 0 & a_{(2)}(x, D) \end{pmatrix}_{MN} + \text{smoothing remainder,}$$

microlocally around (x_0, ξ_0).

This implies that if the multiplicity of a particular eigenvalue $A_M^{\mathrm{prin}}(x, \xi)$ is constant, then the system can be transformed such that the part related to this eigenvalue decouples from the rest of the system, modulo a smoothing remainder. In this survey we will assume that at least some of the modes decouple (microlocally). This is stated as Assumption 1 below.

We now give an alternative characterization of the quantities $A_M^{\mathrm{prin}}(x, \xi)$ and $Q_{iM}^{\mathrm{prin}}(x, \xi)$. The values $\tau = \pm\sqrt{A_M^{\mathrm{prin}}(x, \xi)}$ are precisely the solutions to the equation

$$\det P_{il}^{\mathrm{prin}}(x, \xi, \tau) = 0. \qquad (2\text{--}9)$$

The multiplicity of $A_M^{\text{prin}}(x, \xi)$ is equal to the multiplicity of the corresponding root of (2–9). The columns of $Q_{iM}^{\text{prin}}(x, \xi)$ satisfy

$$Q_{iM}^{\text{prin}} \in \ker P_{il}^{\text{prin}}\left(x, \xi, \sqrt{A_M^{\text{prin}}(x, \xi)}\ \right).$$

Because $P_{il}^{\text{prin}}(x, \xi, \tau)$ is homogeneous in (ξ, τ), one may choose to use the *slowness vector* $-\tau^{-1}\xi$ instead of the cotangent or wave vector ξ in the calculations. The set of $-\tau^{-1}\xi$ such that (2–9) holds is called the slowness surface, which can be easily visualized. A (section of the) slowness surface for the case of a transversely isotropic medium in $n = 3$ dimensions is given in Figure 5(a).

The slowness surface consists of n sheets each corresponding to a mode of propagation. The innermost sheet is convex and is associated with the qP wave. The other sheets need not be convex. The multiplicity of the eigenvalues changes at the points (directions) where the different sheets intersect. In seismology it is quite common to use a parametrization of the slowness surface that differs from the stiffness tensor that directly controls the geometry (shape) of the different sheets. Examples are the Lamé parameters for isotropic media and the Thomsen parameters [100] for transversely isotropic media. For a general insight into such parametrizations; see Tsvankin [103].

The second-order equations (2–7) clearly describe the decoupling of the original system into different elastic modes. These equations inherit the symmetries of the original system, such as time-reversal invariance and reciprocity. Time-reversal invariance follows because the operators $Q_{iM}(x, D), A_M(x, D)$ can be chosen in such a way that $Q_{iM}(x, \xi) = -\overline{Q_{iM}(x, -\xi)}$, $A_M(x, \xi) = \overline{A_M(x, \xi)}$. Then Q_{iM}, A_M are real-valued. Reciprocity for the causal Green's function $G_{ij}(x, x_0, t - t_0)$ means that $G_{ij}(x, x_0, t - t_0) = G_{ji}(x_0, x, t - t_0)$. Such a relationship also holds (modulo smoothing operators) for the Green's function $G_M(x, x_0, t - t_0)$ associated with (2–7). This follows because the transpose operator $Q(x, D)_{Mi}^t$ (obtained by interchanging x, x_0 and i, M in the distribution kernel $Q_{iM}(x, x_0)$ of $Q_{iM}(x, D)$) is also a pseudodifferential operator, with principal symbol $Q^{\text{prin}}(x, \xi)_{Mi}^t$. As noted before for the principal symbol, it follows from the fact that $A_{ij}^t = A_{ij}$ that we can choose Q orthogonal, which is to say, such that $Q(x, D)_{iM}Q(x, D)_{Mj}^t = \delta_{ij}$. From the fact that

$$G_M(x, x_0, t - t_0) = Q(x, D)_{Mi}^{-1}G_{ij}(x, x_0, t - t_0)Q(x_0, D_{x_0})_{jM}$$

it then follows that microlocally G_M is reciprocal, i.e.,

$$G_M(x, x_0, t - t_0) = G_M(x_0, x, t - t_0) \quad \text{modulo smoothing operators.}$$

Up to principal symbols, the equation above represents rotations at the receiver (i) and the source (j) side. In seismology this is referred to as the Alford rotation [3].

REMARK 2.2. We already observed that if an eigenvalue $A_M^{\text{prin}}(x, \xi)$ has constant multiplicity $m_M > 1$ say, then u_M is an m_M-dimensional vector and (2–7) is a

$m_M \times m_M$ system, with scalar principal symbol. For such a system a microlocal solution can be constructed in the same way as for scalar systems. In this case all kinematic quantities, such as bicharacteristics, phase functions, and canonical relations depend only on M. Other quantities such as u_M and $Q_{iM}(x, D)$ will have multiple components. The Green's function G_M and its amplitude A_M, to be introduced just before (2–20), are then $m_M \times m_M$ matrices. To simplify notation we do not take this into account explicitly.

2B. The Green's function. To evaluate the Green's function we use the first-order system for u_M that is equivalent to (2–7),

$$\frac{\partial}{\partial t} \begin{pmatrix} u_M \\ \partial u_M/\partial t \end{pmatrix} = \begin{pmatrix} 0 & 1 \\ -A_M(x, D) & 0 \end{pmatrix} \begin{pmatrix} u_M \\ \partial u_M/\partial t \end{pmatrix} + \begin{pmatrix} 0 \\ f_M \end{pmatrix}. \qquad (2\text{–}10)$$

This system can be decoupled also. Let $B_M(x, D) = \sqrt{A_M(x, D)}$, which is a pseudodifferential operator of order 1 that exists because $A_M(x, D)$ is positive definite. The principal symbol of $B_M(x, D)$ is given by $B_M^{\mathrm{prin}}(x, \xi) = \sqrt{A_M^{\mathrm{prin}}(x, \xi)}$. We find then that (2–10) is equivalent to the two first-order equations

$$\left(\frac{\partial}{\partial t} \pm \mathrm{i} B_M(x, D) \right) u_{M,\pm} = f_{M,\pm} \qquad (2\text{–}11)$$

under the transformations

$$\begin{aligned} u_{M,\pm} &= \tfrac{1}{2} u_M \pm \tfrac{1}{2} \mathrm{i} B_M(x, D)^{-1} \frac{\partial u_M}{\partial t}, \\ f_{M,\pm} &= \pm \tfrac{1}{2} \mathrm{i} B_M(x, D)^{-1} f_M. \end{aligned} \qquad (2\text{–}12)$$

We construct operators $G_{M,\pm}$ with Lagrangian distribution kernel $G_{M,\pm}(x, x_0, t)$ that solve the initial value problem for (2–11). Then using Duhamel's principle we find that

$$u_{M,\pm}(x, t) = \int_0^t \int_X G_{M,\pm}(x, x_0, t - t_0) f_{M,\pm}(x_0, t_0) \, \mathrm{d}x_0 \, \mathrm{d}t_0$$

solves (2–11). It follows from (2–12) that the Green's function for the second-order decoupled equation is given by

$$G_M(x, x_0, t)$$
$$= \tfrac{1}{2} \mathrm{i} G_{M,+}(x, x_0, t) B_M(x_0, D_{x_0})^{-1} - \tfrac{1}{2} \mathrm{i} G_{M,-}(x, x_0, t) B_M(x_0, D_{x_0})^{-1}. \qquad (2\text{–}13)$$

The operators $G_{M,\pm}$ are Fourier integral operators. Their construction is well known; see for example Duistermaat [43], Chapter 5. Singularities are propagated along the bicharacteristics, that are determined by Hamilton's equations generated by the principal symbol (factor i divided out) $\tau \pm B_M^{\mathrm{prin}}(x, \xi)$ of (2–11),

$$\begin{aligned} \frac{\partial x}{\partial \lambda} &= \pm \frac{\partial}{\partial \xi} B_M^{\mathrm{prin}}(x, \xi), & \frac{\partial t}{\partial \lambda} &= 1, \\ \frac{\partial \xi}{\partial \lambda} &= \mp \frac{\partial}{\partial x} B_M^{\mathrm{prin}}(x, \xi), & \frac{\partial \tau}{\partial \lambda} &= 0. \end{aligned} \qquad (2\text{–}14)$$

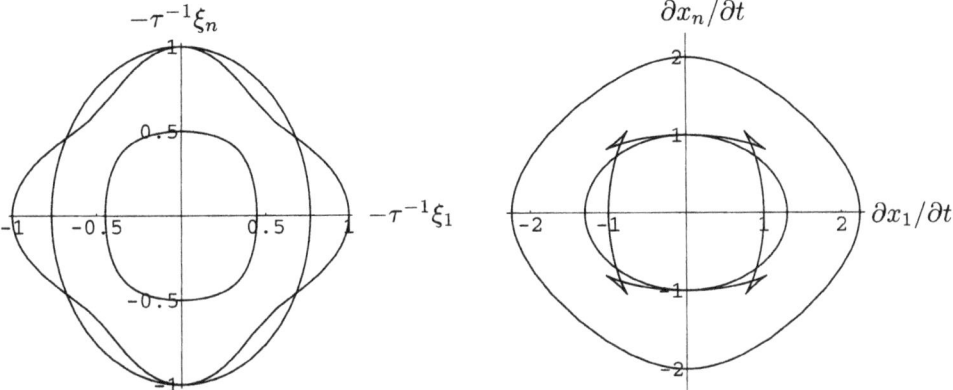

(a) Cotangent: $B_M^{\mathrm{prin}}(x, \tau^{-1}\xi) = 1$ (b) Tangent: $v_M = \partial B_M^{\mathrm{prin}}/\partial \xi\big|_{B_M^{\mathrm{prin}}(x,\tau^{-1}\xi)=1}$

Figure 5. (a) Section of a slowness surface (the characteristic surface) for a transversely isotropic medium in $n = 3$ dimensions. (b) Set of velocities associated to the slowness surface in a). Note the caustics that occur due to the fact that one of the (shear wave) sheets is not convex.

Solving these equations is what seismologists call *ray tracing* [27]. The solution may be parametrized by t. We denote the solution of (2–14) with the $+$ sign and initial values x_0, ξ_0 by $(x_M(x_0, \xi_0, t), \xi_M(x_0, \xi_0, t))$. The solution with the $-$ sign is found upon reversing the time direction; in other words, it is given by $(x_M(x_0, \xi_0, -t), \xi_M(x_0, \xi_0, -t))$. For the later analysis we also use the direction $\alpha = \|\xi_0\|^{-1}\xi_0$ and τ combined to replace ξ_0 in the initial values of the bicharacteristic solution: $\xi_0 = \boldsymbol{\xi}_0(x_0, \alpha, \tau)$.

The first equality in (2–14) represents the velocity $\partial x/\partial t$ of the bicharacteristic identified as the group velocity. Because B_M^{prin} is homogeneous in ξ and Euler's relation, $\langle \xi, \partial_\xi B_M^{\mathrm{prin}} \rangle = B_M^{\mathrm{prin}} = \mp \tau$ it follows directly that the group velocity is orthogonal to the slowness surface. Solving (2–14) reveals the formation of caustics. Caustics may form progressively in the presence of heterogeneities, or instantaneously in the presence of anisotropy even in the absence of heterogeneity. An example of the latter is shown in Figure 5(b).

A complete view of the propagation of singularities is provided by the canonical relation of the operator $G_{M,\pm}$, given by

$$C_{M,\pm} = \{(x_M(x_0, \xi_0, \pm t), t, \xi_M(x_0, \xi_0, \pm t), \mp B_{M,\pm}(x_0, \xi_0); x_0, \xi_0)\}. \quad (2\text{–}15)$$

A convenient choice of phase function is described in Maslov and Fedoriuk [73]. They state that one can always use a subset of the cotangent vector components as phase variables. Let us choose coordinates for $C_{M,+}$ of the form

$$(x_I, x_0, \xi_J, \tau), \quad (2\text{–}16)$$

where $I \cup J$ is a partition of $\{1, \ldots, n\}$. It follows from Theorem 4.21 in Maslov and Fedoriuk [73] that there is a function $S_{M,+}(x_I, x_0, \xi_J, \tau)$, such that locally

$C_{M,+}$ is given by

$$x_J = -\frac{\partial S_{M,+}}{\partial \xi_J}, \qquad t = -\frac{\partial S_{M,+}}{\partial \tau},$$

$$\xi_I = \frac{\partial S_{M,+}}{\partial x_I}, \qquad \xi_0 = -\frac{\partial S_{M,+}}{\partial x_0}. \tag{2-17}$$

Here we take into account the fact that $C_{M,+}$ is a canonical relation, which introduces a minus sign for ξ_0. A nondegenerate phase function for $C_{M,+}$ is then found to be

$$\phi_{M,+}(x, x_0, t, \xi_J, \tau) = S_{M,+}(x_I, x_0, \xi_J, \tau) + \langle \xi_J, x_J \rangle + \tau t. \tag{2-18}$$

In case $J = \varnothing$, the generating function $S_{M,+}$ reduces to frequency, τ, times the negative of travel time.

On the other hand, the canonical relation $C_{M,-}$ is given by

$$C_{M,-} = \big\{ (x, t, -\xi, -\tau; x_0, -\xi_0) : (x, t, \xi, \tau; x_0, \xi_0) \in C_{M,+} \big\}.$$

Thus a phase function for $C_{M,-}$ is

$$\phi_{M,-}(x, x_0, t, \xi_J, \tau) = -\phi_{M,+}(x, x_0, t, -\xi_J, -\tau).$$

We may define the canonical relation for G_M as $C_M = C_{M,+} \cup C_{M,-}$ and a phase function $\phi_M = \phi_{M,-}$ if $\tau > 0$, $\phi_M = \phi_{M,+}$ if $\tau < 0$.

We have to assume that the decoupling is valid microlocally around the bicharacteristic. In that case Theorem 5.1.2 of Duistermaat [43] implies that the operator $G_{M,\pm}$ is microlocally a Fourier integral operator of order $-\frac{1}{4}$. Hence, microlocally we have an expression for $G_{M,\pm}$ in the form of an oscillatory integral

$$G_{M,\pm}(x, x_0, t) =$$

$$(2\pi)^{-(|J|+1)/2-(2n+1)/4} \int \mathcal{A}_{M,\pm}(x_I, x_0, \xi_J, \tau) e^{i\phi_{M,\pm}(x,x_0,t,\xi_J,\tau)} \, d\xi_J \, d\tau. \tag{2-19}$$

The factors of (2π) in front of the integral are according to the convention of Treves [102] and Hörmander [62]. In the special case of $J = \varnothing$ and considering the amplitude $\mathcal{A}_{M,\pm}$ up to leading order, the integral reduces to the leading order term of the Debye series expansion in geometrical optics.

The amplitude $\mathcal{A}_{M,\pm}(x_I, x_0, \xi_J, \tau)$ satisfies a transport equation along the bicharacteristics $(x_M(x_0, \xi_0, \pm t), \xi_M(x_0, \xi_0, \pm t))$. Properties of amplitudes are described for instance in Treves [102], Section 8.4. The amplitude is an element of $M_{C_M} \otimes \Omega^{1/2}(C_M)$, the tensor product of the Keller–Maslov bundle M_{C_M} and the half-densities on the canonical relation C_M. If the subprincipal part of $A_M(x, D)$ is a matrix, then the amplitude is also a matrix; see Remark 2.2. The Keller–Maslov bundle gives a factor i^k, where k is an index, which we will absorb in the amplitude. The index keeps track of the passage through caustics.

It is possible to choose a Maslov phase function with a different set of phase variables, for instance $\xi_{\tilde{J}}$ (and not τ), where $\tilde{I} \cup \tilde{J}$ is a partition of $\{1, \ldots, n\}$ and

$C_{M,\pm}$ is parametrized by $(x_{\bar{I}}, x_0, t, \xi_{\bar{J}})$. In that case the transformed amplitude $\tilde{A}_{M,\pm}(x_{\bar{I}}, x_0, t, \xi_{\bar{J}})$ contains a Jacobian factor to the power one half, i.e.

$$\left| \tilde{A}_{M,\pm}(x_{\bar{I}}, x_0, t, \xi_{\bar{J}}) \right| = |A_{M,\pm}(x_I, x_0, \xi_J, \tau)| \left| \frac{\partial(x_I, x_0, \xi_J, \tau)}{\partial(x_{\bar{I}}, x_0, t, \xi_{\bar{J}})} \right|^{1/2}, \qquad (2\text{--}20)$$

where in the Jacobian both sets of variables are coordinates on $C_{M,\pm}$.

We calculate the left-hand side of (2–20). For this purpose, consider the Green's function $G_{M,\pm}(x, x_0, t - t_0)$ with t and $t_0 = 0$ fixed. This function can be be viewed as the kernel of an invertible Fourier integral operator, mapping the displacement at $t = 0$, $u|_{t=0} \in \mathcal{E}'(X)$ to the displacement at t, $u|_t \in \mathcal{D}'(X)$, with phase $\tilde{\phi}_{M,\pm}(x, x_0, t, \xi_{\bar{J}})$ and amplitude $\tilde{A}_{M,\pm}(x_{\bar{I}}, x_0, t, \xi_{\bar{J}})$. To highest order the energy at time t is given by

$$\int |B_M(x, D)u_{M,\pm}(x, t)|^2 \, dx.$$

Conservation of this quantity is reflected by the relation

$$G_{M,\pm}(t)^* B_M(x, D)^* B_M(x, D) G_{M,\pm}(t) = B_{M,\pm}(x_0, D_{x_0})^* B_{M,\pm}(x_0, D_{x_0}),$$

where the left-hand side denotes a composition of Fourier integral operators and $*$ denotes the adjoint. Since the left-hand side is a product of invertible Fourier integral operators, we can use the theory of Section 8.6 in Treves [102]. We find that to highest order

$$\left| (2\pi)^{-1/4} \tilde{A}_{M,\pm}(x_{\bar{I}}, x_0, t, \xi_{\bar{J}}) \right|^2 = \left| \det \frac{\partial \xi_0}{\partial(x_{\bar{I}}, \xi_{\bar{J}})} \right| \left| \frac{B_M(x_0, \xi_0)}{B_M(x, \xi)} \right|^2.$$

The value of $B_M(x, \xi)$ equals the frequency τ and is conserved along the bicharacteristic. Recall that (x_0, ξ_0, t) are valid coordinates for $C_{M,\pm}$ (cf. (2–15)). The Jacobian $|\partial(x_0, \xi_0, t)/\partial(x_I, x_0, t, \xi_J)|$ is equal to the factor $|\det \partial \xi_0 / \partial(x_I, \xi_J)|$, the reciprocal of which describes the geometrical spreading. It follows that to highest order

$$\left| \tilde{A}_{M,\pm}(x_{\bar{I}}, x_0, t, \xi_{\bar{J}}) \right| = (2\pi)^{1/4} \left| \det \frac{\partial(x_0, \xi_0, t)}{\partial(x_{\bar{I}}, x_0, t, \xi_{\bar{J}})} \right|^{1/2}. \qquad (2\text{--}21)$$

From (2–20) it now follows that

$$|A_{M,\pm}(x_I, x_0, \xi_J, \tau)| = (2\pi)^{1/4} \left| \det \frac{\partial(x_0, \xi_0, t)}{\partial(x_I, x_0, \xi_J, \tau)} \right|^{1/2}. \qquad (2\text{--}22)$$

We give our result about the Green's function for (2–7), collecting the results of this section, and using equations (2–12) and (2–22) to obtain a statement about the amplitude. We will assume that microlocally around the relevant bicharacteristics the decoupling is valid. Let $\mathrm{Char}(P_M)$ be the characteristic set of $P_M(x, D, D_t)$ given by $\{(x, t, \xi, \tau) : P_M(x, \xi, \tau) = 0\}$. The Green's function is such that precisely the singularities of f_M at $\mathrm{Char}(P_M)$ propagate (see Hörmander [61], Theorem 23.2.9). Thus we have

ASSUMPTION 1. *On a neighborhood of the bicharacteristic the multiplicity of the eigenvalue $A_M^{\text{prin}}(x, \xi)$ in (2–8) is constant.*

LEMMA 2.3. *Suppose that for the bicharacteristics through* $\mathrm{WF}(f_M) \cap \mathrm{Char}(P_M)$ *Assumption 1 is satisfied. Then u_M is given microlocally, away from* $\mathrm{WF}(f_M)$, *by*

$$u_M(x, t) = \int G_M(x, x_0, t - t_0) f_M(x_0, t_0) \, dx_0 \, dt_0, \qquad (2\text{–}23)$$

where $G_M(x, x_0, t)$ is the kernel of a Fourier integral operator with canonical relation C_M and order $-1\frac{1}{4}$, mapping functions of x_0 to functions of (x, t). It can be written as

$$G_M(x, x_0, t) = (2\pi)^{-(|J|+1)/2-(2n+1)/4} \int \mathcal{A}_M(x_I, x_0, \xi_J, \tau) e^{i\phi_M(x, x_0, t, \xi_J, \tau)} \, d\xi_J \, d\tau.$$
$$(2\text{–}24)$$

For the amplitude $\mathcal{A}_M(x_I, x_0, \xi_J, \tau)$ we have, to highest order,

$$|\mathcal{A}_M(x_I, x_0, \xi_J, \tau)| = (2\pi)^{1/4} \frac{1}{2} |\tau|^{-1} \left| \det \frac{\partial(x_0, \xi_0, t)}{\partial(x_I, x_0, \xi_J, \tau)} \right|^{1/2}. \qquad (2\text{–}25)$$

The elastic system for generic elastic media has been investigated by Braam and Duistermaat [20]. The set of singular points is generically of codimension three (thus one lower than one would expect naively), and is of conical form in the neighborhood of the singular point. Braam and Duistermaat give a normal form for such systems and investigate the behavior of its associated bicharacteristics and polarization spaces. In this case the system cannot be decoupled. However, in a generic elastic medium there cannot be an open set of bicharacteristics that pass through a singular point, because the singular points form a set of codimension 3. In this sense the set of bicharacteristics that is to be excluded is small. An analysis of conical refraction has been carried out by Melrose and Uhlmann [74] and Uhlmann [104].

When the elastic tensor (for $n = 3$) has symmetries it is determined by less than 21 coefficients. (The classification and analysis of the characteristic sets of such media can be found in the book by Musgrave [77].) In this case the singularities can be of different types. For example, in some classes of media, such as transversely isotropic media, the determinant decomposes into smooth factors. Then the multiplicities of the eigenvalues $A_M^{\text{prin}}(x, \xi)$ can vary on a larger (codimension 2) subset of $T^*X \setminus 0$. Because the bicharacteristics are curves on a codimension 1 surface, Assumption 1 can be violated on an open set of bicharacteristics.

In seismology, representations of the type (2–24) have been used by Chapman and Drummond [29] and Kendall and Thomson [69].

REMARK 2.4. In isotropic media, when Assumption 1 is certainly satisfied, the Hopf–Rinow theorem guarantees that any two points in the domain probed by the waves can be connected by at least one characteristic in each mode of propagation. In anisotropic media, in general, this is no longer true because there will not be an associated smooth Riemannian metric.

2C. Sources. In exploration seismology, a vibrator source is modeled as a point body force that is of the form

$$f_i(x_0, t_0) = d_i \delta(x_0 - s) \mathcal{W}(t_0),$$

acting at s with signature $\mathcal{W}$; the direction $d \in S^{n-1}$. Typically, $\mathcal{W}(t_0)$ is viewed as a regularization of $\delta(t_0)$.

An earthquake in global seismology is modeled with a symmetric moment tensor M_{ij} of rank 2; then

$$f_i(x_0, t_0) = -M_{ij}\partial_{x_0, j}\delta(x_0 - s)\, H(t_0 - t_s),$$

where s denotes the hypocentral location and t_s the origin time. We set $t_s = 0$. In $n = 3$ dimensions, depending on the eigenvalues $\lambda_{1,2,3}$ of M, pure shear faults ($\lambda_1 = -\lambda_3$, $\lambda_2 = 0$), pure tension cracks ($\lambda_1 \neq 0$, $\lambda_2 = \lambda_3 = 0$), explosive sources ($\lambda_1 = \lambda_2 = \lambda_3 \neq 0$), or compensated linear dipoles ($\lambda_1 \neq 0$, $\lambda_2 = \lambda_3 = -\frac{1}{2}\lambda_1$) can be simulated. Substituting the body force representation into (2–23) together with (2–6) leads to the product of operators $G_M Q(x_0, D_{x_0})_{Mi}^{-1} M_{ij}\tau^{-1}\partial/\partial x_{0,j}$. This product is a Fourier integral operator with the same phase as G_M, and amplitude that to highest order (by integrating by parts in x_0) equals the product $\mathcal{A}_M(x_I, x_0, \xi_J, \tau) Q(x_0, \boldsymbol{\xi}_0)_{Mi}^{-1} M_{ij} i\tau^{-1}\boldsymbol{\xi}_{0,j}$, where $\boldsymbol{\xi}_0 = \boldsymbol{\xi}_0(x_I, x_0, \xi_J, \tau)$.

3. High-Frequency Born Modeling and Imaging

Modeling under the Born approximation can be obtained as the leading term in a forward scattering series, while inversion in this approximation can be likewise obtained from the inverse scattering series; see Moses [76] and Razavy [83]. These series originate from a contrast formulation in the medium coefficients assuming a background. This formulation leads to the Lipmann–Schwinger–Dyson equation for the scattered field. We choose the background to be smooth and the contrast to be singular, viewing the contrast as a perturbation from the background. This is important in its own right, and it will also be a motivation for our approach to the model with smooth jumps introduced in the Kirchhoff approximation.

Microlocal analysis of the Born approximation has been discussed by a number of authors. In the absence of caustics, for the acoustic case, see Beylkin [10]; for the isotropic elastic case, see Beylkin and Burridge [12]; for the anisotropic elastic case, see De Hoop, Spencer and Burridge [38]. In the acoustic case, allowing for multipathing (caustics), see Rakesh [82] and Hansen [56]. For the acoustic problem with (non)maximal acquisition geometry, see Nolan and Symes [80]. For the elastic case with maximal acquisition geometry, see De Hoop and Brandsberg–Dahl [33] and Stolk and De Hoop [91].

3A. Scattering: perturbation of the Green's function. In the contrast formulation the total value of the medium parameters c_{ijkl}, ρ is written as the sum of a smooth background constituent $\rho(x), c_{ijkl}(x)$ plus a singular perturbation $\delta\rho, \delta c_{ijkl}$, namely

$$c_{ijkl} + \delta c_{ijkl}, \qquad \rho + \delta\rho.$$

This decomposition induces a perturbation of P_{il} (cf. (2–4)),

$$\delta P_{il} = \delta_{il} \frac{\delta\rho}{\rho} \frac{\partial^2}{\partial t^2} - \frac{\partial}{\partial x_j} \frac{\delta c_{ijkl}}{\rho} \frac{\partial}{\partial x_k}.$$

We denote the causal Green's operator associated with (2–2) by G_{il} and its distribution kernel by $G_{il}(x, x_0, t - t_0)$. The first-order perturbation δG_{il} of G_{il} is derived by demanding that the leading-order term in $(P_{ij} + \delta P_{ij})(G_{jk} + \delta G_{jk})$ vanishes. This results in the representation

$$\delta G_{il}(\hat{x}, \tilde{x}, t) = -\int_0^t \int_X G_{ij}(\hat{x}, x_0, t - t_0) \underbrace{\delta P_{jk}(x_0, D_{x_0}, D_{t_0})}_{\text{−linearized contrast source}} G_{kl}(x_0, \tilde{x}, t_0) \, dx_0 \, dt_0,$$

$$\tag{3-1}$$

which is the Born approximation. Here $\tilde{x}$ denotes a source location, $\hat{x}$ a receiver location, and x_0 a scattering point. Because the background model is smooth the operator δG_{il} contains only the single scattered field.

We apply the decoupling given by equation (2–6). Omitting the factors $Q_{iM}(\hat{x}, D_{\hat{x}}), Q(\tilde{x}, D_{\tilde{x}})^{-1}_{Nl}$ at the beginning and end of the product, we obtain an expression for the perturbation of the Green's function $\delta G_{MN}(\hat{x}, \tilde{x}, t)$ for the pair of modes M (scattered) and N (incident):

$$\delta G_{MN}(\hat{x}, \tilde{x}, t) = -\int_0^t \int_X G_M(\hat{x}, x_0, t - t_0) Q(x_0, D_{x_0})^{-1}_{Mi}$$

$$\times \left(\delta_{il} \frac{\partial}{\partial t_0} \frac{\delta\rho}{\rho} \frac{\partial}{\partial t_0} - \frac{\partial}{\partial x_{0,j}} \frac{\delta c_{ijkl}}{\rho} \frac{\partial}{\partial x_{0,k}} \right) Q(x_0, D_{x_0})_{lN} G_N(x_0, \tilde{x}, t_0) \, dx_0 \, dt_0. \quad (3\text{--}2)$$

Microlocally we can write G_M as in (2–24), with appropriate substitutions for its arguments. For G_N we use in addition the reciprocity relation $G_N(x_0, \tilde{x}, t_0) = G_N(\tilde{x}, x_0, t_0)$. The product of operators

$$G_M Q(x_0, D_{x_0})^{-1}_{Mi} \frac{\partial}{\partial x_{0,j}}$$

is a Fourier integral operator with the same phase as G_M, and amplitude that to highest order equals the product $A_M(\hat{x}_{\hat{f}}, x_0, \hat{\xi}_{\hat{f}}, \tau) Q(x_0, \hat{\xi}_0)^{-1}_{Mi} i \hat{\xi}_{0,j}$, where $\hat{\xi}_0 = \xi_0(\hat{x}_{\hat{f}}, x_0, \hat{\xi}_{\hat{f}}, \tau)$.

Assuming that the medium perturbation vanishes around $\hat{x}$ and $\tilde{x}$ a cut-off is introduced for t_0 near 0 and t. In the resulting expression one of the two frequency variables $\hat{\tau}, \tilde{\tau}$ can now be eliminated using the integral over t_0 (see for instance Duistermaat [43], Section 2.3). In this case the result can be obtained readily by noting that the integral over t_0 can be extended to the whole of $\mathbb{R}$ (the phase is not stationary for t_0 outside $[0,t]$), and then using that $\int_{-\infty}^{\infty} e^{it_0(\hat{\tau}-\tilde{\tau})}\,dt_0 = 2\pi\delta(\hat{\tau}-\tilde{\tau})$. The resulting formula for δG_{MN} is, modulo lower-order terms in the amplitude,

$$\delta G_{MN}(\hat{x},\tilde{x},t) = (2\pi)^{-(3n+1)/4-(|\hat{J}|+|\tilde{J}|+1)/2}\int \mathcal{B}_{MN}(\hat{x}_{\hat{I}},\hat{\xi}_{\hat{j}},\tilde{x}_{\tilde{I}},\tilde{\xi}_{\tilde{j}},x_0,\tau)$$

$$\times\left(w_{MN;ijkl}(\hat{x}_{\hat{I}},\tilde{x}_{\tilde{I}},x_0,\hat{\xi}_{\hat{j}},\tilde{\xi}_{\tilde{j}},\tau)\frac{\delta c_{ijkl}(x_0)}{\rho(x_0)}+w_{MN;0}(\hat{x}_{\hat{I}},\tilde{x}_{\tilde{I}},x_0,\hat{\xi}_{\hat{j}},\tilde{\xi}_{\tilde{j}},\tau)\frac{\delta\rho(x_0)}{\rho(x_0)}\right)$$

$$\times\, e^{i\Phi_{MN}(\hat{x},\tilde{x},t,x_0,\hat{\xi}_{\hat{j}},\tilde{\xi}_{\tilde{j}},\tau)}\,dx_0\,d\hat{\xi}_{\hat{j}}\,d\tilde{\xi}_{\tilde{j}}\,d\tau. \tag{3-3}$$

Here (see (2–18) for the construction of ϕ_M,ϕ_N),

$$\Phi_{MN}(\hat{x},\tilde{x},t,x_0,\hat{\xi}_{\hat{j}},\tilde{\xi}_{\tilde{j}},\tau) = \phi_M(\hat{x},x_0,t,\hat{\xi}_{\hat{j}},\tau)+\phi_N(\tilde{x},x_0,t,\tilde{\xi}_{\tilde{j}},\tau)-\tau t. \tag{3-4}$$

The amplitude factors $\mathcal{B}_{MN}$ are given by

$$\mathcal{B}_{MN}(\hat{x}_{\hat{I}},\tilde{x}_{\tilde{I}},x_0,\hat{\xi}_{\hat{j}},\tilde{\xi}_{\tilde{j}},\tau) = (2\pi)^{-(n-1)/4}\,\mathcal{A}_M(\hat{x}_{\hat{I}},x_0,\hat{\xi}_{\hat{j}},\tau)\,\mathcal{A}_N(\tilde{x}_{\tilde{I}},x_0,\tilde{\xi}_{\tilde{j}},\tau). \tag{3-5}$$

We will refer to the factors $w_{MN;ijkl},w_{MN;0}$ as the radiation patterns. These are given by

$$w_{MN;ijkl}(\hat{x}_{\hat{I}},\tilde{x}_{\tilde{I}},x_0,\hat{\xi}_{\hat{j}},\tilde{\xi}_{\tilde{j}},\tau) = Q_{iM}(x_0,\hat{\xi}_0)Q_{lN}(x_0,\tilde{\xi}_0)\,\hat{\xi}_{0,j}\tilde{\xi}_{0,k},$$

$$w_{MN;0}(\hat{x}_{\hat{I}},\tilde{x}_{\tilde{I}},x_0,\hat{\xi}_{\hat{j}},\tilde{\xi}_{\tilde{j}},\tau) = -Q_{iM}(x_0,\hat{\xi}_0)Q_{iN}(x_0,\tilde{\xi}_0)\,\tau^2,$$

where $\hat{\xi}_0 = \xi_0(\hat{x}_{\hat{I}},x_0,\hat{\xi}_j,\tau)$, $\tilde{\xi}_0 = \xi_0(\tilde{x}_{\tilde{I}},x_0,\tilde{\xi}_j,\tau)$. The scattering is depicted in Figure 6. We illustrate a couple of radiation patterns in Figure 7.

We investigate the map

$$\left(\frac{\delta c_{ijkl}}{\rho},\frac{\delta\rho}{\rho}\right)\mapsto\delta G_{MN}(\hat{x},\tilde{x},t)$$

induced by (3–3). We use the notation C_{ϕ_M} to indicate the subset of the global canonical relation C_M that is associated with a phase function ϕ_M; cf. (2–15).

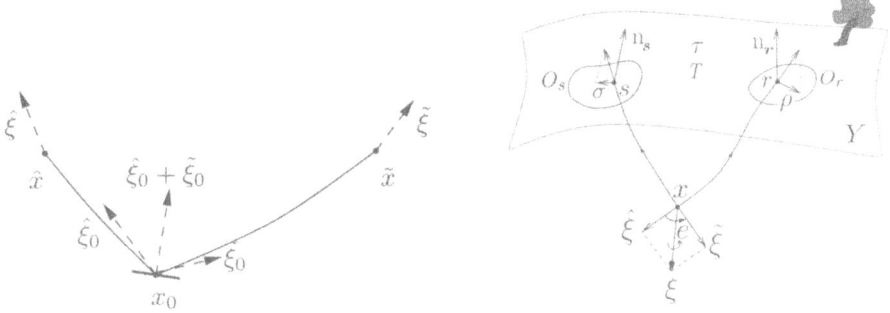

Figure 6. The scattering cotangent vectors ($\Lambda_{0,MN}$, left) and parametrization of Λ_{MN} (right).

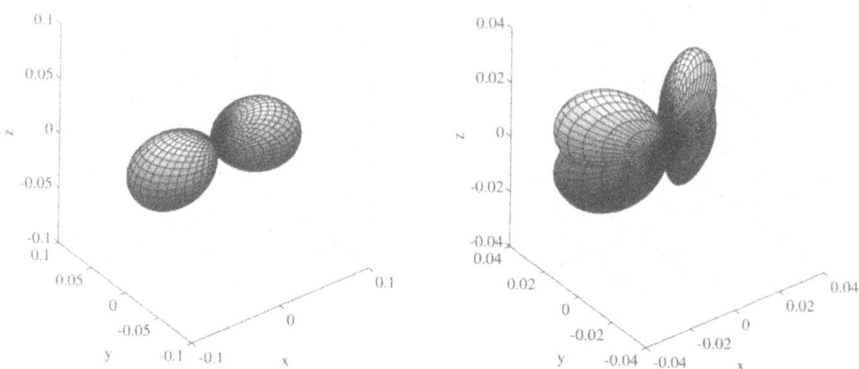

Figure 7. Radiation patterns ($n = 3$) for δc_{1313} (left) and δc_{1112} (right) for qP-qP scattering in an isotropic background. Magnitude is given as as a function of scattering angle and azimuth; cf. (4–4).

LEMMA 3.1. *Assume that if* $(\hat{x}, \hat{t}, \hat{\xi}, \tau; x_0, \hat{\xi}_0) \in C_{\phi_M}$, $(\tilde{x}, \tilde{t}, \tilde{\xi}, \tau; x_0, \tilde{\xi}_0) \in C_{\phi_N}$ *then* $\hat{\xi}_0 + \tilde{\xi}_0 \neq 0$. *Then the map*

$$\left(\frac{\delta c_{ijkl}}{\rho}, \frac{\delta \rho}{\rho} \right) \mapsto \delta G_{MN}(\hat{x}, \tilde{x}, t)$$

of (3–3) *is a Fourier integral operator* $\mathcal{E}'(X) \to \mathcal{D}'(X \times X \times (0, T))$. *Its canonical relation is*

$$\Lambda_{0,MN} = \big\{ (\hat{x}, \tilde{x}, \hat{t} + \tilde{t}, \hat{\xi}, \tilde{\xi}, \tau; x_0, \hat{\xi}_0 + \tilde{\xi}_0) :$$
$$(\hat{x}, \hat{t}, \hat{\xi}, \tau; x_0, \hat{\xi}_0) \in C_{\phi_M}, (\tilde{x}, \tilde{t}, \tilde{\xi}, \tau; x_0, \tilde{\xi}_0) \in C_{\phi_N} \big\}. \quad (3\text{–}6)$$

PROOF. We show that $\Phi_{MN}(\hat{x}_{\hat{j}}, \tilde{x}_{\tilde{j}}, t, x_0, \hat{\xi}_{\hat{j}}, \tilde{\xi}_{\tilde{j}}, \tau)$ is a nondegenerate phase function. The derivatives with respect to the phase variables are

$$\frac{\partial \Phi_{MN}}{\partial \tau} = -\hat{t}(\hat{x}_{\hat{j}}, x_0, \hat{\xi}_{\hat{j}}, \tau) - \tilde{t}(\tilde{x}_{\tilde{j}}, x_0, \tilde{\xi}_{\tilde{j}}, \tau) + t,$$

$$\frac{\partial \Phi_{MN}}{\partial \hat{\xi}_{\hat{j}}} = -\hat{x}_{\hat{j}}(\hat{x}_{\hat{j}}, x_0, \hat{\xi}_{\hat{j}}, \tau) + \hat{x}_{\hat{j}}, \qquad \frac{\partial \Phi_{MN}}{\partial \tilde{\xi}_{\tilde{j}}} = -\tilde{x}_{\tilde{j}}(\tilde{x}_{\tilde{j}}, x_0, \tilde{\xi}_{\tilde{j}}, \tau) + \tilde{x}_{\tilde{j}},$$

where $\hat{x}_{\hat{j}}(\hat{x}_{\hat{j}}, x_0, \hat{\xi}_{\hat{j}}, \tau), \tilde{x}_{\tilde{j}}(\tilde{x}_{\tilde{j}}, x_0, \tilde{\xi}_{\tilde{j}}, \tau)$ are as defined in (2–17), for the receiver side and the source side respectively. The derivatives of these expressions with respect to the variables $(\hat{x}_{\hat{j}}, \tilde{x}_{\tilde{j}}, t)$ are linearly independent, so Φ_{MN} is nondegenerate. From expression (3–4) it follows that the canonical relation of this operator is given by (3–6). By the hypothesis the canonical relation contains no elements with $\hat{\xi}_0 + \tilde{\xi}_0 = 0$, hence it is continuous as a map $\mathcal{E}'(X) \to \mathcal{D}'(X \times X \times (0, T))$. $\square$

The condition in Lemma 3.1 is violated if and only if $M = N$ and there exists a "direct" bicharacteristic from $\tilde{x}, \tilde{\xi}$ to $\hat{x}, -\hat{\xi}$. From the symmetry of the bicharacteristic under the transformation $\xi \to -\xi, t \to -t$ it follows that indeed in this case the condition is violated. On the other hand, we have $B_M(x_0, \hat{\xi}_0) = B_N(x_0, \tilde{\xi}_0) = \pm\tau$. If $\hat{\xi}_0 = -\tilde{\xi}_0$, then we must have $M = N$, because $B_M(x_0, \hat{\xi}_0) = B_M(x_0, -\hat{\xi}_0)$ and the condition that the eigenvalues in (2–8) are different for different modes. If $M = N$ and $\hat{\xi}_0 = -\tilde{\xi}_0$ then we have the mentioned direct bicharacteristic.

3B. Restriction: acquisition. Data are measurements of the scattered wave field which we relate here to the Green's function perturbation in (3–2). These data are assumed to be representable by $\delta G_{MN}(\hat{x}, \tilde{x}, t)$ for $(\hat{x}, \tilde{x}, t)$ in some acquisition manifold, which contains the source and receiver points and time. To make this explicit, let $y \mapsto (\hat{x}(y), \tilde{x}(y), t(y))$ be a coordinate transformation, such that $y = (y', y'')$ and the acquisition manifold Y is given by $y'' = 0$. Assume that the dimension of y'' is $2 + c$, where c is the codimension of the geometry (the 2 enforces "remote sensing"). Then the data are modeled by

$$\delta G_{MN}(\hat{x}(y', 0), \tilde{x}(y', 0), t(y', 0)). \tag{3–7}$$

It follows that the map $(\frac{\delta c_{ijkl}}{\rho}, \frac{\delta \rho}{\rho})$ to the data may be seen as the compose of the map of Lemma 3.1 with the restriction operator to $y'' = 0$. The restriction operator that maps a function $f(y)$ to $f(y', 0)$ is a Fourier integral operator with canonical relation $\Lambda_r = \{(y', \eta'; (y', y''), (\eta', \eta'')) \in T^*Y \times T^*\check{Y} : y'' = 0\}$, where $\check{Y} = X \times X \times (0, T)$. The composition of the canonical relations $\Lambda_{0,MN}$ and Λ_r is well defined if the intersection of $\Lambda_r \times \Lambda_{0,MN}$ with $T^*Y \backslash 0 \times \text{diag}(T^*\check{Y}\backslash 0) \times T^*X \backslash 0$ is transversal [43]. In this case we must have that the intersection of $\Lambda_{0,MN}$ with the manifold $y'' = 0$ is transversal. For the later analysis, the source and receiver points s, r are defined through $(s, r, t) = (\hat{x}(y', 0), \tilde{x}(y', 0), t(y', 0))$; instead, we will shortcut the coordinate transformation and identify

$$y' = (s, r, t) \quad \text{and} \quad \eta' = (\sigma, \rho, \tau),$$

see Figure 6. The tangential slownesses then follow as

$$p^s = \tau^{-1}\sigma, \quad p^r = \tau^{-1}\rho. \tag{3–8}$$

Let's repeat our assumptions:

ASSUMPTION 2. *There are no elements $(y', 0, \eta', \eta'') \in T^*Y \backslash 0$ such that there is a direct bicharacteristic from $(\hat{x}(y', 0), \hat{\xi}(y', 0, \eta', \eta''))$ to $(\tilde{x}(y', 0), -\tilde{\xi}(y', 0, \eta', \eta''))$ with arrival time $t(y', 0)$.*

ASSUMPTION 3. *The intersection of $\Lambda_{0,MN}$ with the manifold $y'' = 0$ is transversal, that is,*

$$\frac{\partial y''}{\partial(x_0, \hat{\xi}_0, \tilde{\xi}_0, \hat{t}, \tilde{t})} \text{ has maximal rank.} \tag{3–9}$$

In the following theorem we parametrize (3–6) by $(x_0, \hat{\xi}_0, \tilde{\xi}_0, \hat{t}, \tilde{t})$ using the parametrization of C_{ϕ_M} given by (2–15). Thus we let $\tau = \mp B_M(x_0, \hat{\xi}_0)$ and

$$\hat{x} = x_M(x_0, \hat{\xi}_0, \pm\hat{t}), \qquad \tilde{x} = x_N(x_0, \tilde{\xi}_0, \pm\tilde{t}),$$
$$\hat{\xi} = \xi_M(x_0, \hat{\xi}_0, \pm\hat{t}), \qquad \tilde{\xi} = \xi_N(x_0, \tilde{\xi}_0, \pm\tilde{t}).$$

We suppose that $\big(y'(x_0, \hat{\xi}_0, \tilde{\xi}_0, \hat{t}, \tilde{t}),\ \eta'(x_0, \hat{\xi}_0, \tilde{\xi}_0, \hat{t}, \tilde{t})\big)$ is obtained by transforming $(\hat{x}, \tilde{x}, \hat{t} + \tilde{t}, \hat{\xi}, \tilde{\xi}, \tau)$ to (y, η) coordinates.

THEOREM 3.2. [91] *If Assumptions 2 and 3 are satisfied, the operator $F_{MN;ijkl}$ (resp. $F_{MN;0}$) that maps the medium perturbation $\delta c_{ijkl}/\rho$ (resp. $\delta\rho/\rho$) to the data as a function of y' (3–7) is microlocally a Fourier integral operator with canonical relation*

$$\Lambda_{MN} = \big\{(y'(x_0, \hat{\xi}_0, \tilde{\xi}_0, \hat{t}, \tilde{t}), \eta'(x_0, \hat{\xi}_0, \tilde{\xi}_0, \hat{t}, \tilde{t}); x_0, \hat{\xi}_0 + \tilde{\xi}_0) :$$
$$B_M(x_0, \hat{\xi}_0) = B_N(x_0, \tilde{\xi}_0) = \pm\tau, y''(x_0, \hat{\xi}_0, \tilde{\xi}_0, \hat{t}, \tilde{t}) = 0\big\}. \quad (3\text{–}10)$$

The order is $(n-1+c)/4$. The amplitude is given to highest order (in coordinates (y'_I, η'_J, x_0) for Λ_{MN}, where I, J is a partition of $\{1, \ldots, 2n-1-c\}$) by the products $\mathcal{B}_{MN}(y'_I, \eta'_J, x_0)w_{MN;ijkl}(y'_I, \eta'_J, x_0)$ and $\mathcal{B}_{MN}(y'_I, \eta'_J, x_0)w_{MN;0}(y'_I, \eta'_J, x_0)$ respectively, where

$$\big|\mathcal{B}_{MN}(y'_I, \eta'_J, x_0)\big| = \tfrac{1}{4}\tau^{-2}(2\pi)^{-(n+1+c)/4}$$
$$\times \left|\det \frac{\partial(\hat{x}, \tilde{x}, t)}{\partial y}\right|^{-1/2} \left|\det \frac{\partial(x_0, \hat{\xi}_0, \tilde{\xi}_0, \hat{t}, \tilde{t})}{\partial(x_0, y'_I, y'', \eta'_J, \Delta\tau)}\right|^{1/2}_{\Delta\tau=0, y''=0}. \quad (3\text{–}11)$$

Here we define $\Delta\tau = \hat{\tau} - \tilde{\tau}$, so that the first constraint in (3–10) reads $\Delta\tau = 0$. The map $(x_0, \hat{\xi}_0, \tilde{\xi}_0, \hat{t}, \tilde{t}) \mapsto (x_0, y'_I, y'', \eta'_J, \Delta\tau)$ is bijective.

PROOF. The first statement has been argued above. The order of the operator is given by

$$\chi + \frac{K}{2} - \frac{\dim X + \dim Y'}{4},$$

where χ is the degree of homogeneity of the amplitude and K is the number of phase variables. The factors $\{w_{MN;ijkl}, w_{MN;0}\}$ are homogeneous of order 2 in the ξ and τ variables; the degree of homogeneity of the factor $\mathcal{B}_{MN}$ follows from (2–22). We find

$$\text{order } F_{MN;ijkl} = 2 + \big(-2 - \tfrac{1}{2}(|\hat{J}| + |\tilde{J}| + 2) + n\big) + \tfrac{1}{2}(|\hat{J}| + |\tilde{J}| + 1) - \tfrac{1}{4}(3n - 1 - c)$$
$$= \tfrac{1}{4}(n - 1 + c).$$

We calculate now the amplitude of the Fourier integral operator in Lemma 3.1. The factor $w_{MN;ijkl}$ is simply multiplicative. Suppose we choose coordinates on $\Lambda_{0,MN}$ to be $(\hat{x}_{\hat{I}}, \hat{\xi}_{\hat{J}}, \tilde{x}_{\tilde{I}}, \tilde{\xi}_{\tilde{J}}, \hat{\tau}, \tilde{\tau}, x_0)$, with ultimately $\hat{\tau} = \tilde{\tau}$ and define $\tau =$

$(\hat{\tau} + \tilde{\tau})/2$, $\Delta\tau = \hat{\tau} - \tilde{\tau}$. Using (2–25) and (3–5) we find that the amplitude $\mathcal{B}_{MN}(x_0, \hat{x}_{\hat{I}}, \hat{\xi}_{\hat{J}}, \tilde{x}_{\tilde{I}}, \tilde{\xi}_{\tilde{J}}, \tau)$ is given by

$$\left|\mathcal{B}_{MN}(\hat{x}_{\hat{I}}, \hat{\xi}_{\hat{J}}, \tilde{x}_{\tilde{I}}, \tilde{\xi}_{\tilde{J}}, \tau, x_0)\right| = \tfrac{1}{4}\tau^{-2}(2\pi)^{-(n-1)/4}\left|\det \frac{\partial(x_0, \hat{\xi}_0, \tilde{\xi}_0, \hat{t}, \tilde{t})}{\partial(\hat{x}_{\hat{I}}, \hat{\xi}_{\hat{J}}, \tilde{x}_{\tilde{I}}, \tilde{\xi}_{\tilde{J}}, \tau, x_0, \Delta\tau)}\right|^{1/2}.$$

The transformation from $(\hat{x}, \tilde{x}, t)$ to y coordinates in Fourier integral (3–7) induces an additional factor $\left|\det \partial(\hat{x}, \tilde{x}, t)/\partial y\right|^{-1/2}$ (note that for the Fourier integral operators it would be more natural to transform as a half-density). The amplitude transforms as a half-density on the canonical relation, and we obtain the factor

$$\left|\det \frac{\partial(y_I', y'', \eta_J')}{\partial(\hat{x}_{\hat{I}}, \hat{\xi}_{\hat{J}}, \tilde{x}_{\tilde{I}}, \tilde{\xi}_{\tilde{J}}, \tau)}\right|^{1/2}.$$

The additional factor $(2\pi)^{-(2+c)/4}$ arises from the normalization. We find (3–11).
□

Natural coordinates for the canonical relation are given by $(x_0, \hat{\xi}_0, \tilde{\xi}_0, \hat{t}, \tilde{t})$ such that $B_M(x_0, \hat{\xi}_0) - B_N(x_0, \tilde{\xi}_0) = 0, y''(x_0, \hat{\xi}_0, \tilde{\xi}_0, \hat{t}, \tilde{t}) = 0$. There is a natural density directly associated with this set, the quotient density. The Jacobian in (3–11) reveals that the amplitude factor $|\mathcal{B}_{MN}(y_I', \eta_J', x_0)|$ is in fact given by the associated half-density times $\tfrac{1}{4}\tau^{-2}(2\pi)^{-(n+1+c)/4}|\partial(\hat{x}, \tilde{x}, t)/\partial y|^{-1/2}$.

REMARK 3.3. If $c = 0$ and there are no rays tangent to the acquisition manifold, that is, if

$$\operatorname{rank} \frac{\partial y''}{\partial(\hat{t}, \tilde{t})} = 2, \qquad . \qquad (3\text{–}12)$$

then a convenient way to parametrize the canonical relation is found using the phase directions $\hat{\alpha} = \hat{\xi}_0/\|\hat{\xi}_0\|$, $\tilde{\alpha} = \tilde{\xi}_0/\|\tilde{\xi}_0\| \in S^{n-1}$ and the frequency τ. See also Figure 6 (right).

3C. Imaging. We collect the medium perturbations into the column matrix

$$g_\alpha = \left(\frac{\delta c_{ijkl}}{\rho}, \frac{\delta\rho}{\rho}\right).$$

The Born approximate forward operator $(F_{MN;ijkl}, F_{MN;0})$ (cf. Theorem 3.2) is then represented by $F_{MN;\alpha}$.

The imaging operator is the adjoint $F_{MN;\alpha}^*$ of $F_{MN;\alpha}$. If $F_{MN;\alpha}$ is a Fourier integral operator then the imaging operator is a Fourier integral operator also. In a geometrical context, seismologists view imaging as "moving" the singular support of the data restricted to a given offset (offset is the difference between the source s and receiver r points) to the singular support of the medium perturbation. Seismologists refer to this process as *migration*.

4. Linearized (Born) Inversion

4A. Imaging-inversion: the least-squares approach. The standard procedure to deal with the fact that the seismic inverse problem is overdetermined is to use the method of least squares. Let us consider data from a single pair of modes (M, N). The normal operator $N_{MN;\alpha\beta}$ is defined as the compose of $F_{MN;\beta}$ and its adjoint $F^*_{MN;\alpha}$,

$$N_{MN;\alpha\beta} = F^*_{MN;\alpha} F_{MN;\beta} \tag{4-1}$$

(no summation over M, N). If $N_{MN;\alpha\beta}$ is invertible (as a matrix-valued operator with indices $\alpha\beta$), then

$$F^{-1}_{MN;\alpha} = (N_{MN})^{-1}_{\alpha\beta} F^*_{MN;\beta} \tag{4-2}$$

(no summation over M, N) is a left inverse of $F_{MN;\alpha}$ that is optimal in the sense of least squares[1].

The properties of the composition in (4–1) depend on those of Λ_{MN}. Let π_Y, π_X be the projection mappings of Λ_{MN} to $T^*Y \setminus 0$, $T^*X \setminus 0$ respectively. We show that under the following assumption $N_{MN;\alpha\beta}$ is a pseudodifferential operator, so that the problem of inverting $N_{MN;\alpha\beta}$ reduces to a finite-dimensional problem for each $(x, \xi) \in \pi_X(\Lambda_{MN})$.

ASSUMPTION 4. (*Guillemin* [51]) *The projection π_Y of Λ_{MN} on $T^*Y \setminus 0$ is an embedding.*

This assumption is also known as the Bolker condition. Because Λ_{MN} is a canonical relation that projects submersively (under Assumption 1) on the subsurface variables (x, ξ), the projection of (3–10) on $T^*Y \setminus 0$ is immersive [62, Lemma 25.3.6 and (25.3.4)]. Therefore only the injectivity in the assumption need be verified [99]. In fact, it is precisely the injectivity condition that has been tacitly assumed in what seismologists call *map migration*.

Note that the "proper" part of Assumption 4 is always satisfied: The definition of proper is that the pre-image of a compact set is a compact set. Let us assume we have a compact subset of $T^*Y \setminus 0$. The pre-image consists of elements of Λ_{MN} corresponding to those "points" where the source and receiver rays intersect. The set of these points can be written as a set on which some continuous function vanishes. Therefore this set is closed. It is also bounded, and hence it is compact.

Assumption 4 implies that the image of π_Y is a submanifold, $\mathcal{L}$ say, of $T^*Y \setminus 0$. Using that Λ_{MN} is a canonical relation we have

---

[1]Equation (4–1) is for the case where one minimizes the difference with the data δG_{MN} in L^2 norm $\|\delta G_{MN} - F_{MN;\alpha} g_\alpha\|$. It can easily be adapted to the case where one minimizes a Sobolev norm of different order, or a weighted L^2 norm. This would introduce extra factors in the amplitude.

LEMMA 4.1. [91] *The projection π_Y of Λ_{MN} on $T^*Y \setminus 0$ is an immersion if and only if the projection π_X of Λ_{MN} on $T^*X \setminus 0$ is a submersion. In this case the image of π_Y is locally a coisotropic submanifold of $T^*Y \setminus 0$.*

As a consequence of immersivity implied by the embedding in Assumption 4, we can use $(x, \xi) \in T^*X \setminus 0$ as the first $2n$ (local) coordinates on Λ_{MN}. In addition, we need to parametrize the subsets of the canonical relation given by $(x, \xi) =$ constant; we denote such parameters by e. The new parametrization of Λ_{MN} is (identifying x_0 in (3–10) with x)

$$\Lambda_{MN} = \{(y'(x, \xi, e), \eta'(x, \xi, e); (x, \xi))\}. \tag{4-3}$$

The results do not depend on the precise definition of e. As noted before, if the variables $(\hat{t}, \tilde{t})$ can be solved from the second constraint in (3–10) (cf. equation (3–12)), then Λ_{MN} can be parametrized using $(x, \hat{\alpha}, \tilde{\alpha}, \tau)$, where $(\hat{\alpha}, \tilde{\alpha})$ are phase directions. In that case (x, ξ, e) should be related by a coordinate transformation to $(x, \hat{\alpha}, \tilde{\alpha}, \tau)$. In isotropic media with $M = n$ (where $\|\hat{\xi}_0\| = \|\tilde{\xi}_0\|$) a suitable choice is the pair scattering angle/azimuth, given by [38]

$$e(x, \hat{\alpha}, \tilde{\alpha}) = \Big(\underbrace{\arccos(\hat{\alpha} \cdot \tilde{\alpha})}_{\theta}, \frac{-\hat{\alpha} + \tilde{\alpha}}{2\sin(\arccos(\hat{\alpha} \cdot \tilde{\alpha})/2)}\Big) \in (0, \pi) \times S^{n-2}. \tag{4-4}$$

The azimuth, the second component, defines together with ξ the plane spanned by $(\hat{\alpha}, \tilde{\alpha})$. It is not very difficult to show that in elastic media the scattering angle (the first component) can be used as a coordinate when the slowness sheets are convex, but not always when one of the slowness sheets fails to be convex. In the canonical relation Λ_{MN} the range of ξ values at x such that y' yields a data point (after the application of mentioned pseudodifferential cutoff) controls the spatial resolution of the reconstruction of $g_\alpha(x)$ (we refer to this as illumination or insonification). The construction of this range is illustrated in Figure 8 and dates back to Ewald [45]; see also Devaney [40].

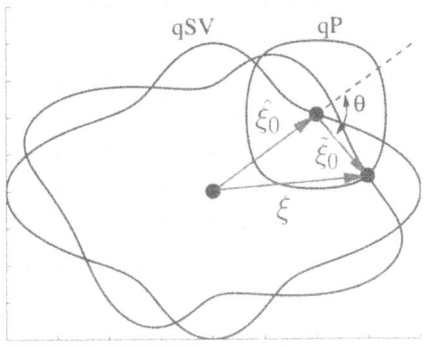

Figure 8. Generalization of the Ewald sphere. Here $M = $ qSV, $N = $ qP in a hexagonal medium; e (here θ) and τ are fixed.

REMARK 4.2. We show that the first part of Assumption 4 implies that

$$\frac{\partial B_M}{\partial \xi}(x, \hat{\xi}_0) + \frac{\partial B_N}{\partial \xi}(x, \tilde{\xi}_0) \neq 0;$$

in other words, the group velocities at the scattering point do not add up to 0. We have seen in Theorem 3.2 that Λ_{MN} may be parametrized by $(x, \hat{\xi}_0, \tilde{\xi}_0, \hat{t}, \tilde{t})$, where $(\hat{\xi}_0, \tilde{\xi}_0)$ are such that

$$B_M(x_0, \hat{\xi}_0) = B_N(x_0, \tilde{\xi}_0) = \pm \tau$$

(and we have the additional constraint $y''(x_0, \hat{\xi}_0, \tilde{\xi}_0, \hat{t}, \tilde{t}) = 0$). The projection π_X is given by $(x, \hat{\xi}_0 + \tilde{\xi}_0)$. Consider tangent vectors to Λ_{MN} given by vectors $v_{\hat{\xi}_0}, v_{\tilde{\xi}_0}$. These must satisfy

$$v_{\hat{\xi}_0} \cdot \frac{\partial B_M}{\partial \xi}(x, \hat{\xi}_0) = v_{\tilde{\xi}_0} \cdot \frac{\partial B_N}{\partial \xi}(x, \tilde{\xi}_0) = \pm v_\tau. \qquad (4\text{--}5)$$

Thus, if $\frac{\partial B_M}{\partial \xi}(x, \hat{\xi}_0) = -\frac{\partial B_N}{\partial \xi}(x, \tilde{\xi}_0)$, (4–5) implies that $(v_{\hat{\xi}_0} + v_{\tilde{\xi}_0}) \cdot \frac{\partial B_M}{\partial \xi}(x, \hat{\xi}_0)$ vanishes, implying that the projection of Λ_{MN} on $T^*X \setminus 0$ is not submersive.

If $c = 0$, and rank $\partial y''/\partial(\hat{t}, \tilde{t}) = 2$ (no tangent rays), then the constraint $y'' = 0$ may be used to solve for the parameters $\hat{t}, \tilde{t}$ and (4–5) is the only condition on $(\hat{\xi}_0, \tilde{\xi}_0)$. In that case $(\partial B_M/\partial \xi)(x, \hat{\xi}_0) \neq -(\partial B_N/\partial \xi)(x, \tilde{\xi}_0)$ implies that the projection is submersive. The solutions to $y''(x_0, \hat{\xi}_0, \tilde{\xi}_0, \hat{t}, \tilde{t}) = 0$ in (3–10) are then denoted as

$$\tilde{t} = T_N(x_0, \tilde{\alpha}), \quad \hat{t} = T_M(x_0, \hat{\alpha}), \quad \text{while} \quad \tilde{t} + \hat{t} = T_{MN}(x_0, \hat{\alpha}, \tilde{\alpha}).$$

For later notational convenience, we write the associated solutions for source and receiver points in y' as (s_N, r_M, T_{MN}) and their cotangent vectors in η' as (σ_N, ρ_M, τ) with

$$s_N(x_0, \tilde{\alpha}), \quad \sigma_N(x_0, \tilde{\alpha}, \tau) = \tau p_N^s, \quad r_M(x_0, \hat{\alpha}), \quad \rho_M(x_0, \hat{\alpha}, \tau) = \tau p_M^r,$$

where p^s and p^r are defined in (3–8). In other cases, the set of $(\hat{\xi}_0, \tilde{\xi}_0)$ is in general a smaller subset of $T_x^*X \setminus 0 \times T_x^*X \setminus 0$.

When constructing the compose (4–1) there is a subtlety that we have to take into account, namely that the linearized forward operator is only *microlocally* a Fourier integral operator. To make it globally a Fourier integral operator, we apply a pseudodifferential cutoff $\psi(y', D_{y'})$ (which seismologists would call a *tapered mute*) with compact support. Due to the fact that an embedding is proper, the forward operator is then a finite sum of local Fourier integral operators.

THEOREM 4.3. [91] *Let $\psi(y', D_{y'})$ be a pseudodifferential cutoff with conically compact support in $T^*Y \setminus 0$, such that for the set*

$$\{(y', \eta'; x_0, \xi_0) \in \Lambda_{MN} : (y', \eta') \in \text{supp } \psi\} \qquad (4\text{--}6)$$

Assumptions 2, 3, and 4 are satisfied. Then

$$F^*_{MN;\beta}\psi(y', D_{y'})^*\psi(y', D_{y'})F_{MN;\alpha} \tag{4-7}$$

is a pseudodifferential operator of order $n - 1$, with principal symbol

$$N_{MN;\beta\alpha}(x, \xi) = \tfrac{1}{16}(2\pi)^{-n}$$

$$\times \int |\psi(y'(x, \xi, e), \eta'(x, \xi, e))|^2 \tau^{-4} \overline{w_{MN;\beta}(x, \xi, e)} w_{MN;\alpha}(x, \xi, e)$$

$$\times \left| \det \frac{\partial(\hat{x}, \tilde{x}, t)}{\partial y} \right|^{-1} \left| \det \frac{\partial(x, \hat{\xi}_0, \tilde{\xi}_0, \hat{t}, \tilde{t})}{\partial(x, \xi, e, y'', \Delta\tau)} \right|_{\substack{\Delta\tau=0 \\ y''=0}} de, \tag{4-8}$$

where $\tau = \tau(x, \xi, e)$.

PROOF. We use the clean intersection calculus for Fourier integral operators (see Treves [102], for example) to show that (4–7) is a Fourier integral operator. The canonical relation of $F^*_{MN;\alpha}$ is given by

$$\Lambda^*_{MN} = \{(x, \xi; y', \eta') : (y', \eta'; x, \xi) \in \Lambda_{MN}\}.$$

Let $L = \Lambda^*_{MN} \times \Lambda_{MN}$ and $M = T^*X \setminus 0 \times \mathrm{diag}(T^*Y \setminus 0) \times T^*X \setminus 0$. We have to show that the intersection $L \cap M$ is clean, that is,

$$L \cap M \text{ is a manifold,} \tag{4-9}$$

$$TL \cap TM = T(L \cap M). \tag{4-10}$$

It follows from Assumption 4 (injectivity) that $L \cap M$ must be given by

$$L \cap M = \{(x, \xi, y', \eta', y', \eta', x, \xi) : (y', \eta'; x, \xi) \in \Lambda_{MN}\}. \tag{4-11}$$

Because Λ_{MN} is a manifold this set satisfies (4–9). The property (4–10) follows from the assumption that the map π_Y is immersive. The excess is given by

$$e = \dim(L \cap M) - (\dim L + \dim M - \dim T^*X \setminus 0 \times T^*Y \setminus 0 \times T^*Y \setminus 0 \times T^*X \setminus 0)$$

$$= n - 1 - c.$$

Taking into account the pseudodifferential cutoff $\psi(y', D_{y'})$, it follows that (4–7) is a Fourier integral operator. The canonical relation $\Lambda^*_{MN} \circ \Lambda_{MN}$ of $F^*_{MN;\beta}\psi^*\psi F_{MN;\alpha}$ is contained in the diagonal of $T^*X \setminus 0 \times T^*X \setminus 0$, so it is a pseudodifferential operator. The order is given by $2\,\mathrm{order}\,F_{MN;\alpha} + e/2 = n - 1$ (note that the codimension c drops out).

We write

$$\psi(y', D_{y'})^*\psi(y', D_{y'}) = \sum_i \chi^{(i)}(y', D_{y'}),$$

where the symbols $\chi^{(i)}(y', \eta')$ have small enough support, so that the distribution kernel of $\chi^{(i)}(y', D_{y'})F_{MN;\alpha}$ can be written as the oscillatory integral

$$\chi^{(i)}(y', D_{y'})\mathcal{F}_{MN;\alpha}(y', x) = (2\pi)^{-(3n-1-c)/4-|J|/2} \int \chi^{(i)}(y_I', \eta_J', x)$$

$$\times \mathcal{B}_{MN}(y_I', \eta_J', x)w_{MN;\alpha}(y_I', \eta_J', x)e^{i(S_{MN}^{(i)}(y_I', x, \eta_J')+\langle \eta_J', y_J'\rangle)}\,d\eta_J',$$

where $\psi^{(i)}(y_I', \eta_J', x) = \psi^{(i)}(y_I', y_J'(y_I', \eta_J', x), \eta_I'(y_I', \eta_J', x), \eta_J')$. We have written $\Phi_{MN}^{(i)}(y', x, \eta_J') = S_{MN}^{(i)}(y_I', x, \eta_J') + \langle \eta_J', y_J'\rangle$, (cf. (2–18) and (3–4)). We do not indicate the dependence of J on i explicitly. The distribution kernel of the normal operator is then given by a sum of terms

$$\int \overline{(\psi(y', D_{y'})\mathcal{F}_{MN;\beta}(y', x))}(\psi(y', D_{y'})\mathcal{F}_{MN;\alpha}(y', x_0))\,dy'$$

$$= (2\pi)^{-(3n-1-c)/2-|J|} \sum_i \int \chi^{(i)}(y_I', \eta_{0,J}', x_0)$$

$$\times \overline{\mathcal{B}_{MN}(y_I', \eta_J', x)}\mathcal{B}_{MN}(y_I', \eta_{0,J}', x_0)\overline{w_{MN;\beta}(y_I', \eta_J', x)}w_{MN;\alpha}(y_I', \eta_{0,J}', x_0)$$

$$\times e^{i(S_{MN}^{(i)}(y_I', x_0, \eta_{0,J}')-S_{MN}^{(i)}(y_I', x, \eta_J')+\langle \eta_{0,J}', y_J'\rangle-\langle \eta_J', y_J'\rangle)}\,d\eta_{0,J}'\,d\eta_J'\,dy'.$$

We now apply the method of stationary phase to integrate out the variables $(y_J', \eta_{0,J}')$. For the remaining variables we use the Taylor expansion,

$$S_{MN}^{(i)}(y_I', x_0, \eta_J') - S_{MN}^{(i)}(y_I', x, \eta_J') = \langle x - x_0, \xi(y_I', \eta_J', x_0)\rangle + O(|x - x_0|^2).$$

Thus we find (to highest order)

$$(2\pi)^{-(3n-1-c)/2} \sum_i \int \chi^{(i)}(y_I', \eta_J', x)^2 |\mathcal{B}_{MN}(y_I', \eta_J', x)|^2$$

$$\times \overline{w_{MN;\beta}(y_I', \eta_J', x)}w_{MN;\alpha}(y_I', \eta_J', x)e^{i\langle x-x_0, \xi(y_I', \eta_J', x_0)\rangle}\,d\eta_J'\,dy_I'.$$

We now change variables $(x, y_I', \eta_J') \to (x, \xi, e)$, and use (3–11). We sum over i and arrive at

$$\mathcal{N}_{MN;\beta\alpha}(x, x_0) = \tfrac{1}{16}(2\pi)^{-2n}$$

$$\times \int |\psi(y'(x, \xi, e), \eta'(x, \xi, e))|^2 \tau^{-4}\overline{w_{MN;\beta}(x, \xi, e)}w_{MN;\alpha}(x, \xi, e)$$

$$\times \left|\det \frac{\partial(\hat{x}, \tilde{x}, t)}{\partial y}\right|^{-1} \left|\det \frac{\partial(x, \hat{\xi}_0, \tilde{\xi}_0, \hat{t}, \hat{t})}{\partial(x, \xi, e, y'', \Delta\tau)}\right|_{\substack{\Delta\tau=0 \\ y''=0}} e^{i\langle x-x_0, \xi\rangle}\,d\xi\,de.$$

It follows that the principal symbol of $\mathcal{N}_{MN;\beta\alpha}$ is given by (4–8). $\square$

In the inverse, $(N_{MN})_{\alpha\beta}^{-1}F_{MN;\beta}^*$, seismologists distinguish the action of the parametrix of the normal operator from the imaging operator: They refer to the first action as *amplitude versus angles* (AVA) inversion, when e is given by (4–4).

REMARK 4.4. So far, we have discussed the inversion of data from one pair of modes (M, N). Often data will be available for some subset S of all possible pairs of modes. Define the normal operator for this case as

$$N_{\alpha\beta} = \sum_{(M,N)\in S} F^*_{MN;\alpha} F_{MN;\beta} = \sum_{(M,N)\in S} N_{MN;\alpha\beta}.$$

If all the $N_{MN;\alpha\beta}$ are pseudodifferential operators, so is $N_{\alpha\beta}$. A left inverse is now given by $N_{\alpha\beta}^{-1} F^*_\beta$, where F^*_β is the vector of Fourier integral operators containing the $F^*_{MN;\beta}$, $(M, N) \in S$.

4B. Parameter resolution. In practice, the parametrix $(N_{MN})^{-1}_{\alpha\beta}$ is replaced by a regularized inverse, $\langle (N_{MN})^{-1} \rangle_{\alpha\beta}$ say. Following the Backus–Gilbert approach [4], we subject the symbol matrix $N_{MN;\alpha\beta}(x, \xi)$ for given (x, ξ) and mode pair MN to a singular value decomposition and invoke thresholding. The thresholding yields the regularization and limits the set of parameters that can be resolved. This is apparent in the symbol resolution matrix,

$$\langle (N_{MN})^{-1} \rangle (x, \xi) N_{MN} (x, \xi),$$

namely through its deviation from the identity matrix: One can read off the linear combinations of parameters that can be resolved. An example is shown in Figure 9, where the background was assumed to be isotropic and $M = N = P$. (There we employ the Voigt notation, C_{IJ}, that replaces the tensor notation c_{ijkl}.)

Seismologists rarely attempt to reconstruct the stiffness tensor components directly. The resolution analysis defines a hierarchy of linear combinations of stiffness tensor components that can actually be extracted from the data.

5. Phantom Images, Artifacts of Type I

We investigate some of the consequences of a violation of the Bolker condition. To this end, we reconsider the composition $F^*_{MN;\beta} F_{MN;\alpha}$ defining the normal operator for a mode pair MN. The canonical relation associated with the imaging operator $F^*_{MN;\beta}$ is given by (cf. (3–10))

$$\Lambda^*_{MN} = \{(x, \xi; y', \eta') : (y', \eta'; x, \xi) \in \Lambda_{MN}\}.$$

Let $L = \Lambda^*_{MN} \times \Lambda_{MN}$ and $M = T^*X \setminus 0 \times \text{diag}(T^*Y \setminus 0) \times T^*X \setminus 0$ as before. The compose $\Lambda^*_{MN} \circ \Lambda_{MN}$ is given by the projection $L \cap M$ on $T^*X \setminus 0 \times T^*X \setminus 0$, namely

$$\Lambda^*_{MN} \circ \Lambda_{MN} = \{(z, \zeta; x, \xi) : (y', \eta'; z, \zeta) \in \Lambda_{MN}, (y', \eta'; x, \xi) \in \Lambda_{MN}\}.$$

If the intersection $L \cap M$ is clean (cf. (4–9)-(4–10)) the compose $\Lambda^*_{MN} \circ \Lambda_{MN}$ is again a canonical relation [62, Theorem 21.2.14]. (Such result was obtained for transversal intersection by Hörmander [59] and refined to clean intersection by Duistermaat and Guillemin [42].) The intersection $L \cap M$ is clean precisely if L

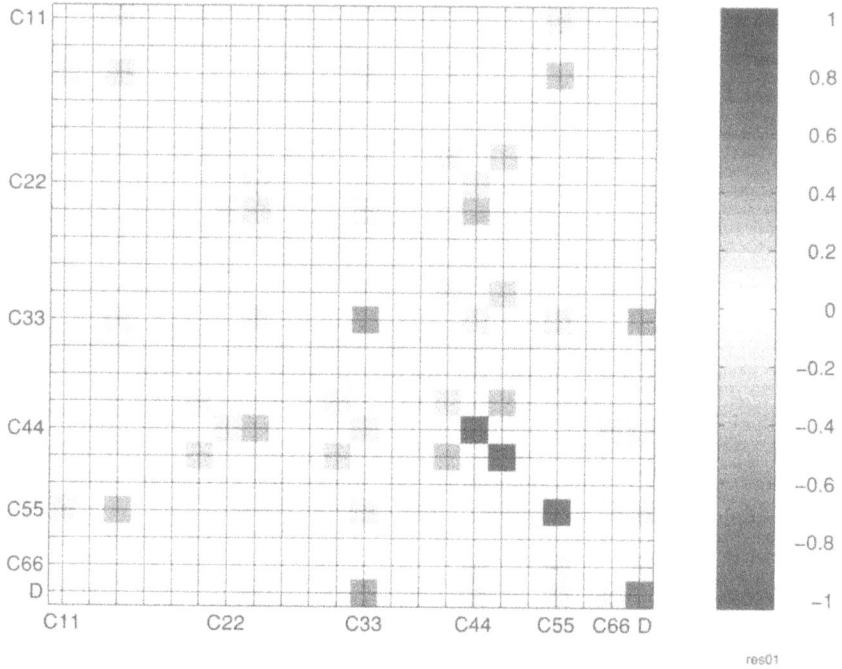

Figure 9. Normal operator symbol resolution matrix; $n = 3$, ξ points in the 3 direction, eigenvalues with a magnitude of less than 0.1 times the primary eigenvalue are zeroed (the matrix has rank 4).

and M intersect transversally in a submanifold of $T^*X\backslash 0 \times T^*Y\backslash 0 \times T^*Y\backslash 0 \times T^*X\backslash 0$.

5A. The composition. Employing the parametrization of Λ_{MN} given in Remark 3.3 and below (4–3), the intersection solves the equation

$$y'(x, \hat{\alpha}, \tilde{\alpha}, \tau) = y'(z, \hat{\beta}, \tilde{\beta}, \tau), \qquad \eta'(x, \hat{\alpha}, \tilde{\alpha}, \tau) = \eta'(z, \hat{\beta}, \tilde{\beta}, \tau). \qquad (5\text{--}1)$$

(Here we have already substituted the immediate equality of frequencies, τ.) These equations describe the following geometry:

(i) x and z lie on the ray in mode M determined by (r, ρ); at x this ray has phase or cotangent direction $\hat{\alpha}$ and at z this ray has phase or cotangent direction $\hat{\beta}$;

(ii) x and z lie on the ray in mode N determined by (s, σ); at x this ray has (phase) cotangent direction $\tilde{\alpha}$ and at z this ray has (phase) cotangent direction $\tilde{\beta}$;

(iii) since

$$\underbrace{T_M(x, \hat{\alpha}) + T_N(x, \tilde{\alpha})}_{T_{MN}(x, \hat{\alpha}, \tilde{\alpha})} = T_M(z, \hat{\beta}) + T_N(z, \tilde{\beta}),$$

if $z \neq x$, $T_M(z, \hat{\beta}) > T_M(x, \hat{\alpha})$ implies $T_N(x, \tilde{\alpha}) > T_N(z, \tilde{\beta})$ and vice versa, in which case the ray originating at r reaches x prior to reaching z while the ray originating at s reaches z prior to reaching x. Because $T_M(z, \hat{\beta}) - T_M(x, \hat{\alpha}) = T_N(x, \tilde{\alpha}) - T_N(z, \tilde{\beta})$, the rays originating at z with initial phase directions $\hat{\beta}$ and $-\tilde{\beta}$ intersect in x at the same time.

From these geometrical observations, it follows that (5–1) can be recast in the form, for some t_{MN}:

$$x_M(z, \hat{\beta}, \tau, t_{MN}) - x_N(z, \tilde{\beta}, \tau, -t_{MN}) = 0, \tag{5-2}$$

and

$$x = x_M(z, \hat{\beta}, \tau, t_{MN}), \tag{5-3}$$

$$\hat{\alpha} = \alpha_M(z, \hat{\beta}, \tau, t_{MN}) := \frac{\xi_M(z, \hat{\beta}, \tau, t_{MN})}{\|\xi_M(z, \hat{\beta}, \tau, t_{MN})\|}, \tag{5-4}$$

$$\tilde{\alpha} = \alpha_N(z, \tilde{\beta}, \tau, -t_{MN}) := \frac{\xi_N(z, \tilde{\beta}, \tau, -t_{MN})}{\|\xi_N(z, \tilde{\beta}, \tau, -t_{MN})\|}, \tag{5-5}$$

so that the compose $\Lambda_{MN}^* \circ \Lambda_{MN}$ is given by the set

$\{(z, \zeta; x, \xi) : (5\text{--}2)\text{--}(5\text{--}3)$ are satisfied, with

$$\xi = \hat{\xi}_0(x, \hat{\alpha}, \tau) + \tilde{\xi}_0(x, \tilde{\alpha}, \tau), \ \zeta = \hat{\xi}_0(z, \hat{\beta}, \tau) + \tilde{\xi}_0(z, \tilde{\beta}, \tau)\}. \tag{5-6}$$

Compare (3–10). In these equations we used the solution representation of the Hamilton system in (2–14). So far, the composition has not been subjected to any assumptions.

5B. Transversal intersection $(c = 0)$. In case $t_{MN} = 0$ then the solution to equations (5–3)-(5–3) is found to be $(z, \hat{\beta}, \tilde{\beta}) = (x, \hat{\alpha}, \tilde{\alpha})$. In the context of the discussion in the previous subsection, we call these solutions the *reciprocal-ray* solutions: the ray originating at s and terminating at r coincides with the ray originating at r and terminating at s. These are the only solutions under the Bolker condition as discussed in Theorem 4.3. It was noted that then L and M intersect cleanly with excess $\mathbf{e} = n - 1 - c$.

In the more general situation when there are solutions with $t_{MN} \neq 0$ (that correspond with the violation of travel time injectivity [99]) the normal operator will attain a contribution that is in general not microlocal (corresponding to *nonreciprocal-ray* solutions). To ensure that this contribution can be represented by a Fourier integral operator, we invoke the condition that L and M still intersect transversally (cleanly with excess 0). This typically occurs in configurations with waveguiding behavior. In seismological terms, then the diffraction surfaces associated with x and z have higher-order contact simultaneously in "common shot" and in "common receiver" data gathers.

Upon linearizing (5–1) the transversal intersection condition can be formulated as the condition that (for the notation, see Remark 4.2)

$$
M = \begin{pmatrix}
\dfrac{\partial r_M}{\partial x} & \dfrac{\partial r_M}{\partial \hat{\alpha}} & 0 & \dfrac{\partial r_M}{\partial z} & \dfrac{\partial r_M}{\partial \hat{\beta}} & 0 \\[2.2ex]
\dfrac{\partial p_M^r}{\partial x} & \dfrac{\partial p_M^r}{\partial \hat{\alpha}} & 0 & \dfrac{\partial p_M^r}{\partial z} & \dfrac{\partial p_M^r}{\partial \hat{\beta}} & 0 \\[2.2ex]
\dfrac{\partial s_N}{\partial x} & 0 & \dfrac{\partial s_N}{\partial \tilde{\alpha}} & \dfrac{\partial s_N}{\partial z} & 0 & \dfrac{\partial s_N}{\partial \tilde{\beta}} \\[2.2ex]
\dfrac{\partial p_N^s}{\partial x} & 0 & \dfrac{\partial p_N^s}{\partial \tilde{\alpha}} & \dfrac{\partial p_N^s}{\partial z} & 0 & \dfrac{\partial p_N^s}{\partial \tilde{\beta}} \\[2.2ex]
\dfrac{\partial T_{MN}}{\partial x} & \dfrac{\partial T_M}{\partial \hat{\alpha}} & \dfrac{\partial T_N}{\partial \tilde{\alpha}} & \dfrac{\partial T_{MN}}{\partial z} & \dfrac{\partial T_M}{\partial \hat{\beta}} & \dfrac{\partial T_N}{\partial \tilde{\beta}}
\end{pmatrix} \quad \text{has maximal rank.}
$$

Following the simplification of (5–1) to (5–2), in [89] it is shown that this rank is maximal if and only if the matrix C defined by

$$
\left(\left. \frac{\partial x_M}{\partial z} \right|_{(z,\hat{\beta},t)} - \left. \frac{\partial x_N}{\partial z} \right|_{(z,\tilde{\beta},-t)} \quad \left. \frac{\partial x_M}{\partial \hat{\beta}} \right|_{(z,\hat{\beta},t)} - \left. \frac{\partial x_N}{\partial \tilde{\beta}} \right|_{(z,\tilde{\beta},-t)} \quad \left. \frac{\partial x_M}{\partial t} \right|_{(z,\hat{\beta},t)} + \left. \frac{\partial x_N}{\partial t} \right|_{(z,\tilde{\beta},-t)} \right)
$$

has maximal rank.

The derivation follows. In view of observation (iii) below (5–1), the map

$$
(z, \hat{\beta}) \rightarrow (r_M(z, \hat{\beta}), p_M^r(z, \hat{\beta}), T_M(z, \hat{\beta}))
$$

is equal to the composition of maps

$$
(z, \hat{\beta}) \rightarrow (x_M(z, \hat{\beta}, \tau, t), \alpha_M(z, \hat{\beta}, \tau, t))
$$

$$
(x, \hat{\alpha}) \rightarrow (r_M(x, \hat{\alpha}), p_M^r(x, \hat{\alpha}), T_M(x, \hat{\alpha}) + t).
$$

Using the chain rule, it follows that

$$
\left. \frac{\partial(r_M, p_M^r, T_M)}{\partial(z, \hat{\beta})} \right|_{(z,\hat{\beta})} = \left. \frac{\partial(r_M, p_M^r, T_M)}{\partial(x, \hat{\alpha})} \right|_{(x,\hat{\alpha})} \left. \frac{\partial(x_M, \alpha_M)}{\partial(z, \hat{\beta})} \right|_{(z,\hat{\beta},t)}. \tag{5–7}
$$

In a similar fashion, we obtain

$$
\left. \frac{\partial(s_N, p_N^s, T_N)}{\partial(z, \tilde{\beta})} \right|_{(z,\tilde{\beta})} = \left. \frac{\partial(s_N, p_N^s, T_N)}{\partial(x, \tilde{\alpha})} \right|_{(x,\tilde{\alpha})} \left. \frac{\partial(x_N, \alpha_N)}{\partial(z, \tilde{\beta})} \right|_{(z,\tilde{\beta},-t)}. \tag{5–8}
$$

Using these relations, the matrix M can then be factorized as

$$
M = AS,
$$

in which

$$
A = \begin{pmatrix} \dfrac{\partial(r_M, p_M^r)}{\partial(x, \hat{\alpha})} & 0 \\[2ex] 0 & \dfrac{\partial(s_N, p_N^s)}{\partial(x, \tilde{\alpha})} \\[2ex] \dfrac{\partial T_M}{\partial(x, \hat{\alpha})} & \dfrac{\partial T_N}{\partial(x, \tilde{\alpha})} \end{pmatrix}
$$

is a $(4n - 3) \times (4n - 2)$ matrix, contains derivatives of the mappings from the subsurface to the acquisition manifold, and has full rank since

$$
\left. \frac{\partial(r_M, p_M^r, T_M)}{\partial(x, \hat{\alpha})} \right|_{(x,\hat{\alpha})} \quad \text{and} \quad \left. \frac{\partial(s_N, p_N^s, T_N)}{\partial(x, \tilde{\alpha})} \right|_{(x,\tilde{\alpha})}
$$

are invertible in view of Liouville's theorem, while

$$
S = \begin{pmatrix} I_n & 0 & 0 & \left.\dfrac{\partial x_M}{\partial z}\right|_{(z,\hat{\beta},t)} & \left.\dfrac{\partial x_M}{\partial \hat{\beta}}\right|_{(z,\hat{\beta},t)} & 0 \\[2ex] 0 & I_{n-1} & 0 & \left.\dfrac{\partial \alpha_M}{\partial z}\right|_{(z,\hat{\beta},t)} & \left.\dfrac{\partial \alpha_M}{\partial \hat{\beta}}\right|_{(z,\hat{\beta},t)} & 0 \\[2ex] I_n & 0 & 0 & \left.\dfrac{\partial x_N}{\partial z}\right|_{(z,\tilde{\beta},-t)} & 0 & \left.\dfrac{\partial x_N}{\partial \tilde{\beta}}\right|_{(z,\tilde{\beta},-t)} \\[2ex] 0 & 0 & I_{n-1} & \left.\dfrac{\partial \alpha_N}{\partial z}\right|_{(z,\tilde{\beta},-t)} & 0 & \left.\dfrac{\partial \alpha_N}{\partial \tilde{\beta}}\right|_{(z,\tilde{\beta},-t)} \end{pmatrix}.
$$

Following the factorization, we have

$$
\operatorname{rank} M = \operatorname{rank} S - \dim(\operatorname{range} S \cap \ker A). \tag{5-9}
$$

The kernel of A follows from the observation (see [33, Appendix A]) that

$$
\left.\frac{\partial(r_M, p_M)}{\partial(x,\hat{\alpha})}\right|_{(x,\hat{\alpha})} \begin{pmatrix} v_M(x,\hat{\alpha}) \\ 0 \end{pmatrix} = 0, \quad \left.\frac{\partial T_M}{\partial(x,\hat{\alpha})}\right|_{(x,\hat{\alpha})} \begin{pmatrix} v_M(x,\hat{\alpha}) \\ 0 \end{pmatrix} = -1;
$$

compare Figure 5, where $v_M(x,\hat{\alpha}) := v_M(x, \tau^{-1}\xi_M(x,\hat{\alpha}))$. A similar observation holds for (s_N, p_N, T_N) with $-v_N(x,\tilde{\alpha})$ replacing $v_M(x,\hat{\alpha})$. We conclude that

$$
\ker A = \operatorname{span}\{v_M(x,\hat{\alpha}), 0, -v_N(x,\tilde{\alpha}), 0\}.
$$

To the range of S the following reasoning applies. Writing $V = [V_{x,\hat{}}, V_{\hat{\alpha}}, V_{x,\tilde{}}, V_{\tilde{\alpha}}]$ for an element of range S we find, upon subtracting rows, from the structure of the matrix S that

$$
V_{x,\hat{}} - V_{x,\tilde{}} \in \operatorname{range} C'
$$

where

$$
C' = \left(\left.\frac{\partial x_M}{\partial z}\right|_{(z,\hat{\beta},t)} - \left.\frac{\partial x_N}{\partial z}\right|_{(z,\tilde{\beta},-t)} \quad \left.\frac{\partial x_M}{\partial \hat{\beta}}\right|_{(z,\hat{\beta},t)} \quad - \left.\frac{\partial x_N}{\partial \tilde{\beta}}\right|_{(z,\tilde{\beta},-t)} \right).
$$

Accounting for the identity submatrices in S, it then follows that $\operatorname{rank} S = 3n - 2 + \operatorname{rank} C'$. Following the subtraction of rows, we find that

$$\dim(\operatorname{range} S \cap \ker A) = \dim(\operatorname{range} C' \cap \underbrace{\operatorname{span}\{v_M(x, \hat{\alpha}) + v_N(x, \tilde{\alpha})\}}_{\text{one-dimensional}}). \quad (5\text{--}10)$$

If $\operatorname{span}\{v_M(x, \hat{\alpha}) + v_N(x, \tilde{\alpha})\} \subset \operatorname{range} C'$, $\operatorname{rank} C = \operatorname{rank} C'$; otherwise $\operatorname{rank} C = \operatorname{rank} C' + 1$, i.e.

$$\operatorname{rank} C = \operatorname{rank} C' + 1 - \dim(\operatorname{range} C' \cap \operatorname{span}\{v_M(x, \hat{\alpha}) + v_N(x, \tilde{\alpha})\}). \quad (5\text{--}11)$$

Combining (5--10), (5--11) with (5--9) yields

$$\operatorname{rank} M = \operatorname{rank} C + 3n - 3.$$

Hence, $\operatorname{rank} M$ is maximal if and only if $\operatorname{rank} C$ is maximal.

5C. Nonmicrolocal contribution to the normal operator ($c = 0$).

The projection π_Y of Λ_{MN} on $T^*Y \setminus 0$ is immersive. (This follows from Lemma 4.1 and the fact that the projection π_X of Λ_{MN} on $T^*X \setminus 0$ is submersive under Assumption 1.) Hence, for any $(y', \eta'; x, \xi) \in \Lambda_{MN}$ there is a small neighborhood $\Lambda_0 \subset \Lambda_{MN}$ such that $\pi_Y : \Lambda_0 \to T^*Y \setminus 0$ is an embedding. Then

$$(\Lambda_0^* \times \Lambda_0) \cap M = \{(x, \xi, y', \eta'; y', \eta', x, \xi) : (x, \xi, y', \eta') \in \Lambda_0\}.$$

Applying this argument for arbitrary $(y', \eta'; x, \xi)$ in the canonical relation, we can find subsets $\Lambda_k \subset \Lambda_{MN}$, $k \in K$ for some set K, such that $\pi_Y : \Lambda_k \to T^*Y \setminus 0$ is an embedding. Let

$$L_{\text{mloc}} := \bigcup_{k \in K} \Lambda_k^* \times \Lambda_k,$$

then $L_{\text{mloc}} \cap M$ consists of points corresponding with the *reciprocal* intersection solutions. For each $k \in K$ the intersection

$$(\Lambda_k^* \times \Lambda_k) \cap M$$

is transversal in the submanifold $T^*X \setminus 0 \times \operatorname{diag}(\pi_Y(\Lambda_k)) \times T^*X \setminus 0$.

Complementary to the microlocal part, $L_{\text{nmloc}} = L \setminus L_{\text{mloc}}$, $L_{\text{nmloc}} \cap M$ consists of points corresponding with the *nonreciprocal* intersection solutions. This intersection is transversal if and only if the following assumption is satisfied:

ASSUMPTION 5 (STOLK). *The matrix* C *given by*

$$\left(\frac{\partial x_M}{\partial z}\Big|_{(z, \hat{\beta}, t)} - \frac{\partial x_N}{\partial z}\Big|_{(z, \tilde{\beta}, -t)} \quad \frac{\partial x_M}{\partial \hat{\beta}}\Big|_{(z, \hat{\beta}, t)} - \frac{\partial x_N}{\partial \tilde{\beta}}\Big|_{(z, \tilde{\beta}, -t)} \quad \frac{\partial x_M}{\partial t}\Big|_{(z, \hat{\beta}, t)} + \frac{\partial x_N}{\partial t}\Big|_{(z, \tilde{\beta}, -t)} \right)$$

has maximal rank.

If $\Lambda_{MN;c} \subset \Lambda_{MN}$ is a conically compact subset of Λ_{MN}, and we replace L by $L_c = \Lambda_{MN;c}^* \times \Lambda_{MN;c}$, the microlocal and the nonmicrolocal parts of $L_c \cap M$ are both conically compact. It follows that in the parameters $(x, \hat{\alpha}, \tilde{\alpha}, \tau)$ the microlocal and nonmicrolocal parts are separated.

If $\Lambda_{MN;c}$ is a compact subset of Λ_{MN}, there is a finite collection $\{\Lambda_k\}_{k \in \check{K}}$ such that $\Lambda_{MN;c} \subset \bigcup_{k \in \check{K}} \Lambda_k$ and $\Lambda_k^* \circ \Lambda_l$ is either diagonal, empty, or accounts for a nonmicrolocal contributrion. Upon applying appropriate pseudodifferential cutoffs, we obtain $\psi(y', D_{y'}) F_{MN;\alpha} = F_{MN;\alpha;k}$ where $F_{MN;\alpha;k}$ has canonical relation Λ_k; in view of the previous observation, we can then write $F_{MN;\alpha}$ as a finite sum of $F_{MN;\alpha;k}$.

To each of the compositions, $F_{MN;\alpha;k}^* F_{MN;\alpha;l}$, the calculus of Fourier integral operators applies. Their orders follow to be $(n-1)/2 + e/2$. With Assumption 5, the excess of the nonmicrolocal contributions is zero whence their contributions to the normal operator have order $\frac{n-1}{2}$ (while for the microlocal part the excess was found to be $e = n - 1$).

The nonmicrolocal contribution to the normal operator is at most as singular as its pseudodifferential contribution:

THEOREM 5.1 (STOLK). *The operator* $F_{MN;\alpha} : H_s \to H_{s-(n-1)/2}$ *is continuous, and hence* $N_{MN;\alpha\beta} : H_s \to H_{s-n+1}$ *is continuous.*

PROOF. The modeling operator $F_{MN;\alpha}$ can be written as the finite sum

$$\sum_{k \in \check{K}} F_{MN;\alpha;k}$$

by "partitioning" its canonical relation: $\Lambda_k \subset \Lambda_{MN}$ is sufficiently small such that it satisfies the Bolker condition. Then $(F_{MN;\alpha;k})^* F_{MN;\beta;k}$ is pseudodifferential of order $n - 1$ and hence continuous as a mapping $H_s \to H_{s-(n-1)}$. We find that $F_{MN;\alpha;k}$ is continuous as a mapping $H_s \to H_{s-(n-1)/2}$.

The nonmicrolocal contribution to the normal operator is thus associated with off-diagonal terms of the form $(F_{MN;\alpha;k})^* F_{MN;\beta;l}$. These can be simply estimated from the fact that $\|F_{MN;\alpha;k} u - F_{MN;\beta;l} u\|^2 \geq 0$, namely,

$$2 \langle F_{MN;\alpha;k} u, F_{MN;\beta;l} u \rangle \leq \|F_{MN;\alpha;k} u\|^2 + \|F_{MN;\beta;l} u\|^2.$$

Since $F_{MN;\alpha;k}$ is continuous as a mapping $H_s \to H_{s-(n-1)/2}$, it follows that $(F_{MN;\alpha;k})^* F_{MN;\beta;l}$ is continuous as a mapping $H_s \to H_{s-(n-1)}$. $\square$

As a consequence, this theorem implies that the normal operator may not be invertible in the sense that $(N_{MN;\text{mloc}})_{\alpha\beta}^{-1} F_{MN;\beta}^* F_{MN;\gamma}$ (cf. equation (4–2)) attains a nonmicrolocal contribution as singular as the identity. The nonmicrolocal contribution, in this case, generates a phantom image, which we also call an artifact of type I; see Figure 10.

The transversal intersection condition for nonreciprocal ray pairs was investigated in [99; 33]; the simplified analysis presented here can be found in [89].

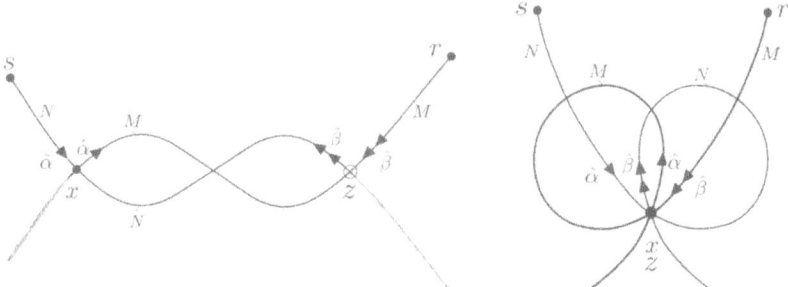

Figure 10. Nonmicrolocal contribution - $z \neq x$ (phantom separated in position; left) and $\zeta \neq \xi$ (phantom separated in orientation; right).

5D. The case $c \geq 1$. Artifacts of type I occur, subject to Assumption 5, in acquisition geometries with $c = 0$ though not generically. Two cases are illustrated in Figure 10. The situation $c \geq 1$ is more of a concern. In principle, neither Assumption 4 nor Assumption 5 are necessarily satisfied. In particular, if the acquisition geometry renders a codimension $c \geq 1$, assumptions of this type may be violated. The precise characterization of the Lagrangian of the normal operator, other than the diagonal, will then yield the artifacts (or phantom images). It is expected that then the calculus of singular Fourier integral operators due to Greenleaf and Uhlmann [48; 50; 49] still applies.

6. Symplectic Geometry of Seismic Data

The wavefront set of the modeled data is not arbitrary. This is a consequence of the fact that data consist of multiple experiments designed to provide a degree of redundancy, which we explain here. Under the Born approximation, subject to the restriction to the acquisition manifold, the singular part of the medium parameters is a function of n variables, while the data are a function of $2n - 1 - c$ variables. This redundancy is exploited in the parameter reconstruction, and is important in the reconstruction of the background medium (or the medium above the interface in the case of a smooth jump) as well.

We consider again the canonical relation Λ_{MN}. Suppose Assumption 4 is satisfied. In this section, denote by Ω the map

$$\Omega : (x, \xi, e) \mapsto (y'(x, \xi, e), \eta'(x, \xi, e)) : T^*X \setminus 0 \times E \to T^*Y \setminus 0$$

introduced above (4–3); seismologists refer to it as *map demigration*. This map conserves the symplectic form of $T^*X \setminus 0$. That is, if $w_{x_i} = \partial(y', \eta')/\partial x_i$ and similarly for w_{ξ_i}, w_{e_i}, we have

$$\sigma_Y(w_{x_i}, w_{x_j}) = \sigma_Y(w_{\xi_i}, w_{\xi_j}) = 0,$$
$$\sigma_Y(w_{\xi_i}, w_{x_j}) = \delta_{ij},$$
$$\sigma_Y(w_{e_i}, w_{x_j}) = \sigma_Y(w_{e_i}, w_{\xi_j}) = \sigma_Y(w_{e_i}, w_{e_j}) = 0. \tag{6–1}$$

The (x, ξ, e) are *symplectic coordinates* on the projection of Λ_{MN} on $T^*Y \setminus 0$, which is a subset $\mathcal{L}$ of $T^*Y \setminus 0$.

The image $\mathcal{L}$ of the map Ω is coisotropic, as noted in Lemma 4.1. The sets $(x, \xi) = $ const. are the isotropic fibers of the fibration of Hörmander [61], Theorem 21.2.6; see also Theorem 21.2.4. Duistermaat [43] calls them characteristic strips (see Theorem 3.6.2). We have sketched the situation in Figure 11. The wavefront set of the data is contained in $\mathcal{L}$ and is a union of fibers.

Using the following result we can extend the coordinates (x, ξ, e) to symplectic coordinates on an open neighborhood of $\mathcal{L}$.

LEMMA 6.1. [91] *Let $\mathcal{L}$ be an embedded coisotropic submanifold of $T^*Y \setminus 0$, with coordinates (x, ξ, e) such that (6-1) holds. Denote $\mathcal{L} \ni (y', \eta') = \Omega(x, \xi, e)$. We can find a homogeneous canonical map G from an open part of $T^*(X \times E) \setminus 0$ to an open neighborhood of $\mathcal{L}$ in $T^*Y \setminus 0$, such that $G(x, e, \xi, \varepsilon = 0) = \Omega(x, \xi, e)$.*

PROOF. The e_i can be viewed as (coordinate) functions on $\mathcal{L}$. We first extend them to functions on the whole $T^*Y \setminus 0$ such that the Poisson brackets $\{e_i, e_j\}$ satisfy

$$\{e_i, e_j\} = 0, \qquad 1 \le i, j \le m - n, \tag{6-2}$$

where $m = \dim Y = 2n - c - 1$. This can be done successively for $e_1, \ldots, e_{m-n}$ by the method that we describe in the sequel, see Treves [102, Chapter 7, proof of Theorem 3.3] or Duistermaat [43, proof of Theorem 3.5.6]. Suppose we have extended $e_1, \ldots, e_l$, we extend e_{l+1}. In order to satisfy (6-2) e_{l+1} has to be a solution u of

$$H_{e_i} u = 0, \qquad 1 \le i \le l,$$

where H_{e_i} is the Hamilton field associated with the function e_i, with initial condition on some manifold transversal to the H_{e_i}. For any $(y', \eta') \in \mathcal{L}$ the covectors de_i, $1 \le i \le l$ restricted to $T_{(y', \eta')}\mathcal{L}$ are linearly independent, so the H_{e_i} are transversal to $\mathcal{L}$ and they are linearly independent modulo $\mathcal{L}$. So we can give the initial condition $u|_{\mathcal{L}} = e_{l+1}$ and even prescribe u on a larger manifold, which leads to nonuniqueness of the extensions e_i.

We now have $m - n$ commuting vectorfields H_{e_i} that are transversal to $\mathcal{L}$ and linearly independent on some open neighborhood of $\mathcal{L}$. The Hamilton systems with parameters ε_i read

$$\frac{\partial y'_j}{\partial \varepsilon_i} = \frac{\partial e_i}{\partial \eta'_j}(y', \eta'), \qquad \frac{\partial \eta'_j}{\partial \varepsilon_i} = -\frac{\partial e_i}{\partial y'_j}(y', \eta'), \qquad 1 \le i, j \le m - n.$$

Let $G(x, e, \xi, \varepsilon)$ be the solution for (y', η') of the Hamilton systems combined with initial value $(y', \eta') = \Omega(x, \xi, e)$ with *flowout parameters* ε. This gives a diffeomorphic map from a neigborhood of the set $\varepsilon = 0$ in $T^*(X \times E) \setminus 0$ to a neighborhood of $\mathcal{L}$ in $T^*Y \setminus 0$. One can check from the Hamilton systems that this map is homogeneous.

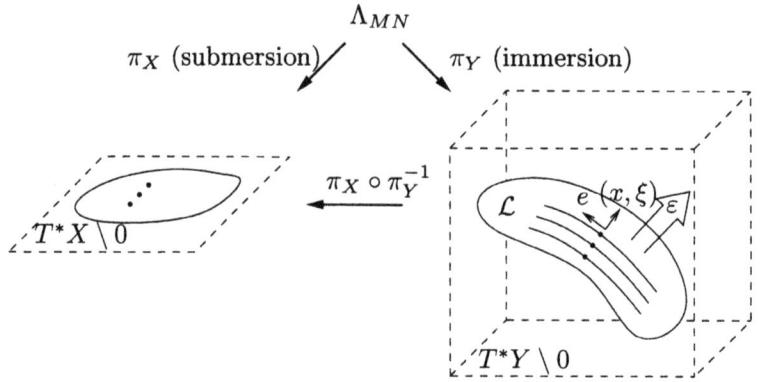

Figure 11. Visualization of the symplectic structure of Λ_{MN} (cone structure omitted).

It remains to check the commutation relations. The relations (6–1) are valid for any ε, because the Hamilton flow conserves the symplectic form on $T^*Y \setminus 0$. The commutation relations for $\partial(y', \eta')/\partial\varepsilon_i$ follow, using that $\partial(y', \eta')/\partial\varepsilon_i = H_{e_i}$.
$\square$

Let M_{MN} be the canonical relation associated with the map G we just constructed, i.e. $M_{MN} = \{(G(x, e, \xi, \varepsilon); x, e, \xi, \varepsilon)\}$. We now construct a Maslov-type phase function for M_{MN} that is directly related to a phase function for Λ_{MN}. Suppose (y_I', η_J', x) are suitable coordinates for Λ_{MN} ($\varepsilon = 0$). For ε small, the constant-ε subset of M_{MN} allows the same set of coordinates, thus we can use coordinates $(y_I', \eta_J', x, \varepsilon)$ on M_{MN}. Now there is (see Theorem 4.21 in Maslov and Fedoriuk [73]) a function $S_{MN}(y_I', x, \eta_J', \varepsilon)$ such that M_{MN} is given by

$$y_J' = -\frac{\partial S_{MN}}{\partial \eta_J'}, \qquad \eta_I' = \frac{\partial S_{MN}}{\partial y_I'}, \qquad \xi = -\frac{\partial S_{MN}}{\partial x}, \qquad e = \frac{\partial S_{MN}}{\partial \varepsilon}.$$

Thus a phase function for M_{MN} is given by

$$\Psi_{MN}(y', x, e, \eta_J', \varepsilon) = S_{MN}(y_I', x, \eta_J', \varepsilon) + \langle \eta_J', y_J' \rangle - \langle \varepsilon, e \rangle. \tag{6–3}$$

A Maslov-type phase function for Λ_{MN} then follows as

$$\Phi_{MN}(y', x, \eta_J') = \Psi_{MN}\left(y', x, \frac{\partial S_{MN}}{\partial \varepsilon}\bigg|_{\varepsilon=0}, \eta_J', 0\right) = S_{MN}(y_I', \eta_J', x, 0) + \langle \eta_J', y_J' \rangle.$$

The variable ε also plays a role in the formulation of the generalized Radon transform in Section 9.

7. Modeling and Inversion under the Kirchhoff Approximation

Another way to model the subsurface is to assume that it consists of different regions (*layers*) separated by smooth interfaces. The medium parameters, stiffness c_{ijkl} and density ρ, are assumed to vary smoothly on each region, and smoothly extendible across each interface, but they vary discontinuously at an interface. Here we model the reflection of waves at a smooth interface between two such layers with smoothly varying medium parameters.

7A. Reflection at an interface: Microlocal analysis of the "Kirchhoff" approximation.

The amplitude of the scattered waves is determined essentially by the reflection coefficients, and, implicitly, also by the curvature of the reflecting interface. Expressions for these coefficients are well known for the case of two constant coefficient media separated by a plane interface (see e.g. Aki and Richards [2], Chapter 5). In the case of smoothly varying media the reflection coefficients determine the scattering in the high-frequency limit. For a treatment of reflection and transmission of waves using microlocal analysis, see Taylor [98]; for the acoustic case, see also Hansen [56].

Mathematically the reflection and transmission of waves is formulated as a boundary value problem. The displacement u_l must satisfy the elastic wave equation under given initial conditions. In addition the displacement and the normal traction must be continuous at the interface. Denote the normal to the interface by ν. The

$$P_{il}u_l = f_i \text{ away from the interface,} \qquad u_l = 0 \text{ for } t < 0, \qquad (7\text{--}1)$$

must hold, while

$$\rho^{-1/2}u_l \text{ is continuous at the interface,}$$
$$\nu_j c_{ijkl}\frac{\partial}{\partial x_k}(\rho^{-1/2}u_l) \text{ is continuous at the interface.} \qquad (7\text{--}2)$$

Here we have the factors ρ because of our normalization (2–3). We assume that the source vanishes on a neighborhood of the interface. That this is a well-posed problem can be shown using energy estimates; see, for instance, Lions and Magenes [71], Section 3.8.

The solutions to the partial differential equation with $f = 0$ follow from the theory discussed in Section 2. The singularities are propagated along the bicharacteristics, curves in $T^*(X \times \mathbb{R}) \setminus 0$, given by

$$(x_M(x_0,\xi_0,\pm t), t, \xi_M(x_0,\xi_0,\pm t), \mp B_M(x_0,\xi_0)).$$

This is the bicharacteristic associated with the $M, \pm$ constituent of the solution; see Section 2. We define a bicharacteristic to be incoming if its direction is from inside a layer towards the interface for increasing time. We define a bicharacteristic to be outgoing if its direction is away from the interface into a layer for increasing time.

Assume that the incoming bicharacteristic stays inside a layer from $t = 0$ until it hits the interface, then the solution along such a bicharacteristic is determined completely by the partial differential equation and the initial condition. On the other hand, the solution along the outgoing bicharacteristics is not determined by the partial differential equation and the initial condition. We show that

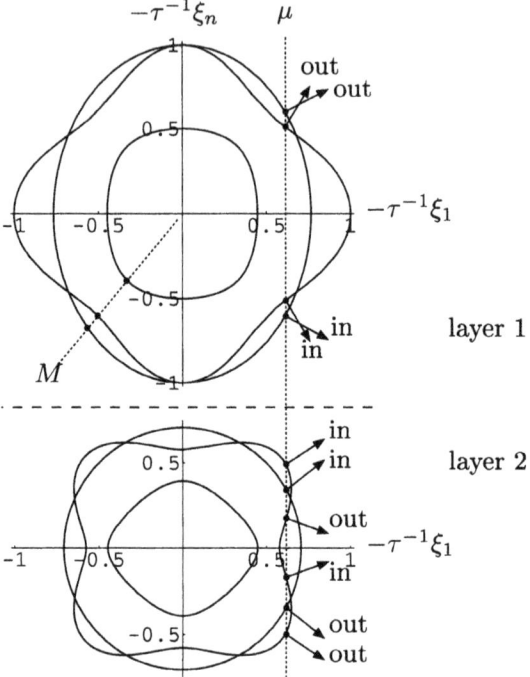

Figure 12. 2-dimensional section of an $n = 3$-dimensional slowness surfaces at some point of the interface, for the medium on both sides of the interface. The slownesses of the modes that interact (i.e. reflect and transmit into each other) are the intersection points with a line that is parallel to the normal of the interface. The group velocity, which is normal to the slowness surface, determines whether the mode is incoming or outgoing.

the solution along the outgoing bicharacteristics is determined by the partial differential equation and the interface conditions in (7–2).

We consider the consequences of the interface conditions. Assume for the moment that the interface is located at $x_n = 0$. We denote $x' = (x_1, \ldots, x_{n-1}), x = (x', x_n)$ and similarly, $\xi = (\xi', \xi_n)$. The wavefront set of the restriction of u_l to $x_n = 0$ satisfies

$$\mathrm{WF}(u_l|_{x_n=0}) = \big\{(x', t, \xi', \tau) : \text{there is } \xi_n \text{ with } (x', 0, t, \xi', \xi_n, \tau) \in \mathrm{WF}(u_l)\big\}.$$

It follows that a solution traveling along a bicharacteristic that intersects the boundary at some point $(x', 0, t)$ interacts with any other such solution as long as the associated values for ξ', τ in their wavefront sets coincide (Snell's law). This is depicted in Figure 12.

Depending upon the boundary coordinate x' and the "tangential" slowness $-\tau^{-1}\xi'$, the number of interacting bicharacteristics may vary. For large values of $-\tau^{-1}\xi'$ there will be neither incoming nor outgoing modes; for small values there are n incoming and n outgoing modes. The situation where the vertical line in Figure 12 is tangent to the slowness surface corresponds to rays tangent to the interface. Such rays are associated with head-waves and are not treated in our analysis. Equation (2–9) implies that the incoming and the outgoing modes correspond to the real solutions ξ_n of

$$\det P_{il}(x', 0, \xi', \xi_n, \tau) = 0.$$

This equation has $2n$ real or complex conjugated roots. The complex roots correspond to evanescent wave constituents. To number the roots we use an index μ.

In the following theorem we show that if none of the rays involved is tangent, there exists a pseudodifferential operator-type relation between the different modes restricted to the surface $x_n = 0$; we calculate its principal symbol in the proof. Let $x \mapsto z(x) : \mathbb{R}^n \to \mathbb{R}^n$ be a coordinate transformation such that the interface is given by $z_n = 0$. The corresponding cotangent vector is denoted by ζ, and satisfies

$$\zeta_i(\xi) = \left(\left(\frac{\partial z}{\partial x} \right)^{-1} \right)^{t_{ij}} \xi_j;$$

moreover the z form coordinates on a manifold Z.

ASSUMPTION 6. *There are no rays tangent to the interface* $z_n = 0$ *microlocally at* (z', t, ζ', τ).

THEOREM 7.1. *Suppose the roots τ of (2–9) have constant multiplicity and Assumption 6 is valid microlocally on some neighborhood in $T^*(Z' \times \mathbb{R}) \setminus 0$. Let $u^{\text{in}}_{N(\nu)}$ be microlocal constituents of a solution describing the "incoming" modes, and suppose $G_{M(\mu)}$ refers to an "outgoing" Green's function (2–19). Microlocally, the single reflected/transmitted constituent of the solution is given by*

$$u_{M(\mu)}(x, t) = \int_{z_n = 0} G_{M(\mu)}(x, x(z), t - t_0) 2 i D_{t_0} \left(R_{\mu\nu}(z, D_{z'}, D_{t_0}) u^{\text{in}}_{N(\nu)}(x(z), t_0) \right)$$

$$\times \left| \det \frac{\partial x}{\partial z} \right| \left\| \frac{\partial z_n}{\partial x} \right\| \, dz' \, dt_0, \quad (7\text{–}3)$$

where $R_{\mu\nu}(z, D_{z'}, D_t)$ is a pseudodifferential operator of order 0.

In the proof we derive the explicit form of $R^{\text{prin}}_{\mu\nu}(z, \zeta', \tau)$; see Remark 7.2. The $|\det \partial x / \partial z| \, \|\partial z_n / \partial x\| \, dz'$ integration is the Euclidean surface integral over the surface $z_n = 0$.

PROOF. (See [91].) For the moment we assume $z(x) = x$, i.e. that we have a reflector at $x_n = 0$, and smooth coefficients on either side. We show that at the interface there is a relation of the type

$$u^{\text{out}}_{M(\mu)}(x', 0, t) = R^0_{\mu\nu}(x', 0, D', D_t) u^{\text{in}}_{N(\nu)}. \quad (7\text{–}4)$$

We use the notation $c_{jk;il} = c_{ijkl}$ and also $(c_{jk})_{il} = c_{ijkl}$ [111]. The partial differential equation (2–1) reads in this notation

$$\left(\rho\delta_{il}\frac{\partial^2}{\partial t^2} - c_{jk;il}\frac{\partial^2}{\partial x_j\partial x_k}\right)(\rho^{-1/2}u_l) + \text{l.o.t.} = 0.$$

This equation can be rewritten as a first-order system in the variable x_n for the vector V_a of length $2n$ that contains both the displacement and the normal traction (normal to the surface $x_n = \text{const.}$)

$$V_a = \begin{pmatrix} \rho^{-1/2}u_i \\ c_{nk;il}\frac{\partial(\rho^{-1/2}u_l)}{\partial x_k} \end{pmatrix}, \qquad i = 1,\dots,n \tag{7-5}$$

in preparation for the boundary value problem (7–1), (7–2). Here a is an index in $\{1,\dots,2n\}$. The first-order system then is

$$\frac{\partial V_a}{\partial x_n} = iC_{ab}(x, D', D_t)V_b,$$

where C_{ab} is a matrix partial differential operator given to highest order by

$$C_{ab}(x, D', D_t) =$$

$$-i \begin{pmatrix} -\sum_{q=1}^{n-1}\sum_{j=1}^{n}(c_{nn})_{ij}^{-1}c_{nq;jl}\frac{\partial}{\partial x_q} & (c_{nn})_{il}^{-1} \\[2ex] -\sum_{p,q=1}^{n-1}b_{pq;il}\frac{\partial^2}{\partial x_p\partial x_q} + \rho\delta_{il}\frac{\partial^2}{\partial t^2} & -\sum_{p=1}^{n-1}\frac{\partial}{\partial x_p}c_{pn;ij}(c_{nn})_{jl}^{-1} \end{pmatrix}_{ab}.$$

Here $b_{pq;il} = c_{pq;il} - \sum_{j,k=1}^{n}c_{pn;ij}(c_{nn})_{jk}^{-1}c_{nq;kl}$ (in this equation we indicate the summations explicitly because the summations over p, q are $1,\dots,n-1$, whereas $j \in \{1,\dots,n\}$).

We next decouple this first-order system microlocally, in a way similar to the decoupling in Section 2A. We want to find scalar pseudodifferential operators $C_\mu(x, D', D_t)$ and a matrix pseudodifferential operator $L_{a\mu}(x, D', D_t)$ such that

$$C_{ab}(x, D', D_t) = L_{a\mu}(x, D', D_t)\,\text{diag}(C_\mu(x, D', D_t))_{\mu\nu}\,L_{\nu b}^{-1}(x, D', D_t).$$

The principal symbols $C_\mu^{\text{prin}}(x, \xi', \tau)$ are the solutions for ξ_n of

$$\det P_{il}^{\text{prin}}(x, (\xi', \xi_n), \tau) = 0. \tag{7-6}$$

In fact, it suffices for the transformed operator (the matrix $\text{diag}\,C_\mu(x, D', D_t)_{\mu\nu}$) to be block-diagonal: a block for each different real root of (7–6), a block with eigenvalues with positive imaginary part, and a block with eigenvalues with negative imaginary part. (This has also been discussed by Taylor [98].) By hypothesis of the theorem this transformation can be realized. This is because when varying ξ', τ, the multiplicity of a real eigenvalue only changes when the multiplicity of the corresponding root of (2–9) changes, or when two real eigenvalues become complex. The number of complex eigenvalues with positive or negative imaginary part changes only when two real eigenvalues become complex or vice versa. The latter case occurs only when there are tangent rays, and is hence excluded.

The $2n \times 2n$ principal symbol $L_{a\mu}^{\mathrm{prin}}$ (the columns appropriately normalized) is given by

$$L_{a\mu}^{\mathrm{prin}}(x, \xi', \tau) = \begin{pmatrix} Q_{iM(\mu)}^{\mathrm{prin}}(x, (\xi', C_{\mu}^{\mathrm{prin}}(x, \xi', \tau))) \\ c_{in;kl}(-\mathrm{i}(\xi', C_{\mu}^{\mathrm{prin}}(x, \xi', \tau))_k) Q_{lM(\mu)}^{\mathrm{prin}}(x, (\xi', C_{\mu}^{\mathrm{prin}}(x, \xi', \tau))) \end{pmatrix}_{a\mu}.$$

(The polarization vector $Q_{iM}(x, \xi)$ can also be defined for complex ξ). We define $V_{\mu} = L(x, D', D_t)_{\mu a}^{-1} V_a$. (The index mapping $\mu \mapsto M(\mu)$ assigns the appropriate mode to the normal component of the wave vector).

If the principal symbol of $C_{\mu}(x, \xi', \tau)$ is real, the decoupled equation for mode μ is of hyperbolic type. It corresponds to an outgoing wave or to an incoming wave, depending on the direction of the corresponding ray. If the principal symbol of $C_{\mu}(x, \xi', \tau)$ is complex, the decoupled operator for mode μ is of elliptic type. Depending on the sign of the imaginary part it corresponds to a mode that grows in the n-direction, a backward parabolic equation, or one that decays, a forward parabolic equation. The growing mode has to be absent, in view of energy considerations; see also Hörmander [62], Section 20.1.

The matrix $L_{a\mu}$ is fixed up to normalization of its columns. For the elliptic modes (Im $C_{\mu}^{\mathrm{prin}}(x, \xi', \tau) \neq 0$) the normalization is unimportant. For the hyperbolic modes the normalization can be such that the vector $V_{\mu} = L(x, D', D_t)_{\mu a}^{-1} V_a$ agrees microlocally with the corresponding mode $u_{M,\pm}$ defined in Section 2. To see this, assume V_{μ} refers to the same mode as $u_{M,\pm}$. In that case there is an invertible pseudodifferential operator $\psi(x, D, D_t)$ of order 0 such that $V_{\mu} = \psi u_{M,\pm}$. Now we can define $V_{\mu,\mathrm{new}} = \psi^{-1} V_{\mu,\mathrm{old}}$. Because ψ may depend on ξ_n, this factor cannot directly be absorbed in L. However, since $V_{\mu,\mathrm{old}}$ satisfies a first-order hyperbolic equation the dependence on ξ_n can be eliminated and the factor ψ^{-1} can be absorbed in L.

In proving this let the in-modes be the modes for which the amplitude is known, which are of the incoming hyperbolic and the growing elliptic type. Denote by $L_{a\mu}^{(1)}, L_{a\mu}^{(2)}$ the matrix $L_{a\mu}$ on either side of the interface. We define the $2n \times 2n$ matrix L^{in} such that it contains the columns related to incoming modes extracted from $L_{a\mu}^{(1)}, L_{a\mu}^{(2)}$, i.e.

$$L_{a\mu}^{\mathrm{in}} = \left(L^{(1),\mathrm{in}} \quad -L^{(2),\mathrm{in}} \right)_{a\mu},$$

and define $L_{a\mu}^{\mathrm{out}}$ similarly (so, here, μ is slightly different). The interface conditions (7–2) now read

$$L_{a\mu}^{\mathrm{out}} V_{\mu}^{\mathrm{out}} + L_{a\mu}^{\mathrm{in}} V_{\mu}^{\mathrm{in}} = 0.$$

If we set $R_{\mu\nu}^0 = -(L^{\mathrm{out}})_{\mu a}^{-1} L_{a\nu}^{\mathrm{in}}$ (for the question whether the inverse exists; see Remark 7.2 after the proof) then the part referring to the hyperbolic modes gives Equation (7–4).

The u_M^{out} are determined at the interface by (7–4). Describing their propagation away from the interface is a (microlocal) initial value problem similar to the problem for $G_{M,\pm}$ above, where now the x_n variable plays the role of time. The solution is again a Fourier integral operator, with canonical relation generated by the bicharacteristics. It follows that we can use $\phi_{M,\pm}(x, t-t_0, x_0, \xi_J, \tau)$ as phase function (taking care that $n \notin J$). The amplitude $\mathcal{A}_{M,\pm}(x_I, x_0, \xi_J, \tau)$ satisfies the transport equation as before. However, the restriction of the Fourier integral operator to the "initial surface" $x_n = 0$ is a pseudodifferential operator that is not necessarily the identity. Let us assume

$$u_M^{\text{out}}(x,t) = \int_{x_{0,n}=0} G_{M,\pm}(x, (x_0', 0), t - t_0)\psi(x, D_{x_0'}, D_{t_0})u_M^{\text{out}}(x_0', 0, t_0)\, dx_0' \, dt_0,$$

$$(7\text{–}7)$$

where $\psi(x, D', D_t)$ is to be found such that the restriction of this representation to $x_n = 0$ is the identity. The $\pm$ sign is chosen such that $G_{M,\pm}$ is the outgoing mode. We can use again Section 8.6 of Treves [102] to find that the principal symbol of this pseudodifferential operator should be

$$\psi(x, \xi', \tau) = \left|\frac{\partial B_M}{\partial \xi_n}(x, \xi', C_\mu^{\text{prin}}(\xi', \tau))\right| = \left|\frac{\partial x_{M,n}}{\partial t}(x, \xi', C_\mu^{\text{prin}}(\xi', \tau), 0)\right|, \quad (7\text{–}8)$$

i.e., the normal component of the velocity of the ray, the group velocity.

We now replace $G_{M,\pm}$ by (the relevant part of) G_M, using the equality $G_M = \frac{1}{2}iG_{M,+}B_M(x, D)^{-1} - \frac{1}{2}iG_{M,-}B_M(x, D)^{-1}$. Taking this and the relation $B_M^{\text{prin}}(x, \xi) = \mp\tau$ into account, we have now obtained (7–3) for the case that $z = x$ (no coordinate transformation).

We argue that (7–3) is also true when $z(x)$ is a general coordinate transformation. This follows from transforming the equations (7–1), (7–2) to z coordinates. In general, to highest order, the symbol of (pseudo)differential operators transforms as $\psi^{\text{transf}}(z, \zeta, \tau) = \psi(x(z), (\partial z/\partial x)^t \zeta, \tau)$. Tracing the steps of the proof we find the following equivalent of (7–4)

$$u_{M(\mu)}^{\text{out}}(x(z', 0), t) = R_{\mu\nu}^0(z', 0, D_{z'}, D_t)u_{N(\nu)}^{\text{in}}(x(z', 0), t). \quad (7\text{–}9)$$

When the interface is at $z_n = 0$ we can obtain (7–7) in z coordinates instead of x coordinates. Transforming G_M, u_M back to x coordinates we find that, for x away from the interface,

$$u_M(x) = \int_{z_n=0} G_M(x, x(z), t - t_0)\left|\frac{\partial z_{M,n}}{\partial t}(z, D_{z'}, D_{t_0})\right|$$

$$\times\, u_M^{\text{out}}(x(z), t_0)\left|\det\frac{\partial x}{\partial z}\right|\, dz' \, dt_0.$$

Here $\left| \dfrac{\partial z_{M,n}}{\partial t}(z, D_{z'}, D_t) \right|$ is the transformed version of (7–8). Expression (7–3) follows, with

$$R_{\mu\nu}(z, \zeta', \tau) = \left| \frac{\partial z_{M,n}}{\partial t}(z, \zeta', \tau) \right| \left\| \frac{\partial z_n}{\partial x} \right\|^{-1} R^0_{\mu\nu}(z, \zeta', \tau). \qquad \square$$

REMARK 7.2. The principal symbol $R^{0,\mathrm{prin}}_{\mu\nu}(z, \zeta', \tau)$ that occurs in the proof is simply the flux-normalized reflection coefficient for the amplitudes. The principal symbol $R^{\mathrm{prin}}_{\mu\nu}(z, \zeta', \tau)$ is obtained by multiplying $R^{0,\mathrm{prin}}_{\mu\nu}$ with the normal component of the velocity of the ray, given (for $z(x) = x$) by (7–8). The reflection coefficients satisfy unitary relations; see Chapman [28] and Kennett [70] (the appendix to Chapter 5). These follow essentially from conservation of energy. It follows that the matrix of reflection coefficients is well defined and in particular that the inverse of $L^{\mathrm{out}}_{a\mu}$ exists. Chapman [28] also gives a direct proof of the reciprocity relations for the reflection coefficients.

REMARK 7.3. We have shown that the reflected/transmitted wave is given by a composition of Fourier integral operators acting on the source. In the proof of Theorem 7.1, one can recognize the elements of the derivation of the Kirchhoff(-Helmholtz) approximate scattering theory for scalar waves [18]. In the case of multiple reflections or transmissions (for instance in a medium consisting of a number of smooth domains separated by smooth interfaces) this is also the case (cf. Frazer and Sen [46]). It follows that microlocally the solution operator describing the reflected solutions is itself a Fourier integral operator, where the canonical relation is given by the generalized bicharacteristics (i.e. the reflected and transmitted bicharacteristics) and the amplitude is essentially the product of the ray amplitudes and the reflection/transmission coefficients. The integration over z' accounts for the effects associated with the curvature of the interface.

7B. Modeling: Kirchhoff versus Born. In this subsection we match the expression for the data modeled using the smooth jump (Kirchhoff) approximation to the expressions for the Born modeled data we obtained in Section 3. The smooth medium above the interface plays the role of the background medium in the Born approximation.

From Theorem 7.1 it follows that reflection of an incident N-mode with covector $\tilde{\xi}_0$ into a scattered M-mode with covector $\hat{\xi}_0$ can take place if the frequencies are equal and $\hat{\xi}_0 + \tilde{\xi}_0$ is normal to the interface. In other words, $\hat{\xi}_0 + \tilde{\xi}_0$ must be in the wavefront set of the singular function of the interface, $\delta(z_n(x))$. Given $\tilde{\xi}_0, \hat{\xi}_0$ one can identify $\mu(M), \nu(N)$, and define (at least to highest order) the reflection coefficient as a function of $(x, \hat{\xi}_0, \tilde{\xi}_0)$, $R^{\mathrm{prin}}_{MN}(x, \hat{\xi}_0, \tilde{\xi}_0) = R^{\mathrm{prin}}_{\mu(M),\nu(N)}(z'(x), \zeta'(\tilde{\xi}_0), \tau)$. This factor can now be viewed as a function either of coordinates (y'_I, x, η'_J) or of coordinates (x, ξ, e) on Λ_{MN} (strictly speaking only defined for x in the interface, and ξ normal to the interface). To highest order, R_{MN} does not depend on $\|\xi\|$ and is simply a function of (x, e). We obtain the following result, which is a generalization of the Kirchhoff approximation. The normalization factor $\|\partial z_n/\partial x\|$ of the δ-function is such that integral

$$\int(\cdots)\left\|\frac{\partial z_n}{\partial x}\right\|\delta(z_n(x))\,dx$$

is an integral over the surface $z_n = 0$ with Euclidean surface measure in x coordinates.

THEOREM 7.4. *Suppose Assumptions* 1, 6, 2, *and* 3 *are satisfied microlocally for the relevant part of the data. Let* $\Phi_{MN}(y',x,\eta'_J),\mathcal{B}_{MN}(y'_I,x,\eta'_J)$ *be the phase and amplitude as in Theorem 3.2, but here for the smooth medium above the interface. Then the data modeled with the smooth jump interface model is given microlocally by*

$$G_{MN}^{\mathrm{refl}}(y')=(2\pi)^{-|J|/2-(3n-1-c)/4}\int\left(\mathcal{B}_{MN}(y'_I,x,\eta'_J)2i\tau(\eta')R_{MN}(y'_I,x,\eta'_J)+\mathrm{l.o.t.}\right)$$

$$\times\,e^{i\Phi_{MN}(y',x,\eta'_J)}\left\|\frac{\partial z_n}{\partial x}\right\|\delta(z_n(x))\,d\eta'_J\,dx;\quad(7\text{–}10)$$

in other words, by a Fourier integral operator with canonical relation Λ_{MN} *and order* $(n-1+c)/4-1$ *acting on the distribution* $\|\partial z_n/\partial x\|\delta(z_n(x))$.

PROOF. We write the distribution kernel of the reflected data (7–3) in a form similar to (3–3). First recall the reciprocal expression for the Green's function (2–24),

$$G_N(x(z),\tilde{x},t_0)=(2\pi)^{-(|\tilde{J}|+1)/2-(2n+1)/4}$$

$$\times\int\mathcal{A}_N(\tilde{x}_{\tilde{I}},x(z),\tilde{\xi}_{\tilde{J}},\tau)e^{i\phi_N(\tilde{x},x(z),t_0,\tilde{\xi}_{\tilde{J}},\tau)}\,d\tilde{\xi}_{\tilde{J}}\,d\tau.$$

By using Theorem 7.1, and doing an integration over a t and a τ variable one finds that the Green's function for the reflected part is given by

$$G_{MN}^{\mathrm{refl}}(\hat{x},\tilde{x},t)=(2\pi)^{-(|\hat{J}|+|\tilde{J}|+1)/2-n}$$

$$\times\int_{z_n=0}\left(2i\tau\mathcal{A}_M(\hat{x}_{\hat{I}},x(z),\hat{\xi}_{\hat{J}},\tau)\mathcal{A}_N(\tilde{x}_{\tilde{I}},x(z),\tilde{\xi}_{\tilde{J}},\tau)R_{\mu(M)\nu(N)}(z,\zeta',\tau)+\mathrm{l.o.t.}\right)$$

$$\times\,e^{i\Phi_{MN}(\hat{x},\tilde{x},t,x(z),\hat{\xi}_{\hat{J}},\tilde{\xi}_{\tilde{J}},\tau)}\left|\det\frac{\partial x}{\partial z}\right|\left\|\frac{\partial z_n}{\partial x}\right\|\,d\hat{\xi}_{\hat{J}}\,d\tilde{\xi}_{\tilde{J}}\,d\tau\,dz',$$

where ζ' depends on $(x(z),\tilde{\xi}_0)$ (the indices μ,ν for the reflection coefficients have been explained in Section 7A). The integration $\int(\cdots)dz'$ is now replaced by $\int(\cdots)\delta(z_n)dz$. The latter can be transformed back to an integral over x. Thus we obtain

$$(2\pi)^{-(|\hat{J}|+|\tilde{J}|+1)/2-n}$$

$$\times \int \left(2\mathrm{i}\tau \mathcal{A}_M(\hat{x}_{\hat{J}}, x, \hat{\xi}_{\hat{J}}, \tau) \mathcal{A}_N(\tilde{x}_{\tilde{J}}, x, \tilde{\xi}_{\tilde{J}}, \tau) R_{\mu(M)\nu(N)}(z(x), \zeta'(\tilde{\xi}_{\tilde{J}}, x), \tau) + \text{l.o.t.}\right)$$

$$\times \, \mathrm{e}^{\mathrm{i}\Phi_{MN}(\hat{x}, \tilde{x}, t, x, \hat{\xi}_{\hat{J}}, \tilde{\xi}_{\tilde{J}}, \tau)} \left\| \frac{\partial z_n}{\partial x} \right\| \delta(z_n(x)) \, \mathrm{d}\hat{\xi}_{\hat{J}} \, \mathrm{d}\tilde{\xi}_{\tilde{J}} \, \mathrm{d}\tau \, \mathrm{d}x.$$

This formula is similar to (3–3), except for the fact that the amplitude is different and $\delta c_{ijkl}(x)/\rho(x)$, $\delta\rho(x)/\rho(x)$ is replaced by the δ-function $\|\partial z_n/\partial x\|\delta(z_n(x))$. Also the factors $w_{MN;ijkl}$, $w_{MN;0}$ depend only on the background medium, while $R_{\mu(M)\nu(N)}$ depends on the total medium. The phase function Φ_{MN} now comes from the smooth medium above the reflector.

The data are modeled by $G_{MN}^{\mathrm{refl}}(\hat{x}, \tilde{x}, t)$ with $(\hat{x}, \tilde{x}, t)$ in the acquisition manifold, as is explained in the text below Lemma 3.1. We follow the approach of Section 3, and do a coordinate transformation $(\hat{x}, \tilde{x}, t) \mapsto (y', y'')$, such that the acquisition manifold is given by $y'' = 0$. It follows that under Assumptions 2 and 3 the data are obtained as the image of a Fourier integral operator acting on $\|\partial z_n/\partial x\|\delta(z_n(x))$ and that it is given by (7–10). $\qquad \square$

We now construct the reflectivity function and the operator that maps it to seismic data. This is done by applying the results of Section 6 to the Kirchhoff modeling formula (7–10), and its equivalent under the Born approximation (3–3).

THEOREM 7.5. [91] *Suppose that microlocally Assumptions 1, 6, 2, 3, and 4 are satisfied. Let $H_{MN} : \mathcal{E}'(X \times E) \to \mathcal{D}'(Y)$ be the Fourier integral operator with canonical relation given by the extended map $(x, \xi, e, \varepsilon) \mapsto (y', \eta')$ constructed in Section 6, and with amplitude to highest order given by*

$$(2\pi)^{n/2}(2\mathrm{i}\tau)\mathcal{B}_{MN}(y'_I, x, \eta'_J, \varepsilon),$$

such that $\mathcal{B}_{MN}(\varepsilon = 0)$ is as given in Theorem 3.2. Then the data, in both Born and Kirchhoff approximations, are given by H_{MN} acting on a distribution $r_{MN}(x, e)$ of the form

$$r_{MN}(x, e) = (\text{pseudo})(x, D_x, e)(\text{distribution})(x). \qquad (7\text{–}11)$$

For the Kirchhoff approximation this distribution equals $\|\partial z_n/\partial x\|\delta(z_n(x))$, while the principal symbol of the pseudodifferential operator equals $R_{MN}(x, e)$, so to highest order $r_{MN}(x, e) = R_{MN}(x, e)\|\partial z_n/\partial x\|\delta(z_n(x))$. For the Born approximation the function $r_{MN}(x, e)$ is given by a pseudodifferential operator acting on

$$\left(\frac{\delta c_{ijkl}}{\rho}, \frac{\delta\rho}{\rho} \right)_\alpha,$$

with principal symbol $(2\mathrm{i}\tau(x, \xi, e))^{-1}w_{MN;\alpha}(x, \xi, e)$; see (3–5).

PROOF. We do the proof for the Kirchhoff approximation using (7–10); for the Born approximation the proof is similar. Because Assumption 4 is satisfied, the projection π_Y of Λ_{MN} into $T^*Y \setminus 0$ is an embedding, and the image is a coisotropic submanifold of $T^*Y \setminus 0$. Therefore we can apply Lemma 6.1. Formula (6–3) implies that the phase factor $e^{i\Phi_{MN}}$ can be written in the form

$$e^{i\Phi_{MN}(y_I',x,\eta_J')} = e^{i(S_{MN}(y_I',x,\eta_J',0)+\langle y_J',\eta_J'\rangle)}$$

$$= (2\pi)^{-(n-1-c)} \int e^{i(S_{MN}(y_I',x,\eta_J',\varepsilon)+\langle y_J',\eta_J'\rangle - \langle e,\varepsilon\rangle)} \, d\varepsilon \, de;$$

we define

$$\Psi_{MN}(y',x,e,\eta_J',\varepsilon) = S_{MN}(y_I',x,\eta_J',\varepsilon) + \langle y_J',\eta_J'\rangle - \langle e,\varepsilon\rangle.$$

Thus the number of phase variables is increased by making use of a stationary phase argument. Let $\mathcal{B}_{MN}(y_I',x,\eta_J',\varepsilon)$ be as described. Then we obtain

$$G_{MN}^{\mathrm{refl}}(y') = (2\pi)^{-(|J|+n-1-c)/2-(2n-1-c)/2}$$

$$\times \int \left((2\pi)^{n/2} 2i\tau(\eta')\mathcal{B}_{MN}(y_I',x,\eta_J',\varepsilon)R_{MN}(x,e) + \text{l.o.t.}\right)$$

$$\times e^{i\Psi_{MN}(y',x,e,\eta_J',\varepsilon)} \left\|\frac{\partial z_n}{\partial x}\right\| \delta(z_n(x)) \, d\eta_J' \, d\varepsilon \, dx \, de.$$

In this formula, the data are represented as a Fourier integral operator acting on $\|\partial z_n/\partial x\|\delta(z_n(x))$ considered as a function of (x,e). Multiplying by H_{MN}^{-1} gives a pseudodifferential operator of the form described acting on $\|\partial z_n/\partial x\|\delta(z_n(x))$. Thus the result follows. $\square$

Figure 13 illustrates the relation between R_{MN} and the Born approximation.

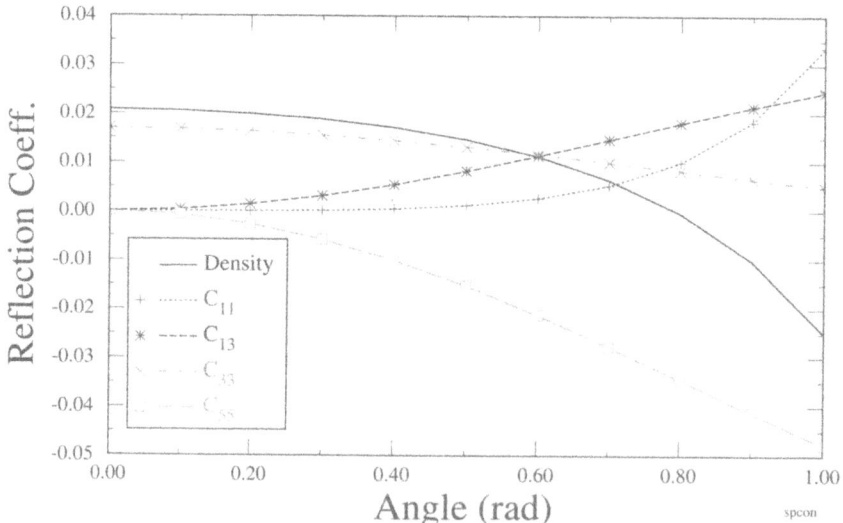

Figure 13. Constituent reflection coefficients linearized in $(\delta c_{ijkl}/\rho, \delta\rho/\rho)_\alpha$ for incident qP and scattered qP in a transversely isotropic medium: The relation between the Born and Kirchhoff approximations.

7C. Inversion. The operator H_{MN} is invertible. A choice of phase function and amplitude for its inverse is given by (see Chapter 8 of Treves [102])

$$-\Psi_{MN}(y', x, e, \eta'_J, \varepsilon), \qquad \mathcal{B}_{MN}(y'_I, x, \eta'_J, \varepsilon)^{-1} \left| \det \frac{\partial(y', \eta')}{\partial(y'_I, x, \eta'_J, \varepsilon)} \right|,$$

respectively. Thus microlocally an explicit expression for $r_{MN}(x, e)$ in terms of the data is given by

$$r_{MN}(x, e) = \int \mathcal{B}_{MN}(y'_I, x, \eta'_J, \varepsilon)^{-1} \left| \det \frac{\partial(y', \eta')}{\partial(y'_I, x, \eta'_J, \varepsilon)} \right|$$
$$\times e^{-i\Psi_{MN}(y', x, e, \eta'_J, \varepsilon)} d_{MN}(y') \, d\eta'_J \, d\varepsilon \, dy'.$$

Because the function $r_{MN}(x, e)$ is to highest order equal to the product of reflection coefficient and the singular function of the reflector surface, this reconstruction of the function $r_{MN}(x, e)$ leads to the following result for Kirchhoff data.

COROLLARY 7.6. *Suppose that the medium above the reflector is given, and that it satisfies Assumptions 1, 6, 2, 3, and 4. Then one can reconstruct the position of the interface and the angle dependent reflection coefficient $R_{\mu\nu}(x, e)$ on the interface.*

REMARK 7.7. In amplitude-versus-angles analysis it is the reflection coefficient obtained from the inversion that is directly subjected to interpretation. Depending on the (medium) coefficients symmetries, the reflection coefficient may be expanded into trigonometric functions of scattering angle and azimuth (e). (The foundation of such expansions dates back to the work of Shuey [88].) The leading order coefficients in such expansion provide information about the P and S phase velocities just below the reflector.

Such information may be used to infer additional properties of the subsurface. For example, the ratio of P and S phase velocities is nearly independent of pore pressure (or effective stress) and hence renders information about the fluid compressibility (and hence fluid mixture) in a porous formation. The S phase velocity as such will reveal primarily the pore pressure. This separation of information is most easily recovered by crossplotting the P and S phase impedances. The analysis is of key importance in time-lapse seismology with application for example to reservoir monitoring.

8. Annihilators of Seismic Data

The results of the previous subsections may be exploited in the problem of estimating the smooth background medium (or, in the Kirchhoff approximation, the smooth medium parameters above/in between the interfaces). If $n-1-c > 0$ there is a redundancy in the data through the variable e. The redundancy in the data manifests itself as a redundancy in images from these data. If the smooth medium parameters (above the interface) are correct, then applying the operator H_{MN}^{-1} of Theorem 7.5 to the data results in a reflectivity function $r_{MN}(x, e)$, such that the position of the singularities does not depend on e. Thus we obtain multiple images of the reflectivity parametrized by e such that their singular supports (in x) should agree. This can be used as a criterion to assess the accuracy of the choice of the background medium.

One way to measure this agreement is by taking a derivative with respect to e. We develop this strategy further. The derivative $\partial r_{MN}(x, e)/\partial e$ is one order less singular if $r_{MN}(x, e)$ depends smoothly on e, as in (7–11), than when it does not (for instance a δ function versus its derivative in the Kirchhoff case). Taking also the factor in front of the δ function of r_{MN} into account — see (7–11 — we obtain that to the highest two orders

$$
\left(R_{MN}(x, e) \frac{\partial}{\partial e} - \frac{\partial R_{MN}^{\mathrm{prin}}}{\partial e}(x, e) \right) r_{MN}(x, e) = 0. \tag{8–1}
$$

If $R_{MN}(x, e)$ is nonzero then the lower order terms can be chosen such that this equation is valid to all orders.

Conjugating the differential operator of (8–1) with the invertible Fourier integral operator H_{MN}, we obtain a pseudodifferential operator on $\mathcal{D}'(Y)$. Thus we obtain the following corollary of Theorem 7.5

COROLLARY 8.1. *Let the pseudodifferential operators* $W_{MN}(y', D_{y'})$ *be given by*

$$
W_{MN}(y', D_{y'}) = H_{MN} \left(R_{MN}(x, e) \frac{\partial}{\partial e} - \frac{\partial R_{MN}}{\partial e}(x, e) \right) H_{MN}^{-1}.
$$

Then for Kirchhoff data $d_{MN}(y')$ *we have to the highest two orders*

$$
W_{MN}(y', D_{y'}) d_{MN}(y') = 0. \tag{8–2}
$$

For values of e where $R_{MN}(x, e) \neq 0$, *the operator* $W_{MN}(y', D_{y'})$ *can be chosen such that (8–2) is valid to all orders.*

We refer to W_{MN} as an annihilator of the data.

REMARK 8.2. In principle, the annihilators $W_{MN}(y', D_{y'})$ can be used to quantify the agreement between the data and the background medium. Symes [95] discusses such criteria for acoustic media using differentiation with respect to the offset coordinate. Upon introducing the seminorm $\|W_{MN}d_{MN}\|$, an error criterion is obtained and an optimization scheme can be derived to minimize the action of the annihilators and to obtain an improved estimate of the background medium coefficients. Such an optimization scheme requires a "gradient" computation combined with a regularization procedure. Here the "gradient" is viewed as a derivative along a curve in the space of background media. We have adapted the Tikhonov regularization that penalizes roughness in the background medium. The gradient computation makes use of ray perturbation theory. The procedure is known to seismologists as *migration velocity analysis*. An example is presented in Section 9E.

9. The Generalized Radon Transform

Honoring the symplectic geometry presented in Section 6, we extract a generalized Radon transform from the modeling and inversion Fourier integral operators developed in Sections 4 and 7. The generalized Radon transform becomes the basis of a processing procedure. It is developed through several intermediate steps by selecting relevant neighborhoods of the canonical relation Λ_{MN}. These steps enable the use of traveltime in the phase function while restricting the imaging operator to a given value of e.

9A. Diffraction stack. To describe the kernel of the operator $F_{MN;\alpha}$ as an oscillatory integral on a neighborhood of the point on Λ_{MN} parametrized by $(x_0, \hat{\alpha}, \tilde{\alpha}, \tau)$, the minimum number of phase variables is given by the corank of the projection

$$D\pi : T\Lambda_{MN} \to T(T^*Y \times T^*X)$$

at $(x_0, \hat{\alpha}, \tilde{\alpha}, \tau)$, which is here given by

$$\text{corank } D\pi = 1 + \text{corank } \frac{\partial s_N}{\partial \tilde{\alpha}}(x, \tilde{\alpha}) + \text{corank } \frac{\partial r_M}{\partial \hat{\alpha}}(x, \hat{\alpha}).$$

This corank is > 1 when s or r is in a caustic point relative to x. Let

$$\Lambda'_{MN} = \Lambda_{MN} \setminus \big(\text{closed neighborhood of } \{\lambda \in \Lambda_{MN} : \text{corank } D\pi > 1\}\big). \quad (9\text{--}1)$$

The subset Λ'_{MN} can be described by phase functions of the *traveltime form*

$$\tau\left(t - T_{MN}^{(m)}(x, s, r)\right)$$

with the only phase variable τ and where $T_{MN}^{(m)}$ is the value of the time variable in (3–10); see Remark 4.2. The index m labels the branches of the multi-valued traveltime function. Thus the set $\{T_{MN}^{(m)}\}_{m \in M}$ describes the canonical relation (3–10) except for a neighborhood of the subset of the canonical relation where mentioned projection is degenerate. Each $T_{MN}^{(m)}$ is defined on a subset $D^{(m)}$ of $X \times O_s \times O_r$ (possibly dependent on MN). We define $F_{MN;\alpha}^{(m)}$ to be a contribution to $F_{MN;\alpha}$ with phase function given by $\tau(t - T_{MN}^{(m)})$, and symbol $A_{MN}^{(m)}$ in a suitable class such that on a subset Λ'_{MN} of the canonical relation where the projection is nondegenerate, $F_{MN;\alpha}$ is given microlocally by $\sum_{m \in M} F_{MN;\alpha}^{(m)}$.

REMARK 9.1. With the phase function $\tau(t - T_{MN}^{(m)})$, the application of the imaging or inversion operator on the data, up to leading order, amounts to a "diffraction stack", which is for the contribution at the image point x, an integration of the analytic extension of the data over (s, r) subjecting the time to $t = T_{MN}^{(m)}(x, s, r)$. Let s, r be restricted to a hyperplane (the earth's surface) and introduce the half offset H and midpoint m such that $s = m - H$ and $r = m + H$. Assume that $M = N$. Upon composing $T_{NN}^{(m)}$ with this map, we obtain $t_N^{(m)}(x, m, H) = T_{NN}^{(m)}(x, m - H, m + H)$. Let $t_0^{(m)} = t_N^{(m)}(x, m, H = 0)$ denote the zero-offset traveltime. It is common practice to write $H = h\varpi$ with

$\varpi \in S^{n-2}$. Since $dt_N^{(m)}/dh|_{h=0} = 0$, expanding $t_N^{(m)}$ into a Taylor series, and squaring the result leads to the expansion

$$(t_N^{(m)})^2 = (t_0^{(m)})^2 + (2h)^2 \underbrace{\langle \varpi, U\varpi \rangle}_{V_{nmo}^2(\varpi)} + \dots,$$

in which $U_{ij} = t_0^{(m)} \partial_{H_i} \partial_{H_j} t_N^{(m)}|_{H=0}/4$ is an $(n-1) \times (n-1)$ matrix, which defines the *normal moveout velocity* along a common midpoint line in the direction ϖ [103]. This hyperboloid was the original shape used in the diffraction stack that was organized in (m, h, ϖ) rather than (s, r). In the generalized Radon transform we can think of $\xi/\|\xi\|$ replacing m and e replacing H.

9B. Common image point gathers in scattering angles. In (4–4) we introduced the scattering angles. Here we introduce in addition the *migration dip* ν, defined as the direction of ξ in Λ_{MN} (cf. (3–10)),

$$\nu_{MN}(x, \hat{\alpha}, \tilde{\alpha}) = \frac{\xi}{\|\xi\|} \in S^{n-1}. \tag{9-2}$$

On $D^{(m)}$ there is a map $(x, \hat{\alpha}, \tilde{\alpha}) \mapsto (x, s, r)$. We define $e_{MN}^{(m)} = e_{MN}^{(m)}(x, s, r)$ as the composition of e_{MN} with the inverse of this map. Likewise, we define $\nu_{MN}^{(m)}$. Note that also $T_{MN}^{(m)}$ is the composition of T_{MN} with mentioned inverse, and that we can introduce $w_{MN;\alpha}^{(m)}$ in a similar manner.

In preparation of the generalized Radon transform (GRT) we define the angles imaging operator, $\check{L}$, via a restriction in $F_{MN;\beta}^*$ of the mapping $e_{MN}^{(m)}$ to a prescribed value e; that is, the distribution kernel of each contribution $(F_{MN;\beta}^{(m)})^*$ is multiplied by $\delta(e - e_{MN}^{(m)}(x, s, r))$. (This restriction transfers over to the construction of the left inverse in (4–2).) Invoking the Fourier representation of this δ, the kernel of L follows as (cf. Section 6)

$$\check{L}(x, e, r, s, t) = \sum_{m \in M} (2\pi)^{-(n-1)} \int \overline{A_{MN}^{(m)}(x, s, r, \tau) w_{MN;\beta}^{(m)}(x, s, r, \tau)}$$
$$\times e^{i\Phi_{MN}^{(m)}(x, e, s, r, t, \varepsilon, \tau)} d\tau d\varepsilon, \tag{9-3}$$

where $A_{MN}^{(m)} = \mathcal{B}_{MN}$ with $|J| = 1$ (and $\eta_J' = \tau$) is a symbol for the m-th contribution to $F_{MN;\beta}$, supported on $D^{(m)}$, and

$$\Phi_{MN}^{(m)}(x, e, s, r, t, \varepsilon, \tau) = \tau(T_{MN}^{(m)}(x, s, r) - t) + \langle \varepsilon, e - e_{MN}^{(m)}(x, s, r) \rangle.$$

As before, ε is the cotangent vector corresponding to e.

Let $\psi_L = \psi_L(D_s, D_r, D_t)$ be a pseudodifferential cutoff such that $\psi_L(\sigma, \rho, \tau) = 0$ on a conic neighborhood of $\tau = 0$. Then $\psi_L \check{L}$ is a Fourier integral operator [90] with canonical relation

$$\Lambda_{\check{L}} = \bigcup_{m \in M} \{(x, e_{MN}^{(m)}, \xi_{MN}^{(m)}, \varepsilon; s, r, T_{MN}^{(m)}, \sigma_N, \rho_M^{(m)}, \tau) : (x, s, r) \in D^{(m)},$$
$$\varepsilon \in \mathbb{R}^{n-1}, \tau \in \mathbb{R} \backslash 0\} \subset T^*(X \times E) \backslash 0 \times T^* Y \backslash 0, \tag{9-4}$$

where $e_{MN}^{(m)}$ and $T_{MN}^{(m)}$ are functions of x, s, r and

$$\xi_{MN}^{(m)} = \xi_{MN}^{(m)}(x, s, r, \tau, \varepsilon) = \partial_x \Phi_{MN}^{(m)} = \tau \partial_x T_{MN}^{(m)}(x, s, r) - \langle \varepsilon, \partial_x e_{MN}^{(m)}(x, s, r) \rangle, \tag{9-5}$$

while similar expressions hold for $\sigma_N^{(m)}$ and $\rho_M^{(m)}$.

Effectively, for each x we select a different subset of the data. This is fundamentally different from the common offset Kirchhoff integral approach which amounts to a straightforward restriction in the acquisition manifold.

9C. Artifacts of type II. With the choice (4–4) for e, the following assumption is implied. However, for other choices of e it needs to be verified.

ASSUMPTION 7. *Consider the mapping*

$$\Xi : \Lambda_{MN} \to T^* X \setminus 0 \times E, \ \lambda(x, \hat{\alpha}, \tilde{\alpha}, \tau) \mapsto (x, \xi, e), \ with \ \xi = \|\xi\| \nu_{MN}.$$

Composing this mapping with the inverse of the mentioned map $(x, \hat{\alpha}, \tilde{\alpha}) \mapsto (x, s, r)$, *yields per branch m a mapping $\Xi^{(m)}$ from (x, s, r, τ) to an element of $T^* X \setminus 0 \times E$. $\Xi^{(m)}$ is locally diffeomorphic, i.e.*

$$\text{rank} \left. \frac{\partial(\xi_{MN}^{(m)}, e_{MN}^{(m)})}{\partial(s, r, \tau)} \right|_{\varepsilon=0} \quad is \ maximal, \ at \ given \ x \ and \ branch \ m.$$

Let d_{MN} be the Born modeled data in accordance with Theorem 3.2. To reveal any artifacts generated by $\check{L}$, i.e. singularities in $\check{L}d_{MN}$ at positions not corresponding to an element of $\text{WF}(g_\alpha)$, we consider the composition $\check{L} F_{MN;\alpha}$. With Assumption 7 this composition is equal to the sum of a smooth e-family of pseudodifferential operators and, in general, a non-microlocal operator. The wavefront set of the non-microlocal operator contains no elements with $\varepsilon = 0$ [90, Theorem 6.1]. The origin of contributions from $\varepsilon \neq 0$ is illustrated in Figure 14. A filter needs to be applied to remove contributions from $|\varepsilon| \geq \varepsilon_0 > 0$; we define the generalized Radon transform L as the Fourier integral operator with canonical relation U_L — also denoted as Λ_L — to be a *neighborhood* of $\Lambda_{\check{L}} \cap \{\varepsilon = 0\}$, which derives from the left inverse (4–2).

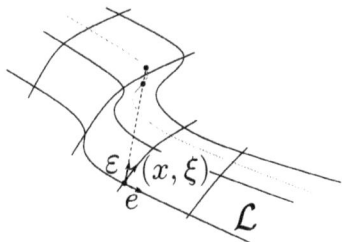

Figure 14. The origin of artifacts generated by the generalized Radon transform. (Inside the $T^* Y \setminus 0$ box of Figure 11.)

The artifacts in the compose of the canonical relation of $\check{L}$ with that of $F_{MN;\alpha}$ can be evaluated by solving the system of equations

$$r = r_M(x, \hat{\alpha}), \tag{9-6}$$

$$s = s_N(x, \tilde{\alpha}), \tag{9-7}$$

$$T_{MN}^{(m)}(z, s, r) = T_M(x, \hat{\alpha}) + T_N(x, \tilde{\alpha}), \tag{9-8}$$

$$\rho_M^{(m)}(z, s, r, \tau, \varepsilon) = -\tau p_M^r(x, \hat{\alpha}), \tag{9-9}$$

$$\sigma_N^{(m)}(z, s, r, \tau, \varepsilon) = -\tau p_N^s(x, \tilde{\alpha}). \tag{9-10}$$

(The frequency is preserved.) Equations (9–6)–(9–8) imply that the image point z must lie on the isochrone determined by (x, s, r). Equations (9–9) and (9–10) enforce a match of slopes (apparent in the appropriate "slant stacks") in the measurement process:

$$- \tau \partial_r T_{MN}^{(m)}(z, s, r) + \langle \varepsilon, \partial_r e_{MN}^{(m)}(z, s, r) \rangle = -\tau p_M^r(x, \hat{\alpha}), \tag{9-11}$$

$$- \tau \partial_s T_{MN}^{(m)}(z, s, r) + \langle \varepsilon, \partial_s e_{MN}^{(m)}(z, s, r) \rangle = -\tau p_N^s(x, \tilde{\alpha}). \tag{9-12}$$

For $\varepsilon \neq 0$ the take-off angles of the pairs of rays at (r, s) may be distinct. Equations (9–11) and (9–12) imply the matrix compatibility relation (upon eliminating ε/τ)

$$\left(\partial_r e_{MN}^{(m)}(z, s, r) \right)^{-1} \left(p_M^r(x, \hat{\alpha}) - \partial_r T_{MN}^{(m)}(z, s, r) \right)$$
$$= \left(\partial_s e_{MN}^{(m)}(z, s, r) \right)^{-1} \left(p_N^s(x, \tilde{\alpha}) - \partial_s T_{MN}^{(m)}(z, s, r) \right). \tag{9-13}$$

Those summarize the geometrical composition equations determining the artifacts: For each $(x, \hat{\alpha}, \tilde{\alpha}) \in K$ solve the $3n - 2$ equations (9–6)–(9–8), (9–13) for the $3n - 2$ unknowns (z, s, r). (From (9–11) we then obtain ε/τ hence ε.)

REMARK 9.2. The generalized Radon transform reconstructs a distribution in $\mathcal{E}'(X)$ smoothly indexed by $e \in E$. Thus, we can carry out the composition $(N_{MN})_{\alpha\beta}^{-1}L$ (no summation over M, N) as in (4–2) to yield the generalized Radon transform *inversion* [33]. Likewise, we can carry out a composition with the modeling operator $F_{MN;\alpha}$ (or $H_{MN;\alpha}$).

9D. Filters. In general, filters need to be applied to the common image point gathers to remove the artifacts of type II. At the same, the generalized Radon transform is a transformation of data as a function of the $(2n - 1)$ variables (s, r, t) to a distribution of the $(2n - 1)$ variables (x, e). After the removal of artifacts, the alignment in the e directions ($\varepsilon = 0$) of the singular support of this distribution represented in the common image point gathers simplifies the task of denoising the final image of the medium perturbation. Treating the artifacts as noise as well, a joint approach based upon non-adaptive wavelet thresholding [41] applies; an analysis of subbands will then aid in the suppression of the artifacts, associated with $\varepsilon \neq 0$.

The removal of the illumination effects (which can be written as the action of a pseudodifferential operator on g_α) is also amenable to the use of wavelets. A possible approach is matching pursuit [72].

9E. An example: Estimating the background model. To get to an example, a discretization both of c_{ijkl}, ρ (background) and of $\delta c_{ijkl}, \delta\rho$ (singular perturbation) has to be chosen. We have represented the background by *cubic splines* the smoothness of which aid in the numerical computation of geometrical spreading.

In Figure 16 we illustrate the performance of an optimization minimizing the annihilators developed with the aid of the generalized Radon transform. In the example, the medium is isotropic and hence we have only two parameters i.e. the P-wave and S-wave speeds. We reconstruct a smooth Gaussian lens (in P-wave speed) in six iterations (the intermediate model on the left and some ray geometry originating at a scattering point on the reflector on the right), starting from an initial model that did not contain a lens whence in the initial model no caustics were formed. Note the alignment in the final common image point gather reflecting the reduncancy in the data. In the final model caustics do occur, as illustrated in Figure 15.

9F. An example: reconstructing the singular perturbation. From a geological point of view, it is attractive to represent the singular perturbation by *fractional splines*.

We apply the generalized Radon transform to multicomponent ocean bottom seismic data acquired over the Valhall field in the Norwegian sector of the North Sea to obtain common image point gathers representing $r_{MN}(x,e)$. In this region, it is believed that the presence of gas in the overburden yields lenses that cause caustics to form. An isotropic elastic background model was obtained;

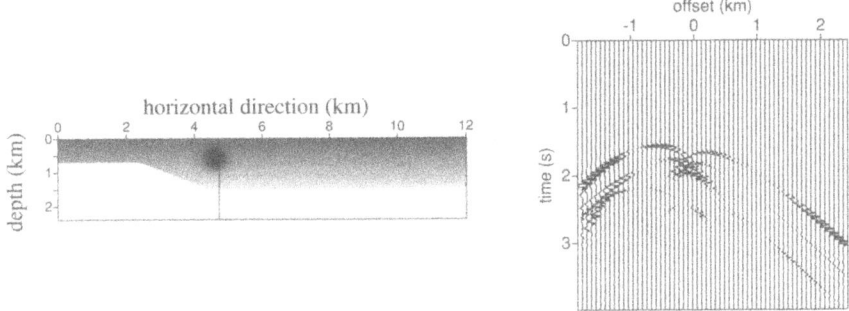

Figure 15. The Gaussian lens model with reflecting surface (left) and the modeled data (vertical component) for a given source position, in an experiment of the type in Figure 1 (right).

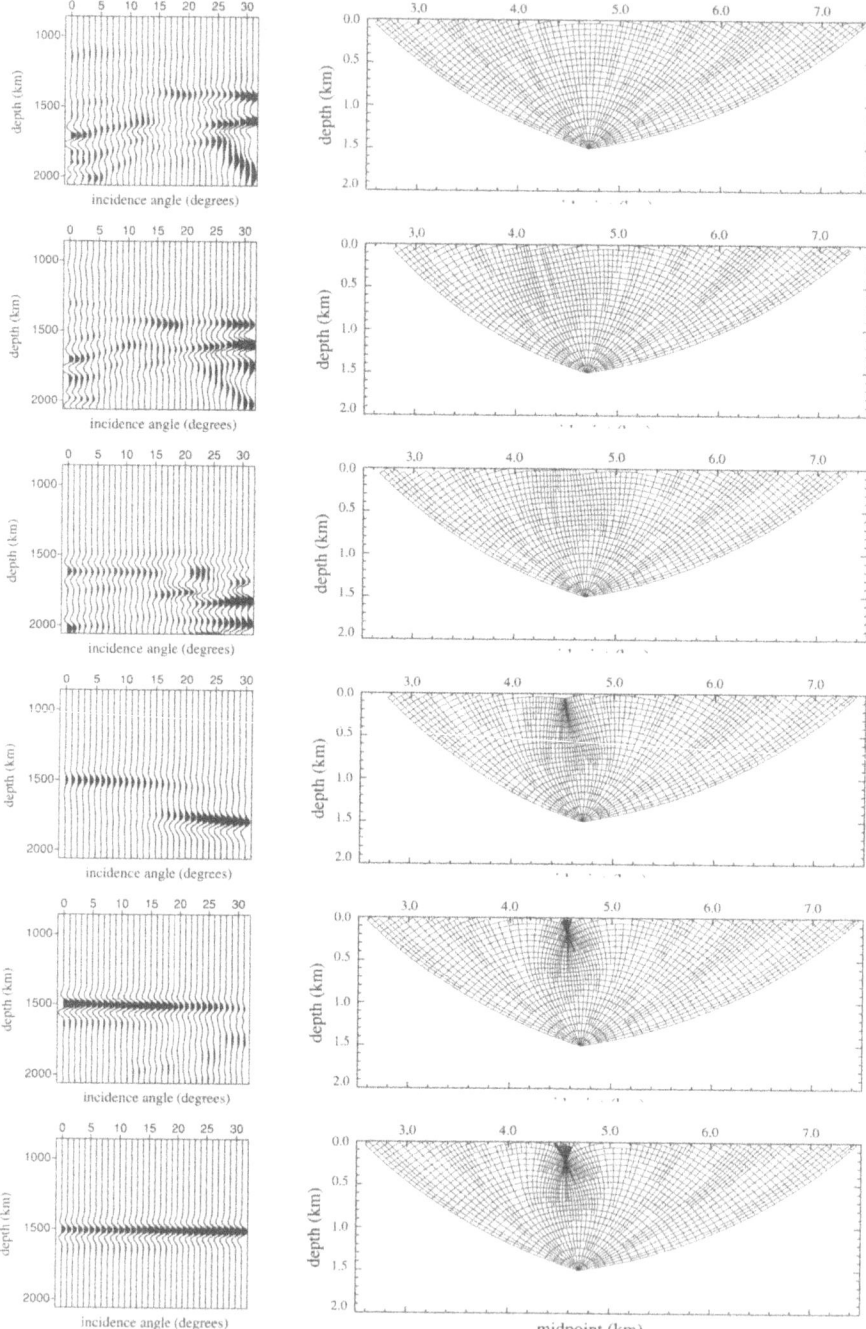

Figure 16. Successive (P, P) common image point gathers (left) and characteristics associated with a particular Green's function originating at the reflector (right) in the process of minimizing the annihilator. A smooth Gaussian lens is being reconstructed (from [22]).

the P-wave speed is shown in Figure 17 (top). Note the presence of the lenses. We will illustrate the data, the action of the generalized Radon transform, the common image point gathers, and images of particular medium parameter combinations in the slice depicted in Figure 17 (bottom). The common image point gathers will be restricted to x lying on the black vertical line and e being the scattering angle. A fan of characteristics originating at the (dark) reservoir layer is shown in Figure 18, which illustrates the formation of caustics in G_M ($M = \mathrm{P}$).

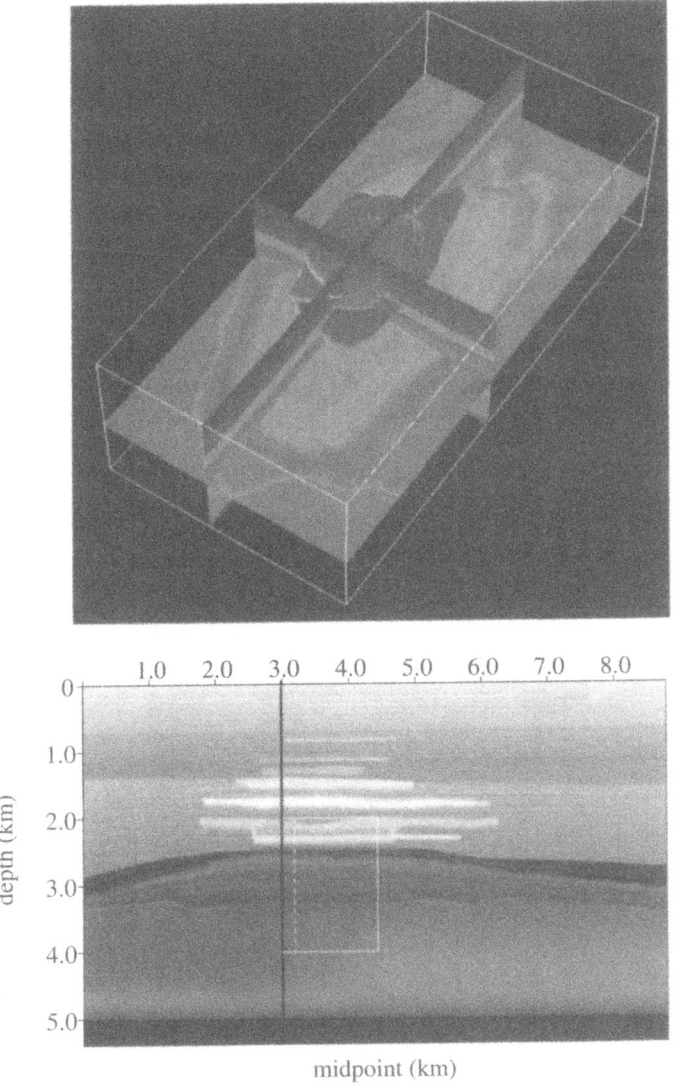

Figure 17. The c_P model in perspective (top) and a slice (bottom). Note the stack of lenses that contain gas. The dark layer represents the reservoir.

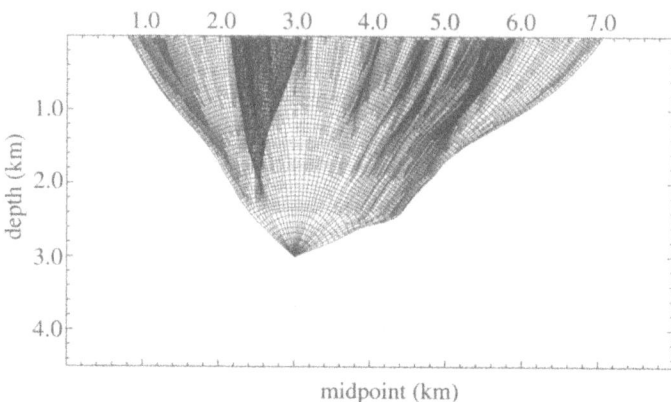

Figure 18. Characteristics in the model of Figure 17 (bottom).

The horizontal component of the data, for given source position, is shown in Figure 19. These illustrations can be thought of as regularizations of the data and medium perturbation distributions.

For the pairs (P, P) and (P, S) the common image point gathers are shown in Figure 20 (left, middle). Observe the alignment of a sequence of (singular) supports at the arrow with scattering angle (here converted to incidence angle) at reservoir depth. The $r_{MN}(x, e)$ for $(N, M) = (P, S)$ corresponding with the

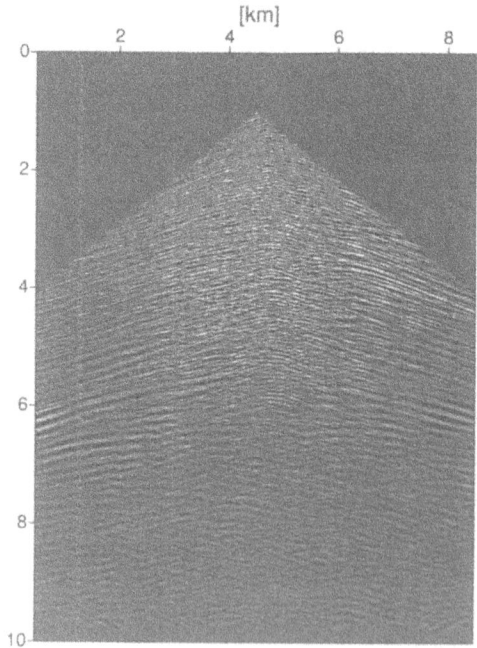

Figure 19. Data: fixed source position, horizontal component (parallel to acquisition surface).

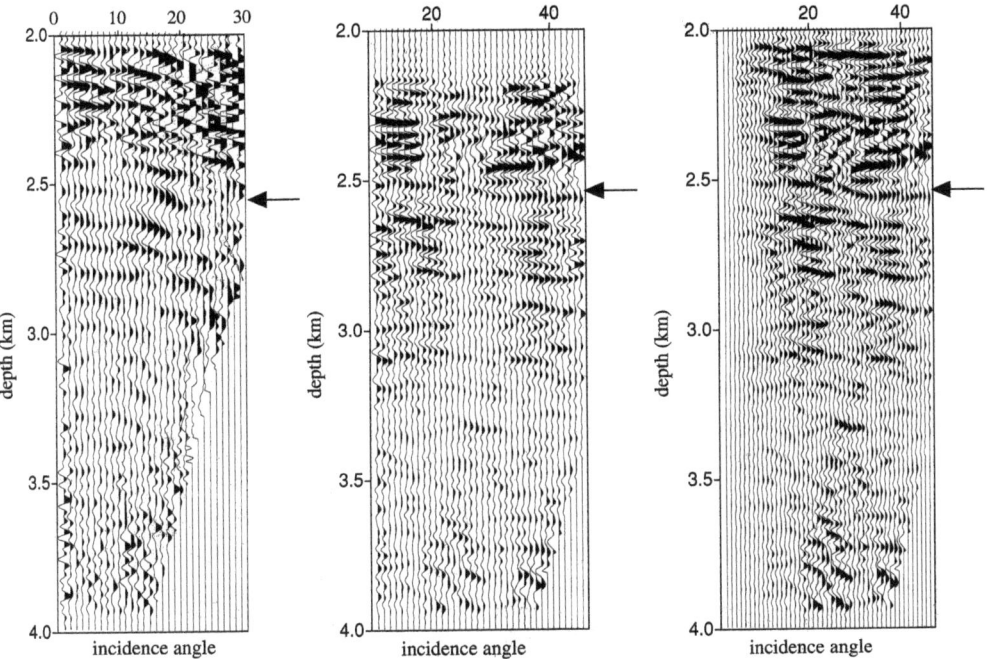

Figure 20. Common image point gathers with x on the black line in Figure 17 (bottom). In the framework of the Born approximation for (P, P) (left) and (P, S) (middle) and in the framework of the Kirchhoff approximation for (P, S) (right); from [21].

middle common image point gather is shown on the right. Note the change of sign of the amplitude at the key reflector, which is an indication of the presence of anisotropy.

In Figure 21, the images of P- and S-phase impedances inside the white box of Figure 17 (bottom) are presented, and compared with a standard seismic image from (P, S) in Figure 22. The use of these images combined was addressed in Remark 7.7.

10. Wavefield "Continuation"

In general, $F_{MN;\alpha} F^*_{MN;\beta}$ cannot be a pseudodifferential operator. However, exploiting carefully the redundancies in the data, L^*L (note that L is an imaging operator itself) attains, under certain conditions, pseudodifferential properties. This observation is at the basis of seismic wavefield "continuation".

In this section, we use a simplified notation: We suppress the subscripts MN in the operators F (originally $F_{MN;\alpha}$) and N (originally $N_{MN;\alpha\beta}$). Also, we

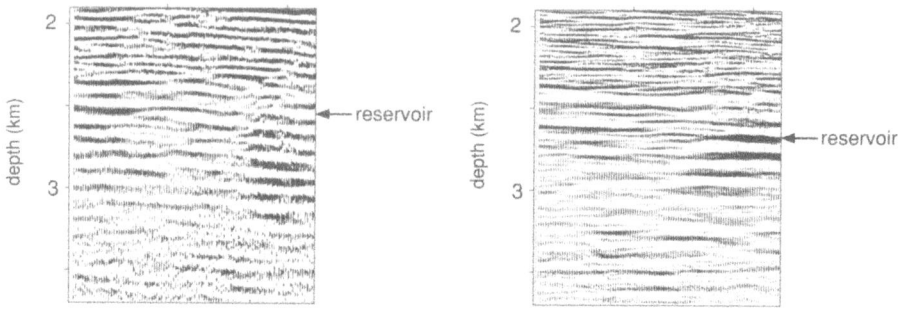

Figure 21. High resolution images of impedance revealing sedimentary layers and faults, from: (P, P) scattered waves (left) and (P, S) scattered waves (right).

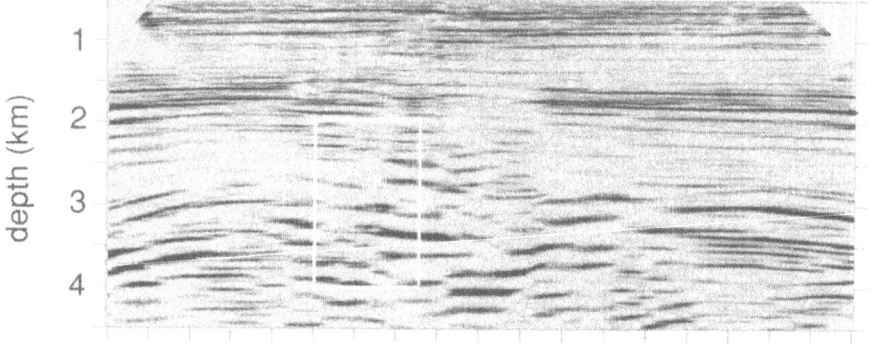

Figure 22. A (P, P) image from hydrophone data obtained with standard seismic processing (transversely isotropic) corresponding with Figure 17 (bottom). The white box corresponds with the image of Figure 21 (left).

re-introduce y replacing y' to enable with the same notation a further restriction to acquisition submanifolds.

10A. Modeling restricted to an acquisition submanifold. Single reflection seismic wavefield continuation aims at generating from reflection data—through the canonical relation (3–10)—associated with $T^*X \setminus 0 \times E_i$, in which E_i is an $(n-1)$-dimensional open neighborhood of e say, reflection data associated with $T^*X \setminus 0 \times E_o$, in which $E_o \supset E_i$. Such continuation, within the acquisition manifold Y, is accomplished through the composition of Fourier integral operators generating an intermediate image of δc_{ijkl} (resp. $\delta\rho$). In the previous section, we analyzed a Fourier integral operator, the generalized Radon transform, that generates (linear combinations of) δc_{ijkl} (resp. $\delta\rho$) from data on $T^*X \setminus 0 \times E_i$. In this section we consider, once data are modeled from δc_{ijkl} (resp. $\delta\rho$) as in Theorem 3.2, the restriction to an acquisition submanifold parametrized by (x, ξ, e) through the canonical relation (3–10), such that $e \in E_o$. In the following subsections, the restriction, modeling and generalized Radon transform imaging operators will be composed to yield the continuation. In this composition, the background coefficients (ρ, c_{ijkl}) are used, but, naturally, δc_{ijkl} (resp. $\delta\rho$) does not appear. The continuation is illustrated in Figure 23.

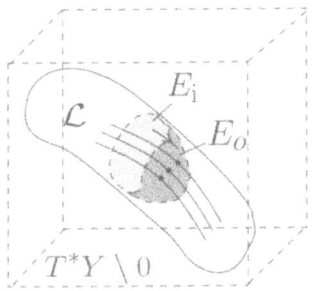

Figure 23. Continuation and characteristic strips. (Inside the $T^*Y \setminus 0$ box of Figure 11.)

Our starting point is the situation where $c = 0$. We consider the restriction to an acquisition submanifold by increasing c to $\tilde{c}$, say. This submanifold is written as $Y^{\tilde{c}} = \Sigma^{\tilde{c}} \times (0, T)$, with $\Sigma^{\tilde{c}} \xrightarrow{\imath} O_s \times O_r$ representing an embedded manifold of codimension $c = \tilde{c} \geq 0$. We reconsider Assumption 3. Let

$$(y'_1, \ldots, y'_{2n-2-c}, \underbrace{y'_{2n-1-c}}_{t}, y''_{2n-c}, \ldots, y''_{2n-1})$$

denote a local coordinate system on Y such that Σ^c is given by $(y''_{2n-c}, \ldots, y''_{2n-1}) = (0, \ldots, 0)$ locally. In the applications under consideration, we order the coordinates such that $y'_{2n-1-c} = t$.

ASSUMPTION 8. *The intersection of Λ_{MN} with the manifold $y'' = 0$ is transversal, i.e.*

$$\frac{\partial y''}{\partial(x, \hat{\alpha}, \tilde{\alpha}, \tau)} \quad \text{has maximal rank.}$$

Applying [43, Thm. 4.2.2] to the pair F and the restriction $\mathcal{R} = \mathcal{R}^c$ from $O_s \times O_r \to \Sigma^c$ with Assumption 8 implies that $\mathcal{R}^c F$ is a Fourier integral operator of order $(n-1+c)/4$ with canonical relation

$$\Lambda^c_{MN} = \{(y', t, \eta', \tau; x, \xi) : \exists\, y', y'', \eta', \eta'' \text{ such that}$$
$$(y', y'', \eta', \eta''; x, \xi) \in \Lambda_{MN} \text{ and } y'' = 0\} \subset T^*Y^c \setminus 0 \times T^*X \setminus 0. \quad (10\text{–}1)$$

(Here $\Lambda^0_{MN} = \Lambda_{MN}$.) We mention two examples: Zero offset, where $c = n - 1$ and $\Sigma^c := \Sigma^{ZO} \subset \operatorname{diag}(\partial X)$ (subject to the $n - 1$ constraints $r = s$ when $\arccos(\hat{\alpha} \cdot \tilde{\alpha}) = 0$ and e_o at x follows from (4–4)), and common azimuth (CA), where $c = 1$ and $\Sigma^c := \Sigma^{CA}$ subject to one constraint typically of the form that the $(n - 1)$st coordinate in $r - s$ is set to zero, while $E_o \ni e$ at x follows from the mapping $e^{(m)}$. We set $Y^{ZO} = \Sigma^{ZO} \times (0, T)$ and $Y^{CA} = \Sigma^{CA} \times (0, T)$.

The restriction to acquisition submanifolds is placed in the context of inversion in [80].

10B. Continuation. We analyze the continuation of multiple finite-offset seismic data — in the absence of knowledge about the singular medium perturbation. The compose FL is a well-defined operator $\mathcal{D}'(Y) \to \mathcal{D}'(Y)$. Its wavefront set is contained in the composition of the wavefront sets of F and L [43, Thm. 1.3.7], hence in the composition of canonical relations,

$$\Lambda_{MN} \circ \Lambda'_L = \big\{ (s_2, r_2, t_2, \sigma_2, \rho_2, \tau_2; s_1, r_1, t_1, \sigma_1, \rho_1, \tau_1) : \exists\, x, \xi, \varepsilon \text{ such that}$$

$$(s_2, r_2, t_2, \sigma_2, \rho_2, \tau_2; x, \xi) \in \Lambda_{MN} \text{ and } (x, e, \xi, \varepsilon; s_1, r_1, t_1, \sigma_1, \rho_1, \tau_1) \in \Lambda_L \big\}$$

$$\subset T^*Y \setminus 0 \times T^*Y \setminus 0, \quad (10\text{–}2)$$

with $\Lambda'_L = \{(x, \xi; s, r, t, \sigma, \rho, \tau) : \exists x, \xi, \varepsilon \text{ such that } (x, e, \xi, \varepsilon; s, r, t, \sigma, \rho, \tau) \in \Lambda_L\}$. (To avoid the introduction of Λ'_L, we could consider the composition of H_{MN} with L; see Theorem 7.5.) Whether the compose is a Fourier integral operator is yet to be investigated.

Using the parametrizations of Λ_{MN} in Remark 3.3 and Λ_L in (9–4), the compose (10–2) can be evaluated through solving a system of equations, the first n being trivial fixing the scattering point $x_0 = x$, the second n equating the cotangent vectors

$$\underbrace{\tau_2 \partial_x T_{MN}(x_0, \hat{\alpha}, \tilde{\alpha})}_{\boldsymbol{\xi}(x, \hat{\alpha}, \tilde{\alpha}, \tau_2)} = \underbrace{\tau_1 \partial_x T^{(m)}_{MN}(x, s, r) - \langle \varepsilon, \partial_x e^{(m)}_{MN}(x, s, r)\rangle}_{\boldsymbol{\xi}^{(m)}_{MN}(x, s, r, \tau_1, \varepsilon)}. \quad (10\text{–}3)$$

Given $e(x, \hat{\alpha}, \tilde{\alpha}) = e$ ($n - 1$ constraints) these constitute n equations with the $2n - 1$ unknowns $(\hat{\alpha}, \tilde{\alpha}, \tau_2)$. (On $D^{(m)}$ the constraints on e can be invoked on s, r instead, namely, via the inverse of the map $(x, \hat{\alpha}, \tilde{\alpha}) \mapsto (x, s, r)$ as before.)

LEMMA 10.1. *With Assumptions 4 and 7 the composition FL yields a smooth family of Fourier integral operators parametrized by e. Their canonical relations are given by*

$$\Lambda_C = \Lambda_{MN} \circ U_L = \{(s_2, r_2, t_2, \sigma_2, \rho_2, \tau_2; s_1, r_1, t_1, \sigma_1, \rho_1, \tau_1)\}$$

parametrized by $(x_0, \hat{\alpha}, s_1, \tau_1, \varepsilon)$, where upon substituting $x = x_0$ and once r_1 is obtained from s_1 through the value e of $e^{(m)}_{MN}$ (which mapping is defined below equation (4–4)), $(s_1, r_1, t_1, \sigma_1, \rho_1)$ are given in (9–4), and, given $(\hat{\alpha}, \varepsilon)$, $(s_2, r_2, t_2, \sigma_2, \rho_2)$ are given in Theorem 3.2 in which $(\tilde{\alpha}, \tau_2)$ are obtained by solving (10–3).

PROOF. First we extend the operator F to act on distributions in $\mathcal{E}'(X \times E)$ by assuming that the action does not depend on $e \in E$. The calculus of Fourier integral operators gives sufficient conditions that the composition of two Fourier integral operators, here F and L, is again a Fourier integral operator. The essential condition is that the composition of canonical relations is transversal, i.e. that $\mathcal{L} = \Lambda_{MN} \times U_L$ and $\mathcal{M} = T^*Y \setminus 0 \times \text{diag}(T^*(X \times E) \setminus 0) \times T^*Y \setminus 0$ intersect transversally. We have

$$
\begin{array}{ccccc}
 & \Lambda_{MN} & & U_L & \\
\swarrow & & \searrow & \swarrow & \searrow \\
T^*Y \setminus 0 & & T^*X \setminus 0 \, (\times E) & & T^*Y \setminus 0
\end{array}
\qquad (10\text{-}4)
$$

where the inner two projections are submersions.

On the other hand, in a neighborhood of a point in Λ_{MN} given by (9–4), Λ_{MN} can be parametrized as in Λ'_{MN}. Using this parametrization one finds that the composition of Λ_{MN} and Λ_L is transversal if and only if the matrix

$$
\frac{\partial}{\partial(s, r, \hat{\alpha}, \tilde{\alpha}, \tau_2, \varepsilon, \tau_1)} \left(\boldsymbol{\xi}(x, \hat{\alpha}, \tilde{\alpha}, \tau_2) - \boldsymbol{\xi}_{MN}^{(m)}(x, s, r, \tau_1, \varepsilon) \right)
$$

has maximal rank (cf. (10–3)). This follows, for example, just from the $\boldsymbol{\xi}$ contribution in view of the submersivity of the projection $\pi_X : \Lambda_{MN} \to T^*X \setminus 0$. However, it follows also from the $\boldsymbol{\xi}_{MN}^{(m)}$ contribution: Parametrizing Λ_L by (x, ξ, ε) and restricting Λ_L to U_L further to $\varepsilon = 0$, results in a parametrization in terms of (x, ξ) (with the artifacts filtered out). Then $\boldsymbol{\xi}_{MN}^{(m)}$ becomes ξ and it follows that the composition of Λ_{MN} and U_L is transversal if and only if

$$
\text{rank} \, \frac{\partial}{\partial(\xi, \hat{\alpha}, \tilde{\alpha}, \tau_2)} \left(\boldsymbol{\xi}(x, \hat{\alpha}, \tilde{\alpha}, \tau_2) - \xi \right) \quad \text{is maximal.}
$$

This is indeed the case.   □

Subjecting the operator F in the composition to the constraint that e (cf. (4–4)) attains a prescribed value, the parameter $\hat{\alpha}$ in the lemma will be eliminated.

REMARK 10.2. Following seismological convention, we have used the terminology wavefield continuation. In fact, this is continuation in the context of continuation theorems also. We consider the continuation of the wavefield in the acquisition manifold from $T^*X \setminus 0 \times E_i$ to $T^*X \setminus 0 \times E_o$. This continuation is unique in the sense that $FLd = 0$ implies $F^*FLd = 0$ and, since $F^*F = N$ is strictly elliptic and pseudodifferential, then $Ld = 0$ so that the image of $(\delta c_{ijkl}, \delta\rho)$ vanishes. In the single scattering approximation this implies that $d = F(\delta c_{ijkl}, \delta\rho) = 0$, all modulo smoothing contributions.

REMARK 10.3. The subject of data regularization is the transformation of measured reflection data, sampled in accordance with the actual acquisition, to data associated with a regular sampling of the acquisition manifold Y. In our approach the operator $\mathcal{R}^c F \int_{E_i} de\langle N^{-1} \rangle L$ replaces the forward interpolation operator in the usual regularization procedures.

10C. Transformation to common azimuth: Azimuth MoveOut. Azimuth MoveOut [13] (AMO) is the process following composing $\mathcal{R}^1 = \mathcal{R}^1_{\mathrm{CA}}$ restricting Y to Y^{CA} with modeling operator F with the imaging generalized Radon transform L centered at a given value of e (conventionally for given value of offset $r - s$); the sing supp of the Lagrangian-distribution kernel of the resulting operator is what seismologists call the *AMO impulse response*. The composition FL has been addressed in Lemma 10.1. The general restriction has been addressed in Section 10A. Here we combine these results in the following

THEOREM 10.4. *With Assumptions* 4, 7 *and* 8 *with* $Y^c = Y^{\mathrm{CA}}$, *the composition* $\mathcal{R}^1_{\mathrm{CA}} F L$ *yields a smooth family of Fourier integral operators parametrized by* e. *The resulting operator is called Azimuth MoveOut.*

The following Bolker-like condition ensures that the restriction to common azimuth is "image preserving". Let $\Lambda^{\mathrm{CA}}_{MN}$ denote the canonical relation of $\mathcal{R}^1_{\mathrm{CA}} F$ in accordance with the analysis of Section 10A,

ASSUMPTION 9. *The projection*

$$\pi_{Y^{\mathrm{CA}}} \; : \; \Lambda^{\mathrm{CA}}_{MN} \to T^* Y^{\mathrm{CA}} \setminus 0$$

is an embedding.

This assumption is most easily verified whether an element in $T^* Y^{\mathrm{CA}} \setminus 0$ uniquely determines an element in $T^* X \setminus 0$ smoothly given the background medium. Using "all" the data (when available), integration over the $(n - 1)$ dimensional e removes the artifacts under the Bolker condition, Assumption 4: We obtain the transformation to common azimuth (TCA)

COROLLARY 10.5. *Let* $\langle N^{-1} \rangle$ *denote the regularized inverse of the normal operator in Theorem* 4.3. *With Assumptions* 2, 3, 4 *and* 8 (*with* $\Sigma^c = \Sigma^{\mathrm{CA}}$), *the composition* $\mathcal{R}^1_{\mathrm{CA}} F \langle N^{-1} \rangle F^* = \int \mathrm{d}e \mathcal{R}^1_{\mathrm{CA}} F \langle N^{-1} \rangle L$ *is a Fourier integral operator,* $\mathcal{D}'(Y) \to \mathcal{D}'(Y^{\mathrm{CA}})$. *With Assumption* 9 *the reduced dataset generates the same image as the original dataset.*

The proof follows that of Theorem 4.3 closely (see [91, Theorem 4.5]).

11. Sampling Canonical Relations: Quasi-Monte Carlo Integration Methods

The canonical relations of the modeling and imaging operators can be optimally sampled through their parametrizations. Here we consider the parametrization in $(x, \hat{\alpha}, \tilde{\alpha}, \tau)$ and the parametrization in (x, ξ, e). In seismic experiments, time is evenly sampled and hence, as far as τ is concerned, Nyquist's criterion (Shannon's law) applies. As far as $(\hat{\alpha}, \tilde{\alpha})$ is concerned, we discuss a sampling approach based upon quasi-Monte Carlo methods (see De Hoop and Spencer [37]).

For quasi-Monte Carlo methods we refer here to one original paper by Nieder-reiter [78]; for a comprehensive treatment of their foundations, see Hammersley and Handscomb [55]. We also apply more recent work by Wózniakowski [112].

11A. The notion of discrepancy. The basic idea underlying Monte Carlo integration is straightforward. The example often quoted in numerical textbooks (e.g., Press *et al.* [81]) is that of determining the volume of a general region E contained in the s-dimensional unit hypercube I^s with $I = [0, 1)$. If N points are chosen at random over the unit hypercube then the volume of $E \subset I^s$, $V(E)$ say, is given by

$$V(E) \approx \frac{1}{N} \sum_{i=1}^{N} c_E(x_i), \qquad (11\text{--}1)$$

where $c_E(x_i)$ is the characteristic function that takes the value 0 if the point x_i is outside E, 1 otherwise. In other words, the volume is computed by simply counting the number of points in I^s that fall within E and dividing by the total number of points.

Likewise, the integral $I_E[f]$ of an integrable function f over E is approximated by the mean

$$I_E[f] = \int_{I^s} f(x) c_E(x) \, dx \approx \frac{1}{N} \sum_{i=1}^{N} f(x_i) c_E(x_i). \qquad (11\text{--}2)$$

From the central limit theorem it can be deduced that the integration error aris-ing from using Eq.(11--2) is Gaussian-like distributed and its expected value is $\mathcal{O}(N^{-1/2})$. The attractive feature of this result is that the order of the error is independent of the dimension s of the problem and, hence, Monte Carlo integra-tion methods become increasingly favorable for higher-dimensional problems.

Monte Carlo methods work as well as they do, because randomly chosen points in s dimensions sample the s-dimensional unit hypercube I^s "fairly". This prop-erty is not uniquely confined to purely random numbers. Any fair, deterministic, distribution will suffice and may be superior. Fairness is here defined based on a deterministic measure of the difference between an estimate of the volume of a region K resulting from the use of the N point samples $x_i, i = 1, 2, \ldots, N$ and the true volume, $V(K)$ say. We assume that K is a Cartesian region contained in I^s, i.e., $K = K_1 \times K_2 \times \cdots \times K_s$ with $K_n = [0, k^n)$, $0 \le k^n \le 1$, $n = 1, 2, \ldots, s$. Let

$$R_N(k; x_1, x_2, \ldots, x_N) = \left| \frac{1}{N} \sum_{i=1}^{N} c_K(x_i) - V(K) \right|, \quad k = \{k^1, k^2, \ldots, k^s\}.$$
$$(11\text{--}3)$$

Given $x_i, i = 1, 2, \ldots, N$, then the *discrepancy* D_N is defined as the supremum,

$$D_N = \sup_{k \in I^s} R_N(k; x_1, x_2, \ldots, x_N),$$

or root-mean-square average or any equivalent measure, of R_N over all $K \subset I^s$ i.e. $k \in I^s$. For example, Wózniakowski [112] uses the L^2 discrepancy T_N of a set of points $x_i, i = 1, 2, \ldots, N$,

$$T_N^2 = \int_{I^s} R_N^2(k; x_1, x_2, \ldots, x_N) \, dk.$$

Finite sets of points with low discrepancies provide valid approximations to a uniform distribution of points.

It is possible to express the error bounds on the integral of f over E in terms of the discrepancy of the point set $x_i, i = 1, 2, \ldots, N$ in I^s and, for example, the Hardy–Krause variation of f on I^s. An error bound can be obtained in which the influence of the regularity of the integrand has been separated from the influence of the uniformity of the distribution of nodes. Hence, the desirability of sampling based on a set of points or nodes with low discrepancy, to give an accurate estimate of the integral.

There exist two approaches to low-discrepancy sets:

(i) Given N, find N points in I^s with small discrepancy D_N (*low-discrepancy point set*).

(ii) Find a set of N points in I^s, such that the first M points of the sequence show low discrepancy D_M for any $M \leq N$ (*low-discrepancy sequence*).

Point sets of dimension s can be derived from sequences of dimension $s-1$ by a method described by Neiderreiter [79]: For $s \geq 2$, let

$$x_i' = \{x_i^1, \ldots, x_i^{s-1}\} \in I^{s-1}, \qquad i = 1, 2, \ldots, N \qquad (11\text{-}4)$$

be a low-discrepancy sequence in I^{s-1}. Let D_M' be the discrepancy of the first $M \leq N$ terms of the sequence. Then, for given N, put

$$x_i = \left\{ \frac{i-1}{N}, x_i' \right\} \in I^s, \quad i = 1, 2, \ldots, N. \qquad (11\text{-}5)$$

Niederreiter [79] has shown that the discrepancy D_N of these points satisfies

$$N \, D_N \leq \max_{1 \leq M \leq N} M \, D_M' + 1, \qquad (11\text{-}6)$$

so that they form a low-discrepancy point set.

11B. Halton sequences and Hammersley point sets. The discrepancy in L^2 and other norms has been extensively studied, and relations with number theory have been established. Halton [54] was the first to demonstrate that it is possible to construct a sequence of points with discrepancy of order $D_N = \mathcal{O}(N^{-1}[\log N]^s)$. These sequences are now known as Halton sequences. If x_i' is a point in such a Halton sequence of N points in I^{s-1}, then

$$x_i = \left\{ \frac{i-1}{N}, x_i' \right\} \quad \text{or} \quad 1 - \left\{ \frac{i-1}{N}, x_i' \right\}, \quad i = 1, 2, \ldots, N$$

is a Hammersley point set in s-dimensional space, with discrepancy $D_N = \mathcal{O}(N^{-1}[\log N]^{s-1})$ (see Eq.(11–6)) . In fact, upon "shifting" the Hammersley points, the L^2 discrepancy can be minimized to yield $T_N = \mathcal{O}(N^{-1}[\log N]^{(s-1)/2})$, which result is optimal (Wózniakowski [112]). The convergence rate of the summation replacing the integration is better for the Hammersley point set than for a set of randomly distributed points.

Several methods for constructing Halton sequences and Hammersley point sets are referred to in Niederreiter [79]. All those methods rely on expansions of integers, to different bases for each of the s coordinates. We will give the construction involving expansions of integers in prime number bases; this construction has been reviewed and exploited by Wózniakowski [112].

Consider the creation of the j'th point in a Halton sequence in an $(s-1)$-dimensional space. Let the first $s-1$ prime numbers be denoted $p_1, p_2, \ldots, p_{s-1}$. Expand the integer j as a power series in each of these $s-1$ prime numbers:

$$\forall_{m \in \{1, \ldots, s-1\}} : \; j = \sum_{\mu=0}^{\lceil \log_{p_m} j \rceil} a_\mu \, (p_m)^\mu, \quad a_\mu \in \{0, \ldots, p_m - 1\}, \qquad (11\text{–}7)$$

where $\lceil \cdot \rceil$ denotes the integral part. By reversing the order of the digits in j, we can uniquely construct a fraction lying between 0 and 1, namely, the radical inverse

$$\phi_{p_m}(j) = \sum_{\mu=0}^{\lceil \log_{p_m} j \rceil} a_\mu \, (p_m)^{-\mu-1}, \qquad (11\text{–}8)$$

which can then be used to assemble a Halton sequence,

$$y'_j = \left\{ \phi_{p_1}(j), \phi_{p_2}(j), \ldots, \phi_{p_{s-1}}(j) \right\}. \qquad (11\text{–}9)$$

Consider the first M terms, $j = 1, 2, \ldots, M$, with $M = (p_1 p_2 \cdots p_{s-1})^h$, $h \in \mathbb{N}$ and extend this sequence *periodically* as

$$y'_{j+M} = y'_j. \qquad (11\text{–}10)$$

Choosing $h = \lceil \log_2 N \rceil + 1$, $N \geq 2$, leads to the desired discrepancy estimate $\mathcal{O}(N^{-1}[\log N]^{s-1})$ for the first N points of the set. The "shifted" Hammersley point set for dimension s is then given by

$$x_i = 1 - z_i, \quad i = 1, 2, \ldots, N, \qquad (11\text{–}11)$$

where

$$z_i = \left\{ \frac{i-1+\eta}{N}, \; y'_i \right\}, \quad \eta \in \mathbb{R} \text{ such that } 0 \leq i-1+\eta < N. \qquad (11\text{–}12)$$

For $\eta = 0$, expressions (11–11) and (11–12) define the original Hammersley point set. The constant shift η was used by Wózniakowski [112] in order to minimize the L^2 discrepancy, though an explicit expression for the optimal shift has not been found, hence its value must be determined by experiment.

11C. Spherical geometry: S^{n-1} **and** $S^{n-1} \times S^{n-1}$**.** For the use of quasi-Monte Carlo methods in the application of Fourier integral operators, we must design an algorithm for integration over double spheres rather than over rectangular domains. In this framework, it is appropriate to consider the construction of low-discrepancy sets over the sphere S^{n-1}. In the application of the generalized Radon transform (for a given value of e) we encounter an integration over $\nu \in S^{n-1}$; in the application of the (imaging-)inversion operator we encounter an integration over $(\hat{\alpha}, \tilde{\alpha}) \in S^{n-1} \times S^{n-1}$ or over $(\nu, e) \in S^{n-1} \times S^{n-1}$. We will discuss and illustrate the case $n = 3$.

The integration over double spheres $S^2 \times S^2$ can be written as the double integral over two spheres defined by the unit vectors $\hat{\alpha}$ and $\tilde{\alpha}$. We consider these spheres separately. The horizontal projection of S^2 onto the tangent cylinder along the equator is an area preserving map; thus we may choose a point on the cylinder and obtain a corresponding point on the sphere. We shall draw $(\hat{\vartheta}, \hat{\varphi})$ from a Hammersley point set in the rectangle $[-\pi, \pi] \times [-1, 1]$ and assign the point $\hat{\alpha} \in S^2$ in accordance with

$$\left(\sqrt{1 - \hat{\varphi}^2} \cos \hat{\vartheta}, \ \sqrt{1 - \hat{\varphi}^2} \sin \hat{\vartheta}, \ \hat{\varphi} \right).$$

(This mapping implies in the case of randomly chosen points a uniform distribution with respect to the natural area measure on S^2.) We apply this procedure also for $\tilde{\alpha}$. Once the sampling in $(\hat{\alpha}, \tilde{\alpha})$ is accomplished, we deduce the sampling in (ν, e) using (4–4) and (9–2).

12. An Outlook on Global Seismology

The primary phases amenable to the application of microlocal analysis are earthquake generated short-period body waves; see, for example, the work of Bostock *et al.* [19]. It was by ray methods that the depth to the core mantle boundary was first estimated, and the existence of an inner core was recognized. Indeed, the phases that interact with the inner core boundary (ICB) and core mantle boundary (CMB) can be modeled with microlocal techniques. In the crust, the Moho discontinuity can be thought of as a conormal distribution reflections off which are detected and interpreted. There are also transitions of a different nature in the deep earth. We mention the ones associated with anomalously large velocity gradients (around 400 km and 600 km depths).

In the analysis of discontinuities in the mantle transition zone, Ps conversions from teleseismic body waves are processed using a delay-and-sum approach [105]. This approach can be mathematically justified using the linearized inversion formulation of Section 4 in a planarly layered background medium. Studies of lithosperic and upper mantle structure, that account for anisotropic elasticity, provide constraints on continental dynamics and evolution.

Here we focus our final discussion on inverse scattering at the CMB and selective neighborhoods. Most seismological research of the CMB region has been based on the scattering [6; 5] and diffraction [57] of relatively high frequency body waves. The dynamic wave group of PKKP fits in the framework of the presented inverse scattering theory. These are waves that propagate through the mantle, refract into the core, bounce back from the underside of the CMB, and refract into the mantle again on their way to the receivers at earth's surface (see Figure 24, which illustrates the formation of caustics).

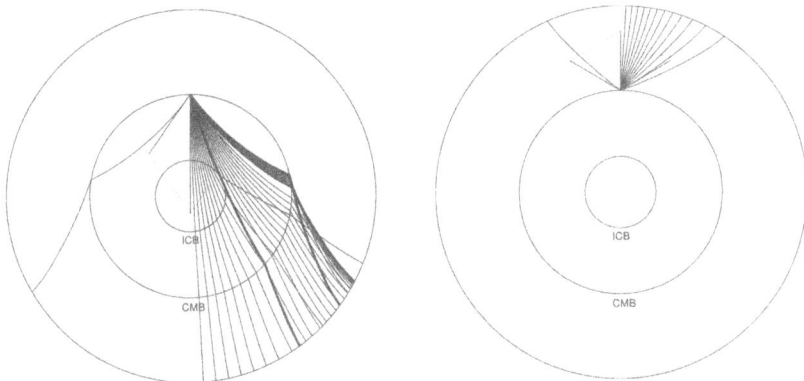

Figure 24. An illustration of characteristics associated with PKKP showing the formation of caustics (left) and characteristics associated with PcP (right).

PKKP is most readily detected in vertical-component short-period records. Often the analysis is restricted to high-frequency data to avoid contamination by long arc surface wave propagation. Other members of this wavegroup are SKKP, PKKS, and SKKS. This multiplicity can provide data redundancy for the study of the CMB near the underside reflections. PKKP and SKKS (and SKKP and PKKS in between them) can best be observed at epicentral distances up to some 100 degrees, and in a time window of some 10 minutes before the arrival of P'P' (PKIKPPKIKP, i.e., a wave that passes through the core and reflects at earth's surface instead of at the underside of the CMB). However, due to inner core attenuation and the small reflection coefficient at near-vertical incidence, the PKKPdf arrivals are typically very weak [44]; hence, most data comes from the caustics.

12A. The core mantle boundary and its vicinity. The core mantle boundary (CMB) is located at about 2880 km depth, where the temperature is about 4000 K and the pressure is about 135 GPa, and marks one of the most dramatic changes in composition and physical properties in our planet. The CMB separates the solid mantle silicates from the liquid iron-alloy in the outer core. From mantle to core, the increase in density is about 4500 kg m^{-3} (compared to the 2700 kg m^{-3} difference between air and crustal rock near the earth's surface), the increase in temperature is about 1000-1500 K, the shear wave speed drops from some 7.2 km sec^{-1} in the lowermost mantle to zero in the liquid outer core, and the compressional wavespeed drops from 13.7 km s^{-1} to 8 km s^{-1}. The CMB also separates two vastly different dynamic regimes. Across it the viscosity drops at least 10 and perhaps as much as 20 orders of magnitude. Mantle convection in the stiff mantle silicates, driven by thermal buoyancy, is slow, a few cm yr^{-1}; in contrast, thermal and compositional buoyancy drives turbulent flow in the liquid core at several mm s^{-1}, that is, some 6 orders of magnitude faster than in the mantle. Another consequence of the enormous viscosity contrast is that the mantle can support lateral variations in density, temperature, and elastic properties, whereas the core is usually considered homogeneous.

12B. Heterogeneity in the outermost core. On the core side, the PKKP underside reflections at high latitude straddle the intersection of the virtual "tangential cylinder", which is an essential feature of the magneto-hydrodynamics of the outer core related to the generation of the earth's magnetic field (e.g. [113; 114]). These reflections allow us to investigate if there are any changes in the character of the inside of the CMB associated with the topology of outer core flow. Of particular interest is the search for any evidence of heterogeneity in the outermost core. Core flow is partly driven by compositional buoyancy, and it is possible that "puddles" of iron that is enriched in light elements form either in topographic highs of the CMB or in certain locations relative to the virtual tangential cylinder or Taylor columns in the outer core.

References

[1] DMO processing. *Geophysics Reprint Series*, 1995.

[2] K. Aki and P. G. Richards. *Quantitative seismology: theory and methods*, volume 1. Freeman, San Francisco, 1980.

[3] R. M. Alford. Shear data in the presence of azimuthal anisotropy. In *Expanded Abstracts*, pages 476–479. Society of Exploration Geophysicists, 1986.

[4] G. Backus and F. Gilbert. The resolving power of gross earth data. *Geophys. J. Roy. Astr. Soc.*, 16:169–205, 1968.

[5] K. Bataille and S. M. Flatté. Inhomogeneities near the core mantle boundary inferred from short-period scattered pkp waves recorded at the global digital seismograph network. *J. Geophys. Res.*, 93:15057–15064, 1988.

[6] K. Bataille and F. Lund. Strong scattering of short-period seismic waves by the core mantle boundary and the p-diffracted wave. *Geophys. Res. Lett.*, 23:2413–2416, 1996.

[7] M. Batzle and Z. Wang. Seismic properties of pore fluids. *Geophysics*, 57:1396–1408, 1992.

[8] J. G. Berryman and G. W. Milton. Exact results for generalized gassmann's equations in composite porous media with two constituents. *Geophysics*, 56:1950–1960, 1991.

[9] G. Beylkin. The inversion problem and applications of the generalized Radon transform. *Comm. Pure Appl. Math.*, XXXVII:579–599, 1984.

[10] G. Beylkin. Imaging of discontinuities in the inverse scattering problem by inversion of a causal generalized Radon transform. *J. of Math. Phys.*, 26:99–108, 1985.

[11] G. Beylkin. Reconstructing discontinuities in multidimensional inverse scattering problems: smooth errors vs small errors. *Applied Optics*, 24:4086–4088, 1985.

[12] G. Beylkin and R. Burridge. Linearized inverse scattering problems in acoustics and elasticity. *Wave Motion*, 12:15–52, 1990.

[13] B. Biondi, S. Fomel, and N. Chemingui. Azimuth moveout for 3-D prestack imaging. *Geophysics*, 63:574–588, 1998.

[14] M. A. Biot. Theory of propagation of elastic waves in a fluid-saturated porous solid. i. low-frequency range. *J. Acoust. Soc. Am.*, 28:168–178, 1956.

[15] M. A. Biot. Theory of propagation of elastic waves in a fluid-saturated porous solid. ii. higher frequency range. *J. Acoust. Soc. Am.*, 28:179–191, 1956.

[16] M. A. Biot. Mechanics of deformation and acoustic propagation in porous media. *J. Appl. Phys.*, 23:1482–2498, 1962.

[17] N. Bleistein. On imaging of reflectors in the earth. *Geophysics*, 52(6):931–942, 1987.

[18] N. Bleistein, J. K. Cohen, and J. W. Jr. Stockwell. *Mathematics of multidimensional seismic imaging, migration and inversion.* Springer-Verlag, New York, 2000.

[19] M. G. Bostock, S. Rondenay, and J. Shragge. Multiparameter two-dimensional inversion of scattered teleseismic body waves. *J. Geophys. Res.*, 106:30771–30782, 2001.

[20] P. J. Braam and J. J. Duistermaat. Normal forms of real symmetric systems with multiplicity. *Indag. Mathem.*, 4:197–232, 1993.

[21] S. Brandsberg-Dahl, M. V. De Hoop, and B. Ursin. Focusing in dip and ava compensation on scattering-angle/azimuth gathers. *Geophysics*, 68:232–254, 2003. in print.

[22] S. Brandsberg-Dahl, B. Ursin, and M. V. De Hoop. Seismic velocity analysis in the scattering-angle/azimuth domain. *Geophys. Prosp.*, in print (July 2003 issue). submitted.

[23] R. J. S. Brown and J. Korringa. On the dependence of the elastic properties of a porous rock on the compressibility of a pore fluid. *Geophysics*, 40:608–616, 1975.

[24] B. Budiansky. On the elastic moduli of some heterogeneous materials. *J. Mech. Phys. Solids*, 13:223–227, 1965.

[25] R. Burridge, M. V. De Hoop, D. Miller, and C. Spencer. Multiparameter inversion in anisotropic media. *Geophys. J. Int.*, 134:757–777, 1998.

[26] R. Burridge and J. B. Keller. Poroelasticity equations derived from microstructure. *J. Acoust. Soc. Am.*, 70:1140–1146, 1981.

[27] V. Červený. *Seismic ray theory.* Cambridge University Press, Cambridge, 2001.

[28] C. H. Chapman. Reflection/transmission coefficient reciprocities in anisotropic media. *Geophysical Journal International*, 116:498–501, 1994.

[29] C. H. Chapman and R. Drummond. Body-wave seismograms in inhomogeneous media using maslov asymptotic theory. *Bull. Seism. Soc. Am.*, 72:277–317, 1982.

[30] J. F. Claerbout. *Fundamentals of Geophysical Data Processing.* McGraw-Hill, 1976.

[31] R. W. Clayton and R. H. Stolt. A born-wkbj inversion method for acoustic reflection data. *Geophysics*, 46:1559–1567, 1981.

[32] M. V. De Hoop and N. Bleistein. Generalized Radon transform inversions for reflectivity in anisotropic elastic media. *Inverse Problems*, 13:669–690, 1997.

[33] M. V. De Hoop and S. Brandsberg-Dahl. Maslov asymptotic extension of generalized Radon transform inversion in anisotropic elastic media:a least-squares approach. *Inverse Problems*, 16:519–562, 2000.

[34] M. V. De Hoop, R. Burridge, C. Spencer, and D. Miller. Generalized Radon transform amplitude versus angle (grt/ava) migration/inversion in anisotropic media. In *Proc SPIE 2301*, pages 15–27. SPIE, 1994.

[35] M. V. de Hoop and A. T. de Hoop. Wavefield reciprocity and optimization in remote sensing. *Proc. R. Soc. Lond. A (Mathematical, Physical and Engineering Sciences)*, 456:641–682, 2000.

[36] M. V. de Hoop, J. H. Le Rousseau, and B. Biondi. Symplectic structure of wave-equation imaging: A path-integral approach based upon the double-square-root equation. *Geoph. J. Int.*, 153:52–74, 2003. submitted.

[37] M. V. De Hoop and C. Spencer. Quasi monte-carlo integration over $s^2 \times s^2$ for migration × inversion. *Inverse Problems*, 12:219–239, 1996.

[38] M. V. De Hoop, C. Spencer, and R. Burridge. The resolving power of seismic amplitude data: An anisotropic inversion/migration approach. *Geophysics*, 64:852–873, 1999.

[39] N. Dencker. On the propagation of polarization sets for systems of real principal type. *Journal of Functional Analysis*, 46:351–372, 1982.

[40] A. J. Devaney. Geophysical diffraction tomography. *IEEE Trans. Geosc. Remote Sens.*, GE-22:3–13, 1984.

[41] D. L. Donoho and I. M. Johnstone. Ideal spatial adaptation via wavelet shrinkage. *Biometrika*, 81:425–455, 1994.

[42] H. H. Duistermaat and V. Guillemin. The spectrum of positive elliptic operators and periodic bicharacteristics. *Inv. Math.*, 29:39–79, 1975.

[43] J. J. Duistermaat. *Fourier integral operators*. Birkhäuser, Boston, 1996.

[44] P. S. Earle and P. N. Shearer. Observations of high-frequency scattered energy associated with the core phase pkkp. *Geophys. Res. Lett.*, 25:405–408, 1998.

[45] P. P. Ewald. Zur begründung der kristalloptik. *Annalen der Physik, IV*, 49:1–38, 1916.

[46] N. L. Frazer and M. K. Sen. Kirchhoff-helmholtz reflection seismograms in a laterally inhomogeneous multi-layered medium – i. theory. *Geophys. J. R. Astr. Soc.*, 80:121–147, 1985.

[47] F. Gassmann. Uber die elastizität poröser medien. *Vierteljahrsschrift der Naturforschenden Gesellschaft in Zürich*, 96:1–23, 1951.

[48] A. Greenleaf and G. Uhlmann. Nonlocal inversion formulas for the x-ray transform. *Duke Math. J.*, 58:205–240, 1989.

[49] A. Greenleaf and G. Uhlmann. Composition of some singular fourier integral operators and estimates for restricted x-ray transforms. *Ann. Inst. Fourier*, 40:443–446, 1990.

[50] A. Greenleaf and G. Uhlmann. Estimates for singular Radon transforms and pseudodifferential operators with singular symbols. *J. Func. Anal.*, 89:202–232, 1990.

[51] V. Guillemin. *Pseudodifferential operators and applications (Notre Dame, Ind., 1984)*, chapter On some results of Gel'fand in integral geometry, pages 149–155. Amer. Math. Soc., Providence, RI, 1985.

[52] A. G. Haas and J. R. Viallix. Krigeage applied to geophysics. *Geophysical Prospecting*, 24:49–69, 1976.

[53] J. G. Hagedoorn. A process of seismic reflection interpretation. *Geophysical Prospecting*, 2:85–127, 1954.

[54] J. H. Halton. On the efficiency of certain quasi-random sequences of points in evaluating multi-dimensional integrals. *Numeric. Math.*, 2:85–90, 1960.

[55] J. M. Hammersley and D. C. Handscomb. *Monte Carlo Methods*. Methuen, London, 1964.

[56] S. Hansen. Solution of a hyperbolic inverse problem by linearization. *Communications in Partial Differential Equations*, 16:291–309, 1991.

[57] D. V. Helmberger, E. J. Garnero, and X.-D. Ding. Modeling 2-d structure at the core-mantle boundary. *J. Geophys. Res.*, 101:13963–13972, 1996.

[58] R. Hill. A self-consistent mechanics of composite matericals. *J. Mech. Phys. Solids*, 13:213–222, 1965.

[59] L. Hörmander. Fourier integral operators. i. *Acta Math.*, 127:79–183, 1971.

[60] L. Hörmander. *The analysis of linear partial differential operators*, volume I. Springer-Verlag, Berlin, 1983.

[61] L. Hörmander. *The analysis of linear partial differential operators*, volume III. Springer-Verlag, Berlin, 1985.

[62] L. Hörmander. *The analysis of linear partial differential operators*, volume IV. Springer-Verlag, Berlin, 1985.

[63] B. E. Hornby. *The elastic properties of shales*. PhD thesis, University of Cambridge, Cambridge, 1994.

[64] B. E. Hornby, L. M. Schwartz, and J. A. Hudson. Anisotropic effective medium modeling of the elastic properties of shales. *Geophysics*, 59:1570–1583, 1994.

[65] J. A. Hudson. Overall properties of a cracked solid. *Math. Proc. Camb. Phil. Soc.*, 88:371–384, 1980.

[66] E. Iversen, H. Gjøystdal, and J. O. Hansen. Prestack map migration as an engine for parameter estimation in ti media. In *70th Ann. Mtg. Soc. Explor. Geoph., Expanded Abstracts*, pages 1004–1007, 2000.

[67] V. Ya. Ivrii. Wave fronts of solutions of symmetric pseudodifferential systems. *Siberian Mathematical Journal*, 20:390–405, 1979.

[68] L. E. A. Jones and H. F. Wang. Ultrasonic velocities in cretaceous shales from the williston basin. *Geophysics*, 46:288–297, 1981.

[69] J. M. Kendall and C. J. Thomson. Maslov ray summation, pseudo-caustics, lagrangian equivalence and transient seismic waveforms. *Geoph. J. Int.*, 113:186–214, 1993.

[70] B. L. N. Kennett. *Seismic Wave Propagation in Stratified Media.* Cambridge University Press, Cambridge, 1983.

[71] J. L. Lions and E. Magenes. *Non-Homogeneous Boundary Value Problems and Applications,* volume 1. Springer-Verlag, Berlin, 1972.

[72] S. Mallat. *A wavelet tour of signal processing.* Academic Press, San Diego, 1997.

[73] V. P. Maslov and M. V. Fedoriuk. *Semi-classical approximation in quantum mechanics.* Reidel Publishing Company, 1981.

[74] R. Melrose and G. Uhlmann. Microlocal structure of involutive conical refraction. *Duke Math. J.,* 46:571–582, 1979.

[75] D. Miller, M. Oristaglio, and G. Beylkin. A new slant on seismic imaging: migration and integral geometry. *Geophysics,* 52(6):943–964, 1987.

[76] H. E. Moses. Calculation of scattering potential from reflection coefficients. *Phys. Rev.,* 102:559–567, 1956.

[77] M. J. P. Musgrave. *Crystal acoustics.* Holden-Day, 1970.

[78] H. Niederreiter. Quasi-monte carlo methods and pseudo-random numbers. *Bull. Am. Math. Soc.,* 84:957–1041, 1978.

[79] H. Niederreiter. *Numerical Integration III. International series of numerical mathematics Vol 85,* chapter Quasi-Monte Carlo methods for multidimensional numerical integration. Birkhauser-Verlag, Berlin, 1988.

[80] C. J. Nolan and W. W. Symes. Global solution of a linearized inverse problem for the wave equation. *Communications in Partial Differential Equations,* 22(5-6):919–952, 1997.

[81] W. H. Press, S. A. Teukolsky, W. T. Vetterling, and B. P. Flannery. *Numerical Recipes in FORTRAN, The art of scientific computing.* Cambridge University Press, 1994.

[82] Rakesh. A linearised inverse problem for the wave equation. *Comm. in Part. Diff. Eqs.,* 13:573–601, 1988.

[83] M. Razavy. Determination of the wave velocity in an inhomogeneous medium from reflection data. *J. Acoust. Soc. Am.,* 58:956–963, 1975.

[84] L. A. Riabinkin. Fundamentals of resolving power of controlled directional reception (cdr) of seismic waves. In Gardner G. H. F. and Lu L., editors, *Slant stack processing,* number 14 in Geophysics Reprint Series, pages 36–60. Society of Exploration Geophysicists, Tulsa, 1957.

[85] F. Rieber. A new reflection system with controlled directional sensitivity. *Geophysics,* I(1):97–106, 1936.

[86] J. Schleicher, M. Tygel, and P. Hubral. 3-D true-amplitude finite-offset migration. *Geophysics,* 58:1112–1126, 1993.

[87] W. A. Schneider. Integral formulation for migration in two and three dimensions. *Geophysics,* 43:49–76, 1978.

[88] R. T. Shuey. A simplication of the zoeppritz equations. *Geophysics,* 50:609–614, 1985.

[89] C. C. Stolk. Microlocal analysis of a seismic linearized inverse problem. *Wave Motion,* 32:267–290, 2000.

[90] C. C. Stolk. Microlocal analysis of the scattering angle transform. Preprint, Dept. of Computational and Applied Mathematics, Rice University, www.caam.rice.edu/ ~cstolk/angle.ps, July 2001.

[91] C. C. Stolk and M. V. De Hoop. Microlocal analysis of seismic inverse scattering in anisotropic, elastic media. *Comm. Pure Appl. Math.*, 55:261–301, 2002.

[92] C. C. Stolk and M. V. De Hoop. Seismic inverse scattering in the "wave equation" approach. *SIAM J. Math. Anal.*, 2002, submitted; available at http://www.msri.org/ publications/preprints/online/2001-047.html.

[93] R. H. Stolt. Migration by fourier transform. *Geophysics*, 43:23–48, 1978.

[94] R. H. Stolt and A. B. Weglein. Migration and inversion of seismic data. *Geophysics*, 50:2456–2472, 1985.

[95] W. W. Symes. A differential semblance algorithm for the inverse problem of reflection seismology. *Comput. Math. Appl.*, 22(4-5):147–178, 1991.

[96] A. Tarantola. Inversion of seismic reflection data in the acoustic approximation. *Geophysics*, 49:1259–1266, 1984.

[97] A. Tarantola. A strategy for nonlinear elastic inversion of seismic reflection data. *Geophysics*, 51:1893–1903, 1986.

[98] M. E. Taylor. Reflection of singularities of solutions to systems of differential equations. *Communications on Pure and Applied Mathematics*, 28:457–478, 1975.

[99] A. P. E. ten Kroode, D.-J. Smit, and A. R. Verdel. A microlocal analysis of migration. *Wave Motion*, 28:149–172, 1998.

[100] L. Thomsen. Weak elastic anisotropy. *Geophysics*, 51(10):1954–1966, 1986.

[101] F. Treves. *Introduction to pseudodifferential and Fourier integral operators*, volume 2. Plenum Press, New York, 1980.

[102] F. Treves. *Introduction to pseudodifferential and Fourier integral operators*, volume 2. Plenum Press, New York, 1980.

[103] I. Tsvankin. *Seismic signatures and analysis of reflection data in anisotropic media*. Elsevier Science, Amsterdam, 2001.

[104] G. Uhlmann. Light intensity distribution in conical refraction. *Comm. Pure Appl. Math.*, XXXV:69–80, 1982.

[105] L. P. Vinnik. Detection of waves converted from P to SV in the mantle. *Phys. Earth Planet. Inter.*, 15:39–45, 1977.

[106] K. Walton. The oblique compression of two elastic spheres. *J. Mech. Phys. Solids*, 26:139–150, 1978.

[107] K. Walton. The effective elastic moduli of a random packing of spheres. *J. Mech. Phys. Solids*, 35:213–226, 1987.

[108] M. Weber. Die bestimmung einer beliebig gekruemmten schichtgrenze aus seismischen reflexionsmessungen. *Geofisica pura e applicata*, 32:7–11, 1955.

[109] D. N. Whitcombe and R. J. Carroll. The application of map migration to 2-d migrated data. *Geophysics*, 59:1121–1132, 1994.

[110] B. S. White, B. Nair, and A. Bayliss. Random rays and seismic amplitude anomalies. *Geophysics*, 53:903–907, 1988.

[111] J. H. Woodhouse. Surface waves in laterally varying layered media. *Geophys. J. R. Astr. Soc.*, 37:461–490, 1974.

[112] H. Wözniakowski. Average case complexity and multivariate integration. *Bull. Am. Math. Soc.*, 24:185–194, 1991.

[113] K. K. Zhang and F. H. Busse. Convection driven magnetohydrodynamic dynamos in rotating spherical-shells. *Geophys. Astro Fluid*, 49(1-4):97–116, 1989.

[114] K. K. Zhang and D. Gubbins. Convection in a rotating spherical fluid shell with an inhomogeneous temperature boundary-condition at infinite prandtl number. *J. Fluid Mech.*, 250:209–232, 1993.

MAARTEN V. DE HOOP
CENTER FOR WAVE PHENOMENA
COLORADO SCHOOL OF MINES
GOLDEN, CO 80401-1887 UNITED STATES
mdehoop@Mines.EDU

Sojourn Times, Singularities of the Scattering Kernel and Inverse Problems

VESSELIN PETKOV AND LUCHEZAR STOYANOV

ABSTRACT. We study inverse problems in the scattering by obstacles in odd-dimensional Euclidean spaces. In general, such problems concern the recovery of the geometric properties of the obstacle from the information related to the scattering amplitude $a(\lambda, \omega, \theta)$, related to the wave equation in the exterior of the obstacle with Dirichlet boundary condition. It turns out that all singularities of the Fourier transform of $a(\lambda, \omega, \theta)$, the so-called scattering kernel, are given by the sojourn (traveling) times of scattering rays in the exterior of the obstacle. Apart from that these sojourn times are a naturally observable data. The purpose of this survey is to describe several results in obstacle scattering obtained in the last twenty years concerning sojourn times of scattering rays, and to motivate further study of related inverse scattering problems.

1. Introduction

The scattering operator $S(\lambda)$ presents a mathematical model for the data observed experimentally in many branches of physics, chemistry and mathematics. The operator $S(\lambda)$ is related to behavior as the time $t \to \pm\infty$ of the solutions of an unperturbed operator L_0 and to its perturbation L. The kernel of $S(\lambda) - I$, the so called *scattering amplitude* $a(\lambda, \omega, \theta)$, contains the information related to the perturbation of L_0 and this kernel is the leading term of the asymptotic of an outgoing solution $v_s(r\theta, \lambda)$ of $Lv_s = 0$ as $|x| = r \to \infty$. Obstacle scattering problems arise in many physical phenomena and concern the perturbation caused by a bounded obstacle K with connected complement Ω. In general the inverse scattering problems deal with recovering geometric properties of K from information related to the scattering amplitude.

Schiffer's result (see [12], [2]) implies that the obstacle K is uniquely determined if we know the scattering amplitude $a(\lambda, \omega, \theta)$ for $\lambda \in (\alpha, \beta) \subset \mathbb{R}^+$ and all $\omega, \theta \in \mathbf{S}^{n-1}$. Some more precise results concerning uniqueness in this inverse scattering problem are known under weaker assumptions (see [2], [7], [11], [26] for more details and references.) On the other hand, in general in experiments one cannot determine the scattering amplitude for all (outgoing) directions $\theta \in \mathbf{S}^{n-1}$ or all (incoming) directions $\omega \in \mathbf{S}^{n-1}$, while the *sojourn times* or *traveling times* of the so-called (ω, θ)-rays in the exterior of the obstacle give a physically observable data. This naturally leads to the consideration of inverse scattering problems involving such rays. In fact, it turns out that all singularities of the Fourier transform $s(t, \omega, \theta)$ of $a(\lambda, \omega, \theta)$, the so-called *scattering kernel*, have the form $-T_\gamma$, where T_γ are sojourn times of (ω, θ)-rays γ. Moreover, for (ω, θ) in a set of full measure in $\mathbf{S}^{n-1} \times \mathbf{S}^{n-1}$ the singularities of $s(t, \omega, \theta)$ are precisely the numbers of the form $-T_\gamma$, that is the so-called *Poisson relation* becomes an equality (see Section 5). This leads to some interesting geometrical observations. The purpose of this survey is to describe several results in obstacle scattering obtained in the last twenty years concerning sojourn times of (ω, θ)-rays, and to motivate further study of related inverse scattering problems.

The scattering amplitude is defined in Section 2. The case of a convex obstacle is then considered in details, and the leading term of the asymptotic of the scattering amplitude as $\lambda \to +\infty$ is derived. Section 3 is devoted to the Fourier transform of the scattering amplitude, the so-called scattering kernel $s(t, \theta, \omega)$, where $t \in \mathbb{R}$ and $\theta, \omega \in \mathbf{S}^{n-1}$. It turns out that the singularities of $s(t, \theta, \omega)$ in t are very much related to the geometry of the obstacle K. Namely, these are given by sojourn (traveling) times of scattering rays in the exterior of the obstacle incoming with direction ω and outgoing with direction θ. This is particularly easy to see in the case of a convex obstacle, where a scattering ray can have at most one reflection at the boundary ∂K of the obstacle. In the general case a typical scattering ray is a mutiply reflecting ray with reflections at ∂K. Moreover there are other, more complicated rays, that have to be taken into account when studying the singularities of the scattering kernel; some of these contain gliding segments on ∂K which are simply geodesics with respect to the metric on ∂K induced by the Euclidean structure. All these are generalized bicharacteristics in the sense of Melrose and Sjöstrand [20]. Their definition is sketched in Section 3, and at the end of that section the leading term of the singularity of $s(t, \theta, \omega)$ at $t \sim -T$ is described, where T is the sojourn time of a scattering ray satisfying some nondegeneracy properties.

Section 4 is purely geometrical. Here we give a simple definition of a reflecting (ω, θ)-ray, and show that for almost all $(\omega, \theta) \in \mathbf{S}^{n-1} \times \mathbf{S}^{n-1}$, the reflecting (ω, θ)-rays in the exterior of K have no tangencies to ∂K and any two of them have different sojourn times. These properties, together with nondegeneracy of the differential cross-sections, play an important role in the analysis of the singularities of the scattering kernel. The latter is dealt with in Section 5. The central point here is the so-called Poisson relation for the scattering kernel, and the first half of Section 5 is devoted to the idea of its proof. We then proceed to discuss the question of how often this relation becomes an equality. One of the problems to do this is to show that (under certain nondegeneracy assumptions about the obstacle) for almost all $(\omega, \theta) \in \mathbf{S}^{n-1} \times \mathbf{S}^{n-1}$, the (ω, θ)-rays in the exterior of K are reflecting rays, i.e. they do not contain gliding segments on the boundary. Combining this with previous results gives that the Poisson relation becomes an equality for almost all $(\omega, \theta) \in \mathbf{S}^{n-1} \times \mathbf{S}^{n-1}$.

In Section 6 we discuss the existence of simply reflecting nondegenerate scattering rays with sojourn times tending to infinity. This leads to some interesting results concerning the behavior of the modified resolvent of the Laplacian.

Finally, in Section 7 the inverse scattering problem is considered of recovering geometric information about the obstacle from its scattering length spectrum, i.e. from the set of sojourn times of scattering rays in the exterior of the obstacle[1]. Pairs of obstacles K, L are considered such that for (almost) all $(\omega, \theta) \in \mathbf{S}^{n-1} \times \mathbf{S}^{n-1}$ the sets of sojourn times of (ω, θ)-rays in the exteriors of K and L are the same. It then turns out that the generalized geodesic flows in the nontrapping parts of the cotangent bundles of the exteriors of K and L are conjugated by a time preserving conjugacy which is almost everywhere smooth and symplectic. Various geometric relationships between K and L are derived.

2. Scattering Amplitude for Strictly Convex Obstacles

Let $K \subset \mathbb{R}^n$, $n \geq 3$, n odd, be a bounded domain with C^∞ boundary ∂K and connected complement $\Omega = \overline{\mathbb{R}^n \setminus K}$. Such K is called an *obstacle* in $\mathbb{R}^n$. Throughout this paper we deal with the Dirichlet problem for the Laplacian but similar considerations can be applied to other boundary value problems. To introduce the scattering amplitude $a(\lambda, \theta, \omega)$, $(\theta, \omega) \in \mathbf{S}^{n-1} \times \mathbf{S}^{n-1}$, consider the *outgoing solution* $v_s = v_s(x, \lambda)$ of the problem

$$\begin{cases} (\Delta + \lambda^2) v_s = 0 & \text{in } \mathring{\Omega}, \\ v_s + e^{-i\lambda\langle x, \omega \rangle} = 0 & \text{on } \partial K \end{cases}$$

satisfying the so-called $(i\lambda)$ - outgoing Sommerfeld radiation condition. This condition means that as $|x| = r \to \infty$ we have

$$v_s(r\theta, \lambda) = \frac{e^{-i\lambda r}}{r^{(n-1)/2}} \left(a(\lambda, \theta, \omega) + O(r^{-1}) \right), \quad x = r\theta.$$

We can interpret $v_i = e^{-i\lambda\langle x, \omega \rangle}$ as an *incoming plane wave*, while $v_s(x, \lambda)$ is the *outgoing wave* obtained after the impact of v_i on ∂K. To obtain a formula for the leading term $a(\lambda, \theta, \omega)$ we apply the Green formula combined with the outgoing condition and deduce the representation

$$v_s(x, \lambda) = \int_{\partial K} \left(E_\lambda(x - y) \frac{\partial v_s}{\partial \nu}(y, \lambda) - \frac{\partial E_\lambda}{\partial \nu}(x - y) v_s(y, \lambda) \right) dS_y, \qquad (2\text{-}1)$$

---

[1] According to the Poisson relation, this is equivalent to trying to obtain information about the obstacle from the singularities of the scattering kernel.

where $E_\lambda(x)$ is the outgoing Green function

$$E_\lambda(x) = \frac{(i\lambda)^{(n-3)/2}}{2(2\pi)^{(n-1)/2}} \frac{e^{-i\lambda r}}{r^{(n-1)/2}} + O\left(r^{-(n+1)/2}\right)$$

and $\nu(x)$ is the unit normal to $x \in \partial K$ pointing into Ω. Next, we multiply (2–1) by $e^{i\lambda r} r^{(n-1)/2}$, put $x = r\theta$, and taking the limit $r \to \infty$, we get

$$a(\lambda,\theta,\omega) = \frac{(i\lambda)^{(n-3)/2}}{2(2\pi)^{(n-1)/2}} \int_{\partial K} \left(i\lambda\langle\nu(x),\theta\rangle e^{i\lambda\langle x,\theta-\omega\rangle} + e^{i\lambda\langle x,\theta\rangle}\frac{\partial v_s}{\partial \nu}(x,\lambda)\right) dS_x,$$

where $\langle\,\cdot\,,\cdot\,\rangle$ denotes the scalar product in $\mathbb{R}^n$.

Following the physical literature, $a(\lambda,\theta,\omega)$ is called the *scattering amplitude*. The analysis of the leading term of its asymptotic as $\lambda \to +\infty$ has a long tradition in mathematical physics. The simplest case to deal with is when $\theta \neq \omega$ and K is a strictly convex obstacle. In this case the integral

$$I(\lambda) = \frac{(i\lambda)^{(n-1)/2}}{2(2\pi)^{(n-1)/2}} \int_{\partial K} \langle\nu(x),\theta\rangle e^{i\lambda\langle x,\theta-\omega\rangle} dS_x$$

is rather easy to study. The phase function $\langle x, \theta-\omega\rangle|_{x\in\partial K}$ has two critical points $x_\pm$ with

$$\langle x_+, \theta-\omega\rangle = \max_{y\in\partial K}\langle y,\theta-\omega\rangle, \quad \langle x_-,\theta-\omega\rangle = \min_{y\in\partial K}\langle y,\theta-\omega\rangle,$$

$$\nu(x_\pm) = \pm\frac{\theta-\omega}{|\theta-\omega|}.$$

Here x^+ denotes the point in the *illuminated region* (see Figure 1)

$$\partial K_+(\omega) = \{y \in \partial K : \langle\nu(y),\omega\rangle < 0\}$$

related to ω, while x^- lies in the *shadow region*

$$\partial K_-(\omega) = \{y \in \partial K : \langle\nu(y),\omega\rangle > 0\},$$

and we have used the convention that the obstacle lies in the half-space

$$\{x \in \mathbb{R}^n : \langle x, \theta-\omega\rangle < 0\}.$$

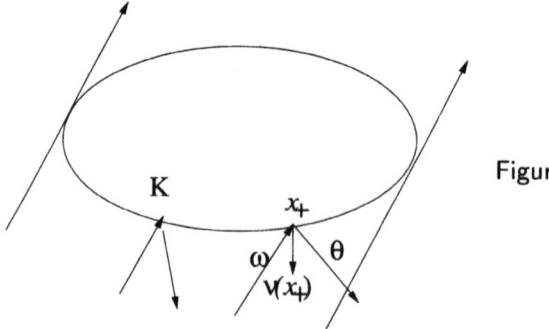

Figure 1.

Applying a stationary phase argument for the integral over $\partial K_+(\omega)$, one gets

$$\frac{(i\lambda)^{(n-1)/2}}{2(2\pi)^{(n-1)/2}} \int_{\partial K_+(\omega)} \langle \nu(x), \theta \rangle e^{i\lambda \langle x, \theta - \omega \rangle} dS_x$$

$$= \frac{1}{2} e^{i\lambda \langle x_+, \theta - \omega \rangle} \mathcal{K}(x_+)^{-1/2} \frac{\langle \nu(x_+), \theta \rangle}{|\theta - \omega|^{(n-1)/2}} + O(|\lambda|^{-1}),$$

$\mathcal{K}(y) > 0$ being the Gauss curvature at $y \in \partial K$. We get a similar expression for the integral over $\partial K_-(\omega)$.

The analysis of the term involving $\partial v_s / \partial \nu$ is more complicated. In mathematical physics many efforts have been concerned with construction of an approximate outgoing solution $w_0(x, \lambda)$ of the problem

$$\begin{cases} (\Delta + \lambda^2) w_0 = f(x, \lambda) & \text{in } \mathring{\Omega}, \\ w_0 + e^{-i\lambda \langle x, \omega \rangle} = g(x, \lambda) & \text{on } \partial K, \end{cases}$$

with $f(x, \lambda) \in C^\infty(\Omega)$ and $g(x, \lambda) \in C^\infty(\partial K)$. This leads to considerable difficulties when one has to describe the form of the solution w_0 in a domain close to the grazing submanifold

$$G(\omega) = \{ y \in \partial K : \langle \nu(y), \omega \rangle = 0 \}.$$

The progress of the microlocal analysis in the seventies led to the investigation of the above problem without a precise information for w_0 in a neighborhood of $G(\omega)$. This was done by Majda [14] exploiting the works of Hörmander [9], Taylor [30] and Melrose [17] for the propagation of the singularities. Below we present the idea of the approach of Majda and refer to [14] for more details.

Consider the boundary problem

$$\begin{cases} (\partial_t^2 - \Delta) u_0 = F(t, x) & \text{in } \mathbb{R} \times \mathring{\Omega}, \\ u_0 + \delta(t - \langle x, \omega \rangle) = G(t, x) & \text{on } \mathbb{R} \times \partial K, \end{cases}$$

where $F(t, x) \in C^\infty(\mathbb{R} \times \Omega)$ vanishes for $t \leq -t_0$, $G(t, x) \in C_0^\infty(\mathbb{R} \times \partial K)$ and t_0 is chosen so that

$$\text{supp}_t \, \delta(t - \langle x, \omega \rangle|_{x \in \partial K}) \subset \{t : |t| \leq t_0\}.$$

Taking a partition of unity $\{\psi_j(t, x)\}_{j=1}^M$ on $[-t_0, t_0] \times \partial K$, we pass to the analysis of the solutions of the localized problems

$$\begin{cases} (\partial_t^2 - \Delta) u_j = F_j(t, x) & \text{in } \mathbb{R} \times \mathring{\Omega}, \\ u_j + \psi_j \delta(t - \langle x, \omega \rangle) = G_j(t, x) & \text{on } \mathbb{R} \times \partial K, \end{cases} \tag{2-2}$$

with $F_j(t, x) \in C^\infty(\mathbb{R} \times \Omega), G_j(t, x) \in C_0^\infty(\mathbb{R} \times \partial K)$ and $F_j = 0$ for $t \leq t_0$. Then using the decay of local energy for strictly convex obstacles we get

$$\frac{\partial v_s}{\partial \nu}\bigg|_{\partial K} = \sum_{j=1}^M \int e^{-i\lambda t} \frac{\partial u_j(t, x)}{\partial \nu}\bigg|_{\mathbb{R} \times \partial K} dt + O(|\lambda|^{-N}) \quad \text{for all } N.$$

The results on the propagation of the wave front set $WF(u_j)$ of the solutions of (2–2) (see [30], [17]) say that

$$WF\left(\frac{\partial u_j}{\partial \nu}\Big|_{\mathbb{R}\times\partial K}\right) \subset WF\left(\psi_j\delta(t-\langle x,\omega\rangle)\right|_{\mathbb{R}\times\partial K}). \qquad (2\text{–}3)$$

In the case when $\operatorname{supp}\psi_j \cap (\mathbb{R}\times G(\omega)) = \varnothing$ the above relation follows from the pseudo-local property of pseudo-differential operators [10] since we have, modulo smooth terms, the representation

$$\frac{\partial u_j}{\partial \nu}\Big|_{\mathbb{R}\times\partial K} = -B_j\left(\psi_j\delta(t-\langle x,\omega\rangle)\right|_{\mathbb{R}\times\partial K}),$$

B_j being a first order pseudo-differential operator. In the case where $\operatorname{supp}\psi_j$ overlaps with $\mathbb{R}\times G(\omega)$ we apply the results of Taylor [30] and Melrose [17] for diffraction problems. Thus we are going to study the expression

$$\sum_j \iint_{\partial K} e^{-i\lambda(t-\langle x,\theta\rangle)}\frac{\partial u_j}{\partial \nu}\, dt\, dS_x, \qquad (2\text{–}4)$$

where the integral is interpreted in the sense of distributions. From the definition of the wave front it is easy to see that the condition

$$(t, y', d_t\Phi, d'_y\Phi) \cap WF(u) = \varnothing \quad \text{for } y' \in D \subset \mathbb{R}^{n-1}$$

implies

$$\int_{\mathbb{R}}\int_D e^{-i\lambda\Phi(y',t)}u(y',t)\, dt\, dy' = O(|\lambda|^{-N}) \quad \text{for all } N.$$

In order to exploit this property, assume that in local coordinates $U_j \cap \partial K$ is given by

$$y_n = g(y'), \quad y' = (y_1,\ldots,y_{n-1}) \in D \subset \mathbb{R}^{n-1}.$$

Then (2–3) yields

$$WF\left(\frac{\partial u_j}{\partial \nu}\Big|_{\mathbb{R}\times\partial K}\right) \subset \big\{(t,y,\tau,\xi) \in T^*(\mathbb{R}\times\partial K): t = \langle y,\omega\rangle$$
$$\text{with } y \in \operatorname{supp}\psi_j\left(y, \langle y,\omega\rangle\right) \text{ and } (\xi,\tau) = \pm\big(-\omega' - \nabla g(y')\omega_n, 1\big)\big\}.$$

Clearly, for the phase function $\Phi = t - \langle y,\theta\rangle|_{y\in U_j\cap\partial K}$ we have

$$d_{y',t}\Phi = (-\theta' - \nabla g(y')\theta_n, 1),$$

which coincides with the directions of the wave front of $(\partial u_j/\partial \nu)\big|_{\mathbb{R}\times\partial K}$ only in the case

$$-\omega' - \nabla g(y')\omega_n = -\theta' - \nabla g(y')\theta_n.$$

Thus we deduce immediately

$$\frac{\theta - \omega}{|\theta - \omega|} = \pm\nu(y', g(y')).$$

The assumption $\theta \neq \omega$ implies that for $y \in G(\omega)$ the last condition is impossible. Moreover, the same argument shows that supp $\psi_j(y, \langle y, \omega \rangle)$ must be included in a small neighborhood $U_\pm$ of $x_\pm$ with $\psi_j(y, \langle y, \omega \rangle) = 1$ in a neighborhood of $x_\pm$.

Since x_- lies in the shadow region, we have $\langle \nu(x_-), \omega \rangle > 0$ and the solution of the wave equation which is smooth for $t < 0$ in a small neighborhood of $(\langle x_-, \omega \rangle, x_-)$ has the form $u_- = -\delta(t - \langle x, \omega \rangle)$. Thus we obtain

$$\frac{\partial v_s}{\partial \nu}\bigg|_{U_- \cap \partial K} = i\lambda \langle \nu, \omega \rangle e^{-i\lambda \langle x, \omega \rangle}|_{U_- \partial K},$$

and replacing $(\partial v_s / \partial \nu)|_{U_- \cap \partial K}$ in expression (2–4), we see that the shadow region makes no contribution to $a(\lambda, \theta, \omega)$ because

$$\langle \nu(x_-), \theta + \omega \rangle = 0.$$

Passing to the illuminated region, denote by ψ_+ and B_+ the cutoff function and the pseudo-differential operator related to U_+. Then for the formally adjoint operators B_+^* we obtain

$$-\int\int_{U_+} B_+^* \left(e^{-i\lambda(t - \langle y', \theta' \rangle - g(y')\theta_n)} \right) \psi_+ \delta \left(t - \langle y', \omega' \rangle - g(y')\omega_n \right) \left(1 + |\nabla g(y')|^2 \right)^{1/2} dt dy'$$

$$= -\lambda \int_{U_+} e^{i\lambda(\langle y', \theta' - \omega' \rangle + g(y')(\theta_n - \omega_n))} b_+(y', \theta) dy' + O(1)$$

with

$$b_+(y', \theta) = -i\beta_+ \left(y', -1, \theta' + \nabla g(y')\theta_n \right) \left(1 + |\nabla g(y')|^2 \right)^{1/2},$$

$i\beta_+$ being the principal symbol of B_+. Thus our task is reduced to the study of an integral having the same form as $I(\lambda)$.

Without loss of generality we can assume that $\nabla g(x'_+) = 0$. From the construction of the asymptotic solution in a neighborhood of x_+ we obtain

$$\beta_+(x'_+, -1, \theta') = \langle \nu(x_+), \theta \rangle > 0$$

and we conclude that

$$\frac{1}{2}\left(\frac{i\lambda}{2\pi}\right)^{(n-1)/2} \int_{U_+} e^{i\lambda(\langle y', \theta' - \omega' \rangle + g(y')(\theta_n - \omega_n))} b_+(y', \theta) dy'$$

$$= \frac{1}{2} e^{i\lambda \langle x_+, \theta - \omega \rangle} \mathcal{K}(x_+)^{-1/2} \frac{\langle \nu(x_+), \theta \rangle}{|\theta - \omega|^{(n-1)/2}} + O(|\lambda|^{-1}).$$

Taking the sum of all contributions, one gets

$$a(\lambda, \theta, \omega) = e^{i\lambda \langle x_+, \theta - \omega \rangle} \mathcal{K}(x_+)^{-1/2} \langle \nu(x_+), \theta \rangle |\theta - \omega|^{(1-n)/2} + O(|\lambda|^{-1}).$$

Finally, in the illuminated region we have

$$\frac{\langle \nu(x_+), \theta \rangle}{|\theta - \omega|} = \frac{\langle \theta - \omega, \theta \rangle}{|\theta - \omega|^2} = \frac{1}{2}.$$

and
$$a(\lambda, \theta, \omega) = \tfrac{1}{2} e^{i\lambda \langle x_+, \theta - \omega \rangle} \mathcal{K}(x_+)^{-1/2} |\theta - \omega|^{(3-n)/2} + O(|\lambda|^{-1}).$$

Thus from the limit
$$|a(\omega, \theta)| = \lim_{\lambda \to \infty} |a(\lambda, \omega, \theta)|$$

we can determine the Gauss curvature $\mathcal{K}(x_+)$ at x_+. When (ω, θ) runs over a set
$$V \subset \mathbf{S}^{n-1} \times \mathbf{S}^{n-1} \setminus \{(\omega, \omega) : \omega \in \mathbf{S}^{n-1}\},$$

we can recover the Gauss curvature $\mathcal{K}(y)$ at every point $y \in \partial K$, provided the map
$$V \ni (\omega, \theta) \to \frac{\theta - \omega}{|\theta - \omega|} \in \mathbf{S}^{n-1}$$

is onto. On the other hand, the knowledge of the Gauss curvature at all points of ∂K determines uniquely ∂K (see [14] for more details).

The case $\omega = \theta$ is more complicated since the singularities associated to diffracted rays must be taken into account. See [19] and [31] for results in this direction.

3. Singularities of the Scattering Kernel

Throughout this section we assume that $\theta \neq \omega$. To study the general case of nonconvex obstacles it is more convenient to consider the *scattering kernel* $s(t, \theta, \omega)$ defined as the Fourier transform of the scattering amplitude:
$$s(t, \theta, \omega) = \mathcal{F}_{\lambda \to t} \left(\left(\frac{\lambda}{2\pi i} \right)^{(n-1)/2} \overline{a(\lambda, \theta, \omega)} \right),$$

where $(\mathcal{F}_{\lambda \to t} \varphi)(t) = (2\pi)^{-1} \int e^{it\lambda} \varphi(\lambda) \, d\lambda$ for functions $\varphi \in \mathcal{S}(\mathbb{R})$. Let $V(t, x; \omega)$ be the solution of the problem
$$\begin{cases} (\partial_t^2 - \Delta) V = 0 & \text{in } \mathbb{R} \times \mathring{\Omega}, \\ V + \delta(t - \langle x, \omega \rangle) = 0 & \text{on } \mathbb{R} \times \partial K, \\ V|_{t < -t_0} = 0. \end{cases}$$

Then we have
$$s(\sigma, \theta, \omega) = (-1)^{(n+1)/2} 2^{-n} \pi^{1-n} \int_{\partial K} \partial_t^{n-2} \partial_\nu V(\langle x, \theta \rangle - \sigma, x; \omega) \, dS_x,$$

where the integral is interpreted in the sense of distributions. Our aim will be to examine the singularities of $s(t, \theta, \omega)$ with respect to t.

First we define the so-called reflecting (ω, θ)-rays. Given two directions θ, ω in $\mathbf{S}^{n-1} \times \mathbf{S}^{n-1}$, consider a curve $\gamma \in \Omega$ having the form
$$\gamma = \bigcup_{i=0}^{m} l_i, \qquad m \geq 1,$$

where $l_i = [x_i, x_{i+1}]$ are finite segments for $i = 1, \ldots, m-1$, $x_i \in \partial K$, and l_0 (resp. l_m) is the infinite segment starting at x_1 (resp. at x_m) and having direction $-\omega$ (resp. θ). The curve γ is called a *reflecting* (ω, θ)-ray in Ω if for $i = 0, 1, \ldots, m-1$ the segments l_i and l_{i+1} satisfy the law of reflection at x_{i+1} with respect to ∂K. The points $x_1, \ldots, x_m$ are called *reflection points* of γ and this ray is called *ordinary reflecting* (or *simply reflecting*) if γ has no segments tangent to ∂K.

Next, we define two important notions related to (ω, θ)-rays (also-called *scattering rays*). Fix an arbitrary open ball U_0 with radius $a > 0$ containing K. For $\xi \in \mathbf{S}^{n-1}$ introduce the hyperplane Z_ξ orthogonal to ξ and such that ξ is pointing into the interior of the open half space H_ξ with boundary Z_ξ containing U_0. Let $\pi_\xi : \mathbb{R}^n \to Z_\xi$ be the orthogonal projection. For a reflecting (ω, θ)-ray γ in Ω with successive reflecting points $x_1, \ldots, x_m$ the *sojourn time* T_γ of γ is defined by

$$T_\gamma = \|\pi_\omega(x_1) - x_1\| + \sum_{i=1}^{m-1} \|x_i - x_{i+1}\| + \|x_m - \pi_{-\theta}(x_m)\| - 2a.$$

Obviously, $T_\gamma + 2a$ coincides with the length of this part of γ which lies in $H_\omega \cap H_{-\theta}$ (see Figure 2). In fact, the sojourn time T_γ does not depend on the choice of the ball U_0 since it follows easily that

$$\|\pi_\omega(x_1) - x_1\| = a + \langle x_1, \omega \rangle, \|x_m - \pi_{-\theta}(x_m)\| = a - \langle x_m, \theta \rangle ,$$

therefore

$$T_\gamma = \langle x_1, \omega \rangle + \sum_{i=1}^{m-1} \|x_i - x_{i+1}\| - \langle x_m, \theta \rangle.$$

Given an ordinary reflecting (ω, θ)-ray γ set $u_\gamma = \pi_\omega(x_1)$. There exists a small neighborhood W_γ of u_γ in Z_ω such that for every $u \in W_\gamma$ there are a unique direction $\theta(u) \in \mathbf{S}^{n-1}$ and points $x_1(u), \ldots, x_m(u)$ which are the successive reflection points of a reflecting $(u, \theta(u))$-ray in Ω with $\pi_\omega(x_1(u)) = u$. This defines a smooth map

$$J_\gamma : W_\gamma \ni u \to \theta(u) \in \mathbf{S}^{n-1}$$

and $dJ_\gamma(u_\gamma)$ is called a *differential cross section* related to γ. We say that γ is *nondegenerate* if

$$\det dJ_\gamma(u_\gamma) \neq 0.$$

The notion of sojourn time as well as that of differential cross section are well known in the physical literature. The definitions given above are due to Guillemin [5].

For strictly convex obstacles all (nontrivial) reflecting rays have only one reflection point x_1 and the corresponding sojourn time is equal to $\langle x_1, \omega - \theta \rangle$.

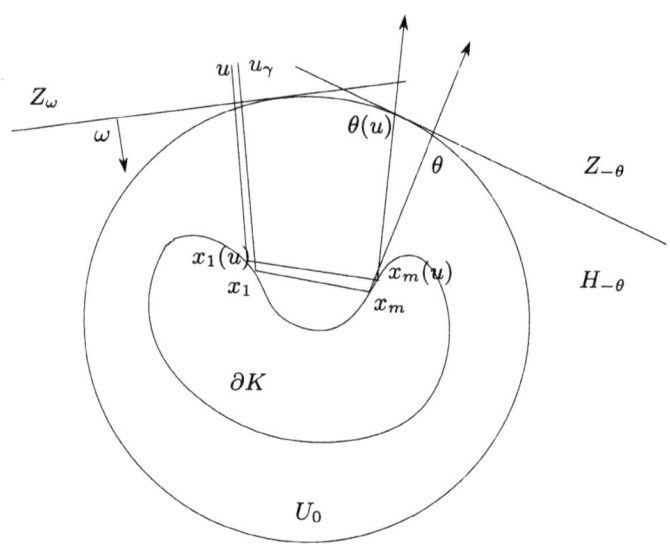

Figure 2.

Moreover, the stationary phase argument of the previous section implies that $\overline{a(\lambda, \omega, \theta)}$ has a complete asymptotic expansion

$$\overline{a(\lambda, \omega, \theta)} = e^{i\langle x_+, \omega - \theta\rangle} \sum_{j=0}^{N} c_j \lambda^{-j} + O(|\lambda|^{-N-1}) \quad \text{for all } N \in \mathbb{N},$$

which gives

$$\operatorname{sing\,supp} s(t, \theta, \omega) = \{-T_+\},$$

$T_+ = \langle x_+, \omega - \theta\rangle$ being the sojourn time of the (ω, θ)-ray γ_+ reflecting at x_+. A simple geometric argument implies that

$$|\det dJ_{\gamma_+}(u_{\gamma_+})| = 4|\theta - \omega|^{(n-3)}\mathcal{K}(x_+),$$

and for t close to $-T_+$ we have

$$s(t, \theta, \omega) = \left(\frac{-1}{2\pi}\right)^{(n-1)/2} |dJ_{\gamma_+}(u_{\gamma_+})|^{-1/2} \delta^{(n-1)/2}(t + T_+) + \text{l.o.s.}$$

(the abbreviation stands for "lower order singularities").

For strictly convex obstacles T_+ is an isolated singularity of $s(t, \theta, \omega)$ related to an ordinary reflecting ray. This situation can be generalized for generic obstacles if we consider the back scattering direction $\theta = -\omega$. Without loss of the generality we may assume that K lies in the half space $\{x \in \mathbb{R}^n : \langle x, \omega\rangle > 0\}$. Then the function

$$\partial K \ni x \to \langle x, \omega\rangle \in \mathbb{R}^+$$

has a positive minimum $\rho(\omega)$ and there exists at least one reflecting $(\omega, -\omega)$-ray γ with sojourn time $T_\gamma = 2\rho(\omega)$. Of course we could have many $(\omega, -\omega)$-rays with the same minimal sojourn time. A geometric argument based on Sard's theorem shows that there exists a subset $B \subset \mathbf{S}^{n-1}$ with full measure such that for every $\omega \in B$ we have only a finite number of reflecting (ω, θ)-rays with sojourn time $2\rho(\omega)$. Moreover, each of these rays $\gamma_1, \ldots, \gamma_M$, has only one reflection point $x_k \in \partial K, k = 1, \ldots, M$, and ∂K has a nonvanishing Gauss curvature $\mathcal{K}(x_k) \neq 0$ for every $k = 1, \ldots, M$. Thus, repeating the argument from Section 2, it follows that for $\omega \in B$ the sojourn time $T = -2\rho(\omega)$ is an isolated singularity of the scattering kernel $s(t, -\omega, \omega)$, and for such ω we have

$$\max \operatorname{sing} [\operatorname{supp}_t s(t, -\omega, \omega)] = -2\rho(\omega),$$

and for t close to $-2\rho(\omega)$,

$$s(t, -\omega, \omega) = 2^{1-n} \left(-\frac{1}{\pi}\right)^{(n-1)/2} \sum_{k=1}^{M} |\mathcal{K}(x_k)|^{-1/2} \delta^{(n-1)/2}(t + 2\rho(\omega)) + \text{l.o.s.}$$

This result is due to Majda [15]. From the maximal singularity of the back scattering kernel one obtains that the *convex hull* of the obstacle is given by

$$\hat{K} = \bigcap_\omega \{x : \langle x, \omega \rangle \geq \rho(\omega)\}.$$

Thus one can recover the geometry of a convex obstacle.

It is much more complicated to get similar results in the case of nonconvex obstacles. Now the information obtained by means of rays having only one reflection is no longer sufficient. One needs to consider multiple reflecting (ω, θ)-rays leading to isolated singularities of $s(t, \theta, \omega)$. Roughly speaking, the singularities of the scattering kernel are amongst the sojourn times of (ω, θ)-rays, however now one has to consider not only simply reflecting (ω, θ)-rays but all generalized geodesics *incoming* with direction ω and *outgoing* with direction θ (see [22, Chapter 9] and [18]); these are simply called (ω, θ)-rays. In general, there exist (ω, θ)-rays with grazing or gliding segments (see Figure 3).

The precise definition of an (ω, θ)-ray is based on the notion of a generalized bicharacteristic of the operator $\Box = \partial_t^2 - \Delta_x$ given as trajectories of the generalized Hamilton flow $\mathcal{F}_t$ in Ω generated by the symbol $\sum_{i=1}^{n} \xi_i^2 - \tau^2$ of $\Box$ (see [20] for a precise definition). In general, $\mathcal{F}_t$ is not smooth and in some cases there may exist two different integral curves issued from the same point in the phase space (see [30] for an example). To avoid this situation we assume that the following generic condition is satisfied.

($\mathcal{G}$) If for $(x, \xi) \in T^*(\partial K)$ the normal curvature of ∂K vanishes of infinite order in direction ξ, then ∂K is convex at x in direction ξ.

We will now sketch the definition of a generalized bicharacteristics of $\Box$. Let $p(x, \xi)$ be the restriction of the principal symbol of $\Box$ to the level surface $\tau = 1$ (this is the case of motion with unit speed along geodesics). Notice that in this case the so-called *zero bicharacteristic set* $\Sigma = p^{-1}(0)$ coincides with the

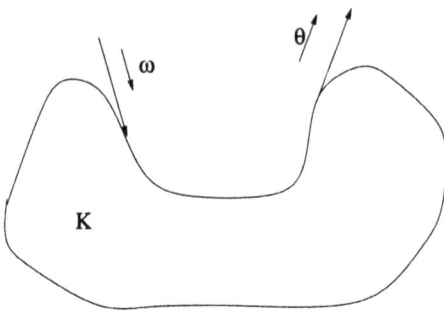

Figure 3.

cosphere bundle $S^*(\Omega)$ of Ω. Given a point $x \in \partial K$, we choose local coordinates

$$x = (x_1, \ldots, x_n), \xi = (\xi_1, \ldots, \xi_n)$$

in $T^*(\mathbb{R}^n)$ so that locally ∂K is given by $x_1 = 0$ and Ω by $x_1 \geq 0$. The coordinates (x, ξ) can be chosen so that, up to a nonzero smooth factor, $p(x, \xi)$ has the form

$$p(x, \xi) = \xi_1^2 - r(x, \xi')$$

with $x' = (x_2, \ldots, x_n), \xi' = (\xi_2, \ldots, \xi_n)$ and $r(x, \xi')$ homogeneous of order two in ξ'. Introduce the sets

$$\Sigma_0 = \{(x, \xi) \in T^*(\mathbb{R}^n) \setminus \{0\} : x_1 > 0\},$$
$$H = \{(x, \xi) \in \Sigma : x_1 = 0, r(0, x', \xi') > 0\},$$
$$G = \{(x, \xi) \in \Sigma : x_1 = 0, r(0, x', \xi') = 0\}.$$

The sets H and G are called *hyperbolic* and *glancing* set, respectively. Next consider the symbols

$$r_0(x', \xi') = r(0, x', \xi'), r_1(x', \xi') = \frac{\partial r}{\partial x_1}(0, x', \xi'),$$

and define the *diffractive* and *gliding* sets by

$$G_d = \{(x, \xi) \in G : r_1(x', \xi') > 0\},$$
$$G_g = \{(x, \xi) \in G : r_1(x', \xi') < 0\},$$

respectively. The generalized bicharacteristics are related to the Hamilton vector fields

$$H_p = \sum_{j=1}^{n} \left(\frac{\partial p}{\partial \xi_j} \cdot \frac{\partial}{\partial x_j} - \frac{\partial p}{\partial x_j} \cdot \frac{\partial}{\partial \xi_j} \right),$$

$$H_{r_0} = \sum_{j=2}^{n} \left(\frac{\partial r_0}{\partial \xi_j} \cdot \frac{\partial}{\partial x_j} - \frac{\partial r_0}{\partial x_j} \cdot \frac{\partial}{\partial \xi_j} \right).$$

We have $d_\xi p(x, \xi) \neq 0$ on $S^*(\Omega)$ and $d_{\xi'} r_0(x', \xi') \neq 0$ on G. Moreover, the above definitions are independent on the choice of the local coordinates. Using the above local coordinates the generalized bicharecteristics of $\square$ are defined as follows.

Let $I \subset \mathbb{R}$ be an open interval. A curve $\gamma : I \to S^*(\Omega)$ is called a *generalized bicharacteristic* of $\square$ if there exists a discrete subset $B \subset I$ such that the following conditions hold:

(i) If $\gamma(t_0) \in \Sigma_0 \cup G_d$ for some $t_0 \in I \backslash B$, then γ is differentiable at t_0 and

$$\frac{d}{dt} \gamma(t_0) = H_p(\gamma(t_0)).$$

(ii) If $\gamma(t_0) \in G \backslash G_d$ for some $t_0 \in I \backslash B$, then

$$\gamma(t) = (x_1(t), x'(t), \xi_1(t), \xi'(t))$$

is differentiable at t_0 and

$$\frac{dx_1}{dt}(t_0) = \frac{d\xi_1}{dt}(t_0) = 0, \qquad \frac{d}{dt}(x'(t), \xi'(t))_{|t=t_0} = H_{r_0}(\gamma(t_0)).$$

(iii) If $t_0 \in B$, then $\gamma(t) \in \Sigma_0$ for all $t \neq t_0, t \in I$ with $|t - t_0|$ sufficiently small. Moreover, in this case for $\xi_1^\pm(x', \xi') = \pm\sqrt{r_0(x', \xi')}$ we have

$$\lim_{t \to t_0, \pm(t-t_0)>0} \gamma(t) = (0, x'(t), \xi_1^\pm(x'(t_0)), \xi'(t_0)) \in H.$$

The functions $x(t)$, $\xi'(t)$, $|\xi_1(t)|$ are continuous on I, while the function $\xi_1(t)$ has a jump discontinuity at any point $t \in B$. Finally, under the condition $(\mathcal{G})$ a generalized bicharacteristic $\gamma : \mathbb{R} \to S^*(\Omega)$ of $\square$ is *uniquely extendible* in the sense that for each $t \in \mathbb{R}$ the only generalized bicharacteristic (up to the change of parameter t) passing through $\gamma(t)$ is γ ([20]; see also [10, vol. III]).

More generally, working with the restriction of the principal symbol of $\square$ to a level surface $\tau = \tau_0 \neq 0$, one defines generalized bicharacetristics on the set $\dot{T}^*(\Omega)$ of all $(x, \xi) \in T^*(\Omega)$ such that $\xi \neq 0$. Given $\sigma = (x, \xi) \in \dot{T}^*(\Omega)$, there exists a unique generalized bicharacteristic $(x(t), \xi(t)) \in \dot{T}^*(\Omega)$ such that $x(0) = x$ and $\xi(0) = \xi$. Set $\mathcal{F}_t(x, \xi) = (x(t), \xi(t))$ for all $t \in \mathbb{R}$. This defines a flow $\mathcal{F}_t : \dot{T}^*(\Omega) \to \dot{T}^*(\Omega)$ ([20]) which is sometimes called the *generalized geodesic flow* on $\dot{T}^*(\Omega)$. Obviously, it leaves the *cosphere bundle* $S^*(\Omega)$ invariant. At points of transversal reflection at $\dot{T}^*_{\partial K}(\Omega)$ the flow $\mathcal{F}_t$ is discontinuous. To make it continuous, consider the *quotient space* $\dot{T}^*_b(\Omega) = \dot{T}^*(\Omega)/\sim$ of $\dot{T}^*(\Omega)$ with respect to the following equivalence relation: $\rho \sim \sigma$ if and only if $\rho = \sigma$ or $\rho, \sigma \in T^*_{\partial K}(\Omega)$ and either $\lim_{t \nearrow 0} \mathcal{F}_t(\rho) = \sigma$ or $\lim_{t \searrow 0} \mathcal{F}_t(\rho) = \sigma$. Let $S^*_b(\Omega)$ be the image of $S^*(\Omega)$ in $\dot{T}^*_b(\Omega)$. Melrose and Sjöstrand ([20]) proved that the natural projection of $\mathcal{F}_t$ on $\dot{T}^*_b(\Omega)$ is continuous.

After these definitions a curve $\gamma = \{x(t) \in \Omega : t \in \mathbb{R}\}$ is called an (ω, θ)-*ray* if there exist real numbers $t_1 < t_2$ such that

$$\tilde{\gamma}(t) = (x(t), \xi(t)) \in S^*(\Omega)$$

is a *generalized bicharacteristic* of $\square$ and

$$\xi(t) = \begin{cases} \omega & \text{for } t \leq t_1, \\ \theta & \text{for } t \geq t_2, \end{cases}$$

provided that the time t increases when we move along $\tilde{\gamma}$. Denote by $\mathcal{L}_{\omega,\theta}(\Omega)$ the set of all (ω, θ)-rays in Ω. The *sojourn time* T_δ of $\delta \in \mathcal{L}_{\omega,\theta}(\Omega)$ is defined as the length of the part of δ lying in $H_\omega \cap H_{-\theta}$.

Turning to the problem of the behavior of $s(t, \theta, \omega)$ near singularities, assume that γ is a fixed *nondegenerate ordinary reflecting* (ω, θ)-ray such that

$$T_\gamma \neq T_\delta \quad \text{for every } \delta \in \mathcal{L}_{\omega,\theta}(\Omega) \setminus \{\gamma\}. \tag{3-1}$$

By using the continuity of the generalized Hamilton flow, it is easy to show that

$$(-T_\gamma - \varepsilon, -T_\gamma + \varepsilon) \cap \text{sing supp } s(t, \theta, \omega) = \{-T_\gamma\}$$

for $\varepsilon > 0$ sufficiently small. The singularity of $s(t, \theta, \omega)$ at $t = -T_\gamma$ can be investigated using a global construction of an asymptotic solution as a Fourier integral operator ([6], [21], Chapter 9 in [22]).

THEOREM 3.1 [21]. *Under the assumption (3–1) we have*

$$-T_\gamma \in \text{sing supp } s(t, \theta, \omega)$$

and for t close to $-T_\gamma$ the scattering kernel has the form

$$s(t, \theta, \omega) = \left(\frac{1}{2\pi i}\right)^{(n-1)/2} (-1)^{m_\gamma - 1} \exp\left(i\frac{\pi}{2}\beta_\gamma\right)$$
$$\times \left| \frac{\det dJ_\gamma(u_\gamma)\langle \nu(q_1), \omega \rangle}{\langle \nu(q_m), \theta \rangle} \right|^{-1/2} \delta^{(n-1)/2}(t + T_\gamma) + \text{l.o.s.,} \tag{3-2}$$

where m_γ is the number of reflections of γ and q_1, q_m are the first and last reflection points, respectively, of γ and $\beta_\gamma \in \mathbb{Z}$.

For strictly convex obstacles we have $\beta_\gamma = -(n-1)/2$, $q_1 = q_m$ and $\theta - \omega$ is parallel to $\nu(q_1)$.

4. Properties of Reflecting (ω, θ)-Rays

To apply the result of the previous section we need the condition (3–1) and it is desirable to prove that there exists a subset $\mathcal{S} \subset \mathbf{S}^{n-1} \times \mathbf{S}^{n-1}$ with zero Lebesgue measure such that for all directions $(\omega, \theta) \in \mathbf{S}^{n-1} \times \mathbf{S}^{n-1} \setminus \mathcal{S}$ the corresponding (ω, θ)-rays satisfy (3–1). Here one has to deal with all (generalized) (ω, θ)-rays and this makes the problem rather difficult. We start with a result concerning the *ordinary reflecting* (ω, θ)-rays only.

THEOREM 4.1 [23]. *For every $\omega \in \mathbf{S}^{n-1}$ there exists a set $S(\omega) \subset \mathbf{S}^{n-1}$ the complement of which is a countable union of compact subsets of $\mathbf{S}^{n-1}$ of measure zero such that if $\theta \in S(\omega)$, then any two different ordinary reflecting (ω, θ)-rays in Ω have distinct sojourn times.*

SKETCH OF PROOF. Let U_0 be an open ball with center 0 and radius a containing K and let $Z = Z_\omega$ be the hyperplane introduced in Section 3. Given an integer $k \geq 1$, denote by U_k the set of those $u \in Z$ for which the trajectory $\gamma(u)$ of the generalized Hamiltonian flow starting in u with direction ω is an ordinary reflecting ray with exactly k reflection points. Let $J_k(u) \in \mathbf{S}^{n-1}$ be the direction of $\gamma(u)$ after the last reflection. Obviously, U_k is open in Z and the map

$$J_k : U_k \ni z \rightarrow J_k(u) \in \mathbf{S}^{n-1}$$

is smooth.

Now let us fix two arbitrary integers $k \geq 1, s \geq 1$. For $u \in U_k$ denote by $f(u)$ the sojourn time of the scattering ray determined by $\gamma(u)$. In the same way denote by $g(v)$ the sojourn time of the scattering ray with s reflections determined by $v \in V_s$. The functions $f : U_k \rightarrow \mathbb{R}, g : V_s \rightarrow \mathbb{R}$ are smooth.

For $u \in U_k$ denote by $x_1(u), \ldots, x_k(u)$ the successive reflection points of $\gamma(u)$. The corresponding maps $x_i : U_k \rightarrow \partial K$ are smooth and for every $y \in \partial K$ we denote by $N(y)$ the unit normal to ∂K pointing into Ω. Thus for $u \in U_k$ we obtain

$$J_k(u) = \frac{x_k(u) - x_{k-1}(u)}{\|x_k(u) - x_{k-1}(u)\|} - 2 \left\langle \frac{x_k(u) - x_{k-1}(u)}{\|x_k(u) - x_{k-1}(u)\|}, N(x_k(u)) \right\rangle N(x_k(u)),$$

and

$$f(u) = \sum_{i=0}^{k-1} \|x_{i+1}(u) - x_i(u)\| + t - 2a,$$

with the convention that $x_0(u)$ (resp. $x_{k+1}(u)$) denotes the orthogonal projection of $x_1(u)$ (resp. $x_k(u)$) on Z (resp. $Z_{-J_k(u)}$), and where $t = \|x_k(u) - x_{k+1}(u)\|$. We obtain easily $t = a - \langle J_k(u), x_k \rangle$, so

$$f(u) = \sum_{i=0}^{k-1} \|x_{i+1}(u) - x_i(u)\| - \langle x_k(u), J_k(u) \rangle - a.$$

For $v \in V_s$ the successive reflection points of $\gamma(v)$ will be denoted by $y_1(v), \ldots, y_s(v)$. Next we set $y_0(v) = v$ and we define $y_{s+1}(v)$ in the same way as $x_{k+1}(u)$. Now denote by $W(k, s)$ the set of those $(u, v) \in U_k \times V_s$ for which

$$J_k(u) = J_s(v), f(u) = g(v)$$

and

$$\operatorname{rank} dJ_k(u) = \operatorname{rank} dJ_s(v) = n - 1.$$

LEMMA 4.2. $W(k, s)$ is a smooth $(n-2)$-dimensional submanifold of $U_k \times U_s$.

PROOF OF LEMMA 4.2.   Consider a point $w_0 = (u_0, v_0) \in W(k, s)$. Since $\operatorname{rank} dJ_k(u_0) = \operatorname{rank} dJ_s(v_0) = n - 1$, there exists a neighborhood U of w_0 in $U_k \times V_s$ such that for every $(u, v) \in U$ we have

$$\operatorname{rank} dJ_k(u) = \operatorname{rank} dJ_s(v) = n - 1.$$

Define the map $L : U \to \mathbb{R}^n$ by

$$L(u, v) = \left(\lambda(u, v), \ (\chi^{(j)}(u, v))_{1 \leq j \leq n-1} \right)$$

with

$$\lambda(u, v) = f(u) - g(v), \chi(u, v) = J_k(u) - J_s(v).$$

Clearly, $W(k, s) \cap U \subset L^{-1}(0)$ and to prove that $W(k, s)$ is a smooth $(n-2)$-dimensional submanifold of $U_k \times V_s$ it is sufficient to show that L is a submersion at any point w_0 of $L^{-1}(0)$. For this purpose we assume without loss of generality that $\theta_n \neq 0$. Suppose that

$$\sum_{j=1}^{n-1} A_j \operatorname{grad} \chi^{(j)}(w_0) + C \operatorname{grad} \lambda(w_0) = 0$$

with some constants A_j, C. Calculating the derivatives involved above and using the geometrical meaning of f, g, J_k and J_s, one derives $A_1 = \cdots = A_{n-1} = C = 0$. Thus L is a submersion at w_0. See [23] for more details. $\square$

Consider the map $\varphi : U_k \times V_s \to \mathbf{S}^{n-1}$ given by $\varphi(u, v) = J_k(u)$. This map is smooth and $\dim W(k, s) = n - 2$ shows that $\varphi(W(k, s))$ is a countable union of compact subsets of $\mathbf{S}^{n-1}$ of measure zero. Clearly

$$F_k = \{u \in U_k : \operatorname{rank} dJ_k(u) \leq n - 2\}$$

is a countable union of compact subsets. By Sard's theorem, $J_k(F_k)$ has measure zero in $\mathbf{S}^{n-1}$ for all k, so $F = \bigcup_k J_k(F_k)$ also has measure zero in $\mathbf{S}^{n-1}$. Hence the subset

$$S(\omega) = \mathbf{S}^{n-1} \setminus \left(F \cup \bigcup_{k,s} J_k(W(k, s)) \right)$$

of $\mathbf{S}^{n-1}$ has the desired properties. This concludes the proof of Theorem 4.1. $\square$

Setting $\mathcal{S} = \{(\omega, \theta) \in \mathbf{S}^{n-1} \times \mathbf{S}^{n-1} : \theta \in S(\omega)\}$, we see that for $(\omega, \theta) \in \mathcal{S}$ any two different ordinary reflecting rays in Ω have distinct sojourn times and the complement of $\mathcal{S}$ in $\mathbf{S}^{n-1} \times \mathbf{S}^{n-1}$ has measure zero.

To deal with reflecting rays with tangent segments, we introduce a more general type of trajectories. A curve γ in $\mathbb{R}^n$ is called an (ω, θ)-*trajectory* for Ω if it has the form $\gamma = \bigcup_{i=0}^{s} l_i$, where $l_i = [x_i, x_{i+1}]$, i ranges from 1 through $s - 1$, $x_i \in \partial K$ for $i = 1, \ldots, s$, while l_0 (resp. l_s) is the infinite ray starting at x_1 (resp. x_s) with direction $-\omega$ (resp. θ) and, for every $i = 0, 1, \ldots, s - 1$, l_i and l_{i+1} satisfy the law of reflection at x_i with respect to ∂K. It is clear that every reflecting (ω, θ)-ray is an (ω, θ)-trajectory, but the converse is not true in general since some (ω, θ)-trajectory may intersect transversally ∂K. On the other hand, every (ω, θ)-reflecting ray with tangent segment is an (ω, θ)-trajectory. We have the following.

THEOREM 4.3 [23]. *There exists $\mathcal{T} \subset \mathbf{S}^{n-1} \times \mathbf{S}^{n-1}$ the complement of which is a countable union of compact subsets of measure zero in $\mathbf{S}^{n-1} \times \mathbf{S}^{n-1}$ such that for $(\omega, \theta) \in \mathcal{T}$ all (ω, θ)-trajectories for Ω are ordinary.*

PROOF. We follow the idea of the proof of Theorem 4.1. For simplicity set $\partial K = X$. Fix two integers k and s so that $s \geq 1, 0 \leq k \leq s$. Let $M(s,k)$ be the set of those

$$\zeta = (\omega; x; y; \theta) \in M_s = \mathbf{S}^{n-1} \times X^{(s)} \times X \times \mathbf{S}^{n-1}$$

with $x = (x_1, \ldots, x_s)$ such that there exists an (ω, θ)-trajectory for X with successive transversal reflection points $x_1, \ldots, x_s$, the segment $[x_k, x_{k+1}]$ of which is tangent to X at $y \in (x_k, x_{k+1})$. Here

$$X^{(s)} = \{(x_1, \ldots, x_s) \in X^s : x_i \neq x_j, i \neq j\}$$

and x_0 (resp. x_{s+1}) is the orthogonal projection of x_1 on Z_ω (resp. of x_s on $Z_{-\theta}$).

The main step in the proof is to show that $M(s,k)$ is a smooth submanifold of M_s of dimension $2n - 3$. This follows from a specially adapted parametrization of $M(s,k)$; see [23] for details. Using this one obtains Theorem 4.3 easily. Consider the projection

$$\pi_s : M_s = \mathbf{S}^{n-1} \times X^{(s)} \times X \times \mathbf{S}^{n-1} \to \mathbf{S}^{n-1} \times \mathbf{S}^{n-1}$$

given by

$$\pi_s(\omega; x; y; \theta) = (\omega, \theta),$$

and introduce the open subsets of M_s

$$U_r(s,k) = \{(\omega; x; y; \theta) \in M_s : x_k^{(r)} \neq x_{k+1}^{(r)}\}, r = 1, \ldots, n.$$

Then $M_r(s,k) = M(s,k) \cap U_r(s,k)$ is a smooth submanifold of M_s of dimension $2n - 3 < \dim(\mathbf{S}^{n-1} \times \mathbf{S}^{n-1})$. Since π_s is smooth, the set

$$L_r(s,k) = \pi_s(M_r(s,k)) \subset \mathbf{S}^{n-1} \times \mathbf{S}^{n-1}$$

has measure zero. Consequently, for the covering $M_r(s,k) = \bigcup_{j=1}^\infty K_j$ with K_j compact, one gets that

$$L_r(s,k) = \bigcup_{j=1}^\infty \pi_s(K_j)$$

is a countable union of compact subsets of $\mathbf{S}^{n-1} \times \mathbf{S}^{n-1}$ of measure zero. Setting

$$\mathcal{T} = \mathbf{S}^{n-1} \times \mathbf{S}^{n-1} \setminus \bigcup_{0 \leq k \leq s} \bigcup_{r=1}^\infty L_r(s,k),$$

completes the proof of Theorem 4.3.   □

Finally, we find a subset $\mathcal{U} \subset \mathbf{S}^{n-1} \times \mathbf{S}^{n-1}$ such that for $(\omega, \theta) \in \mathcal{T} \cap \mathcal{U}$ all reflecting (ω, θ)-rays are ordinary and nondegenerate. So there exists a subset $\mathcal{A} = \mathcal{T} \cap \mathcal{U} \cap \mathcal{S}$ of $\mathbf{S}^{n-1} \times \mathbf{S}^{n-1}$ of full measure so that for every $(\omega, \theta) \in \mathcal{A}$ the corresponding (ω, θ)-reflecting rays are ordinary, nondegenerate and with distinct sojourn times.

The study of the generalized (ω, θ)-rays leads to many difficulties. However it is quite natural to expect that for almost all (ω, θ) in $\mathbf{S}^{n-1} \times \mathbf{S}^{n-1}$ there are no generalized (ω, θ)-rays different from reflecting ones. This will be discuss in details in the next section.

5. Poisson Relation for the Scattering Kernel

Let K be an obstacle in $\mathbb{R}^n, n \geq 3, n$ odd, with C^∞ boundary ∂K so that

$$K \subset \{x \in \mathbb{R}^n : |x| \leq \rho_0\}$$

and let $\Omega = \overline{\mathbb{R}^n \setminus K}$. In what follows we assume that K satisfies the condition $(\mathcal{G})$ from Section 3. Let $\pi : T^*(\mathbb{R} \times \Omega) \to \Omega$ be the natural projection.

The following result of [21], [1] (see also [22, Chapter 8] and [18]) shows that for $\omega \neq \theta$ all singularities in t of $s(t, \theta, \omega)$ are given by (negative) sojourn times.

THEOREM 5.1 [21], [1]. *For $\omega \neq \theta$ we have*

$$\text{sing supp } s(t, \theta, \omega) \subset \{-T_\gamma : \gamma \in \mathcal{L}_{\omega, \theta}(\Omega)\}. \tag{5–1}$$

In analogy with the well-known Poisson relation for the Laplacian on Riemannian manifolds, (5–1) is called the *Poisson relation for the scattering kernel*, while the set of all T_γ, where $\gamma \in \mathcal{L}_{\omega, \theta}(\Omega), (\omega, \theta) \in \mathbf{S}^{n-1} \times \mathbf{S}^{n-1}$, is called the *scattering length spectrum* of K.

SKETCH OF PROOF. The proof uses results on the propagation of singularities along generalized bicharactaristics, and some properties of oscillatory integrals. Consider a fixed t_0 so that

$$-t_0 \notin \{-T_\gamma : \gamma \in \mathcal{L}_{(\omega, \theta)}(\Omega)\}.$$

Take $T > 0$ with $|t_0| < T$ and introduce the set

$$\Gamma_T = \{T_\gamma : |T_\gamma| \leq T, \gamma \in \mathcal{L}_{(\omega, \theta)}(\Omega)\}.$$

The continuity of the generalized Hamiltonian flow implies that Γ_T is closed, so we can choose $\varepsilon_0 > 0$ so that

$$T_\gamma \notin [t_0 - \varepsilon_0, t_0 + \varepsilon_0] \quad \text{for all } \gamma \in \mathcal{L}_{(\omega, \theta)}(\Omega).$$

Let $\rho(t) \in C_0^\infty(\mathbb{R}), \rho(t) = 1$ for $|t| \leq 1/2, \rho(t) = 0$ for $|t| \geq 1$. Set $\rho_\delta(t) = \rho(t/\delta)$ for $0 < \delta \leq \varepsilon_0/2$. To prove that $t_0 \notin \text{sing supp } s(t, \theta, \omega)$, it is sufficient to show that the integral

$$J(\lambda) = \langle s(t, \theta, \omega), \rho_\delta(t + t_0) e^{-i\lambda t} \rangle$$

$$= \sum_{k=0}^{n-2} c_k (-i\lambda)^{n-2-k} \int_{\mathbb{R}} \int_{\partial\Omega} e^{i\lambda(t - \langle x, \theta \rangle)} \frac{d^k \rho_\delta}{dt^k} (\langle x, \theta \rangle - t + t_0) \frac{\partial w}{\partial \nu}(t, x; \omega) \, dt \, dS_x,$$

with c_k a constant, is rapidly decreasing with respect to λ. Here $w(t, x; \omega) = V(t, x; \omega) + \delta(t - \langle x, \omega \rangle)$, where $V(t, x; \omega)$ is defined in Section 3. Let us treat the term with $k = 0$, the other ones can be examined by a similar argument.

Without loss of generality we may assume that $\omega = (0, \ldots, 0, 1)$. Set

$$Z(\tau) = \{x \in \mathbb{R}^n : x_n = \tau\},$$

where $\tau < -\rho_0$ and let $\mathbb{R}_\tau^+ = \{t \in \mathbb{R} : t > \tau\}$. To localize the problem, introduce a partition of unity on $Z(\tau)$ given by functions

$$\varphi_j(x') \in C_0^\infty(\mathbb{R}^{n-1}), x' = (x_1, \ldots, x_{n-1}).$$

Consider the problems

$$\begin{cases} \Box v_j = 0 \quad \text{in } \mathbb{R}_\tau^+ \times \mathbb{R}_x^n, \\ v_j(\tau, x) = \varphi_j(x')\delta(\tau - x'), \\ \dfrac{\partial v_j}{\partial t}(\tau, x) = \varphi_j(x')\delta'(\tau - x_n), \end{cases}$$

$$\begin{cases} \Box W_j = 0 \quad \text{in } \mathbb{R} \times \mathring{\Omega}, \\ W_j = 0 \quad \text{on } \mathbb{R} \times \partial\Omega, \\ W_j(\tau, x) = \varphi_j(x')\delta(\tau - x'), \\ \dfrac{\partial W_j}{\partial t}(\tau, x) = \varphi_j(x')\delta'(\tau - x_n). \end{cases}$$

Clearly, there exists a compact set $F_0' \subset \mathbb{R}^{n-1}$ such that if $\operatorname{supp} \varphi_j \cap F_0' = \varnothing$, then the straight lines issued from $(x', \tau), x' \in \operatorname{supp} \varphi_j$, with direction ω do not meet $\partial\Omega$. For such j and $\omega \neq \theta$ we have

$$WF\left(\left(\frac{\partial W_j}{\partial\nu}\right)_{|\mathbb{R} \times \partial\Omega}\right) \cap \{(t, x, 1, -\theta_{|T_x(\partial\Omega)}) : |t| \leq T + \rho_0 + 1, x \in \partial\Omega\} = \varnothing. \tag{5-2}$$

This implies easily

$$\int_{\mathbb{R}} \int_{\partial\Omega} e^{i\lambda(t - \langle x, \theta \rangle)} \rho_\delta(\langle x, \theta \rangle - t + t_0) \frac{\partial W_j}{\partial\nu} \, dt \, dS_x = O(|\lambda|^{-m}) \quad \text{for all } m \in \mathbb{N}. \tag{5-3}$$

Now set $F_0 = \{x \in \mathbb{R}^n : x' \in F_0', x_n = \tau\}$ and denote by $l(u_0)$ the straight line passing through $u_0 \in F_0$ with direction ω. There are three cases:

(i) $\varnothing \neq l(u_0) \cap \overline{K} \subset \partial\Omega$;
(ii) $l(u_0)$ meets transversally $\partial\Omega$ at $x_1(u_0)$;
(iii) $l(u_0)$ is tangent to $\partial\Omega$ at $x_1(u_0)$ and ω is an asymptotic direction for $\partial\Omega$ at $x_1(u_0)$.

In the case (i) the generalized bicharecteristic γ_0 with $\operatorname{Im}(\pi \circ \gamma_0) = l(u_0)$ is uniquely extendible, and results on propagation of singularities lead to (5–2) which in turn gives (5–3). To deal with the case (ii), set $t_1(u) = |u - x_1(u)|, u \in F_0$. The solution v_j with such j is given by an oscillatory integral and $WF(v_j)$ is included in the set of all $(t, x, \pm\sigma, \mp\omega) \in T^*(\mathbb{R}^{n+1}) \setminus \{0\}$ such that $\sigma > 0$ and there exist $\hat{x} \in Z(\tau), \hat{x}' \in \operatorname{supp}\varphi_j, s \geq 0$ with $t = \tau \pm \sigma s, x = \hat{x} \pm \sigma s \omega$. We modify v_j on the intersection of a small neighborhood of $x_1(u_0)$ with the interior of K so that the modified function $\tilde{v}_j$ satisfies (for some $\varepsilon > 0$) the properties

$$\tilde{v}_j = \begin{cases} v_j & \text{for } t < t_1 + \varepsilon, \\ 0 & \text{for } t > t_1 + 2\varepsilon. \end{cases}$$

Here $t_1 = \max\{t_1(u) : u \in O(u_0)\}$, where $O(u_0)$ is a sufficiently small neighborhood of u_0 with supp $\varphi_j \subset O(u_0)$ and ε is small enough. Moreover, we preserve the condition

$$\Box \tilde{v}_j = 0 \quad \text{in } \mathbb{R}^+_\tau \times \mathring{\Omega}.$$

Set $h_j = (\tilde{v}_j)_{|\mathbb{R}^+_\tau \times \partial\Omega}$ and notice that $h_j = 0$ for t sufficiently close to τ. We extend h_j as 0 for $t < \tau$ and consider the solution w_j of the problem

$$\begin{cases} \Box w_j = 0 & \text{in } \mathbb{R} \times \mathring{\Omega}, \\ w_j + h_j = 0 & \text{on } \mathbb{R} \times \partial\Omega, \\ w_j = 0 & \text{for } t < \tau. \end{cases}$$

We have $(\partial/\partial t)(w_j + \tilde{v}_j)_{|\mathbb{R}^+_\tau \times \partial\Omega} = 0$ and we are going to study the integrals

$$I_{j,\delta}(\lambda) = \int_{\mathbb{R}} \int_{\partial\Omega} e^{i\lambda(t - \langle x, \theta\rangle)} \rho_\delta\big(\langle x, \theta\rangle - t + t_0\big) \left(\frac{\partial}{\partial\nu} - \langle\nu, \theta\rangle \frac{\partial}{\partial t}\right) \tilde{v}_j \, dt \, dS_x,$$

$$J_{j,\delta}(\lambda) = \int_{\mathbb{R}} \int_{\partial\Omega} e^{i\lambda(t - \langle x, \theta\rangle)} \rho_\delta\big(\langle x, \theta\rangle - t + t_0\big) \left(\frac{\partial}{\partial\nu} - \langle\nu, \theta\rangle \frac{\partial}{\partial t}\right) w_j \, dt \, dS_x.$$

This study is based on certain information about the generalized wave front set

$$WF_b(v) \subset T^*(\mathbb{R} \times \mathring{\Omega}) \cup T^*(\mathbb{R} \times \partial\Omega) = \tilde{T}^*(\mathbb{R} \times \Omega),$$

where the map $\sim$ is the one introduced in Section 3 (see [20] for the properties of $WF_b(u)$). For $x \in \partial\Omega$ we have

$$\sim \; : T^*(\mathbb{R} \times \Omega) \ni (t, x, \tau, \xi) \longrightarrow (t, x, \tau, \xi_{|T_x(\partial\Omega)}) \in T^*(\mathbb{R} \times \partial\Omega).$$

The crucial step in the analysis of $I_{j,\delta}(\lambda)$ and $J_{j,\delta}(\lambda)$ is the following.

PROPOSITION 5.2. *Set* $T_1 = \rho_0 + |t_0| + 1$ *and suppose that there exists* $\eta > 0$ *such that*

$$WF_b(w_j) \cap \{\mu \in \tilde{T}^*(\mathbb{R} \times \Omega) : \mu =_\sim (t, x, 1, -\theta), T_1 + \eta \leq t \leq T_1 + 2\eta\} = \varnothing,$$

$$WF_b(\tilde{v}_j) \cap \{\mu \in \tilde{T}^*(\mathbb{R} \times \Omega) : \mu =_\sim (t, x, 1, -\theta), T_1 + \eta \leq t \leq T_1 + 2\eta\} = \varnothing.$$

Then

$$I_{j,\delta}(\lambda) = O(|\lambda|^{-m}), \; J_{j,\delta}(\lambda) = O(|\lambda|^{-m}) \quad \text{for all } m \in \mathbb{N}.$$

A similar argument can be applied in case (iii), which completes the proof of Theorem 5.1.   □

While in general the relation (5–1) is not an equality, it turns out that there exists a set $\mathcal{R}$ of full measure in $\mathbf{S}^{n-1} \times \mathbf{S}^{n-1}$ such that for $(\omega, \theta) \in \mathcal{R}$ the Poisson relation becomes an equality. This is rather important for some inverse scattering problems.

It is proved in [27] that for each $T > 0$, $S^*(\Omega)$ can be represented as a countable union of Borel subsets S_i such that on each S_i , $\{\mathcal{F}_t\}_{0 \leq t \leq T}$ coincides with the restriction of an one-parameter family $\mathcal{G}_t^{(i)}$ of Lipschitz maps defined in a neighborhood of S_i in $\dot{T}^*(\Omega)$, taking values in $T^*(\mathbb{R}^n)$ and such that for all but finitely many t, $\mathcal{G}_t^{(i)}$ is smooth and its restriction to smooth local cross-sections is a contact transformation. As a consequence of this regularity property one gets the following.

THEOREM 5.3 [27]. *The generalized geodesic flow $\mathcal{F}_t$ preserves the Hausdorff dimension of Borel subsets of $S^*(\Omega)$.*

This would have been a trivial fact if the maps $\mathcal{F}_t$ were Lipschitz. However, it is well-known and easy to see that this not the case. Locally near a point $\rho \in S^*(\Omega)$, the map $\mathcal{F}_t$ is Lipschitz on a neighborhood of ρ for small $|t|$ when $\rho \notin S_{\partial K}^*(\Omega)$ or ρ is a transversal reflection point. Whenever $\rho \in G$, the map $\mathcal{F}_t$ is not Lipschitz (see [20] or [10, vol. III]). For example, in the simplest case of a diffractive tangent point $\rho \in G_d$, the map $\mathcal{F}_t$ has a singularity of "square root type" at ρ, so it is clearly not Lipschitz.

Let $\Gamma : I \to S^*(\Omega)$ be a generalized geodesic in Ω. We say that Γ is *gliding* on ∂K if the set of those $t \in I$ such that $\Gamma(t) \in G_g$ is dense in I. In this case the trajectory $\{\Gamma(t) : t \in I\}$ is called a *gliding segment* on ∂K.

Given $T > 0$, denote by $\mathcal{T}_T$ *the set of those $\rho \in S^*(\Omega)$ such that $\{\mathcal{F}_t(\rho) : 0 \leq t \leq T\} \cap G_g \neq \varnothing$,* that is, the trajectory $\{\mathcal{F}_t(\rho) : 0 \leq t \leq T\}$ contains a nontrivial gliding segment on ∂K.

LEMMA 5.4. ([27]) *Let $\mathcal{L}_0$ be an isotropic submanifold of $S^*(\Omega) \backslash S_{\partial K}^*(\Omega)$ of dimension $n-1$ such that $H_p(\rho)$ is not tangent to $\mathcal{L}_0$ at any $\rho \in \mathcal{L}_0$. Then for every $T > 0$ we have $\dim_H \mathcal{F}_T(\mathcal{T}_T \cap \mathcal{L}_0) \leq n-2$. Moreover, if for a given T we have $\mathcal{F}_T(\mathcal{L}_0) \subset S^*(\Omega) \backslash S_{\partial K}^*(\Omega)$, then there exists a countable family $\{\mathcal{I}_m\}$ of smooth $(n-2)$-dimensional isotropic submanifolds of $S^*(\Omega)$ such that $\mathcal{F}_T(\mathcal{T}_T \cap \mathcal{L}_0) \subset \bigcup_m \mathcal{I}_m$.*

Using Theorems 3.1, 4.1, 4.3, 5.1 and Lemma 5.4, one obtains:

THEOREM 5.5 [27]. *There exists a subset $\mathcal{R}$ of full Lebesgue measure in $\mathbf{S}^{n-1} \times \mathbf{S}^{n-1}$ such that for each $(\omega, \theta) \in \mathcal{R}$ the only (ω, θ)-rays in Ω are reflecting (ω, θ)-rays and*

$$\text{sing supp } s(t, \theta, \omega) = \{-T_\gamma : \gamma \in \mathcal{L}_{\omega, \theta}(\Omega)\}.$$

SKETCH OF PROOF. It follows from the results of Melrose and Sjöstrand [20] (see also Theorem 24.3.9 in [10], vol. III) that every (ω, θ)-ray γ in Ω that does not contain gliding segments is a reflecting (ω, θ)-ray, that is, it consists of finitely many straight line segments in Ω (see Section 3).

We will show that there exists a subset $\mathcal{R}$ of full Lebesgue measure in $\mathbf{S}^{n-1} \times \mathbf{S}^{n-1}$ such that for each $(\omega, \theta) \in \mathcal{R}$ the only (ω, θ)-rays in Ω are reflecting (ω, θ)-rays.

As before, denote by $U_0 = \{x \in \mathbb{R}^n : |x| < \rho_0\}$ an open ball in $\mathbb{R}^n$ containing the obstacle K and let C be the boundary sphere of $\mathcal{U}_0$. Fix $\omega \in \mathbf{S}^{n-1}$, $x_0 \in C$ and consider the generalized geodesic $(x(t), \xi(t)) = \mathcal{F}_t(x_0, \omega)$. Let $T > 0$ be such that $x(T) \in C$. Set

$$S_0 = \big\{(x, \xi) \in S^*(\Omega) : x \in C, \ \xi \text{ is transversal to } C\big\}.$$

Since $\Sigma = p^{-1}(0) = S^*(\Omega)$, using the notation $S_C^*(\Omega) = \big\{(x,\xi) \in S^*(\Omega) : x \in C\big\}$, we have

$$S_0' = S_0 \cap \Sigma = \big\{(x, \xi) \in S_C^*(\Omega) : \xi \text{ is transversal to } C\big\}.$$

Then S_0' is a symplectic submanifold of S. Let $\mathcal{P} : S_0 \to S_0$ be the local map defined in a neighborhood of (x_0, ω) using the shift along the flow $\mathcal{F}_t$; then $\mathcal{P}(S_0') \subset S_0'$. Consider the Lagrangian submanifold

$$\mathcal{L}_0 = \big\{(x, \xi) \in S_0' : \xi = \omega\big\}$$

of S_0'. Setting $\mathcal{T} = \mathcal{T}_T$ and applying Lemma 5.4 to $\mathcal{L}_0$ gives that $\mathcal{F}_T(\mathcal{L}_0 \cap \mathcal{T})$ is contained in a countable union of isotropic $(n-2)$-dimensional submanifolds of S. Since locally near (x_0, ω) the map $\mathcal{F}_T : S_0 \to \mathcal{F}_T(S_0)$ is smooth, $\mathcal{F}_T(S_0)$ is a $(2n-1)$-dimensional submanifold of S transversal to the flow $\mathcal{F}_t$ at $\mathcal{F}_T(x_0, \omega)$. Consequently, locally near $\mathcal{F}_T(x_0, \omega) \in \mathcal{F}_T(S_0) \cap S_0$ the shift $\mathcal{Q}$ along $\mathcal{F}_t$ from $\mathcal{F}_T(S_0)$ to S_0 (forwards or backwards) is a smooth map. Moreover $\mathcal{Q}$ maps $\mathcal{F}_T(S_0')$ into S_0' (since $p^{-1}(0)$ is invariant under the flow $\mathcal{F}_t$), the restriction $\mathcal{Q} : \mathcal{F}_T(S_0') \to S_0'$ is a local symplectic map, and $\mathcal{P} = \mathcal{Q} \circ \mathcal{F}_T$. Hence the set $\mathcal{P}(\mathcal{L}_0 \cap \mathcal{T}) = \mathcal{Q}(\mathcal{F}_T(\mathcal{L}_0 \cap \mathcal{T}))$ is contained in a countable union of isotropic $(n-2)$-dimensional submanifolds of S. The projection $j : S_0' \to \mathbf{S}^{n-1}$, $j(x, \xi) = \xi$, is smooth, so Sard's theorem gives now that the set $j(\mathcal{P}(\mathcal{L}_0 \cap \mathcal{T}))$ has Lebesgue measure zero in $\mathbf{S}^{n-1}$. Hence there exists a neighborhood U of x_0 in C and a subset $\mathcal{R}_\omega(U) = \mathbf{S}^{n-1} \setminus j(\mathcal{P}(\mathcal{L} \cap \mathcal{T}))$ of full Lebesgue measure in $\mathbf{S}^{n-1}$ such that for $x \in U$ every generalized (ω, θ)-ray in Ω passing through x with $\theta \in \mathcal{R}_\omega(U)$ is a reflecting (ω, θ)-ray. Covering C by a finite family of neighborhoods U_i, we find a subset $\mathcal{R}_\omega = \bigcap_i \mathcal{R}_\omega(U_i)$ of full Lebesgue measure in $\mathbf{S}^{n-1}$ such that every (ω, θ)-ray in Ω with $\theta \in \mathcal{R}_\omega$ is a reflecting (ω, θ)-ray. It now follows from Fubini's theorem that

$$\mathcal{R}' = \big\{(\omega, \theta) \in \mathbf{S}^{n-1} \times \mathbf{S}^{n-1} : \theta \in \mathcal{R}_\omega\big\}$$

is a subset of full Lebesgue measure in $\mathbf{S}^{n-1} \times \mathbf{S}^{n-1}$. Moreover it is clear that for $(\omega, \theta) \in \mathcal{R}'$, all (ω, θ)-rays in Ω are reflecting ones.

According to Theorems 4.1 and 4.3 above, there exists a subset $\mathcal{R}'' = \mathcal{T} \cap S$ of full Lebesgue measure in $\mathbf{S}^{n-1} \times \mathbf{S}^{n-1}$ such that for $(\omega, \theta) \in \mathcal{R}''$ every reflecting (ω, θ)-ray in Ω has no tangencies to ∂K and $T_\gamma \neq T_\delta$ whenever γ and δ are different reflecting (ω, θ)-rays in Ω. Then $\mathcal{R} = \mathcal{R}' \cap \mathcal{R}''$ has full Lebesgue measure in $\mathbf{S}^{n-1} \times \mathbf{S}^{n-1}$. Given $(\omega, \theta) \in \mathcal{R}$, it follows from Theorem 3.1 that $-T_\gamma \in \operatorname{sing\,supp} s(t, \theta, \omega)$ for all $\gamma \in \mathcal{L}_{\omega, \theta}(\Omega)$. Combining this with Theorem 5.1 completes the proof of the theorem. $\qquad \square$

Using Theorem 5.5 we will now derive a simple but rather important property of obstacles ([12]; see also [27, Proposition 2.3]): most rays incoming from infinity

are not trapped by the obstacle K. Here it is essential that we consider points in the set
$$S_C^*(\Omega) = \{(x,\xi) \in S^*(\Omega) : x \in C\},$$
where C as before is the boundary sphere of an open ball U_0 containing K. In general it is not true that the trapped points $(x,\xi) \in S^*(\Omega_K)$ with x near K form a set of Lebesgue measure zero in $S^*(\Omega_K)$. Example 7.1 below, due to M. Livshitz, shows that in some cases the set of trapped points may even contain a nontrivial open subset of $S^*(\Omega_K)$.

PROPOSITION 5.6.   *The set of those* $(x,\xi) \in S_C^*(\Omega)$ *such that the trajectory* $\{\mathcal{F}_t(x,\xi) : t \geq 0\}$ *is bounded has Lebesgue measure zero in* $S_C^*(\Omega)$.

PROOF.   For $(x,\omega) \in S_C^*(\Omega)$, let $\delta(x,\omega)$ be the generalized geodesic in Ω_K issued from x in direction ω. Assume that there exists a subset W of positive Lebesgue measure in $S_C^*(\Omega)$ such that $\delta(x,\omega) \subset \mathcal{U}_0$ for all $(x,\omega) \in W$. According to Theorem 4.3 and to an argument from the proof of Theorem 5.5 above (or using Lemma 5.4 directly), we may assume that for all $(x,\omega) \in W$ the generalized geodesic $\delta(x,\omega)$ does not contain gliding segments on ∂K and has only transversal reflections at ∂K. Given $(x,\omega) \in W$, denote by x' the first common point of $\delta(x,\omega)$ with ∂K and by ω' the reflected direction of $\delta(x,\omega)$ at x', i.e. $\omega' = \omega - 2\langle\omega, \nu(x')\rangle\nu(x')$, where $\nu(x')$ is the outer unit normal to K at x'. Then the set $W' = \{(x',\omega') \in S_{\partial K}^*(\Omega) : (x,\omega) \in W\}$ is a subset of positive Lebesgue measure in $S_{\partial K}^*(\Omega)$.

Denote by $M \subset S_{\partial K}^*(\Omega)$ the set of those $(y,\eta) \in S_{\partial K}^*(\Omega)$ for which the standard billiard ball map B is well-defined. The map B (as a local map) preserves the so-called Liouville's measure μ on M which is absolutely continuous with respect to the Lebesgue measure on $S_{\partial K}^*(\Omega)$.

Next, we use the argument from the proof of the Poincaré Recurrence Theorem in ergodic theory. It follows from the definition of W' that $B^k(W') \subset M$ and $\mu(B^k(W')) = \mu(W') > 0$ for all $k = 0, 1, 2, \ldots$. On the other hand, in the situation under consideration we clearly have $\mu(\bigcup_{k=0}^\infty B^k(W')) < \infty$. Therefore there exist nonnegative integers $k < m$ with $B^k(W') \cap B^m(W') \neq \varnothing$. Since B is invertible, this means that there exists $(x',\omega') \in W' \cap B^{m-k}(W')$. Then $(x',\omega') = B(y,\eta)$ for some $(y,\eta) \in B^{m-k-1}(W') \subset M$. Now the choice of W and the definition of W' show that W' has no common points with $B(M)$. This is a contradiction which proves the proposition.   □

6. Existence of Scattering Rays with Sojourn Times Tending to Infinity

In this section we study the existence of (ω, θ)-rays for trapping obstacles. The image $S_b^*(\Omega) = \sim(S^*(\Omega))$ of the characteristic set $S^*(\Omega)$ is called the *compressed characteristic set* and the image $\tilde{\gamma} =\sim (\gamma)$ of a generalized bicharacteristic defined in Section 3 is called a *compressed generalized bicharacteristic*.

Let again U_0 be an open ball containing K and C be its boundary sphere. Given a point $z = (x, \xi) \in S_b^*(\Omega)$, consider the compressed generalized bicharacteristic

$$\gamma_z(t) = \big(x(t), \xi(t)\big) \in S_b^*(\Omega)$$

parametrized by the time t and passing through z for $t = 0$. Denote by $T(z) \in \mathbb{R}^+ \cup \infty$ the maximal $T > 0$ such that $x(t) \in U_0$ for $0 \le t \le T(z)$. We introduce the *trapping set*

$$\Sigma_\infty = \big\{ (x, \xi) \in S_b^*(\Omega) : x \in C, T(z) = \infty \big\}.$$

It follows from the continuity of the generalized Hamiltonian flow that Σ_∞ is closed in Σ. The obstacle K is called *trapping* if $\Sigma_\infty \ne \varnothing$. We have the following.

THEOREM 6.1 [23]. *Let the obstacle K be trapping and satisfy the condition $(\mathcal{G})$. Then there exists a sequence of ordinary reflecting nondegenerate scattering rays γ_m with sojourn times $T_{\gamma_m} \to \infty$.*

PROOF. It is easy to see that $\Sigma_\infty \ne S_b^*(\Omega)$, hence the boundary $\partial \Sigma_\infty$ of Σ_∞ in $S_b^*(\Omega)$ is not empty. Take a point $\hat{z} \in \partial \Sigma_\infty$. Since $S_b^*(\Omega) \setminus \Sigma_\infty \ne \varnothing$, there exists a sequence $z_m = (x_m, \xi_m) \in S_b^*(\Omega), x_m \in C$, such that $z_m \notin \Sigma_\infty$ for all m and $z_m \to \hat{z}$. Consider the compressed generalized bicharacteristics $\gamma_{z_m}(t) = (z_m, \xi_m(t))$ passing through z_m for $t = 0$ and such that $T(z_m) < \infty$. The sequence $\{T(z_m)\}$ is unbounded, since otherwise we will have $T(\hat{z}) < \infty$ in contradiction with $\hat{z} \in \Sigma_\infty$. Thus we may assume that $\lim_{m \to \infty} T(z_m) = +\infty$. Set $y_m = x_m(T(z_m)) \in C$, $\omega_m = \xi_m(T(z_m)) \in \mathbf{S}^{n-1}$. Taking a subsequence, we may assume that $y_m \to u \in C$ and $\omega_m \to \omega \in \mathbf{S}^{n-1}$. For the generalized bicharacteristics $\gamma_\mu(t) = (y(t), \xi(t))$ issued from $\mu = (u, \omega)$ we have $T(\mu) = \infty$ and $y(t) \in U_0$ for $t \ge 0$.

Let Z_ω be the hyperplane passing through u and orthogonal to ω and let Z_∞ be the set of those points $y \in Z_\omega$ for which the generalized bicharacteristic γ_{μ_y} passing through $\mu_y = (y, \omega)$ has the property $T(\mu_y) = \infty$. The set Z_∞ is closed in Z_ω, $Z_\omega \ne \varnothing$ and $Z_\infty \ne Z_\omega$. Thus there exists a sequence of points $u_m \to y_0$ for some $y_0 \in Z_\omega$ with $u_m \in Z_\omega \setminus Z_\infty$ such that $T(\mu_{u_m}) < \infty$ for all m and $T(\mu_{u_m}) \to \infty$. Applying Proposition 5.6, we can approximate γ_{u_m} by ordinary reflecting rays γ_{δ_m} with sojourn times going to infinity and by a second approximation we may choose the ordinary reflecting rays γ_{δ_m} to be nondegenerate.

Now consider a fixed ordinary reflecting (ω_m', θ_m')-ray with sojourn time T_m which is nondegenerate. In general it is possible to have other (generalized) (ω_m', θ_m')-rays with the same sojourn time and T_m could be a nonisolated point in sing supp $s(t, \omega_m', \theta_m')$. Let $\mathcal{A} \subset \mathbf{S}^{n-1} \times \mathbf{S}^{n-1}$ be the set introduced at the end of Section 4 and let $\mathcal{R} \subset \mathbf{S}^{n-1} \times \mathbf{S}^{n-1}$ be the set of Theorem 5.5. Let

$$\Xi = \mathcal{R} \cap \mathcal{A} \subset \mathbf{S}^{n-1} \times \mathbf{S}^{n-1}.$$

Then for $(\omega, \theta) \in \Xi$ each (ω, θ)-ray is ordinary reflecting and nondegenerate. By applying the inverse mapping theorem, it is easy to see that we may approximate (ω_m', θ_m') by a pair $(\omega_m'', \theta_m'') \in \Xi$ sufficiently close to (ω_m', θ_m') so that there exist ordinary reflecting nondegenerate (ω_m'', θ_m'')-rays with sojourn times $T_m'' \to \infty$ (see [23] for more details). $\square$

The sojourn times T_m'' are isolated points in sing supp $s(t, \omega_m'', \theta_m'')$ and the argument of Section 3 based on (3–2) implies that following.

THEOREM 6.2. *Under the assumptions of Theorem 6.1 there exists a sequence* $(\omega_m, \theta_m) \in \mathbf{S}^{n-1} \times \mathbf{S}^{n-1}$ *and ordinary reflecting nondegenerate* (ω_m, θ_m)-rays with *sojourn times* $T_m \to \infty$ *so that*

$$-T_m \in \text{sing supp } s(t, \omega_m, \theta_m) \quad \text{for all } m \in \mathbb{N}. \tag{6-1}$$

Relation (6–1) was called property (S) in [24], and there we conjectured that every trapping obstacle has the property (S). The above result shows that for generic obstacles this conjecture is true. Moreover, the above argument implies that for each $m \in \mathbb{N}$ there exists a set $\Pi_m \subset \mathbf{S}^{n-1} \times \mathbf{S}^{n-1}$ with positive measure $\varepsilon_m > 0$ so that the (ω, θ)-rays with $(\omega, \theta) \in \Pi_m$ produce singularities $-\tau_m \le -m$ of the scattering kernel $s(t, \omega, \theta)$. Thus for obstacles satisfying (S) some sojourn times can be observed after a sufficiently long time.

The property (S) leads to some interesting results concerning the behavior of the modified resolvent of the Laplacian [23]. For Im $\lambda > 0$ consider the outgoing resolvent $R(\lambda) = (-\Delta - \lambda^2)^{-1}$ of the Laplacian in Ω with Dirichlet boundary conditions on $\partial \Omega$. The outgoing condition means that for $f \in C_0^\infty(\Omega)$ there exists $g(x) \in C_0^\infty(\mathbb{R}^n)$ so that we have

$$R(\lambda) f(x) = R_0(\lambda) g(x) \quad \text{as } |x| \to \infty,$$

where

$$R_0(\lambda) = (-\Delta - \lambda^2)^{-1} : L^2_{\text{comp}}(\mathbb{R}^n) \to H^2_{\text{loc}}(\mathbb{R}^n)$$

is the outgoing resolvent of the free Laplacian in $\mathbb{R}^n$ related to the outgoing Green function introduced in Section 2. The operator

$$R(\lambda) : L^2_{\text{comp}}(\Omega) \ni f \to R(\lambda) f \in H^2_{\text{loc}}(\Omega)$$

has a meromorphic continuation in $\mathbb{C}$ with poles λ_j such that Im $\lambda_j < 0$, called *resonances* ([12], [25]). Let $\chi_1(x), \chi_2(x) \in C_0^\infty(\mathbb{R}^n)$ be cutoff functions such that $\chi_1(x) = \chi_2(x) = 1$ on a neighborhood of K and $\chi_1(x) = 1$ on supp $\chi_2(x)$. It is easy to see that the *modified resolvent*

$$\tilde{R}(\lambda) = \chi_1 R(\lambda) \chi_2$$

has a meromorphic continuation in $\mathbb{C}$. The poles of $\tilde{R}(\lambda)$ are independent of the choice of χ_i and they coincide with their multiplicities with those of the resonances (see [12], [25]). On the other hand, the scattering amplitude $a(\lambda, \omega, \theta)$ also admits a meromorphic continuation in $\mathbb{C}$ and the poles of this continuation and their multiplicities are the same as those of the resonances (see [12]). From the general results on propagation of singularities ([20]) it follows that if K is nontrapping, there exist $\varepsilon > 0$ and $d > 0$ so that $\tilde{R}(\lambda)$ has no poles in the domain

$$U_{\varepsilon,d} = \{\lambda \in \mathbb{C} : d - \varepsilon \log(1 + |\lambda|) \leq \operatorname{Im} \lambda \leq 0\}.$$

For trapping obstacles we expect to have poles in all domains $U_{\varepsilon,d}$. For the moment this is an open problem and we have a weaker result.

THEOREM 6.3 [23]. *Assume that there exists a sequence of ordinary reflecting* (ω_m, θ_m)*-rays in* Ω *with sojourn times* $T_m \to \infty$. *Let* $\Phi \in C_0^\infty(\mathbb{R})$ *be such that* $supp\ \Phi \subset (-1, 1)$ *and* $\Phi(t) = 1$ *for* $|t| \leq \frac{1}{2}$. *Assume that there exists a sequence* $\gamma_m \to 0$ *of nonzero real numbers and an integer k independent on m such that*

$$\left| \mathcal{F}_{t \to \lambda} \left(\Phi\left(\frac{t + T_m}{\gamma_m} \right) s(t, \omega_m, \theta_m) \right) \right| \geq (c_m - o_m(1)) |\lambda|^k \quad as\ |\lambda| \to \infty,$$

where $c_m > 0$. *Then there are two possibilities:*

(i) *For each $\varepsilon > 0$ and each $d > 0$, the modified resolvent $\tilde{R}(\lambda)$ has poles in the domain $U_{\varepsilon,d}$.*

(ii) *For some $\varepsilon > 0$ and $d > 0$ the modified resolvent $\tilde{R}(\lambda)$ is holomorphic in $U_{\varepsilon,d}$ but for all $\alpha \geq 0, p \in \mathbb{N}, k \in \mathbb{N}$ we have*

$$\sup_{\substack{\lambda \in U_{\varepsilon,d} \\ \|\varphi\|_{H^k(\Omega)} = 1}} (1 + |\lambda|)^{-p} e^{-\alpha |\operatorname{Im} \lambda|} \|\tilde{R}(\lambda)\varphi\|_{H^1(\Omega)} = +\infty.$$

It is natural to make the conjecture that under the assumptions of Theorem 6.3, condition (i) always takes place.

7. Rigidity of the Scattering Length Spectrum

Fix again a large open ball U_0 in $\mathbb{R}^n$, $n \geq 3$, n odd[2], and let $C = \partial U_0$. Throughout this section we consider obstacles K in $\mathbb{R}^n$ contained in U_0 with smooth boundaries ∂K that satisfy the condition $(\mathcal{G})$ from Section 3 and such that $\gamma_K(\sigma)$ is a nondegenerate simply reflecting ray for almost all $\sigma \in S_C^*(\Omega)$ such that $\gamma_K(\sigma) \cap \partial K \neq \varnothing$. Denote by $\mathcal{K}_0$ the class of obstacles with these properties. One can derive from [22] (see Chapter 3 there) that $\mathcal{K}_0$ is of second Baire category (with respect to the C^∞ Whitney topology; see [8]) in the class of all obstacles with smooth boundaries.

Since in this section we deal with more than one obstacle, it is convenient to replace the notation Ω, $\mathcal{F}_t$, $s(t, \omega, \theta)$, $\dot{T}_b^*(\Omega)$ and $S_b^*(\Omega)$ used so far (see Section 3 for the latter two) by Ω_K, $\mathcal{F}^{(K)}_t$, $s_K(t, \omega, \theta)$, $\dot{T}_b^*(\Omega_K)$ and $S_b^*(\Omega_K)$, respectively.

A point $\sigma = (x, \omega) \in \dot{T}^*(\Omega_K)$ is called a *trapped point* if at least one of the curves $\{\operatorname{pr}_1(\mathcal{F}^{(K)}_t(\sigma)) : t \leq 0\}$ and $\{\operatorname{pr}_1(\mathcal{F}^{(K)}_t(\sigma)) : t \geq 0\}$ in Ω_K is bounded.

---

[2]In fact, most of the considerations in this section are purely geometrical and apply also in the case when n is even, $n \geq 2$.

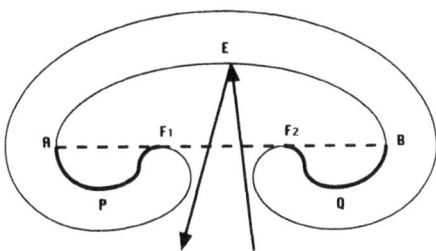

Figure 4. Livshits' example. Adapted from [18, Chapter 5].

Here we use the notation $\text{pr}_1(y, \eta) = y$ and $\text{pr}_2(y, \eta) = \eta$. Denote by $\text{Trap}\,\Omega_K$ the *set of all trapped points* in $\dot{T}^*(\Omega_K)$. Notice that the set Σ_∞ used in Section 6 coincides with $\text{Trap}\,\Omega_K \cap S_C^*(\Omega_K)$. It is easy to see that $\Sigma_\infty \neq \varnothing$ if and only if $\text{Trap}\,\Omega_K \neq \varnothing$. So, if $\text{Trap}\,\Omega_K = \varnothing$, then K is a nontrapping obstacle. It is known for example that all star-shaped obstacles are nontrapping.

The *scattering length spectrum* (SLS) of K is by definition the family of sets of real numbers $SL_K = \{SL_K(\omega, \theta)\}_{(\omega, \theta)}$ where (ω, θ) runs over $\mathbf{S}^{n-1} \times \mathbf{S}^{n-1}$ and $SL_K(\omega, \theta)$ is the set of sojourn times T_γ of all (ω, θ)-rays γ in Ω_K. Thus, SL_K is a map which assigns to each pair of directions (ω, θ) a set $SL_K(\omega, \theta)$ of real numbers.

In this section we discuss the problem of recovering information about the geometry of the obstacle K from its SLS. Two obstacles K and L in $\mathbb{R}^n$ are said to have *almost the same SLS* if there exists a subset $\mathcal{R}$ of full Lebesgue measure in $\mathbf{S}^{n-1} \times \mathbf{S}^{n-1}$ such that $SL_K(\omega, \theta) = SL_L(\omega, \theta)$ for all $(\omega, \theta) \in \mathcal{R}$. We will say that a property P of obstacles in $\mathbb{R}^n$ can be recovered by the SLS of the obstacle if whenever K and L have almost the same SLS and K has property P, then L has property P as well.

It follows from results of A. Majda [15] (see also Majda and Ralston [16]) and P. Lax and R. Phillips [13] that the convex hull $\hat{K}$ of K can be recovered from SL_K. Consequently, in the class of convex obstacles and also in the class of connected obstacles with real analytic boundaries, K is completely determined by its SLS.

EXAMPLE 7.1. The following example of M. Livshits (Chapter 5 in [18]) shows that in general SL_K does not determine K uniquely. Here the part E is half an ellipse with foci F_1 and F_2. The ellipse has the property that any ray intersecting the segment connecting the foci, after reflection at the boundary, intersects the same segment again. It is now clear that no scattering ray in the exterior of the obstacle K has a common point with the parts P and Q, so these two "pockets" cannot be recovered from the SLS of the obstacle. It should be mentioned that this example is in $\mathbb{R}^2$ and no examples like this in higher dimensions are known to the authors.

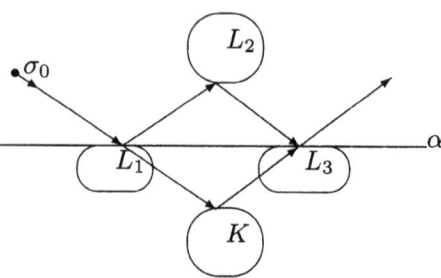

Figure 5.

The problem considered at the beginning of this section is of a global nature. The following simple example shows that in the corresponding local problem there is no uniqueness (unless possibly some nondegeneracy conditions are imposed).

EXAMPLE 7.2. Consider two obstacles K and $L = L_1 \cup L_2 \cup L_3$ in $\mathbb{R}^n$, $n \geq 2$, as shown in Figure 5. Here K and L_2 are (strictly) convex domains, while L_1 and L_3 are convex domains. Moreover K and L_2 are symmetric with respect to the hyperplane α containing the flat "top parts" of ∂L_1 and ∂L_3. The rays on the figure are generated by some σ_0 (far from K and L). For any σ close to σ_0 we have $\mathcal{F}^{(K)}{}_t(\sigma) = \mathcal{F}^{(L)}{}_t(\sigma)$ for $t \gg 0$ and both trajectories have common points with the corresponding obstacles (and are nondegenerate). On the other hand, $K \cap L = \varnothing$. It should be mentioned however that the obstacles K and L in this example do not satisfy the condition $\mathcal{G}$. Whether such examples exist with K and L satisfying $\mathcal{G}$ is an open problem.

It turns out that if two obstacles K and L have almost the same SLS, then their generalized geodesic flows are conjugate with a time preserving conjugacy on the nontrapping parts of their phase spaces.

THEOREM 7.3 [29]. *If the obstacles* $K, L \in \mathcal{K}_0$ *have almost the same SLS, then there exists a homeomorphism*

$$\Phi : \dot{T}^*_b(\Omega_K) \setminus \operatorname{Trap}\Omega_K \to \dot{T}^*_b(\Omega_L) \setminus \operatorname{Trap}\Omega_L$$

with the following properties:

(i) Φ *defines a symplectic map on an open dense subset of* $\dot{T}^*_b(\Omega_K) \setminus \operatorname{Trap}\Omega_K$;
(ii) Φ *maps* $S^*_b(\Omega_K) \setminus \operatorname{Trap}\Omega_K$ *onto* $S^*_b(\Omega_L) \setminus \operatorname{Trap}\Omega_L$;
(iii) $\mathcal{F}^{(L)}{}_t \circ \Phi = \Phi \circ \mathcal{F}^{(K)}{}_t$ *for all* $t \in \mathbb{R}$;
(iv) $\Phi(x, \xi) = (x, \xi)$ *for any* $(x, \xi) \in \dot{T}^*_b(\Omega_K) \setminus \operatorname{Trap}\Omega_K = \dot{T}^*_b(\Omega_L) \setminus \operatorname{Trap}\Omega_L$ *such that* $x \notin U_0$.

Conversely, it is not difficult to show that if $K, L \in \mathcal{K}_0$ are two obstacles for which there exists a homeomorphism $\Phi : S^*_b(\Omega_K) \setminus \operatorname{Trap}\Omega_K \to S^*_b(\Omega_L) \setminus \operatorname{Trap}\Omega_L$ such that $\mathcal{F}^{(L)}{}_t \circ \Phi = \Phi \circ \mathcal{F}^{(K)}{}_t$ for all $t \in \mathbb{R}$ and $\Phi = \operatorname{id}$ on $S^*(\mathbb{R}^n \setminus U_0) \setminus \operatorname{Trap}\Omega_K$, then K and L have the same SLS ([29]).

There is a clear analogy between the property described above and the *lens equivalence* of geodesic flows on Riemannian manifolds without boundary (see [3] and the references there).

SKETCH OF PROOF OF THEOREM 7.3. Assume that the obstacles K and L have almost the same SLS. The existence of the conjugacy Φ follows easily from the following main lemma.

LEMMA 7.4. *Assume that $\sigma \in S^*(\mathbb{R}^n \setminus U_0)$ and $t \in \mathbb{R}$ with $\mathcal{F}^{(K)}{}_t(\sigma) \in S^*(\mathbb{R}^n \setminus U_0)$. Then $\mathcal{F}^{(K)}{}_t(\sigma) = \mathcal{F}^{(L)}{}_t(\sigma)$.*

Given $\sigma \in \dot{T}^*(\Omega) \setminus \operatorname{Trap} \Omega_K$, take $t \in \mathbb{R}$ so large that $\mathcal{F}^{(K)}{}_t(\sigma) \in S^*(\mathbb{R}^n \setminus U_0)$. Then define $\Phi(\sigma) = \mathcal{F}^{(L)}{}_{-t} \circ \mathcal{F}^{(K)}{}_t(\sigma)$. It follows from the above lemma that the definition of Φ is correct and moreover $\mathcal{F}^{(L)}{}_t \circ \Phi = \Phi \circ \mathcal{F}^{(K)}{}_t$ for all $t \in \mathbb{R}$ and $\Phi(\sigma) = \sigma$ for $\sigma \in \dot{T}^*(\mathbb{R}^n \setminus U_0) \setminus \operatorname{Trap} \Omega_K$. Clearly Φ is a homeomorphism and it follows from the properties of the generalized geodesic flows ([20]) that it is a symplectic map on an open dense subset of $\dot{T}_b^*(\Omega_K) \setminus \operatorname{Trap} \Omega_K$. This shows how Theorem 7.3 is derived from Lemma 7.4.

PROOF OF LEMMA 7.4. Fix for a moment an arbitrary $(\omega_0, \theta_0) \in \mathbf{S}^{n-1} \times \mathbf{S}^{n-1}$, and let δ be a nondegenerate simply reflecting (ω_0, θ_0)-ray in Ω_K with reflection points $x_1, \ldots, x_k$ $(k \geq 1)$ and δ' is a nondegenerate simply reflecting (ω_0, θ_0)-ray in Ω_L with reflection points $y_1, \ldots, y_m$ $(m \geq 1)$. Using the nondegeneracy of δ and the Inverse Mapping Theorem one derives the existence of a neighborhood U of (ω_0, θ_0) in $\mathbf{S}^{n-1} \times \mathbf{S}^{n-1}$ such that for each $(\omega, \theta) \in U$ there are a unique reflecting (ω, θ)-ray $\delta(\omega, \theta)$ in Ω_K with reflection points $x_1(\omega, \theta), \ldots, x_k(\omega, \theta)$ close to $x_1, \ldots, x_k$, respectively, and a unique reflecting (ω, θ)-ray $\delta'(\omega, \theta)$ in Ω_L with reflection points $y_1(\omega, \theta), \ldots, y_m(\omega, \theta)$ close to $y_1, \ldots, y_m$, respectively.

LEMMA 7.5. *Under the preceding assumptions, suppose in addition that $T_{\delta(\omega,\theta)} = T_{\delta'(\omega,\theta)}$ for all $(\omega, \theta) \in U$. Then for each $(\omega, \theta) \in U$ there exist real numbers $\lambda(\omega, \theta)$ and $\mu(\omega, \theta)$ such that*

$$y_1(\omega, \theta) = x_1(\omega, \theta) + \lambda(\omega, \theta)\omega, \qquad y_m(\omega, \theta) = x_k(\omega, \theta) + \mu(\omega, \theta)\theta. \qquad (7\text{-}1)$$

PROOF. Let $(\omega, \theta) = (\omega(u), \theta(v))$, $(u, v) \in \mathbb{R}^{n-1} \times \mathbb{R}^{n-1}$, be a smooth parametrization of U and set $x_j(u, v) = x_j(\omega(u), \theta(v))$ and $y_j(u, v) = y_j(\omega(u), \theta(v))$. For the functions

$$f(u, v) = \langle \omega(u), x_1(u, v) \rangle + \sum_{i=1}^{k-1} \|x_i(u, v) - x_{i+1}(u, v)\| - \langle x_k(u, v), \theta(v) \rangle,$$

$$g(u, v) = \langle \omega(u), y_1(u, v) \rangle + \sum_{i=1}^{m-1} \|y_i(u, v) - y_{i+1}(u, v)\| - \langle y_m(u, v), \theta(v) \rangle,$$

we have $f(u,v) = g(u,v)$ for all (u,v), therefore the derivatives of these two functions coincide. A simple calculation gives

$$\frac{\partial f}{\partial u_j}(u) = \left\langle \frac{\partial \omega}{\partial u_j}, x_1 \right\rangle + \left\langle \omega, \frac{\partial x_1}{\partial u_j} \right\rangle + \sum_{i=1}^{k-1} \left\langle \frac{x_{i+1} - x_i}{\|x_{i+1} - x_i\|}, \frac{\partial x_{i+1}}{\partial u_j} - \frac{\partial x_i}{\partial u_j} \right\rangle - \left\langle \frac{\partial x_k}{\partial u_j}, \theta \right\rangle.$$

Using the notation $e_i = \dfrac{x_{i+1} - x_i}{\|x_{i+1} - x_i\|}$ and the reflection law at the points $x_1, \ldots, x_{k-1}$, we find

$$\frac{\partial f}{\partial u_j}(u) = \left\langle \frac{\partial \omega}{\partial u_j}, x_1 \right\rangle + \left\langle \omega - e_1, \frac{\partial x_1}{\partial u_j} \right\rangle + \left\langle e_1 - e_2, \frac{\partial x_2}{\partial u_j} \right\rangle + \cdots$$

$$+ \left\langle e_{k-2} - e_{k-1}, \frac{\partial x_{k-1}}{\partial u_j} \right\rangle + \left\langle e_{k-1} - \theta, \frac{\partial x_k}{\partial u_j} \right\rangle$$

$$= \left\langle \frac{\partial \omega}{\partial u_j}, x_1 \right\rangle.$$

In the same way one gets $\dfrac{\partial g}{\partial u_j} = \left\langle \dfrac{\partial \omega}{\partial u_j}, y_1 \right\rangle$. Hence

$$\left\langle \frac{\partial \omega}{\partial u_j}, x_1 \right\rangle = \left\langle \frac{\partial \omega}{\partial u_j}, y_1 \right\rangle$$

for all $j = 1, \ldots, n-1$, so $y_1 - x_1 = \lambda \omega$ for some $\lambda \in \mathbb{R}$.

Similarly, $y_m = x_k + \mu \theta$ for some $\mu \in \mathbb{R}$.   □

We continue the proof of Lemma 7.4. As usual, we denote by $\overset{\circ}{M}$ the *interior* (largest open subset) of a subset M of $\mathbb{R}^n$. Let $\mathcal{R}$ be a subset of full Lebesgue measure in $\mathbf{S}^{n-1} \times \mathbf{S}^{n-1}$ such that

$$SL_K(\omega, \theta) = SL_L(\omega, \theta), \quad (\omega, \theta) \in \mathcal{R}. \tag{7-2}$$

Shrinking $\mathcal{R}$ if necessary, we will assume that $(\omega, \omega) \notin \mathcal{R}$ for any $\omega \in \mathbf{S}^{n-1}$. Then for $(\omega, \theta) \in \mathcal{R}$, any (ω, θ)-ray in Ω_K (and in fact in the exterior of any obstacle) must have at least one reflection point. Furthermore, using Theorems 4.3, 5.1 and 5.5 above, we may assume that the set $\mathcal{R}$ is chosen in such a way that: (i) for $(\omega, \theta) \in \mathcal{R}$ all (ω, θ)-rays in Ω_K (resp. Ω_L) are nondegenerate simply reflecting (ω, θ)-rays; (ii) if $(\omega, \theta) \in \mathcal{R}$ and γ and δ are (ω, θ)-rays in Ω_K (resp. Ω_L), then $T_\gamma \neq T_\delta$.

It follows from [13] and [15] (see also [16]) that $\hat{K} = \hat{L}$.

Let $\sigma_0 = (u_0, \omega_0) \in S^*(\overset{\circ}{\Omega}_{\hat{K}})$ and $t_0 \in \mathbb{R}$ be such that $\mathcal{F}^{(K)}{}_{t_0}(\sigma_0) \notin S^*(\overset{\circ}{\Omega}_{\hat{K}})$. We will show that $\mathcal{F}^{(K)}{}_{t_0}(\sigma_0) = \mathcal{F}^{(L)}{}_{t_0}(\sigma_0)$. Using various results from [20], [23] and [29], one derives that it is enough to consider the case when σ_0 is nontrapped and $(\omega_0, \theta_0) \in \mathcal{R}$. Then $\delta = \gamma_K(\sigma_0)$ is a nondegenerate simply reflecting (ω_0, θ_0)-ray in Ω_K.

The essential case to consider is when $\gamma_K(\sigma_0) \cap \partial K \neq \varnothing$. Then there exists $s_0 \in \mathbb{R}$ with $\mathcal{F}^{(K)}{}_{s_0}(\sigma_0) = (x_0, \xi_0)$, $x_0 \in \partial K$, and without loss of generality we will assume $s_0 > 0$ and moreover that s_0 is the minimal positive number with $\mathrm{pr}_1(\mathcal{F}^{(K)}{}_{s_0}(\sigma_0)) \in \partial K$. Let $x_1 = x_0, x_2, \ldots, x_k$ be the successive reflection points of δ. According to (7–2), there exists a reflecting (ω_0, θ_0)-ray δ' in Ω_L with $T_{\delta'} = T_\delta$. Let $y_1, \ldots, y_m$ be the successive reflection points of δ'. The choice of $\mathcal{R}$ and $(\omega_0, \theta_0) \in \mathcal{R}$ imply that δ' is nondegenerate. From the latter one derives that there exist a neighborhood U of (ω_0, θ_0) in $\mathbf{S}^{n-1} \times \mathbf{S}^{n-1}$ and a neighborhood U_i of x_i in ∂K for each $i = 1, \ldots, k$ such that for every $(\omega, \theta) \in U$ there is a unique reflecting (ω, θ)-ray $\delta(\omega, \theta)$ in Ω_K with reflection points $x_1(\omega, \theta) \in U_1, \ldots, x_k(\omega, \theta) \in U_k$ smoothly depending on (ω, θ). Similarly, there exists a neighborhood U_j' of y_j in ∂L for each $j = 1, \ldots, m$ such that for every $(\omega, \theta) \in U$ there is a unique reflecting (ω, θ)-ray $\delta'(\omega, \theta)$ in Ω_L with reflection points $y_1(\omega, \theta) \in U_1', \ldots, y_m(\omega, \theta) \in U_m'$ smoothly depending on (ω, θ). Moreover $\delta(\omega_0, \theta_0) = \delta$ and $\delta'(\omega_0, \theta_0) = \delta'$.

According to (7–2), for each $(\omega, \theta) \in \mathcal{R} \cap U$ there exists a unique reflecting (ω, θ)-ray $\delta''(\omega, \theta)$ in Ω_L with

$$T_{\delta''(\omega,\theta)} = T_{\delta(\omega,\theta)}. \tag{7–3}$$

Assuming U is small enough, it then follows that $\delta''(\omega, \theta) = \delta'(\omega, \theta)$ for each $(\omega, \theta) \in \mathcal{R} \cap U$. Otherwise there exists a sequence $\{(\omega_p, \theta_p)\}_{p=1}^\infty \subset \mathcal{R} \cap U$ converging to (ω_0, θ_0) such that $\delta''(\omega_p, \theta_p) \neq \delta'(\omega_p, \theta_p)$ for all p. Let $Z = Z_{\omega_0}$. Denote by u_p the (incoming) intersection point of $\delta''(\omega_p, \theta_p)$ with Z; then $\delta''(\omega_p, \theta_p) = \gamma_L(u_p, \omega_p)$. Considering an appropriate subsequence, we may assume that $u_p \to u \in Z$ as $p \to \infty$. Then $\delta'' = \gamma_L(u, \omega_0)$ is an (ω_0, θ_0)-ray in Ω_L and clearly $T_{\delta''} = \lim_p T_{\delta''(\omega_p, \theta_p)} = T_{\delta''(\omega_0, \theta_0)}$. Now (7–3) implies $T_{\delta''} = T_{\delta(\omega_0,\theta_0)} = T_\delta$ and therefore $T_{\delta''} = T_{\delta'(\omega_0,\theta_0)} = T_{\delta'}$. This and $(\omega_0, \theta_0) \in \mathcal{R}$ give $\delta'' = \delta'$. Hence u belongs to $\delta' = \delta'(\omega_0, \theta_0)$ and therefore for large p, the ray $\delta''(\omega_p, \theta_p)$ has m reflection points belonging to the neighborhoods U_j', respectively. From the choice of U and the uniqueness of the (ω, θ)-rays $\delta'(\omega, \theta)$ for $(\omega, \theta) \in U$, it now follows that $\delta''(\omega_p, \theta_p) = \delta'(\omega_p, \theta_p)$. This is a contradiction with the choice of the sequence $\{(\omega_p, \theta_p)\}_p$ which proves that $\delta''(\omega, \theta) = \delta'(\omega, \theta)$ for all $(\omega, \theta) \in \mathcal{R} \cap U$. Hence

$$T_{\delta'(\omega,\theta)} = T_{\delta(\omega,\theta)} \tag{7–4}$$

for $(\omega, \theta) \in \mathcal{R} \cap U$. This gives that (7–4) holds for all $(\omega, \theta) \in U$, and then Lemma 7.5 implies that equations (7–1) hold for some real numbers $\lambda(\omega, \theta)$ and $\mu(\omega, \theta)$ for all $(\omega, \theta) \in U$. In particular, $\delta' = \gamma_L(\sigma_0)$.

Let $\mathcal{F}^{(K)}_{t_0}(\sigma_0) = (z, \zeta)$. Then either $\zeta = \omega_0$ and $z = x_1 + s\omega_0$ for some $s < 0$, or $\zeta = \theta_0$ and $z = x_k + s\theta_0$ for some $s > 0$. The same holds for $\mathcal{F}^{(L)}_{t_0}(\sigma_0) = (z', \zeta')$. In both cases (7–1) and (7–4) imply $(z, \zeta) = (z', \zeta')$, i.e. $\mathcal{F}^{(K)}_{t_0}(\sigma_0) = \mathcal{F}^{(L)}_{t_0}(\sigma_0)$. $\square$

Using the existence of the conjugacy Φ and the fact that it is measure preserving with respect to the canonical measures on $S_b^*(\Omega_K)$ and $S_b^*(\Omega_L)$, one derives the following.

COROLLARY 7.6. *Let the obstacles K and L have almost the same SLS. If the sets of trapped points of both K and L have Lebesgue measure zero, then* $\mathrm{Vol}\,K = \mathrm{Vol}\,L$.

Livshits' example shows that the above conclusion is not true without any assumption about the sets of trapped points. Notice that far from the obstacle the trapping set is relatively small. For example, if C is a large sphere in $\mathbb{R}^n$ (i.e. it contains K in its interior), a slight modification of the proof of Proposition 5.6 shows that $\dim\big(S_C^*(\Omega_K) \cap \mathrm{Trap}\,\Omega_K\big) \leq 2n - 3$. On the other hand, in some cases (as in Livshits' example) we have $\dim\big(\mathrm{Trap}\,\Omega_K \cap S_b^*(\Omega_K)\big) = 2n - 1 = \dim\big(S_b^*(\Omega_K)\big)$.

Another simple consequence of Theorem 7.3 concerns backscattering rays. Denote by $\mathrm{Trap}^{(n)}\,\partial K$ the set of those $x \in \partial K$ such that $(x, \nu_K(x)) \in \mathrm{Trap}\,\Omega_K$, where $\nu_K(x)$ is the *outward unit normal* to ∂K at x.

Suppose that K and L are obstacles with almost the same SLS. Let Φ be the conjugacy from Theorem 7.3. Given $x \in \partial K \setminus \mathrm{Trap}^{(n)}\,\partial K$, take an arbitrary $t > 0$ such that $(z, \zeta) = \mathcal{F}^{(K)}{}_t(x, \nu_K(x)) \in S^*(\mathbb{R}^n \setminus U_0)$. Then $\mathcal{F}^{(K)}{}_t(z, -\zeta) = (x, \nu_K(x))$ and $\mathcal{F}^{(K)}{}_{2t}(z, -\zeta) = (z, \zeta)$. Therefore

$$(z, \zeta) = \Phi(z, \zeta) = \Phi \circ \mathcal{F}^{(K)}{}_{2t}(z, -\zeta) = \mathcal{F}^{(L)}{}_{2t} \circ \Phi(z, -\zeta) = \mathcal{F}^{(L)}{}_{2t}(z, -\zeta),$$

so for $(y, \eta) = \mathcal{F}^{(L)}{}_t(z, -\zeta)$ we must have $y \in \partial L$ and $\eta \perp \partial L$ at y. Thus, $\Phi(x, \nu_K(x)) = (y, \nu_L(y))$ for some $y \in \partial L \setminus \mathrm{Trap}^{(n)}\,\partial L$. Setting $\varphi(x) = y$, one gets a homeomorphism

$$\varphi : \partial K \setminus \mathrm{Trap}^{(n)}\,\partial K \to \partial L \setminus \mathrm{Trap}^{(n)}\,\partial L$$

such that $\varphi(x) = y$ whenever $\Phi(x, \nu_K(x)) = (y, \nu_L(y))$. In particular, assuming that $\dim \mathrm{Trap}^{(n)}\,\partial K < n - 2$ and $\dim \mathrm{Trap}^{(n)}\,\partial L < n - 2$, it follows that K and L must have the same number of connected components.

Here we denote by $\dim X$ the *topological dimension* of X (see [4], for example). Since $\dim X \leq \dim_H X$, where $\dim_H X$ is the Hausdorff dimension of the metric space X (see [4], for example), all assumptions of the form $\dim X \leq a$ can be replaced by $\dim_H X \leq a$.

It seems natural to conjecture that in the case of nontrapping obstacles the SLS uniquely determines the obstacle. While this is still an open problem, using Theorem 7.3 and backscattering rays as above, one can prove this conjecture at least for star-shaped obstacles (as mentioned above, these are necessarily nontrapping).

PROPOSITION 7.7 [29]. *Let K and L have almost the same SLS. If K is star-shaped, $L = K$.*

Even though the trapping set is relatively small far from the obstacle, in general it may be big enough to *topologically divide* $S_C^*(\Omega_K)$, i.e. it may happen that $S_C^*(\Omega_K) \setminus \mathrm{Trap}\,\Omega_K$ has more than one connected component.

We will denote by $\partial K^{(\mathrm{ob})}$ the union of all connected components of ∂K that have a common point with at least one scattering ray in Ω_K, and call it the *observable part* of the boundary ∂K. The obstacle K will be called *observable*, if $\partial K = \partial K^{(\mathrm{ob})}$.

THEOREM 7.8 [28]. *Let K, L be obstacles in $\mathbb{R}^n$ with real analytic boundaries that have almost the same SLS. If K is such that $\operatorname{Trap}\Omega_K$ does not topologically divide $S_C^*(\Omega_K)$, then $\partial K^{(\mathrm{ob})} = \partial L^{(\mathrm{ob})}$. If in addition both K and L are observable, then $K = L$.*

The idea of the proof of Theorem 7.8 is rather simple. Let Y be the union of all connected components of $\partial K^{(\mathrm{ob})}$ that do not coincide with connected components of L. Assuming $Y \neq \varnothing$, one finds $\sigma \in S^*(\mathbb{R}^n \setminus U)$ such that $\gamma_K(\sigma) = \{\operatorname{pr}_1(\mathcal{F}^{(K)}{}_t(\sigma)) : t \in \mathbb{R}\}$ has a common point with Y. Consider a smooth curve $\sigma(s)$ in $S^*(\mathbb{R}^n \setminus U)$ that connects σ to a point $\sigma(0) = \sigma_0$ generating a *free ray*, i.e. a ray without common points with K. After some regularization of the curve $\sigma(s)$ (imposing some transversality conditions on it), we choose the smallest s with $\gamma_K(\sigma(s)) \cap Y \neq \varnothing$. For $\rho = \sigma(s)$, the scattering ray $\gamma_K(\rho)$ has only one common point y with Y which is a tangent point, and all transversal reflection of its occur at connected components of ∂K that coincide with connected components of L. Then we show that $y' \in \partial L$ for a dense set of points y' in a neighborhood of y in Y. Thus, $\partial K = \partial L$ near y which is a contradiction with the definition of Y. See [28] for details.

It is not clear how restrictive the condition that $\operatorname{Trap}\Omega_K$ does not topologically divide $S_C^*(\Omega_K)$ is. It turns out ([28]) that this condition is satisfied when K is a finite disjoint union of strictly convex domains with C^∞ boundaries. This and Theorem 7.8 imply the following.

COROLLARY 7.9. ([28]) *If K is a finite disjoint union of strictly convex domains, K and L have almost the same SLS and both ∂K and ∂L are real analytic, then $K = L$.*

It is an open problem whether the statement of Corollary 7.9 remains true for obstacles with C^∞ boundaries ∂K and ∂L.

Next, we describe a few results from [29] involving scattering rays having tangencies to the boundary.

Denote by $\mathcal{K}^{(fin)}$ the class of obstacles $K \in \mathcal{K}_0$ such that the normal curvature of K does not vanish of infinite order. From now on until the end of this section we assume that $K, L \in \mathcal{K}^{(fin)}$.

Consider an arbitrary scattering ray γ in Ω_K and let X and Y be arbitrary cross-sections of the incoming and outgoing rays of γ. Define the cross-sectional map $\mathcal{P}_K : S_X^*(\mathbb{R}^n) \to S_Y^*(\mathbb{R}^n)$ by the shift along the flow $\mathcal{F}^{(K)}{}_t$. Now assume that the obstacle K and L have almost the same SLS. It then follows from Theorem 7.3 that $\mathcal{P}_K = \mathcal{P}_L$. In particular the singularities of $\mathcal{P}_K$ and $\mathcal{P}_L$ are the same, and this implies that for any $\sigma_0 = (x_0, \xi_0) \in S^*(\mathbb{R}^n \setminus U_0) \setminus \text{Trap}\,\Omega_K$, the ray $\gamma_K(\sigma_0)$ contains a point of tangency to ∂K if and only if $\gamma_L(\sigma_0)$ contains a point of tangency to ∂L.

Next, suppose that $\sigma(s)$, $s \in [0, a]$, is a continuous curve in $S^*(\Omega_K)$ consisting of nontrapped points. Using an idea of Melrose and Sjöstrand [20] involving winding numbers, one shows that if $\gamma_K(\sigma(s))$ is simply reflecting for each s, then the number of reflection points of $\gamma_K(\sigma(s))$ is the same for all $s \in [0, a]$. Now assume that $\sigma = \sigma(0)$ generates a ray $\gamma_K(\sigma)$ containing a gliding segment on ∂K. If $s_k \searrow 0$ are such that each $\gamma_K(\sigma(s_k))$ is simply reflecting, it follows from [20] that the number of reflection points of $\gamma_K(\sigma(s_k))$ tends to ∞. Hence there must be infinitely many $s \in (0, a]$ such that $\mathcal{F}^{(K)}{}_t(\sigma(s)) \in S^*(\partial K)$ for some $t = t(s)$. On the other hand if $\gamma_K(\sigma)$ is tangent to ∂K but does not contain a gliding segment, then it is not difficult to construct a continuous curve $\sigma(s)$ ($0 \leq s \leq a$, $a > 0$) in $S^*(\Omega_K)$ with $\sigma(0) = \sigma$ such that $\gamma(\sigma(s))$ is a simply reflecting ray for all $s \in (0, a]$.

These observation yield that from the SLS of an obstacle one can determine which points $\sigma \in S^*(\Omega_K) \setminus \text{Trap}\,\Omega_K$ generate rays containing gliding segments on ∂K.

COROLLARY 7.10. ([29]) *Let* K, L *have almost the same SLS. If there exists a scattering ray containing a gliding segment in* Ω_K, *then* Ω_L *has the same property. Consequently, if* K *is a finite disjoint union of convex domains in* $\mathbb{R}^n$ *and* $\dim \text{Trap}\,\Omega_L \cap S^*(\partial L) < 2n-3$, *then* L *is also a finite disjoint union of convex domains, moreover* K *and* L *must have the same number of connected components and are therefore diffeomorphic.*

A point $\sigma \in S_C^*(\Omega_K)$ will be called *accessible* if it belongs to a connected component of $S_C^*(\Omega_K) \setminus \text{Trap}\,\Omega_K$ containing a point that generates a free ray. Presumably the SLS provides more substantial information about the behavior of the flow $\mathcal{F}^{(K)}{}_t$ near accessible points $\rho \in S_C^*(\Omega_K)$ and correspondingly about parts of ∂K that can be reached by rays generated by accessible points. The following result shows for example that the SLS determines uniquely the number of reflection points of simply reflecting rays $\gamma_K(\sigma)$ generated by accessible points σ.

PROPOSITION 7.11 [29]. *Let* K, L *have almost the same SLS. For every connected component* W *of* $S_C^*(\Omega_K) \setminus \text{Trap}\,\Omega_K$ *there exists an integer* $m = m(K, L, W)$ *such that*

$$\#(\gamma_K(\sigma) \cap \partial K) = \#(\gamma_L(\sigma) \cap \partial L) + m$$

for all $\sigma \in W \cap U^{(K)}$. *Whenever* W *is accessible,* $m = 0$; *that is,*

$$\#(\gamma_K(\sigma) \cap \partial K) = \#(\gamma_L(\sigma) \cap \partial L)$$

for any accessible point σ.

See [29] for further results concerning relationship between obstacles having almost the same SLS.

References

[1] F. Cardoso, V. Petkov, L. Stoyanov, *Singularities of the scattering kernel for generic obstacles*, Ann. Inst. H. Poincaré (Physique théorique) **53** (1990), 445–466.

[2] D. Colton and R. Kress, *Inverse Acoustic and Electromagnetic Scattering Theory*, 2nd edition, Appl. Math. Sciences, Springer, Berlin, 1998.

[3] C. Croke, *Rigidity and the distance between boundary points*, J. Diff. Geometry **36**, (1991), 445–464.

[4] G. Edgar, *Measure, Topology and Fractal Geometry*, Berlin, Springer, 1990.

[5] V. Guillemin, *Sojourn time and asymptotic properties of the scattering matrix*, Publ. RIMS Kyoto Univ. **12** (1977), 69–88.

[6] V. Guillemin and R. Melrose, *The Poisson summation formula for manifolds with boundary*, Adv. in Math. **32** (1979), 128–148.

[7] V. Isakov, *Inverse problems for partial differential equations*, Appl. Math. Sci., **127**, Springer, Berlin, 1998.

[8] M. Hirsch, *Differential Topology*, Berlin, Springer 1976.

[9] L. Hörmander, *Fourier integral operators I*, Acta Math. **127** (1971), 79–183.

[10] L. Hörmander, *The Analysis of Linear Partial Differential Operators*, Springer, Berlin, 1983–1985 (4 volumes).

[11] A. Kirsch and R. Kress, *Uniqueness in inverse scattering*, Inverse Problems **9** (1993), 285–299.

[12] P. Lax and R. Phillips, *Scattering Theory*, 2nd Edition, Academic Press, New York, 1989.

[13] P. Lax and R. Phillips, *The scattering of sound waves by an obstacle*, Comm. Pure Appl. Math. **30** (1977), 195–233.

[14] A. Majda, *High Frequency Asymptotics for the Scattering Matrix and the Inverse Problem of Acoustical Scattering*, Comm. Pure Appl. Math. **29** (1976), 261–291.

[15] A. Majda, *A representation formula for the scattering operator and the inverse problem for arbitrary bodies*, Comm. Pure Appl. Math. **30** (1977), 165–194.

[16] A. Majda and J. Ralston, *An analogue of Weyl's formula for unbounded domains*, Duke Math. J. **45** (1978), 183–196.

[17] R. Melrose, *Microlocal parametrices for diffractive boundary value problems*, Duke Math. J. **42** (1975), 605–635.

[18] R. Melrose, *Geometric Scattering Theory*, Cambridge Univ. Press, Cambridge, 1994.

[19] R. Melrose and M. Taylor, *Near peak scattering and the corrected Kirchoff approximation*, Adv. in Math. **55** (1985), 242–315.

[20] R. Melrose and J. Sjöstrand, *Singularities in boundary value problems*, I, II, Commun. Pure Appl. Math. **31** (1978), 593–617 and **35** (1982), 129–168.

[21] V. Petkov, *High frequency asymptotics of the scattering amplitude for non-convex bodies*, Commun. PDE. **5** (1980), 293–329.

[22] V. Petkov and L. Stoyanov, *Geometry of Reflecting Rays and Inverse Spectral Problems*, John Wiley & Sons, Chichester, 1992.

[23] V. Petkov and L. Stoyanov, *Sojourn times of trapping rays and the behavior of the modified resolvent of the Laplacian*, Ann. Inst. H. Poincaré (Physique théorique) **62** (1995), 17–45.

[24] V. Petkov and L. Stoyanov, *Singularities of the scattering kernel for trapping obstacles*, Ann. Scient. Ec. Norm. Sup. **29** (1996), 737–756.

[25] J. Sjöstrand and M. Zworski, *Complex scaling and the distribution of scattering poles*, Journal of AMS, **4** (1991), 729–769.

[26] P. Stefanov and G. Uhlmann, *Local uniqueness for the fixed energy fixed angle inverse problem in obstacle scattering*, Preprint, 2002.

[27] L. Stoyanov, *Generalized Hamiltonian flow and Poisson relation for the scattering kernel*, Ann. Scient. Ec. Norm. Sup. **33** (2000), 361–382.

[28] L. Stoyanov, *On the scattering length spectrum for real analytic obstacles*, J. Funct. Anal., **177** (2000), 459–488.

[29] L. Stoyanov, *Rigidity of the scattering length spectrum*, Math. Ann. **324** (2002), 743–771.

[30] M. Taylor, *Grazing rays and reflection of singularities of solutions to wave equations*, Comm. Pure Appl. Math. **29** (1976), 1–37.

[31] M. Zworski, *High frequency scattering by a convex obstacle*, Duke Math. J. **61** (1990), 543–634.

VESSELIN PETKOV
DÉPARTEMENT DE MATHÉMATIQUES APPLIQUÉES
UNIVERSITÉ BORDEAUX I
351, COURS DE LA LIBÉRATION
33405 TALENCE
FRANCE
petkov@math.u-bordeaux.fr

LUCHEZAR STOYANOV
DEPARTMENT OF MATHEMATICS AND STATISTICS
UNIVERSITY OF WESTERN AUSTRALIA
PERTH 6709
AUSTRALIA
stoyanov@maths.uwa.edu.au

Geometry and Analysis in Many-Body Scattering

ANDRÁS VASY

ABSTRACT. This chapter explains in relatively nontechnical terms recent results in many-body scattering and related topics. Many results in the many-body setting should be understood as new results on the propagation of singularities, here understood as lack of decay of wave functions at infinity, with much in common with real principal type propagation (wave phenomena). Classical mechanics plays the role that geometric optics has in the study of the wave equation, but even at this point quantum phenomena emerge. Propagation of singularities has immediate applications to the structure of scattering matrices and to inverse scattering; these topics are addressed here. The final section studies a problem very closely related to many-body scattering, namely scattering on higher rank noncompact symmetric spaces.

1. Introduction

This chapter is an effort to explain in relatively nontechnical terms recent results in many-body scattering and related topics. Thus, many results in the many-body setting should be understood as new results on the propagation of singularities, here understood as lack of decay of wave functions at infinity, with much in common with real principal type propagation, i.e. wave phenomena. Motivated by this, I first briefly describe propagation of singularities for the wave equation. This is a remarkable relationship between geometric optics (the particle view of light) and the solutions of the wave equation (the wave view).

Next, in Section 3, I explain the geometry of many-body scattering, which includes both that of the configuration space and phase space. This geometry is closely related to classical mechanics, playing the role of geometric optics, but even at this point quantum phenomena emerge. This leads to the analytic results, namely the propagation of singularities connecting classical and quantum mechanics.

This work is partially supported by NSF grant #DMS-0201092, a Fellowship from the Alfred P. Sloan Foundation, and the Université de Nantes, where these lectures were originally given.

Much as for the wave equation, such a result has immediate applications, including the description of the scattering matrices and of the scattering phase. Slightly stronger versions can even lead to inverse results, a topic covered in the following section.

After so explaining the results, in Sections 5-6, I will try to at least give a flavor of how they are proved. This uses a many-body pseudodifferential algebra and positive commutator estimates, so these are discussed. We remark that these techniques are closely related to the proofs of the propagation of singularities for the wave equation, but there are significant differences as well, mostly arising from bound states of particles, which have no analogues for the wave equation. The pseudodifferential algebra itself is very interesting from the viewpoint of noncommutative geometry: there is a hierarchy of operator valued symbols at infinity.

Asymptotic completeness was the main focus of work in many-body scattering for a long period. In Section 7, I briefly explain how it relates to the microlocal estimates.

There is another area that is very closely related to many-body scattering, namely scattering on higher rank noncompact symmetric spaces. Here, in Section 8, we only discuss rank two, which corresponds to three-body scattering, since this is the only part that has been properly written up, but it is expected that very soon these results will extend to all higher rank spaces.

I hope that these notes will make many of these results more accessible, the connections more transparent, and explain the motivation behind them. Many-body scattering has a long history, and here I can only talk about the most recent developments. An excellent overview of results known in the early 1990s can be found in Hiroshi Isozaki's lecture notes [30]. Indeed, in some sense, the current notes continue where [30] left off. I introduce a fully microlocal picture, motivated by the geometric approach of Richard Melrose [42], and emphasize the results these give, but the basic spectral and scattering results follow from a simpler 'partial' microlocalization, which is one of the subjects of [30].

Acknowledgements. The notes were originally prepared for a mini-course at the Université de Nantes at the invitation of Professor Xue-Ping Wang, whose hospitality I gratefully acknowledge. The analytic continuation of the resolvent on symmetric spaces is a more recent development, but it was fueled by a discussion during the visit of Rafe Mazzeo, my collaborator, to Nantes. Over an espresso, Gilles Carrón mentioned that the existence of the analytic continuation was not known, something that was hard to believe, but we immediately realized that our methods should yield such a continuation rather directly. I also thank Gunther Uhlmann for urging me to write up these notes: without him, they may never have been written up, and Rafe Mazzeo for a careful reading of the manuscript.

2. Geometric Optics and the Wave Equation

According to the rules of geometric optics, light propagates in straight lines, and reflects/refracts from surfaces according to the Snell–Descartes law. That is to say, considering light as a stream of billiard balls, the energy as well as the tangential component of the momentum (tangential to the surface hit) is conserved upon hitting the surface.

But light satisfies the wave equation, i.e. if $u = u(x,t)$ is the electromagnetic field on $\Omega_x \times \mathbb{R}_t$, $\Omega \subset \mathbb{R}^n$, then $Pu = 0$ where P is the wave operator $c^2\Delta - D_t^2$, and a boundary condition also holds (say, Dirichlet), if Ω is not the whole space. (Here $D_t = \frac{1}{i}\frac{\partial}{\partial t}$ and $\Delta = \sum_j D_{x_j}^2$ is the *positive* Laplacian.) How are these two viewpoints related?

One can phrase the connection in different ways. The most usual one in physics is that the billiard ball picture is accurate in the high frequency, i.e. low wave length, limit. That is to say, for high frequency light, geometric optics is accurate up to a 'small' error. A slightly different way of looking at this, which however does not involve approximations, is that the location of singularities of the solution of the wave equation is *exactly* predicted by geometric optics. Here singularities are understood as lack of smoothness, or possibly lack of analyticity.

Indeed, it is convenient at this point to generalize the setting somewhat. So let (Ω, g) be a Riemannian manifold with corners, $P = c^2\Delta_g - D_t^2$, $c > 0$. The speed of light, c, may be absorbed in the metric g, of course, we keep the notation in analogy with the usual wave equation.

For simplicity of notation in this paragraph we assume that $M_z = \Omega_x \times \mathbb{R}_t$ is boundaryless; in general, the same definitions hold in the interior of M. Thus, we associate a homogeneous real function on T^*M to P, namely its principal symbol: $p = c^2|\xi|_g^2 - \tau^2$, where we write $\zeta = (\xi, \tau)$ as the dual variable of $z = (x,t)$. Now T^*M is a symplectic manifold with symplectic form $\omega = \sum d\zeta_j \wedge dz_j$. Thus, p gives rise to a vector field H_p, called the Hamilton vector field, by requiring that $\omega(V, H_p) = Vp$ for any vector field V on T^*M. Hence H_p is a smooth vector field on T^*M explicitly given by

$$H_p = \frac{\partial p}{\partial \zeta}\frac{\partial}{\partial z} - \frac{\partial p}{\partial z}\frac{\partial}{\partial \zeta}.$$

Note that p is constant along the integral curves of H_p since taking $V = H_p$, $0 = \omega(H_p, H_p) = H_pp$. Null bicharacteristics are the integral curves of H_p inside its characteristic set $\Sigma = p^{-1}(\{0\})$. Thus, if $\gamma : I \to T^*M$ is a null bicharacteristic (here I is an interval), and $z(s) = z(\gamma(s))$, $\zeta(s) = \zeta(\gamma(s))$, then these solve the ODE's $dz/ds = \partial p/\partial \zeta$, $d\zeta/ds = -\partial p/\partial z$. Hence, when $M = \Omega \times \mathbb{R}$, $\Omega \subset \mathbb{R}^n$, $p = c^2|\xi|^2 - \tau^2$ as above, we deduce that ξ and τ are constant along the integral curves of H_p, hence their projection to M consists of straight line segments. More generally, the projection of null-bicharacteristics to Ω are geodesics of g.

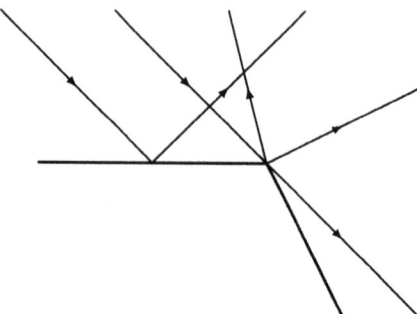

Figure 1. Projection of broken bicharacteristics to Ω. When rays hit the boundary hypersurfaces, the tangential component of the momentum and the kinetic
energy are conserved, but the normal component may change. At the corner,
there is no tangential component (though there would be if the time variable
were not projected out), so the only constraint is the conservation of kinetic
energy.

There is an appropriate extension of this at boundary surfaces and even at
corners, called generalized broken bicharacteristics, see [43; 36], which I will not
explain in full generality, though I remark that many-body scattering, discussed
in the next section in detail, is rather similar. However, a somewhat typical
example is that of broken bicharacteristics. These are piecewise bicharacteristics,
i.e. there is a sequence s_j, j in a subset of integers, such that for each j, $\gamma|_{(s_j,s_{j+1})}$
is a null bicharacteristic in the sense described above, the projection $z \circ \gamma$ of γ to
M is continuous, and $\gamma(s_j+) - \gamma(s_j-)$ is conormal to the smallest dimensional
boundary face containing $z(\gamma(s_j))$. Thus, the tangent vectors to $z \circ \gamma|_{(s_j,s_{j+1})}$ and
$z \circ \gamma|_{(s_{j-1},s_j)}$ differ by a vector normal to the smallest boundary face containing
$z(\gamma(s_j))$. This expresses that the normal component of the momentum may
change, while the tangential component is conserved, when a light ray hits a
boundary.

Now one can describe the singularities of u using null bicharacteristics. Let o
be the zero section of T^*M. The location of the singularities is described by an
object

$$\mathrm{WF}(u) \subset T^*M \setminus o = \{(z,\zeta) : \zeta \neq 0\}$$

that is conic in ζ, i.e. $(z,\zeta) \in \mathrm{WF}(u)$ if and only if $(z,r\zeta) \in \mathrm{WF}(u)$ for every
$r > 0$. $\mathrm{WF}(u)$ is called the wave front set of u, and it describes where (in z) and in
which codirection ζ is the distribution u not $\mathcal{C}^\infty$. More precisely, the definition of
$\mathrm{WF}(u)$ is that $(z_0,\zeta_0) \notin \mathrm{WF}(u)$ if and only if there exists $\phi \in \mathcal{C}_c^\infty(M)$, $\phi(z_0) \neq 0$
such that the Fourier transform $\mathcal{F}(\phi u)$ of ϕu is rapidly decreasing in an open
cone around ζ_0. Here we assume that M is boundaryless; otherwise we need
to require that ϕ is supported in the interior of M. Again, there is a natural
definition at ∂M which we do not give here. (There are more natural versions of
this definition using pseudodifferential operators that I will describe later.) As

an example, consider the step function: write $z = (z_1, z'')$, $u(z) = 1$ if $z_1 > 0$, $u(z) = 0$ if $z_1 < 0$. Then

$$\mathrm{WF}(u) = N^*\{z_1 = 0\} \setminus o = \{(0, z'', \zeta_1, 0) : \zeta_1 \neq 0\},$$

the conormal bundle of the hypersurface $z_1 = 0$, with its zero section removed. The same statement holds, with $=$ possibly replaced by $\subset$, if we take any $\mathcal{C}^\infty$ function u_0 on M, and then define $u = u_0$ in $z_1 > 0$ and $u = 0$ in $z_1 < 0$. Informally, one might say that u is singular in z_1 at $z_1 = 0$, but it depends smoothly on z''. The wave front set thus pinpoints not only the locations z of singularities (lack of smoothness) in M, but it refines it by also giving the frequencies (or rather direction of frequencies) at which these appear at z.

The theorem we are after is the following. In early versions it goes back to Lax [35], its boundaryless version is due to Hörmander [28], the smooth boundary versions are due to Melrose, Sjöstrand, Taylor and Ivrii [33; 43; 44; 61], and the corner version in the analytic category is due to Lebeau [36] (the $\mathcal{C}^\infty$ version is still not known in the corner setting) while a different extension, to conic points, is due to Melrose and Wunsch [45].

THEOREM 2.1. *Suppose $Pu \in \mathcal{C}^\infty(M)$, and if $\partial M \neq 0$ then $u|_{\partial M} = 0$. Then $\mathrm{WF}(u) \subset \Sigma = p^{-1}(\{0\})$ (microlocal elliptic regularity). Moreover, $\mathrm{WF}(u)$ is a union of maximally extended generalized broken bicharacteristics inside Σ (propagation of singularities).*

This theorem states that if a point $(z, \zeta) \in T^*M \setminus o$ is in $\mathrm{WF}(u)$ and u solves $Pu \in \mathcal{C}^\infty(M)$, and satisfies a boundary condition if appropriate, then there is at least one maximally extended generalized broken bicharacteristic through (z, ζ) that is completely contained in $\mathrm{WF}(u)$. Of course, in the absence of boundaries, and often even in their presence, there is a unique maximally extended generalized broken bicharacteristic through (z, ζ), so the statement is that this bicharacteristic is completely in $\mathrm{WF}(u)$. However, as soon as codimension two or higher corners appear, there is no hope for such uniqueness, and this theorem is the optimal statement.

At least in the nicest settings (no boundaries, or nondegeneracy assumption at the boundaries which are assumed to be smooth), this theorem can be improved significantly to predict not only the location, but also the amplitude of the singularities of u.

3. Propagation in Many-Body Scattering

There is an analogous setup for scattering. Now we want to understand how interacting particles behave. Again, there is a classical mechanical setup (the analogue of geometric optics) and a quantum mechanical setup (the analogue of the wave equation). To focus on the most relevant points, I formulate the prob-

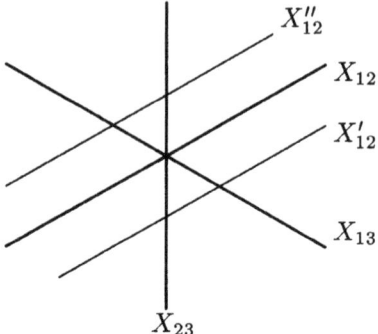

Figure 2. Collision planes X_{12}, X_{13} and X_{23} and translates X'_{12} and X''_{12} of X_{12}. V_{12} is constant along X_{12}, is a (typically different) constant along X'_{12}, etc., so it does not decay at infinity unless it is identically zero.

lem in a time-independent fashion, though it is easy to reformulate everything in a time dependent way. We only do this in a remark following Theorem 3.1.

Thus, we want to understand tempered distributional solutions u of $(H - \lambda)u = 0$; here $\lambda \in \mathbb{R}$ is the energy, and H is the Hamiltonian, i.e. the analogue of $H - \lambda$ is P above. Namely, if we have N particles, each of which is d-dimensional with positions $x_1, \ldots, x_N \in \mathbb{R}^d$, mass $m_1, \ldots, m_N$, and the interaction between particle i and j is given by a potential V_{ij} (which is a function on $\mathbb{R}^d$), then the Hamiltonian describing this system is

$$H = \sum_{i=1}^{N} \frac{1}{2m_i} \Delta_{x_i} + \sum_{i<j} V_{ij}(x_i - x_j) = \Delta + V,$$

which is an operator on (functions on) $\mathbb{R}^n = \mathbb{R}^{Nd}$. Planck's constant $\hbar$ is here taken to be 1; it could be absorbed in the x_i by a simple rescaling.

Now H is elliptic in the standard sense, namely its principal symbol is

$$\sum \frac{1}{2m_i} |\xi|^2,$$

which never vanishes outside the zero section o. Note that the potential is lower order than Δ in the standard sense, so it is not part of the principal symbol. So, by the previous theorem,

$$(H - \lambda)u = 0 \implies \mathrm{WF}(u) = \varnothing \implies u \in C^\infty(\mathbb{R}^n).$$

So the only possibility of interesting behavior for u is at infinity, and this is exactly what we want to understand.

The main feature of many-body problems is that even if V_{ij} decays at infinity on $\mathbb{R}^d$, it does *not* decay at infinity in $\mathbb{R}^n$ since it is a constant along $X_{ij} = \{x_i = x_j\}$, as well as along its translates X'_{ij}, X''_{ij}, so it does not decay if we go to infinity, say, along X_{ij}; see Figure 2. The X_{ij} are called collision planes (as are their intersections) since at X_{ij} particles i and j are at the same place.

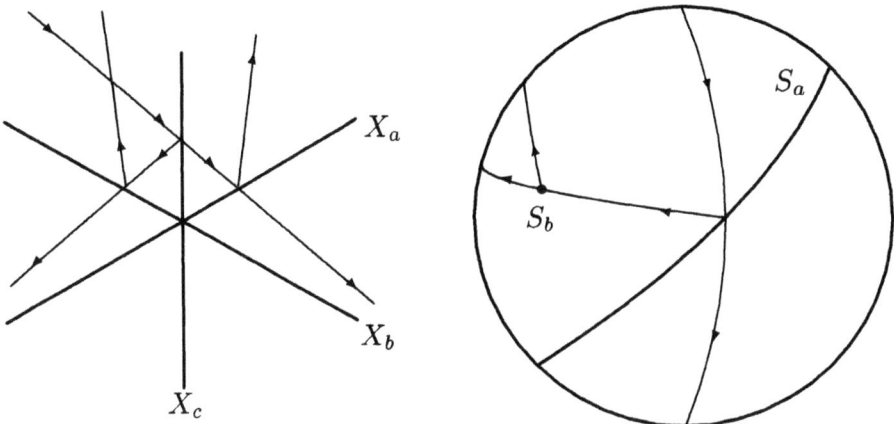

Figure 3. On the left, broken geodesics in $\mathbb{R}^n \setminus \{0\}$, $n = 2$, broken at the collision planes X_a, X_b and X_c. On the right, the projection of broken geodesics in $\mathbb{R}^n \setminus \{0\}$, $n = 3$, emanating from the north pole, to the unit sphere S_0, better understood as the sphere at infinity. The C_a, C_b are the intersection of the collision planes X_a, X_b with S_0; $\dim X_a = 2$, $\dim X_b = 1$.

In the two-body problem one actually has $H = \Delta_{x_1,x_2} + V_{12}(x_1 - x_2)$, i.e. V_{12} still does not decay at infinity, e.g. if one keeps $x_1 = x_2$ but lets $x_1 \to \infty$. However, one can easily remove the center of mass by performing a Fourier transform along X_{12}. This conjugates $H - \lambda$ to $H^{12} + |\xi_{12}|^2 - \lambda$, where ξ_{12} is the variable on X_{12}^*, and $H^{12} = \Delta_{X^{12}} + V_{12}$, X^{12} being the orthocomplement of X_{12}. Thus, one reduces the study of $H - \lambda$ to that of a Hamiltonian on X^{12}, namely $H^{12} - \lambda'$, $\lambda' = \lambda - |\xi_{12}|^2$ being a shifted spectral parameter. Now V_{12} decays at infinity (we are working on X^{12}!), so H^{12} can be considered as a perturbation of $\Delta_{X^{12}}$, hence its analysis is rather simple. Notice that the point spectrum of H^{12} gives rise to a branch of the continuous spectrum of H: this is a phenomenon that is very typical in many-body scattering. The center of mass can also be removed in any actual many-body problem, but one still obtains a Hamiltonian with nondecaying potentials as before.

One can still talk about classical mechanics, just as for the wave equation, using bicharacteristics. These are deterministic – if V is smooth enough (we usually assume that V is C^∞). But much like for corners, there is a compressed description of dynamics near infinity. This is somewhat more complicated than for the wave equation, but only because particles can be bound together. Thus, even the 'classical' description is partly quantum. These two facts, the presence of collision planes and the bound states, are the two crucial features of many-body scattering.

The compressed dynamics in the absence of bound states looks just like in the wave equation setting. One should think of this as a good description when a classical trajectory is uniformly near infinity.

More precisely, it is convenient to introduce Agmon's generalization of the many-body problem, which amounts to using the vector space structure of $\mathbb{R}^n$ as the setting. One can also give geometric generalizations (in the sense of differential geometry) that arose from the work of Melrose [41; 42], and I will do this later.

So we work on the vector space $X_0 = \mathbb{R}^n$, equipped with the Euclidean metric. We are also given a finite collection $\mathcal{X} = \{X_a : a \in I\}$ of linear subspaces X_a of $\mathbb{R}^n$, called the collision planes. We assume that $\mathcal{X}$ is closed under intersections, and $X_0 = \mathbb{R}^n \in \mathcal{X}$, $X_1 = \{0\} \in \mathcal{X}$. We let $X^a = X_a^{\perp}$ be the orthocomplement of X_a in $\mathbb{R}^n$, so $\mathbb{R}^n = X_a \oplus X^a$. (Agmon's generalization is thus that the X_a do not have to come from intersections of the planes $X_{ij} = \{x_i = x_j\}$.) We write the corresponding coordinates as (x_a, x^a), and denote the orthogonal projection to X^a by π^a. A many-body Hamiltonian in potential scattering is an operator of the form

$$H = \Delta + \sum_a (\pi^a)^* V_a,$$

where V_a is a real valued function on X^a in a certain class, for example V_a is a symbol on X^a of negative order: $V_a \in S^{-\rho}(X^a)$, $\rho > 0$. We also assume that $V_0 = 0$ for normalization; note that $X^0 = \{0\}$, so V_0 would simply play the role of the spectral parameter. We sometimes drop the pull-back notation from now on and write $H = \Delta + \sum_a V_a$.

Another useful piece of terminology is the following. We say that V_a is short range if $V_a \in S^{-\rho}(X^a)$ for some $\rho > 1$. We say that V_a is long-range if $V_a \in S^{-\rho}(X^a)$ for some $\rho \in (0, 1]$. The Coulomb potential is thus 'marginally long-range', at least if we ignore its singularity at 0 (which is not a serious problem anyway). Whether V_a is short- or long-range does not make any difference for the propagation phenomena we discuss in this section. However, it does make a major difference for the precise behavior of generalized eigenfunctions at the 'radial sets' which we discuss later. This also shows up in the related issue of asymptotic completeness.

Yet another notation we use on occasion is that of a k-cluster. Physically, a cluster describes particles that are close (or collide), and a k-cluster means that there are k clusters of particles, inside each of which the particles are close to each other. So in N-particle scattering, the N-cluster describes N asymptotically free particles (none is close to any other), hence we say that the collision plane $X_0 = \mathbb{R}^n$ is the N-cluster. On the other hand, if $X_a \neq \{0\}$ is such that $X_b \subsetneq X_a$ implies that $X_b = \{0\}$, then X_a, or rather a, is a 2-cluster. E.g. given five particles, a 2-cluster is where $x_1 = x_2$ and $x_3 = x_4 = x_5$, i.e. the particles 1 and 2, resp. 3, 4 and 5, are close to each other. In general, a k-cluster X_a can be defined by the length of nested chains of collision planes inside X_a.

One need not assume that all interactions between the particles are via potentials. Indeed, V_a may be allowed to be any first order differential operator on the vector space X^a with symbolic coefficients of negative order. Also, one may generalize the metric g in an analogous fashion, as discussed later, which in effect allows V_a to be second order provided that H remains elliptic. To simplify the notation, and due to the traditions, we mostly talk as if V_a were potentials, but the generalization to such higher order perturbations requires only occasional and minor modifications, which will be pointed out.

The subsystem Hamiltonians are defined by

$$H^a = \Delta_{X^a} + \sum_{X_a \subset X_b} V_b.$$

Note that $X_a \subset X_b$ if and only if $X^a \supset X^b$, so above V_b is really the pull-back of V_b from X^b to to X^a by the orthogonal projection. Thus, H^a is an operator on (functions on) X^a, and indeed it is a many-body Hamiltonian.

We also let

$$X_{a,\mathrm{sing}} = \bigcup \{X_b : X_b \subsetneq X_a\} \quad \text{and} \quad X_{a,\mathrm{reg}} = X_a \setminus X_{a,\mathrm{sing}}$$

be the singular and regular parts of X_a. Thus, if X_c is a collision plane and X_a is not a subset of X_c, then $X_a \cap X_c$ is a proper subset of X_a, and is a collision plane (since $\mathcal{X}$ is closed under intersections), so $X_a \cap X_c \subset X_{a,\mathrm{sing}}$. Correspondingly, V_c decays at $X_{a,\mathrm{reg}}$, so

$$H_a = \Delta_{X_a} + H^a,$$

which is an operator on (functions on) $X_0 = \mathbb{R}^n$, has the property that $H - H_a$ is a function that decreases at $X_{a,\mathrm{reg}}$. So H_a should be thought of as a good approximation of H at $X_{a,\mathrm{reg}}$. Note that $X_{a,\mathrm{sing}}$ is a finite union of codimension ≥ 1 submanifolds of X_a, so $X_{a,\mathrm{reg}}$ is in particular an open dense subset of X_a. Also, note that Δ_{X_a} plays a role analogous to the kinetic energy of the center of mass in the two-body setting, but now this description only valid locally, at $X_{a,\mathrm{reg}}$.

Having thus described the configuration space $X = X_0 = \mathbb{R}^n$, the next step is to describe the phase space, as was done first in [65] and [66]. The main goal in the process is to obtain a space on which broken bicharacteristics behave well. We remind the reader that we are concerned with singularities at infinity, hence with bicharacteristics that are uniformly close to infinity. Later we give a compactified description, but here for simplicity we give its homogeneous version, much as for the wave equation where bicharacteristics were integral curves of the homogeneous principal symbol. So we start with T^*X, but we wish to compress it at X_a in such a way that at $X_{a,\mathrm{reg}}$, $T^*_{X_{a,\mathrm{reg}}}X$ is replaced by $T^*_{X_{a,\mathrm{reg}}}X_a = T^*X_{a,\mathrm{reg}}$. For broken bicharacteristics this has the effect that only the X_a-tangential component of the momentum is preserved at $X_{a,\mathrm{reg}}$. So we define the compressed cotangent bundle as

$$\dot{T}^*X = \bigcup_{a \neq 1} T^*X_{a,\mathrm{reg}}.$$

Note that this is at first just a set, equipped with a projection $\dot{T}^*X \to X \setminus \{0\}$ induced by the bundle projections $T^*X_{a,\mathrm{reg}} \to X_{a,\mathrm{reg}}$. There is also a natural $\mathbb{R}^+$-action on $\dot{T}^*X$ via dilation in the configuration variables:

$$\mathbb{R}^+_r \times T^*X_{a,\mathrm{reg}} \ni (r, x_a, \xi_a) \mapsto (rx_a, \xi_a) \in T^*X_{a,\mathrm{reg}}. \qquad (3\text{--}1)$$

We topologize $\dot{T}^*X$ via the projection

$$\pi : T^*_{X \setminus \{0\}} X \to \dot{T}^*X,$$

whose restriction to $T^*_{X_{a,\mathrm{reg}}}X$ is the pull-back of one-forms by the inclusion map $X_{a,\mathrm{reg}} \hookrightarrow X$. Thus, writing (ξ_a, ξ^a) as the momenta dual to (x_a, x^a), π projects out the normal component of the momentum, ξ^a. The topology is then the weakest topology that makes π continuous, i.e. a set C in $\dot{T}^*X$ is closed if and only if $\pi^{-1}(C)$ is closed.

We can now describe the contribution of the bound states to the characteristic sets. As mentioned above, this is one of the most interesting features of many-body scattering that has no analogue for the wave equation. These are conic subsets of $\dot{T}^*X$ (conic with respect to the $\mathbb{R}^+_r$-action in (3–1)). The characteristic sets describe where certain operators are not elliptic, i.e. invertible, at infinity, in a precise sense described in the subsequent sections. They correspond to the 'energy shell', i.e. being on the characteristic set at energy λ means that the particles have total energy λ. We let

$$\mathrm{Char}_0(\lambda) = \{(x, \xi) \in T^*X : g(\xi) = \lambda\}$$

be the free characteristic variety, with g being the metric function on T^*X, and more generally we set

$$\mathrm{Char}_a(\lambda) = \{(x_a, \xi_a) \in T^*X_a : \lambda - g_a(\xi_a) \in \mathrm{spec}_{\mathrm{pp}} H^a\} \subset T^*X_a.$$

Notice that $\lambda = g_a(\xi_a) + \varepsilon_\alpha$, $\varepsilon_\alpha \in \mathrm{spec}_{\mathrm{pp}} H^a$, corresponds to the splitting of the total energy λ to the kinetic energy of the cluster, $g_a(\xi_a)$, plus the energy of the bound state, ε_α. Thus, $\mathrm{Char}_a(\lambda)$ describes that particles may exist in a bound state of H^a, of energy ε_α, along X_a, with kinetic energy $g_a(\xi_a) = \lambda - \varepsilon_\alpha$. Moreover, H^0 is the zero operator on $X^0 = \{0\}$, so if $a = 0$, these two definitions are consistent. If $X_a \subset X_b$, the pull-back of one-forms gives a projection $\pi_{ba} : T^*_{X_a}X_b \to T^*X_a$. Let

$$\dot{\mathrm{Char}}(\lambda) = \bigcup \dot{\mathrm{Char}}_a(\lambda) \subset \dot{T}^*X,$$

$$\dot{\mathrm{Char}}_a(\lambda) = \bigcup_{X_b \supset X_a} \pi_{ba}(\mathrm{Char}_b(\lambda)) \cap T^*X_{a,\mathrm{reg}}.$$

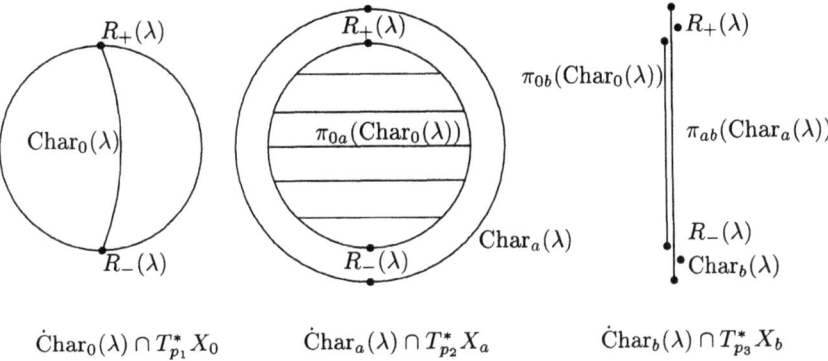

$$\dot{\mathrm{Char}}_0(\lambda) \cap T^*_{p_1} X_0 \qquad \dot{\mathrm{Char}}_a(\lambda) \cap T^*_{p_2} X_a \qquad \dot{\mathrm{Char}}_b(\lambda) \cap T^*_{p_3} X_b$$

Figure 4. The characteristic set of many-body Hamiltonians. Here H is a 4-body Hamiltonian, a is a 3-cluster, b is a 2-cluster, $p_1 \in X_{0,\mathrm{reg}}$, $p_2 \in X_{a,\mathrm{reg}}$, $p_3 \in X_{b,\mathrm{reg}}$. The solid dots are the radial sets, defined below.

In order to understand $\dot{\mathrm{Char}}(\lambda)$ it is important to keep in mind several results on the structure of the eigenvalues of the subsystems. So let

$$\Lambda_a = \bigcup_{b:X^b \subsetneq X^a} \mathrm{spec}_{\mathrm{pp}} H^b$$

be the set of thresholds of H^a. Fundamental results of Perry, Sigal and Simon [53] and of Froese and Herbst [16] show that Λ_a is closed, countable, and the countable set $\mathrm{spec}_{\mathrm{pp}} H^a$ can only accummulate at Λ_a, so

$$\Lambda'_a = \Lambda_a \cup \mathrm{spec}_{\mathrm{pp}} H^a = \bigcup_{b:X^b \subset X^a} \mathrm{spec}_{\mathrm{pp}} H^b$$

is also closed. Hence, $\dot{\mathrm{Char}}(\lambda)$ is a closed subset of $\dot{T}^*X$. In fact, the quotient $\dot{\Sigma}(\lambda)$ of $\dot{\mathrm{Char}}(\lambda)$ by the $\mathbb{R}^+$ action (which can be realized by restricting the various bundles to the unit sphere, $S_0 = \{x \in X_0 : |x| = 1\}$) is compact, and indeed it is metrizable, see [65]. Since compact topological spaces have better properties than noncompact ones, it is quite natural to work with $\dot{\Sigma}(\lambda)$, although we do not follow this route in this section. We also remark that it is much better to think of $\dot{\Sigma}(\lambda)$ lying at the sphere at infinity, rather than at S_0, since it is the dynamics at infinity that is described here. We will take up this approach in later sections.

We also recall another result of Froese and Herbst [15], namely that eigenfunctions ψ_α of H^a with eigenvalue ε_α decay exponentially on X^a, at a specified rate, if $\varepsilon_\alpha \notin \Lambda_a$. This generalizes to higher order perturbations, but requires a somewhat different approach, see [67]. In fact, this is the only place where second order perturbations behave differently from first or zeroth order ones. For the latter, there can be no positive energy bound states, while for the former this has been only proved for small metric perturbations in [67], and it is not clear whether it holds more generally, especially for trapping perturbations. (Note that if H^a, $a \neq 1$, is trapping then H is trapping at infinity!)

A generalized broken bicharacteristic (at energy λ) is then a continuous map $\gamma : I \to \dot{\mathrm{Char}}(\lambda)$, I an interval, such that a Hamilton vector field condition holds. To see what this is, we consider a subset of continuous functions on $\dot{T}^*X$, namely the class of π-invariant C^∞ functions on T^*X. π-invariance means that if $\zeta, \zeta' \in T^*X$ and $\pi(\zeta) = \pi(\zeta')$ then $f(\zeta) = f(\zeta')$. If f is π-invariant then it induces a function f_π on $\dot{T}^*X$ by $f_\pi(q) = f(\zeta)$ if $q = \pi(\zeta)$. Moreover, if f is smooth (or indeed just continuous) then f_π is continuous by the definition of the topology on $\dot{T}^*X$.

Now, if $\tilde{\gamma}$ is a curve in a manifold, one way to put that it is an integral curve of a vector field V is that

$$\frac{d}{ds}(f \circ \tilde{\gamma})|_{s=s_0} = (Vf)(\tilde{\gamma}(s_0))$$

for all smooth functions f. If f is a smooth π-invariant function on T^*X, then f defines a C^∞ function on T^*X_a for all a, so $H_{g_a}f$ makes sense. Here H_{g_a} is the Hamilton vector field of the metric function g_a on T^*X_a, so explicitly, $H_{g_a} = 2\xi_a \cdot \partial_{x_a}$. Now we would like to say that along a generalized broken bicharacteristic γ, $(d/ds)(f_\pi \circ \gamma)|_{s=s_0}$ should be given by $H_{g_b}f$ for some b and some ζ with $\pi(\zeta) = \gamma(s_0)$. The problem is that there are many such points ζ and clusters b, so this statement does not make any sense. However, we may replace the derivative by the lim inf of the difference quotients, i.e. by

$$D_{\pm}h(s_0) = \liminf_{s \to s_0} \frac{h(s) - h(s_0)}{s - s_0},$$

and demand an inequality instead of the equality. That is, we may demand that $D_{\pm}(f_\pi \circ \gamma)(s_0)$ may not be less than the worst possible scenario as we run over all such b and ζ. Thus, the condition for a continuous map $\gamma : I \to \dot{\mathrm{Char}}(\lambda)$ to be a generalized broken bicharacteristic is then that for any $s_0 \in I$, if $\gamma(s_0) \in T^*X_{a,\mathrm{reg}}$ then

$$D_{\pm}(f_\pi \circ \gamma)(s_0) \geq \inf\big\{(H_{g_b}f)(\zeta) : \zeta \in \mathrm{Char}_b(\lambda), \ \pi(\zeta) = \gamma(s_0), \ X_a \subset X_b\big\}.$$

If the set of bound states is discrete, then such a curve γ is piecewise an integral curve of the Hamilton vector field of g_b inside $\mathrm{Char}_b(\lambda)$, where b may of course vary. In particular, if there are no bound states in any proper subsystem, the picture is very similar to wave propagation: the definiton can be reduced to the analogue of Lebeau's [36].

The structure of the generalized broken bicharacteristics, including the above claims, depends on having a large supply of π-invariant functions. But these exist, since the pull-backs of all functions on X to T^*X is π-invariant, so one can localize in X using smooth cutoffs. Moreover, near $X_{a,\mathrm{reg}}$, each component of ξ_a is π-invariant, as is $\xi^a \cdot x^a$. Note that the generalized broken bicharacteristics depend on V, but only via the characteristic set $\dot{\mathrm{Char}}(\lambda)$, i.e. only via the bound states of the subsystem Hamiltonians.

There is also a wave front set associated to many-body scattering which measures the microlocal decay of tempered distributions at infinity. For a tempered distribution u, $\mathrm{WF}_{\mathrm{sc}}(u)$ is a closed conic subset of $\dot{T}^*X$. Apart from u, it depends on $\mathcal{X}$, since $\dot{T}^*X$ depends on $\mathcal{X}$, but we suppress this in the notation, and write

$$\mathrm{WF}_{\mathrm{sc}}(u) = \mathrm{WF}_{\mathrm{sc},\mathcal{X}}(u).$$

Its definition is slightly complicated, and I only refer to [65] for the general definition, which uses the structure of the pseudodifferential algebra, in particular the operator-valued nature of symbols at infinity. However, for generalized eigenfunctions of many-body Hamiltonians it is simple. Namely, suppose that $(H - \lambda)u \in \mathcal{S}(X)$, where $\mathcal{S}(X)$ is the space of Schwartz functions. For $\bar{x} = \bar{x}_a \in X_{a,\mathrm{reg}}$ and $\bar{\xi}_a \in X_a^*$ we say that $(\bar{x}_a, \bar{\xi}_a) \notin \mathrm{WF}_{\mathrm{sc}}(u)$ if there exists $\phi \in \mathcal{C}_c^\infty(X_a^*)$ such that $\phi(\bar{\xi}_a) \neq 0$ and $\mathcal{F}^{-1}\phi\mathcal{F}u$ is rapidly decreasing in an open cone in X around $\bar{x}_a$. Two examples are:

$$\mathrm{WF}_{\mathrm{sc}}(e^{ix\cdot\xi_0}) = \pi(\{(x,\xi_0) : x \neq 0\}), \qquad \xi_0 \in X_0^*,$$
$$\mathrm{WF}_{\mathrm{sc}}(e^{i\alpha|x|}) = \pi(\{(x, \alpha x/|x|) : x \neq 0\}). \quad \alpha \in \mathbb{R}.$$

More generally, if v is a symbol of any order on X_0, say $v \in S^k(X_0)$, and $\phi \in \mathcal{C}^\infty(X_0)$ is homogeneous degree 1 for $|x| > 1$, then

$$\mathrm{WF}_{\mathrm{sc}}(e^{i\phi(x)}v(x)) \subset \pi(\mathrm{graph}\, d\phi) = \pi(\{(x, (d\phi)(x)) : |x| > 1\}).$$

The condition $|x| > 1$ is due to the requirement of the homogeneity of ϕ only for $|x| > 1$; technically we should add a subset of $|x| \leq 1$ to the right hand side to make it conic. The theorem on the propagation of singularities is the following.

THEOREM 3.1. *Suppose that $\lambda \in \mathbb{R}$ and H is a many-body Hamiltonian. If $u \in \mathcal{S}'(\mathbb{R}^n)$ and $(H - \lambda)u \in \mathcal{S}(\mathbb{R}^n)$ then $\mathrm{WF}_{\mathrm{sc}}(u) \subset \dot{\mathrm{Char}}(\lambda)$ (microlocal elliptic regularity). Moreover, $\mathrm{WF}_{\mathrm{sc}}(u)$ is a union of maximally extended generalized broken bicharacteristics of $H - \lambda$ (propagation of singularities).*

REMARK. In the time dependent version, one considers tempered distributional solutions of $(D_t + H)u = 0$ on $X_0 \times \mathbb{R}_t$. Then $D_t + H$ still has the structure of a many-body Hamiltonian, with $D_t + \Delta$ in place of Δ, with collision planes given by $X_a \times \mathbb{R}$, with $\{0\}$ added for the sake of completeness. Thus, t is always a variable along the collision planes, so in particular, its dual variable τ, is π-invariant. Moreover, $\mathrm{Char}_a(\lambda)$ is replaced by

$$\mathrm{Char}_a = \{(x_a, t, \xi_a, \tau) \in T^*(X_a \times \mathbb{R}) : -\tau - g_a(\xi_a) \in \mathrm{spec}_{\mathrm{pp}} H^a\} \subset T^*X_a,$$

so effectively $-\tau$ plays the role of the energy λ. Generalized broken bicharacteristics can be defined as before with H_{g_b} replaced by $H_{\tau+g_b} = \partial_t + 2\xi_b \cdot \partial_{x_b}$. The main additional issue is that they can only be expected to give a good description of propagation at finite energies since $D_t + H$ is not elliptic in the usual sense. So the analogue of Theorem 3.1 is that if $u \in \mathcal{S}'(\mathbb{R}^n \times \mathbb{R})$, $(D_t + H)u = 0$, and $u = \psi(H)u$ for some $\psi \in \mathcal{C}_c^\infty(\mathbb{R})$, then $\mathrm{WF}_{\mathrm{sc}}(u)$ is a subset of the characteristic set, and in fact $\mathrm{WF}_{\mathrm{sc}}(u)$ is a union of maximally extended generalized broken bicharacteristics of $D_t + H$ inside it. The proof of this statement only requires simple modifications of the proof of the theorem.

The interpretation of the theorem is much analogous to that for the wave equation. However, there is a difference which also occurs in the traditional microlocal setting for more general operators (i.e. for operators other than the wave operator), see [21]. Namely, the orbits of the $\mathbb{R}^+$-action may be bicharacteristics, and then the statement of the theorem is empty at the points lying on these orbits since the wave front set is *a priori* conic. This happens for $(x_a, \xi_a) \in T^* X_{a,\mathrm{reg}} \subset \dot{T}^* X$ if and only if there exists some cluster b with $X_b \supset X_a$, and $\zeta \in \mathrm{Char}_b(\lambda)$ such that H_{g_b} at ζ is tangent to the orbits of the $\mathbb{R}^+$-action. This happens, in turn, if and only if ξ_a is parallel to x_a and $\lambda - |\xi_a|^2 \in \mathrm{spec}_{\mathrm{pp}} H^b$. Such points are called radial points, and their collection is denoted by

$$\mathcal{R}(\lambda) = \bigcup_{a \neq 1} \{ (x_a, \xi_a) \in T^* X_{a,\mathrm{reg}} : \exists c \in \mathbb{R},\ \xi_a = c x_a,$$
$$\exists b,\ X_b \supset X_a,\ \lambda - |\xi_a|^2 \in \mathrm{spec}_{\mathrm{pp}} H^b \}.$$

As we discuss in Section 7, $\mathcal{R}(\lambda)$ plays an important role in asymptotic completeness. In many-body scattering it appeared in the work of Sigal and Soffer [55] and was called 'propagation set' because in the time-dependent picture this is where particles end up as time goes to infinity. (In the stationary semiclassical picture, this is where nontrapped classical trajectories starting in a compact region end up.) It is thus unfortunate, in terms of terminology, that this is also the region where there is no real principal type propagation.

REMARK. In the time-dependent problem, the set of radial points is

$$\mathcal{R} = \bigcup_{a} \{ (x_a, t, \xi_a, \tau) \in T^* (X_{a,\mathrm{reg}} \times \mathbb{R}) : x_a = 2t\xi_a,$$
$$\exists b,\ X_b \supset X_a,\ -\tau - |\xi_a|^2 \in \mathrm{spec}_{\mathrm{pp}} H^b \}.$$

In terms of radial points, the difference between threshold energies $\lambda \in \Lambda = \Lambda_0$ and nonthreshold energies is that if $\lambda \in \Lambda$, then there are *constant* generalized broken bicharacteristics, i.e. bicharacteristics whose image is a single point. Namely, if $x_a \in X_{a,\mathrm{reg}}$ and $\lambda \in \mathrm{spec}_{\mathrm{pp}} H^a$, then

$$(x_a, 0) \in \mathrm{Char}_a(\lambda) \cap T^* X_{a,\mathrm{reg}} \subset \dot{\mathrm{Char}}(\lambda),$$

and $H_{g_a} = 2\xi_a \cdot \partial_{x_a}$ vanishes there, so $(x_a, 0)$ is indeed the image of a constant bicharacteristic. While this does not make any difference for the propagation of singularities, it does for the related limiting absorption principle, which in this generality is due to Perry, Sigal and Simon [53].

THEOREM 3.2. *If $\lambda \notin \Lambda$, then the limits $R(\lambda \pm i0) = (H - (\lambda \pm i0))^{-1}$ exists as bounded operators between L_s^2 and H_{-s}^2 for $s > \frac{1}{2}$. Here H_l^m is the weighted Sobolev space $\langle x \rangle^{-l} H^m(\mathbb{R}^n)$, $L_s^2 = H_s^0$.*

In fact, the proofs of the limiting absorption principle and the propagation of singularities are related. Indeed, the statement on propagation of singularities can be strengthened for $R(\lambda + i0)f$, $f \in \mathcal{S}(\mathbb{R}^n)$, by saying that $\mathrm{WF}_{sc}(R(\lambda + i0)f)$ is not only a union of maximally extended generalized broken bicharacteristics, as follows from Theorem 3.1, but in fact it is a union of generalized broken bicharacteristics $\gamma : \mathbb{R}_s \to \dot{T}^* X$ which go to

$$\mathcal{R}_+(\lambda) = \mathcal{R}(\lambda) \cap \bigcup_{a \neq 1} \{(x_a, \xi_a) \in T^* X_{a,\mathrm{reg}} : x_a \cdot \xi_a > 0\}$$

as $s \to -\infty$. That is, the singularities at $\mathcal{R}_+(\lambda)$ (where the statement of Theorem 3.1 is empty) can only leave $\mathcal{R}_+(\lambda)$ in the *forward* direction. The limiting absorption principle is thus strengthened to:

THEOREM 3.3. *If $\lambda \notin \Lambda$, then for $f \in \mathcal{S}(\mathbb{R}^n)$, $\mathrm{WF}_{sc}(R(\lambda + i0)f)$ is a subset of the image of $\mathcal{R}_+(\lambda)$ under the* forward *generalized broken bicharacteristic relation. A similar statement holds for $R(\lambda - i0)f$ with $\mathcal{R}_+(\lambda)$ replaced by*

$$\mathcal{R}_-(\lambda) = \mathcal{R}(\lambda) \cap \bigcup_{a \neq 1} \{(x_a, \xi_a) \in T^* X_{a,\mathrm{reg}} : x_a \cdot \xi_a < 0\},$$

and the forward relation by the backward relation.

In fact, if $u \in \mathcal{S}'(\mathbb{R}^n)$ and $\mathrm{WF}_{sc}(u)$ is disjoint from the image of $\mathcal{R}_-(\lambda)$ under the backward generalized broken bicharacteristic relation, then $R(\lambda + i0)u$ is defined by duality and $\mathrm{WF}_{sc}(R(\lambda + i0)u)$ is a subset of the image of $\mathcal{R}_+(\lambda) \cup \mathrm{WF}_{sc}(u)$ under the forward relation.

REMARK. $\lambda \notin \Lambda$ can be also characterized by $\mathcal{R}(\lambda) = \mathcal{R}_+(\lambda) \cup \mathcal{R}_-(\lambda)$, i.e. that $x_a \cdot \xi_a$ never vanishes on $\mathcal{R}(\lambda) \cap T^* X_{a,\mathrm{reg}}$ for any a.

In the time-dependent setting, $x_a = 2t\xi_a$ on $\mathcal{R}$, so $x \cdot \xi_a = 2t$. So $\mathcal{R}_+$, defined in $\mathcal{R}$ by $x_a \cdot \xi_a > 0$, is the subset of $\mathcal{R}$ where $t > 0$. Hence the 'outgoing' terminology for $R(\lambda + i0)$ and 'incoming' for $R(\lambda - i0)$. In fact, the solution of $(D_t + H)u = 0$ with $u|_{t=0} = \phi$, $\phi \in \mathcal{S}(X_0)$, say, is $u(\cdot, t) = e^{-iHt}\phi$. The time-dependent propagation of singularities shows that $\mathrm{WF}_{sc}(u)$ is a subset of the union of the image of $\mathcal{R}_+$ under the forward broken bicharacteristic relation and the image of $\mathcal{R}_-$ under the backward bicharacteristic relation. Using the spectral measure and Stone's theorem,

$$u(\cdot, t) = \frac{1}{2\pi i} \int_{\mathbb{R}} e^{-i\lambda t} (R(\lambda + i0) - R(\lambda - i0))\phi \, d\lambda$$

Fixing some $\psi \in \mathcal{C}_c^\infty(\mathbb{R})$, for ϕ in the range of $\psi(H)$ we thus deduce that in $t > 0$, $\mathrm{WF}_{sc}(u)$ arises from the $R(\lambda + i0)$ term, and in $t < 0$ from the $R(\lambda - i0)$ term. So the time-dependent and stationary settings are very close: the only difference is that in the latter, λ is a parameter, while in the former, it is a variable, $\lambda = -\tau$.

Again, one can make more precise propagation statements in some circumstances, such as three-body scattering, where the precise nature of the singularities can be analyzed, see [22; 63]. Here we only state the stronger implication for the structure of the scattering matrices, which we proceed to analyze.

4. Scattering Matrices

Physically, the scattering matrices relate incoming and outgoing data in an experiment. In the time independent framework (where $-\lambda$ is the dual variable of time), for short-range potentials an incoming wave of energy λ in channel α (a channel is the choice of a cluster a and an L^2-eigenfunction ψ_α of H^a of energy ε_α) takes the following form in $|x| > 1$:

$$u_{\alpha,-} = e^{-i\sqrt{\lambda-\varepsilon_\alpha}|x_a|}|x_a|^{-(\dim X_a - 1)/2} g_{\alpha,-}\left(\frac{x_a}{|x_a|}\right)\psi_\alpha(x^a) + u'_-$$

Similarly, an outgoing wave has the form

$$u_{\alpha,+} = e^{i\sqrt{\lambda-\varepsilon_\alpha}|x_a|}|x_a|^{-(\dim X_a - 1)/2} g_{\alpha,+}\left(\frac{x_a}{|x_a|}\right)\psi_\alpha(x^a) + u'_+$$

i.e. the sign of the phase has changed. Here $g_{\alpha,\pm}$ may be taken e.g. L^2 functions on S_a, the unit sphere in X_a, or ideally, at least one of them may be taken C^∞. In either case, $u'_\pm$ are 'lower order terms', namely they must be in $L^2_{-1/2}$. (Note that $\langle x_a\rangle^{-(\dim X_a-1)/2} \in L^2_s(X_a)$ for $s < -\frac{1}{2}$ but not for $s = -\frac{1}{2}$.) In fact, for $g_{\alpha,\pm} \in C^\infty_c(S_{a,\mathrm{reg}})$ we may take them to be of the form $e^{-i\sqrt{\lambda-\varepsilon_\alpha}|x_a|}|x_a|^{-(\dim X_a+1)/2}v$ where v is a 0th order symbol, with $S_{a,\mathrm{reg}}$ denoting $X_{a,\mathrm{reg}} \cap S_0$.

One can now produce tempered distribution with given incoming, or alternatively of given outgoing, asymptotics. A typical example is of the form

$$P_{\alpha,+}(\lambda)g_{\alpha,-} = u_{\alpha,-} - (H - (\lambda+i0))^{-1}((H - \lambda)u_{\alpha,-}); \qquad (4\text{--}1)$$

here the lower order terms may be dropped from $u_{\alpha,-}$ without affecting $u = P_{\alpha,+}(\lambda)g_{\alpha,-}$ and $g_{\alpha,-}$ can be specified to be any smooth function on S_a. In general, even if the incoming data are in a single channel α, as in (4–1), the corresponding generalized eigenfunction u of H will have outgoing waves in all channels. The S-matrix $S_{\alpha\beta}(\lambda)$ picks out the component in channel β by projection in a certain sense, see [64]. Thus, $S_{\alpha\beta}(\lambda)$ maps functions on S_a, the unit sphere in X_a, to functions on S_b, by

$$S_{\alpha\beta}(\lambda)g_{\alpha,-} = g_{\beta,+}$$

for u as in (4–1). For example, the free-to-free (i.e. N-cluster to N-cluster in N-body scattering) S-matrix $S_{00}(\lambda)$ maps functions on S_0, the unit sphere in $\mathbb{R}^n$, to functions on S_0, more precisely $S_{00}(\lambda) : L^2(S_0) \to L^2(S_0)$ is bounded.

More precisely, let T_+ be a pseudodifferential operator that is identically 1 on the outgoing radial set and identically 0 on the incoming radial set; see the paragraph of (5–6) for a precise statement. Then

$$S_{\alpha\beta}(\lambda) = \frac{1}{2i\sqrt{\lambda - \varepsilon_\beta}}((H - \lambda)T_+ P_{\beta,-}(\lambda))^* P_{\alpha,+}(\lambda), \qquad (4\text{–}2)$$

i.e. for any $g \in \mathcal{C}^\infty(S_{a,\mathrm{reg}})$, $h \in \mathcal{C}^\infty(S_{b,\mathrm{reg}})$,

$$\langle h, S_{\alpha\beta}(\lambda)g \rangle = \left\langle (H - \lambda)T_+ P_{\beta,-}(\lambda)h, \ \frac{1}{2i\sqrt{\lambda - \varepsilon_\beta}} P_{\alpha,+}(\lambda)g \right\rangle.$$

This is equivalent to the usual wave operator definition in the time-dependent setting, see [64]. An immediate consequence of the propagation of singularities and the definition of the scattering matrices is the following:

THEOREM 4.1. *The wave front relation of $S_{\alpha\beta}(\lambda)$ is given by the broken bicharacteristic relation. In particular, if no proper subsystem of H has bound states, the wave front relation of $S_{00}(\lambda)$ is given by the broken geodesic flow on S_0 at distance π.*

While typically broken bicharacteristics can be continued in many ways when they hit a collision plane, it is important to keep in mind that under suitable assumptions (which rule out geometric complications) the broken bicharacteristic relation is Lagrangian, hence its dimension is the same as if there were no collision planes. The reason is that only a low dimensional family of broken bicharacteristics hits any specified collision plane, with the dimension of the possible continuations of each of these these bicharacteristics compensating to yield the correct dimension for Lagrangian submanifolds.

The significant improvement in the three-body case, as shown by Hassell and the author [63; 22], is that one can pinpoint not only the location of the singularities, but also their precise form. This theorem was motivated by the geometric result of Melrose and Zworski [46], showing that the scattering matrix on asymptotically Euclidean manifolds is a Fourier integral operator.

THEOREM 4.2. *Suppose that H is a three-body Hamiltonian and the V_a are Schwartz on X^a for all a. Then $S_{00}(\lambda)$ is a finite sum of Fourier integral operators (FIOs) associated to the broken geodesic relation on S_0 to distance π. Its canonical relation corresponds to the various collision patterns. The principal symbol of the term corresponding to a single collision at X_a is given by, and in turn determines, the 2-body S-matrix of H^a at energies $\lambda' \in (0, \lambda)$.*

REMARK. This result presumably extends to short range symbolic potentials, using the same methods, though it is technically more complicated to write down the argument in that case, and it has not been done. In fact, it should also extend to the N-body problem, provided that there are no bound states in any proper subsystem. Some assumption on the bound states is necessary, for otherwise the generalized broken bicharacteristic relation can become fairly complicated; see [66]. The reason why one does not need any assumption on bound states in three-body scattering is that for any 2-cluster a, $\text{Char}_a(\lambda) \cap \pi_{0a}(\text{Char}_0(\lambda))$ is either empty (if 0 is not an eigenvalue of H^a) or consists of the boundary of $\pi_{0a}(\text{Char}_0(\lambda))$. In the former case there is no interaction (modulo smoothing terms) between the 0-cluster and the a-cluster dynamics, while in the latter case in the only place they interact, the two dynamics give the same propagation.

It should also be noted that the normalization of $S_{\alpha\beta}(\lambda)$ is not the standard one in many-body scattering (which is based on wave operators), but rather follows the geometric conventions [41]. The difference is that in the wave operator approach, free motion is factored out, so the free scattering matrix is the identity operator. On the other hand, in the geometric approach we describe the asymptotics of generalized eigenfunctions, or alternatively of the Schrödinger equation. Since free particles move to infinity in the opposite direction from which they came, it is reasonable that the two should differ by (a constant multiple of) pullback by the antipodal map, and this is indeed the case, see [64]. The distance π propagation along (not broken!) geodesics on the sphere indeed takes particles to the antipodal point.

An immediate corollary, when combined with two-body results (e.g. analyticity of the S-matrix in λ' and the Born approximation) is the following inverse result.

COROLLARY. *If the V_a decay exponentially and* $\dim X_a \geq 2$ *for all a then* $S_{00}(\lambda)$ *for a single value of λ determines all interactions.*

This result is analogous to the recovery of cracks in a material by directing sound waves at it and observing the singularities of the reflected waves, except the last step which uses two-body results to get the potentials from the two-body S-matrices.

The other extremal scattering matrices are the 2-cluster to 2-cluster ones, and they describe the physically most interesting events. Indeed, it is hard to make more than two particles collide in an accelerator, so the initial state in a physical experiment tends to be a 2-cluster. The following result is due to Skibsted [59], and it also follows from the propagation of singularities and the definition of the S-matrices.

THEOREM 4.3. *Let α and β be two-clusters, and suppose that either $\varepsilon_\alpha \in \text{spec}_d H^a$ and $\varepsilon_\beta \in \text{spec}_d H^b$, or V_c is Schwartz for all c. Then the two-cluster to two-cluster S-matrix $S_{\alpha\beta}(\lambda)$ has C^∞ Schwartz kernel, except if $\alpha = \beta$ in which case the Scwartz kernel of $S_{\alpha\alpha}(\lambda)$ is conormal to the graph of the antipodal map on S_a, corresponding to free motion.*

Thus, principal symbol calculations do not help in this inverse problem. Note that if H is a 3-body Hamiltonian, then $\varepsilon_\alpha \in \text{spec}_d H^a$ and $\varepsilon_\beta \in \text{spec}_d H^b$ holds for any nonthreshold bound state energies. The new result, in a joint project with Gunther Uhlmann, is the following [62].

THEOREM 4.4. *Suppose that H is a 3-body Hamiltonian, a is a 2-cluster, α is a channel of energy $\varepsilon_\alpha < 0$, V_a is a symbol of negative order (i.e. may be long range). For any $\mu > \dim X_a$ there exists $\delta > 0$ such that the following holds.*

Suppose that $\sup |(1 + |x^b|)^\mu V_b(x^b)| < \delta$ for all $b \neq a$. Suppose also that $I \subset (\varepsilon_\alpha, 0)$ is a nonempty open set, and let

$$R = 2\sqrt{\sup I - \varepsilon_\alpha}.$$

Then $S_{\alpha'\alpha''}(\lambda)$ given for all $\lambda \in I$ and for all bound states α', α'' of H^a with $\varepsilon_{\alpha'}, \varepsilon_{\alpha''} < \sup I$, determines the Fourier transform of the effective interaction $V_{\alpha,\mathrm{eff}}$ in the ball of radius R centered at 0.

The effective interaction is the interaction that arises if we consider the 3-body problem as a 2-body problem, i.e. if we regard the two particles forming the cluster a as a single particle. Mathematically, this amounts to projecting to the state ψ_α in X^a and obtaing a new Hamiltonian $\Delta_{X_a} + V_{\alpha,\mathrm{eff}}$ on X_a. Thus, the effective interaction is physically relevant. Moreover, there is no hope for recovering anything better than $V_{\alpha,\mathrm{eff}}$ as shown by the high-energy inverse results of Enss and Weder [11; 13], Novikov [49] and Wang [70; 71].

This theorem says that if the unknown interactions are small then the effective interaction can be determined from the knowledge of all S-matrices with incoming and outgoing data in the cluster a in the relevant energy range. In fact, near-forward information suffices as in two-body scattering, where this was observed recently by Novikov [50]. Also, if one is willing to take small R and α is the ground state of H^a, it suffices to know $S_{\alpha\alpha}(\lambda)$ to recover $\hat{V}_{\alpha,\mathrm{eff}}$ in a small ball.

In case V_b decay exponentially on X^b for all $b \neq a$, then $V_{\alpha,\mathrm{eff}}$ decays exponentially on X_a, hence its Fourier transform is analytic, so $V_{\alpha,\mathrm{eff}}$ itself can be recovered from these S-matrices.

REMARK. It is clear from the proof in [62] that there is a natural extension of this theorem to many-body scattering at low energies.

This result should extend to higher energies, i.e. $\sup I \leq 0$ is *not* expected to be essential. But it is hard to make R greater than $2\sqrt{-\varepsilon_\alpha}$ even then. The reason is that our method relies on the construction of exponential solutions following Faddeev [14], Calderón [6], Sylvester and Uhlmann [60] and Novikov and Khenkin [48], but in the three-body setting. One thus allows complex momenta $\rho \in \mathbb{C}(X_a)$, the complexification of X_a, and one wants to construct solutions of $(H - \lambda)u = 0$ of the form

$$e^{i\rho \cdot x_a}(\psi_\alpha(x^a) + v),$$

where $v = v_\rho$ is supposed to be 'small' in the sense that it goes to 0 as $\rho \to \infty$ in an appropriate fashion. Note that with $v = 0$ these complex plane waves solve $(H_a - \lambda)u = 0$ with

$$\lambda = \rho \cdot \rho + \varepsilon_\alpha; \tag{4-3}$$

this expresses that the total energy λ is the sum of the kinetic energy, $\rho \cdot \rho$, and the potential energy ε_α.

To construct u, we need to find v, and its study reduces to that of the conjugated Hamiltonian

$$e^{-i\rho \cdot x_a}(H - \lambda)e^{i\rho \cdot x_a} = H^a + \Delta_{X_a} + 2\rho \cdot D_{X_a} + I_a - \varepsilon_\alpha$$

with $\rho \in \mathbb{C}(X_a)$ the complex frequency. Here we used (4–3). Now, I_a is considered as a perturbation (this is the reason for the smallness assumption in the theorem), so we really study the model operator,

$$H^a + \Delta_{X_a} + 2\rho \cdot D_{X_a} - \varepsilon_\alpha.$$

Taking the Fourier transform in the X_a variables, one obtains

$$H^a + |\xi_a|^2 + 2\rho \cdot \xi_a - \varepsilon_\alpha.$$

Writing $\rho = z\nu + \rho_\perp$ with $|\nu| = 1$, $\rho_\perp \cdot \nu = 0$, ρ, ν real, $z \in \mathbb{C}$, this operator becomes

$$H^a + |\xi_a|^2 + 2\rho_\perp \cdot \xi_a + 2z\nu \cdot \xi_a - \varepsilon_\alpha = H^a + (\xi_a + \rho_\perp)^2 + 2z\nu \cdot \xi_a - |\rho_\perp|^2 - \varepsilon_\alpha.$$

If ρ is not real, then neither is z, so this operator is invertible if $\nu \cdot \xi_a \neq 0$ since H^a is self-adjoint. On the other hand, if $\nu \cdot \xi_a = 0$, this operator becomes

$$H^a + (\xi_a + \rho_\perp)^2 - |\rho_\perp|^2 - \varepsilon_\alpha,$$

i.e. its invertibility properties correspond to the behavior of the boundary values of the resolvent of H^a at the real axis. If $|\rho_\perp|^2 + \varepsilon_\alpha < 0$, i.e. if $|\rho_\perp| < \sqrt{-\varepsilon_\alpha}$, then the spectral parameter $|\rho_\perp|^2 + \varepsilon_\alpha - (\xi_a + \rho_\perp)^2$ is negative, so only the bound states of H^a contribute to the characteristic variety, i.e. the two-cluster a may not break up. On the other hand, if $|\rho| \geq \sqrt{-\varepsilon_\alpha}$, such a break-up is possible *even if* $\lambda < 0$, i.e. where the break up may not happen for *real* frequencies. The break-up greatly influences analyticity properties, hence one cannot easily use large $\rho_\perp$. On the other hand, one needs such large $\rho_\perp$ to recover $V_{\alpha,\text{eff}}$ on larger balls, hence the limitation in the theorem. This also suggests that the fixed energy problem would be hard, since then one always needs to let $\rho_\perp \to \infty$ to keep $\rho \cdot \rho = |\rho_\perp|^2 + z^2$ fixed and yet have $\rho \to \infty$.

5. Many-body Scattering Pseudo-Differential Operators

I will present the calculus from the compactified point of view. Both the one-step polyhomogeneous (i.e. 'classical') and the nonpolyhomogeneous calculus can be described in noncompact terms, i.e. directly on X_0, but this is more complicated and less natural. Indeed, one of the beauties of compactification is that it exactly captures the structure of many-body Hamiltonians. *We warn the reader here that from now on the Euclidean variable is written as z, rather than* *x in the preceeding sections, for compatibility with previous papers espousing this approach, such as [41; 42].*

To see how the compactification should go, recall first that a classical symbol of order 0 on $\mathbb{R}^n_z$ has an asymptotic expansion

$$a(r\omega) \sim \sum_{j=0}^{\infty} r^{-j} a_j(\omega), \quad a_j \in C^{\infty}(\mathbb{S}^{n-1}),$$

in the polar coordinates (r, ω): $z = r\omega$. The meaning of such an expansion is that, for any k, the difference of a and the sum of the first k terms on the right hand side is a symbol of order $-k$. This expansion is just a Taylor series at $r = \infty$, or rather at '$r^{-1} = 0$'. So we compactify $\mathbb{R}^n$ into a ball $\overline{\mathbb{B}^n}$ by adding points $(0, \omega)$, $\omega \in \mathbb{S}^{n-1}$, and making $(r^{-1}, \omega) = (x, \omega)$ coordinates near these points. The resulting space is called the radial compactification $\overline{\mathbb{R}^n}$ of $\mathbb{R}^n$. Thus, a classical symbol of order 0 is simply a smooth function of $\overline{\mathbb{R}^n}$; the asymptotic expansion at infinity is its Taylor series around the boundary, $x = 0$.

This compactification, whose utility in this context was emphasized by Melrose [41], can also be realized as the closed unit upper hemisphere via a modified stereographic projection. So let $\mathrm{RC} : \mathbb{R}^n \to \mathbb{S}^n_+$ be given by

$$\mathrm{RC}(z) = \left(\frac{1}{\langle z \rangle}, \frac{z}{\langle z \rangle} \right), \quad \text{where } \langle z \rangle = (1 + |z|^2)^{1/2}, \quad z \in \mathbb{R}^n.$$

Then n of the $n + 1$ variables $(1/\langle z \rangle, z/\langle z \rangle)$ give local coordinates on various regions of $\mathbb{S}^n_+$. In particular, in coordinate patches near the equator, which is $\partial \mathbb{S}^n_+$, $1/\langle z \rangle$ (or indeed $x = |z|^{-1}$) and $n - 1$ of $z_j/\langle z \rangle$ (or indeed $\omega_j = z_j/|z|$) can be taken as coordinates, showing that $\mathbb{S}^n_+$ can be identified with the radial compactification $\overline{\mathbb{R}^n}$. A slightly modified version of x (it needs to be smoothed at $z = 0$, where '$x = \infty$'), or $\langle z \rangle^{-1}$, can be taken as a boundary defining function. We will usually write x for this, so $x = |z|^{-1}$ for $|z| \geq 1$, say. (A boundary defining function is a nonnegative function whose zero set is exactly the boundary, and whose differential does not vanish there.)

How can we adapt this to many-body scattering? Let $\bar{X}_a$ denote the closure of X_a in the compactification $\overline{\mathbb{R}^n}$ of $\mathbb{R}^n$, and let $C_a = \partial \bar{X}_a \subset \partial \mathbb{S}^n_+ = C_0$. The closure of any translate of X_a intersects C_0 in the same submanifold (a sphere) as X_a itself. Indeed, writing the coordinates as (z_a, z^a) on $X_0 = X_a \oplus X^a$, local coordinates near C_a are given by $Z^a = z^a/|z|$, $|z|^{-1}$ and $\dim X_a - 1$ of $(z_a)_j/|z|$. Thus, $Z^a \to 0$ as $x \to 0$ along any translate, since z^a is constant along these. So V_a is not even continuous on $\bar{X}_0$, as it takes different values on the different translates of X_a. However, it *is* a negative order symbol (in particular continuous with boundary value 0) on $\bar{X}_0 \setminus C_a$, if V_a is such on X^a; see Figure 5.

So the compactification works for V_a, except at C_a. To remedy this, we blow up C_a. This is an invariant way of introducing polar coordinates about it (i.e. projective coordinates in various charts). That is, curves approaching C_a from various normal directions will correspond to different points on the blown-up

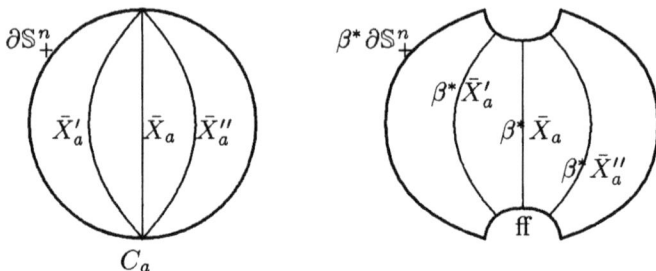

Figure 5. Translates of X_a on $[\bar{X}_0; C_a]$.

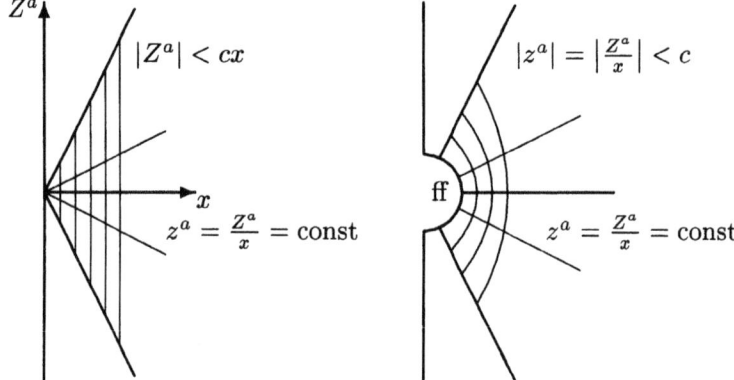

Figure 6. The blow up of C_a, given by $Z^a = 0$, $x = 0$.

space $[\bar{X}_0; C_a]$. Since C_a is given by $x = 0$, $Z^a = 0$, in local coordinates, this means concretely that the components of Z^a/x become coordinate functions on the part of $[\bar{X}_0; C_a]$ where this quotient is finite. (For the sake of completeness, a complete set of coordinates in this region is given by x, the components of Z^a as well as the dim $X_a - 1$ coordinates on the sphere $y_a = z_a/|z_a|$; see Figure 6.) But $Z^a/x = z^a$, so it is now easy to see that for classical symbols V_a on X^a (of negative integer order), V_a is a C^∞ function on $[\bar{X}_0; C_a]$.

In general, there are many collision planes, and we blow them up recursively, starting with ones of the largest codimension, to get $[\bar{X}_0; \mathcal{C}]$,

$$\mathcal{C} = \{C_a : X_a \in \mathcal{X}, \ a \neq 1\}.$$

We refer to [65] for details.

There is no reason at all to take $\bar{X}_0$ as the space we start with. Given any compact manifold with boundary, $\bar{X}$, and a cleanly intersecting family of closed embedded submanifolds $\mathcal{C}$ of $\partial \bar{X}$, we can define $[\bar{X}; \mathcal{C}]$ analogously. For instance, one can start with $\bar{X} = \overline{\mathbb{R}^n} \times \mathbb{S}^k$. The space $[\bar{X}; \mathcal{C}]$ is equipped with boundary fibrations given by the blow-down maps, see [38] for a simpler case where these first appeared explicitly.

Having described the configuration space, we turn to differential operators. X_0 has a nice algebra of differential operators, consisting of operators with symbolic coefficients: $\sum_{|\alpha| \le m} a_\alpha(z) D_z^\alpha$, $a_\alpha \in S^0(X_0)$. We may require instead that a_α is 'classical', i.e. that $a_\alpha \in C^\infty(\bar{X}_0)$. The resulting algebras were denoted $\mathrm{Diff}_{\mathrm{sccl}}(\bar{X}_0)$ and $\mathrm{Diff}_{\mathrm{sc}}(\bar{X}_0)$ by Melrose; he called them 'scattering differential operators'.

This setup generalizes to the geometric set-up as follows. Let (x, y), with $y = (y_1, \ldots, y_{n-1})$, be local coordinates near $\partial \bar{X}_0$. Then the vector fields in $\mathrm{Diff}_{\mathrm{sc}}(\bar{X}_0)$ are linear combinations of $x^2 D_x$ and the $x D_{y_j}$ with coefficients in $C^\infty(\bar{X}_0)$, as can be seen easily by an explicit calculation.

Now, if $\bar{X}$ is a manifold with boundary, $\mathcal{V}_{\mathrm{b}}(\bar{X})$ is the Lie algebra of vector fields tangent to $\partial \bar{X}$, and $\mathcal{V}_{\mathrm{sc}}(\bar{X}) = x \mathcal{V}_{\mathrm{b}}(\bar{X})$, where x is a defining function of $\partial \bar{X}$. $\mathcal{V}_{\mathrm{sc}}(\bar{X})$ is independent of the choice of x. Then $\mathcal{V}_{\mathrm{b}}(\bar{X})$ is spanned by $x \partial_x$ and ∂_y over $C^\infty(\bar{X})$, so $\mathcal{V}_{\mathrm{sc}}(\bar{X})$ is spanned by $x^2 \partial_x$ and $x \partial_y$ over $C^\infty(\bar{X})$. By definition, these generate $\mathrm{Diff}_{\mathrm{sc}}(\bar{X})$. Also, $\mathcal{V}_{\mathrm{sc}}(\bar{X})$ is the set of all smooth sections of a vector bundle over $\bar{X}$, this is denoted by $^{\mathrm{sc}}T\bar{X}$. Its dual bundle is the scattering cotangent bundle, denoted by $^{\mathrm{sc}}T^*\bar{X}$. In the Euclidean setting,

$$^{\mathrm{sc}}T\bar{X}_0 = \bar{X}_0 \times X_0, \quad {}^{\mathrm{sc}}T^*\bar{X}_0 = \bar{X}_0 \times X_0^*.$$

The way to generalize this differential operator algebra to one that includes many-body potentials is to allow singular coefficients $a_\alpha \in C^\infty([\bar{X}; \mathcal{C}])$. Thus,

$$\mathrm{Diff}_{\mathrm{sc}}(\bar{X}; \mathcal{C}) = C^\infty([\bar{X}; \mathcal{C}]) \otimes_{C^\infty(\bar{X})} \mathrm{Diff}_{\mathrm{sc}}(\bar{X}).$$

In particular, if H is a many-body Hamiltonian, with either potential or higher order interactions, then $H \in \mathrm{Diff}_{\mathrm{sc}}^2(\bar{X}_0; \mathcal{C})$.

We let $^{\mathrm{sc}}T[\bar{X}; \mathcal{C}] = \beta^{*\,\mathrm{sc}}T\bar{X}$ and $^{\mathrm{sc}}T^*[\bar{X}; \mathcal{C}] = \beta^{*\,\mathrm{sc}}T^*\bar{X}$, where $\beta : [\bar{X}; \mathcal{C}] \to \bar{X}$ is the blow-down map, and we are pulling back the vector bundles by it. Hence in the Euclidean setting,

$$^{\mathrm{sc}}T[\bar{X}_0; \mathcal{C}] = [\bar{X}_0; \mathcal{C}] \times X_0, \quad {}^{\mathrm{sc}}T^*[\bar{X}_0; \mathcal{C}] = [\bar{X}_0; \mathcal{C}] \times X_0^*.$$

Now it is natural to define pseudodifferential operators using these bundles. Although I restrict the discussion to the Euclidean setting, the construction generalizes to any $\bar{X}$ via localization.

So we consider symbols

$$a \in \langle z \rangle^{-l} \langle \zeta \rangle^m C^\infty([\bar{X}_0; \mathcal{C}] \times \bar{X}_0^*), \tag{5-1}$$

$X_0 = \mathbb{R}^n$, where ζ is the dual variable of z, i.e. the variable on X_0^*. Note that this means that a is a classical symbol of order m in ζ. As usual, we define the Schwartz kernel of the left quantization of a by

$$A = q_L(a) = (2\pi)^{-n} \int_{\mathbb{R}^n} e^{i(z-z') \cdot \zeta} a(z, \zeta) \, d\zeta, \tag{5-2}$$

understood as an oscillatory integral. In particular, for any $f \in \mathcal{S}(\mathbb{R}^n)$,

$$Af(z) = (2\pi)^{-n} \int_{\mathbb{R}^n} \int_{\mathbb{R}^n} e^{i(z-z')\cdot\zeta} a(z,\zeta) f(z') \, d\zeta \, dz',$$

again understood as an oscillatory integral. We write $A \in \Psi_{\mathrm{sc}}(\bar{X}_0; \mathcal{C})$ for this class of operators.

Note that $\langle z \rangle^l a \in S^m_\infty(X_0; X_0^*)$, Hörmander's uniform symbol space [27, Section 18.1], so $A = \langle z \rangle^l \tilde{A}$, $\tilde{A} \in \Psi^m_\infty(X_0)$, the uniform ps.d.o.-algebra arising by quantizing $S^m_\infty(X_0; X_0^*)$ as in (5–2). In particular, since the mapping properties of $\Psi^m_\infty(X_0)$ between weighted Sobolev spaces $H^{r,s}$ are well known, the corresponding properties of A follow. Namely, $A : H^{r,s} \to H^{r-m,s+l}$ for all r, s, where

$$H^{r,s} = \langle z \rangle^{-s} H^r = \langle f \in \mathcal{S}'(\mathbb{R}^n) : \langle z \rangle^s f \in H^r \rangle.$$

Now, $\Psi_{\mathrm{sc}}(\bar{X}_0; \mathcal{C})$ is a $*$-algebra, in particular is closed under composition. Indeed, since $\Psi_{\mathrm{sc}}(\bar{X}_0; \mathcal{C}) \subset \Psi_\infty(X_0)$, and the latter is closed under composition, it suffices to follow the usual proof and make sure that the product is in $\Psi_{\mathrm{sc}}(\bar{X}_0; \mathcal{C})$, rather than merely in $\Psi_\infty(X_0)$. Thus, the key fact is that for any

$$b \in \langle z \rangle^{-l} \langle \zeta \rangle^m C^\infty([\bar{X}_0; \mathcal{C}]_z \times [\bar{X}_0; \mathcal{C}]_{z'} \times (\bar{X}_0^*)_\zeta)$$

there exists a as in (5–1) such that the induced operators

$$B = (2\pi)^{-n} \int_{\mathbb{R}^n} e^{i(z-z')\cdot\zeta} b(z, z', \zeta) \, d\zeta, \tag{5–3}$$

and A as in (5–2) are the same. The proof of this claim is standard. Indeed, we can expand b in Taylor series in z' around $z = z'$ to finite order k. The finite order terms depend on z' only via $(z' - z)^\alpha$, $|\alpha| \leq k$. We rewrite $(z' - z)^\alpha e^{i(z-z')\cdot\zeta}$ as $(-1)^{|\alpha|} D_\zeta^\alpha e^{i(z-z')\cdot\zeta}$, and integrate by parts with respect to ζ. Thus, the α-term is the left quantization of

$$\frac{1}{\alpha!} D_{z'}^\alpha D_\zeta^\alpha b(z, z', \zeta)|_{z'=z}, \tag{5–4}$$

which is of the desired form, i.e. is in $\langle z \rangle^{-l} \langle \zeta \rangle^m C^\infty([\bar{X}_0; \mathcal{C}] \times \bar{X}_0^*)$. In fact, the weight $\langle \zeta \rangle^m$ can be replaced by $\langle \zeta \rangle^{m-|\alpha|}$ due to the symbolic properties of b in $(X_0^*)_\zeta$, but no corresponding change may be made for the z weight. Similarly, the remainder term is of the form

$$K_k(z, z') = (2\pi)^{-n} \int_{\mathbb{R}^n} e^{i(z-z')\cdot\zeta} b_k(z, z', \zeta) \, d\zeta,$$

$$b_k \in \langle z \rangle^{-l} \langle \zeta \rangle^{m-k-1} C^\infty([\bar{X}_0; \mathcal{C}]_z \times [\bar{X}_0; \mathcal{C}]_{z'} \times (\bar{X}_0^*)_\zeta). \tag{5–5}$$

Now we can asymptotically sum the b_α to get a new symbol

$$c \in \langle z \rangle^{-l} \langle \zeta \rangle^m C^\infty([\bar{X}_0; \mathcal{C}] \times \bar{X}_0^*).$$

Let C be the left quantization of c. Then $B - C$ is of the form (5–5) for all k, with b_k replaced by some b_k' with the same properties. It is then straightforward to show that the Schwartz kernel K' of $B - C$ is $\mathcal{C}^\infty$, decays rapidly with all derivatives as $\langle z - z' \rangle \to \infty$, and more precisely it is of the form

$$K' \in \mathcal{C}^\infty([\bar{X}_0; \mathcal{C}]_z \times (\bar{X}_0)_{z-z'})$$

with infinite order vanishing at the boundary of the second factor. Taking its Fourier transform b' in $z - z'$, K' is thus the left quantization of $a = c + b'$, proving the claim, hence in turn that $\Psi_{\mathrm{sc}}(\bar{X}_0; \mathcal{C})$ is closed under composition.

In the two-body setting, where $\mathcal{C} = \varnothing$, there is a principal symbol at infinity. Namely, if $A \in \Psi_{\mathrm{sc}}^{m,l}(\bar{X})$, $A = q_L(a)$, then $\sigma_{m,l}(A)$ is given by the restriction of $\langle z \rangle^l \langle \zeta \rangle^{-m} a \in \mathcal{C}^\infty(\bar{X}_0 \times \bar{X}_0^*)$ to $\partial(\bar{X}_0 \times \bar{X}_0^*) = (\partial \bar{X}_0 \times \bar{X}_0^*) \cup (\bar{X}_0 \times \partial \bar{X}_0^*)$. Of the two boundary hypersurfaces, the restriction to $\bar{X}_0 \times \partial \bar{X}_0^*$ yields the usual principal symbol, while the restriction to $\partial \bar{X}_0 \times \bar{X}_0^*$ is the principal symbol at infinity. More precisely, if $l = 0$, we can indeed define the part of $\sigma_{m,0}(A)$ at infinity to be the restriction of a to $(\partial \bar{X}_0) \times \bar{X}_0^*$. The principal symbol is multiplicative, i.e. $\sigma_{m+m',l+l'}(AB) = \sigma_{m,l}(A)\sigma_{m',l'}(B)$. Thus, $[A, B] \in \Psi_{\mathrm{sc}}^{m+m'-1,l+l'+1}(\bar{X})$, and its principal symbol is given by the Poisson bracket of their symbols, see Section 6.

Since in the many-body setting we do not gain decay in z in (5–4), we cannot expect to have a commutative principal symbol at infinity, i.e. at $\partial[\bar{X}_0; \mathcal{C}]$. For $C \in \Psi_{\mathrm{sc}}^{m,0}(\bar{X}, \mathcal{C})$, $y_a \in C_{a,\mathrm{reg}}$, $\zeta_a \in X_a^*$, we let

$$\hat{C}_a(y_a, \zeta_a) = (2\pi)^{-\dim X^a} \int e^{i(z^a - (z')^a) \cdot \zeta^a} c(y_a, z^a, \zeta)\, d\zeta \in \mathcal{S}'(X^a \times X^a)$$

be the operator valued principal symbol of C at (y_a, ζ_a). Thus, $\hat{C}_a(y_a, \zeta_a)$ is a tempered distribution on $X^a \times X^a$ (denoted by the variables $(z^a, (z^a)')$), and it is in fact a many-body ps.d.o. itself: $\hat{C}_a(y_a, \zeta) \in \Psi_{\mathrm{sc}}^{m,0}(\bar{X}^a, \mathcal{C}^a)$ corresponding to the collision planes $X^a \cap X_b$, with b satisfying $X_b \supset X_a$. We also call it the indicial operator of C to make it clear we are not talking about the standard principal symbol. We also write $\hat{C}_a(z_a, \zeta_a)$ in the same setting, where we extend $\hat{C}_a(y_a, \zeta_a)$ to be homogeneous degree 0 in z_a. It can be easily seen to satisfy

$$\hat{A}_a \hat{B}_a = \widehat{(AB)}_a,$$

where on the left hand side we compose the operators $\hat{A}_a(z_a, \zeta_a)$ and $\hat{B}_a(z_a, \zeta_a)$. Thus, multiplication of operators is only partially commutative, even to top order. This can be observed already from $[D_{z_a}, V_a] = 0$, hence certainly lower order at infinity, while $[D_{z^a}, V_a] \in \mathcal{C}^\infty([\bar{X}_0; \mathcal{C}])$ without any decay at C_a.

This observation has important implications for the positive commutator estimates that we take up in the next section. Namely, H must commute to leading order with the operators we want to microlocalize with. This means that these operators A must have $\hat{A}_a$ commute with $\hat{H}_a$, and the most reasonable way of achieving this is to have $\hat{A}_a$ be a scalar multiple of $\psi(\hat{H}_a)$, where e.g. $\psi \in \mathcal{C}_c^\infty(\mathbb{R})$. This multiple defines a function on $\dot{T}^* \bar{X}_0$; we want this to arise from a smooth π-invariant function for our estimates. On the other hand, $\psi(\hat{H}_a)$ provides localization at the characteristic set.

Here, however, I would like to talk about pseudodifferential constructions first. Namely, if $\lambda \notin \mathbb{R}$, or indeed if $\lambda \in \mathbb{C} \setminus [\inf \Lambda, +\infty)$ then there exists a parametrix $G(\lambda) \in \Psi_{\mathrm{sc}}^{-2,0}(\bar{X}_0; \mathcal{C})$ for $H - \lambda$, i.e. such that

$$(H - \lambda)G(\lambda) - \mathrm{Id}, \; G(\lambda)(H - \lambda) - \mathrm{Id} \in \Psi_{\mathrm{sc}}^{-\infty,\infty}(\bar{X}_0; \mathcal{C}).$$

Then the parametrix identities show that

$$\lambda \in \mathbb{C} \setminus \operatorname{spec} H \implies (H - \lambda)^{-1} \in \Psi_{\mathrm{sc}}^{-2,0}(\bar{X}_0; \mathcal{C}).$$

The parametrix construction proceeds inductively by constructing $(\hat{H}_a - \lambda)^{-1}$ in $\Psi_{\mathrm{sc}}^{-2,0}(\bar{X}_a, \mathcal{C}^a)$ for every $a \neq 1$ and then combining these: there exists a $G_0(\lambda) \in \Psi_{\mathrm{sc}}^{-2,0}(\bar{X}_0; \mathcal{C})$ with specified indicial operators $(\hat{H}_a - \lambda)^{-1}$, hence satisfying

$$(H - \lambda)G_0(\lambda) - \mathrm{Id}, \; G_0(\lambda)(H - \lambda) - \mathrm{Id} \in \Psi_{\mathrm{sc}}^{-1,1}(\bar{X}_0; \mathcal{C}).$$

Then the standard Neumann series argument yields $G(\lambda)$.

The Helffer-Sjöstrand argument [24] then shows that for any $\phi \in \mathcal{C}_c^\infty(\mathbb{R})$,

$$\phi(H) = \frac{-1}{2\pi i} \int_{\mathbb{C}} \bar{\partial}_\lambda \tilde{\phi}(\lambda)(H - \lambda)^{-1} \, d\lambda \wedge d\bar{\lambda},$$

where $\tilde{\phi}$ is an almost analytic extension of ϕ: $\tilde{\phi} \in \mathcal{C}_c^\infty(\mathbb{C})$, $|\bar{\partial}_\lambda \tilde{\phi}| \leq C_k |\operatorname{Im} \lambda|^k$ for all k. We can control $(H - \lambda)^{-1}$ in $\Psi_{\mathrm{sc}}^{-2,0}(\bar{X}, \mathcal{C})$ as $\lambda \to \mathbb{R}$ with semi-norm estimates $\mathcal{O}(|\operatorname{Im} \lambda|^{-j})$ (j depends on the norm), so we conclude that $\phi(H) \in \Psi_{\mathrm{sc}}^{-\infty,0}(\bar{X}; \mathcal{C})$.

We can now explain the precise specifications on T_+ in (4-2). Namely, we require that on a neighborhood of $\mathcal{R}_+(\lambda)$ in $\dot{T}^*X$, the indicial operators $\widehat{T_+}$ should equal $\widehat{\phi(H)}$ for some $\phi \in \mathcal{C}_c^\infty(\mathbb{R})$ identically 1 near λ, and on a neighborhood of $\mathcal{R}_-(\lambda)$ they should vanish. Explicitly this can be arranged by taking any ϕ as above, and any $\chi \in \mathcal{C}^\infty(\mathbb{R})$ identically 1 on $(\sqrt{\lambda}/2, +\infty)$, identically 0 on $(-\infty, -\frac{\sqrt{\lambda}}{2})$. Then let $T_+ = \phi(H)q_R(\chi((z \cdot \zeta)/\langle z \rangle))$, with q_R denoting the 'right quantization' (i.e. where we take $b = \chi((z' \cdot \zeta)/\langle z' \rangle)$ in (5-3)). Although $q_R(\chi((z \cdot \zeta)/\langle z \rangle))$ is not in $\Psi_{\mathrm{sc}}(\bar{X}; \mathcal{C})$, due to the nonsymbolic behavior of b as $\zeta \to \infty$, T_+ is, namely $T_+ \in \Psi_{\mathrm{sc}}^{-\infty,0}(\bar{X}; \mathcal{C})$, since $\phi(H)$ is smoothing: see [65]. Moreover,

$$\widehat{T_{+a}}(z_a, \zeta_a) = \chi\left(\frac{z_a \cdot \zeta_a}{|z_a|}\right) \phi(\hat{H}_a(z_a, \zeta_a)), \tag{5-6}$$

hence has the desired properties.

Our construction of $\phi(H)$ in fact shows that if all potentials are in $S^{-\rho}(X_a)$, $\rho > 0$, and $\chi_a \in \mathcal{C}^\infty(\bar{X}_0)$ is supported away from C_b such that $C_b \supset C_a$ does *not* hold, then $\chi_a(\phi(H) - \phi(H_a)) \in \Psi_{\mathrm{sc}}^{-\infty,\rho}(\bar{X}; \mathcal{C})$, hence trace class if $\rho > \dim X_0$. In the three-body setting this shows that

$$\phi(H) - \phi(H_0) - \sum_{\#b=2} (\phi(H_b) - \phi(H_0))$$

is trace class. Indeed, near C_a this can be written as

$$(\phi(H) - \phi(H_a)) - \sum_{\#b=2,\ b\neq a} (\phi(H_b) - \phi(H_0)),$$

and now all terms in parantheses are in $\Psi_{\mathrm{sc}}^{-\infty,\rho}(\bar{X};\mathcal{C})$ near C_a. So we conclude, with a proof that shows much more, a result of Buslaev and Merkureev:

$$\sigma(\phi) = \mathrm{tr}((\phi(H) - \phi(H_0)) - \sum_{\#b=2} (\phi(H_b) - \phi(H_0)))$$

defines a distribution $\sigma \in C^{-\infty}(\mathbb{R})$. Writing $\sigma = \xi'$ defines the spectral shift function, up to a constant, which in turn, in two-body scattering, is the well-known generalization of the eigenvalue counting function on compact manifolds. These statements, as well as the following theorem, which is joint work with Xue-Ping Wang [68], generalize to arbitrary many-body Hamiltonians (with short-range interactions as indicated).

THEOREM 5.1. *Suppose H is a three-body Hamiltonian with Schwartz interactions: $V_a \in S(X^a)$, and that all interactions are pair interactions (i.e. $V_a \neq 0$ implies that a is a 2-cluster.) Then the spectral shift function σ is C^∞ in $\mathbb{R} \setminus (\Lambda \cup \mathrm{spec}_{\mathrm{pp}} H)$, and it is a classical symbol at infinity (i.e. outside a compact set) with a complete asymptotic expansion:*

$$\sigma(\lambda) \sim \lambda^{(n/2)-3} \sum_{j=0}^{\infty} c_j \lambda^{-j}, \quad c_0 = C_0(n) \sum_{a,b:a\neq b} \int_{\mathbb{R}^n} V_a V_b \, dg.$$

Note that σ decays one order faster than in 2-body scattering, and two orders faster than Weyl's law on compact manifolds. This is because $\phi(H_0) + \sum_{\#b=2}(\phi(H_b) - \phi(H_0))$ is, in a high-energy sense, closer to $\phi(H)$ than $\phi(H_0)$ is to $\phi(H)$ in two body scattering. If not all interactions are pair interactions, the order of the leading term changes, namely becomes $\lambda^{(n/2)-2}$ as in 2-body scattering.

The proof of this theorem relies on the propagation of singularities, applied to the Schwartz kernel of the resolvent, $R(\lambda + i0)$. (In fact, the theorem should generalize to symbolic potentials, but the proof would require a more precise microlocalization than provided by $\mathrm{WF}_{\mathrm{sc}}$.) So we now turn to the positive commutator methods that prove this.

6. Microlocal Positive Commutator Estimates

First I sketch, somewhat vaguely, the idea of positive commutator estimates. So suppose that we want to obtain estimates on the solutions of $Pu = f$, where f is known, and is 'nice', and P is self-adjoint. Suppose that we can construct an operator A which is self-adjoint and is such that

$$i[A, P] = B^*B + E. \tag{6-1}$$

Here B^*B is the positive term, giving the name to the estimate. The point is that we can estimate Bu in terms of Eu. Indeed, at least formally,

$$\langle u, i[A, P]u \rangle = \langle u, B^*Bu \rangle + \langle u, Eu \rangle = \|Bu\|^2 + \langle u, Eu \rangle.$$

On the other hand,

$$\langle u, i[A, P]u \rangle = \langle u, iAPu \rangle - \langle u, iPAu \rangle = \langle u, iAPu \rangle + \langle iAPu, u \rangle = 2\operatorname{Re}\langle u, iAPu \rangle.$$

Combining these yields

$$\|Bu\|^2 \le 2|\operatorname{Re}\langle u, iAPu \rangle| + |\langle u, Eu \rangle|. \tag{6-2}$$

This means that Bu can be estimated in terms of Pu, which is known from the PDE, and Eu, on which we need to make assumptions. The typical application is that E is supported in one region of phase space and B in another, in which case we can *propagate* estimates of u.

In fact, one can also apply this estimate if one does not know a priori that $Bu \in L^2$. Namely, an approximation argument gives that if Pu and Eu are in appropriate spaces so that the right hand side of (6-2) makes sense, then $Bu \in L^2$, and (6-2) holds. Considering pseudodifferential operators A of various orders, this means that we obtain microlocal weighted Sobolev estimates for u. Also, typically one has an error term F, i.e. $i[A, P] = B^*B + E + F$, but F is 'lower order' in some sense. Thus, $|\langle u, Fu \rangle|$ is added to the right hand side of (6-2), but being 'lower order' means that $|\langle u, Fu \rangle|$ automatically makes sense, hence is irrelevant when proving that $Bu \in L^2$.

In fact, this method also yields estimates for the resolvent very directly. Since for $t \in \mathbb{R}$, $i[A, P - it] = i[A, P]$, and

$$\langle u, i[A, P - it]u \rangle = \langle u, iA(P - it)u \rangle - \langle u, i(P - it)Au \rangle$$
$$= \langle u, iA(P - it)u \rangle + \langle iA(P + it)u, u \rangle$$
$$= 2\operatorname{Re}\langle u, iA(P - it)u \rangle - 2t\langle Au, u \rangle.$$

Thus, we deduce

$$\|Bu\|^2 + 2t\langle Au, u \rangle \le 2|\operatorname{Re}\langle u, iAPu \rangle| + |\langle u, Eu \rangle|. \tag{6-3}$$

This is in particular an estimate for $\|Bu\|$ provided that $t \ge 0$ and A is positive. Here we may take $u = u_t = (P - it)^{-1}f$, defined for $t > 0$, say, and we find a uniform estimate for Bu_t as $t \to 0$.

The question is thus how one can produce operators A which have a positive commutator with P as above. First, we recall how this happens in the scattering calculus. Namely, if $A \in \Psi_{\mathrm{sc}}^{m,l}(\bar{X})$, $P \in \Psi_{\mathrm{sc}}^{m',l'}(\bar{X})$ then $[A, P] \in \Psi_{\mathrm{sc}}^{m+m'-1,l+l'+1}(\bar{X})$ and

$$\sigma_{m+m'-1,l+l'+1}(i[A, P]) = H_a p = -H_p a, \quad a = \sigma_{m,l}(A), \quad p = \sigma_{m',l'}(P),$$

where H_a is the Hamilton vector field of a, H_p the Hamilton vector field of p. So modulo lower terms, which I ignore here and which are easy to deal with, we need to arrange that

$$H_p a = -b^2 + e, \qquad (6\text{--}4)$$

and then take B, E with $\sigma(B) = b$, $\sigma(E) = e$. Indeed, under these assumptions (6–2) shows that $\|Bu\|$ can be estimated in terms of Pu and Eu. That is, u microlocally on $\operatorname{supp} b$ is estimated by u on $\operatorname{supp} e$ (and Pu) in the precise sense described in the next paragraph, so we can *propagate* estimates of u from $\operatorname{supp} e$ to $\operatorname{supp} b$. (Incidentally, this is a good example of the F term: only the principal symbols of E and B are specified. Take any E and B with these principal symbols, $F = i[A, P] - B^* B - E$ is lower order.)

This can be used in a very straightforward manner to obtain bounds on $\mathrm{WF_{sc}}(u)$. Namely, one works with 'relative wave front sets', relative to $x^s H^r = H^{r,s}$, that is. Thus, for $\bar{X} = \overline{\mathbb{R}^n}$, $(z, \zeta) \notin \mathrm{WF_{sc}^{r,s}}(u)$ means that there is a cutoff function $\phi \in \mathcal{C}_c^\infty(\mathbb{R}^n)$ with $\phi(\zeta) \neq 0$ such that $\mathcal{F}^{-1} \phi \mathcal{F} u$ is in $H^{r,s}$ in an open cone around z. But this is equivalent to the existence of some $Q \in \Psi_{sc}^{r,-s}(\bar{X})$ such that $\sigma(Q)(z, \zeta) \neq 0$ and $Qu \in L^2$. Note that $\sigma(Q)(z, \zeta) \neq 0$ means that Q is elliptic at (z, ζ). So if we find $A \in \Psi_{sc}^{m,l}(\bar{X})$, and consequently $B \in \Psi_{sc}^{(m-1)/2,(l+1)/2}(\bar{X})$, $E \in \Psi_{sc}^{m-1,l+1}(\bar{X})$ as above, then the conclusion is that (if $Pu \in \dot{\mathcal{C}}^\infty(\bar{X})$)

$$\mathrm{WF_{sc}^{(m-1)/2,-(l+1)/2}}(u) \cap \operatorname{supp} e = \varnothing \implies \mathrm{WF_{sc}^{(m-1)/2,-(l+1)/2}}(u) \cap \operatorname{supp} b = \varnothing.$$

In scattering theory m is usually irrelevant by standard elliptic regularity. Thus, one iteratively reduces l, proving that $\operatorname{supp} b$ is disjoint from the wave front set with respect to more and more decaying Sobolev spaces. (In fact, b is shrunk slightly during the iterative procedure for technical reasons.)

I now illustrate how to prove the propagation of singularities at ∂X for real principal type $P \in \Psi_{sc}^{m,0}(\bar{X})$. For example, we may take $P = H - \lambda$, $\lambda > 0$, and H is a two-body Hamiltonian. (Note that microlocal elliptic regularity is the consequence of the standard microlocal parametrix construction.) We thus want to prove that if $Pu \in \dot{\mathcal{C}}^\infty(\bar{X})$ (or a microlocal version of it holds), $\bar{\xi} \in {}^{sc}T^*_{\partial \bar{X}} \bar{X}$ and there is a point on the backward bicharacteristic through $\bar{\xi}$ which is not in $\mathrm{WF_{sc}}(u)$, then $\bar{\xi} \notin \mathrm{WF_{sc}}(u)$. In fact, by a simple argument it suffices to prove that there exists a neighborhood U of $\bar{\xi}$ such that if there is a point $\tilde{\xi}$ in U which is also on the backward bicharacteristic through $\bar{\xi}$ and which is not in $\mathrm{WF_{sc}}(u)$, then $\bar{\xi} \notin \mathrm{WF_{sc}}(u)$.

The standard proof proceeds via linearization of H_p, see [28]. Thus, first note that $x^{-1} H_p$ is a smooth vector field on ${}^{sc}T^* \bar{X}$ which is tangent to the boundary. (For example, for Euclidean two-body Hamiltonians,

$$x^{-1} H_p = 2|z|\zeta \cdot \partial_z = 2 \sum_j \zeta_j |z| \partial_{z_j},$$

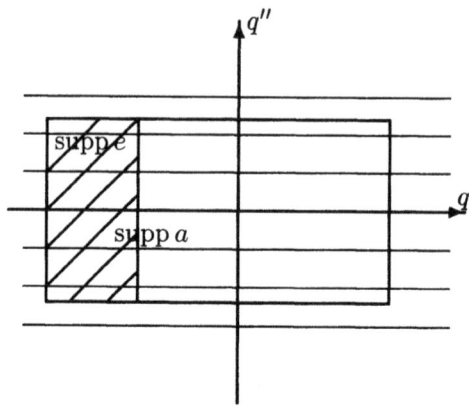

Figure 7. The region $\operatorname{supp} a$ superimposed on the linearized Hamilton flow. $\operatorname{supp} e$ is the shaded region on the left.

and $|z|\partial_{z_j}$ is a smooth vector field tangent to $\partial \bar{X}$, i.e. it is in $\mathcal{V}_b(\bar{X})$.) Thus, given any point $\bar{\xi} \in {}^{sc}T^*_{\partial \bar{X}}\bar{X}$ one can introduce local coordinates $(q_1, \ldots, q_{2n-1}) = (q_1, q'')$ on ${}^{sc}T^*_{\partial \bar{X}}\bar{X}$ centered at $\bar{\xi}$ such that $x^{-1}H_p = \partial_{q_1}$ at ${}^{sc}T^*_{\partial \bar{X}}\bar{X}$. Thus, bicharacteristics at $\partial \bar{X}$ are curves $q'' = $ constant. Now let $\chi_1 \in \mathcal{C}^\infty_c(\mathbb{R}_{q_1})$ and $\chi_2 \in \mathcal{C}^\infty_c(\mathbb{R}^{2n-2}_{q''})$ be smooth functions supported near 0 with the property that

$$\chi_1' = -b_1^2 + e_1,$$

$b_1, e_1 \in \mathcal{C}^\infty_c(\mathbb{R})$, and $\operatorname{supp} e_1 \subset (-\infty, 0)$. Let

$$a = \chi_1 \chi_2^2, \quad b = b_1 \chi_2, \quad e = e_1 \chi_2^2.$$

Then (6–4) holds. In fact, we can even allow weights and take

$$a_s = x^s \chi_1 \chi_2^2, \quad s \in \mathbb{R},$$

since $(x^{-1}H_p x^s)\chi_1$ can be absorbed in $x^s(x^{-1}H_p\chi_1)$ by choosing χ_1' large compared to χ_1. This gives microlocal weighted L^2 estimates in $x^{-s-1/2}L^2$. The iterative argument, in which we gradually let $s \to -\infty$, then allows one to conclude that

$$\operatorname{supp} e \cap \mathrm{WF}_{sc}(u) = \varnothing \wedge \operatorname{supp} a \cap \mathrm{WF}_{sc}(Pu) = \varnothing \implies \{b > 0\} \cap \mathrm{WF}_{sc}(u) = \varnothing.$$

By choosing $\operatorname{supp}\chi_1$ and $\operatorname{supp}\chi_2$ appropriately, we may arrange that e is supported near $\bar{\xi}$ so that $\operatorname{supp} e \cap \mathrm{WF}_{sc}(u) = \varnothing$, and so that $b(\bar{\xi}) > 0$, as shown below.

There are several directions from here. One can use finer notion of regularity, such as Lagrangian regularity, which would correspond to using χ_2 that vanishes simply on a Lagrangian submanifold, or such as regularity at radial points, which is the subject of a joint paper with Andrew Hassell and Richard Melrose [23].

Here I will talk about a rougher version, namely what happens if the bicharacteristic 'flow' is more complicated, e.g. in the presence of boundaries or corners for the wave equation [43; 36] or many-body scattering. In fact, here I will not explain the detailed behavior of bicharacteristics, rather just show how to microlocalize positive commutator estimates in a versatile fashion. This method goes back to the work of Melrose and Sjöstrand [43].

The main point is that if we cannot put the operator P, or at least its Hamilton vector field H_p in a model form, the previous construction will not work. Indeed, unless $H_p\chi_2 = 0$, $H_p(\chi_1\chi_2)$ will always yield a term $\chi_1 H_p\chi_2$, which cannot be controlled by $(H_p\chi_1)\chi_2$: the problem being near the boundary of $\operatorname{supp}\chi_2$. So instead use a different form of localization. First let $\eta \in C^\infty(^{sc}T^*\bar{X})$ be a function with

$$\eta(\bar{\xi}) = 0, \quad H_p\eta(\bar{\xi}) > 0.$$

Thus, η measures propagation along bicharacteristics, e.g. $\eta = q_1$ in the above example would work, but so would many other choices. We will use a function ω to localize near putative bicharacteristics. This statement is deliberately vague; at first we only assume that $\omega \in C^\infty(^{sc}T^*\bar{X})$ is the sum of the squares of C^∞ functions σ_j, $j = 1, \ldots, l$, with nonzero differentials at $\bar{\xi}$ such that $d\eta$ and $d\sigma_j$, $j = 1, \ldots, l$, span $T_{\bar{\xi}}^{sc}T_{\partial\bar{X}}^*\bar{X}$. Such a function ω is nonnegative and it vanishes quadratically at $\bar{\xi}$, i.e. $\omega(\bar{\xi}) = 0$ and $d\omega(\bar{\xi}) = 0$. An example is $\omega = q_2^2 + \ldots + q_{2n-1}^2$ with the notation from before, but again there are many other possible choices. We now consider a family symbols, parameterized by constants $\delta \in (0,1)$, $\varepsilon \in (0, \delta]$, of the form

$$a = \chi_0\left(2 - \frac{\phi}{\varepsilon}\right)\chi_1\left(\frac{\eta + \delta}{\varepsilon\delta} + 1\right),$$

where

$$\phi = \eta + \frac{1}{\varepsilon}\omega, \qquad \chi_0(t) = \begin{cases} 0 & \text{if } t \leq 0, \\ e^{-1/t} & \text{if } t > 0, \end{cases}$$

and $\chi_1 \in C^\infty(\mathbb{R})$ with $\operatorname{supp}\chi_1 \subset [0, +\infty)$ and $\operatorname{supp}\chi_1' \subset [0, 1]$. Although we do not do it explicitly here, weights such as x^s can be accommodated for any $s \in \mathbb{R}$, by replacing the factor $\chi_0(2 - \frac{\phi}{\varepsilon})$ by $\chi_0(A_0^{-1}(2 - \frac{\phi}{\varepsilon}))$ and taking $A_0 > 0$ large.

We analyze the properties of a step by step. First, note that $\phi(\bar{\xi}) = 0$, $H_p\phi(\bar{\xi}) = H_p\eta(\bar{\xi}) > 0$, and $\chi_1((\eta + \delta)/(\varepsilon\delta) + 1)$ is identically 1 near $\bar{\xi}$, so $H_pa(\bar{\xi}) < 0$. Thus, H_pa has the correct sign, and is in particular nonzero, at $\bar{\xi}$.

Next,

$$\xi \in \operatorname{supp}a \implies \phi(\xi) \leq 2\varepsilon \quad \text{and} \quad \eta(\xi) \geq -\delta - \varepsilon\delta.$$

Since $\varepsilon < 1$, we deduce that in fact $\eta = \eta(\xi) \geq -2\delta$. But $\omega \geq 0$, so $\phi = \phi(\xi) \leq 2\varepsilon$ implies that $\eta = \phi - \varepsilon^{-1}\omega \leq \phi \leq 2\varepsilon$. Hence, $\omega = \omega(\xi) = \varepsilon(\phi - \eta) \leq 4\varepsilon\delta$. Since ω vanishes quadratically at $\bar{\xi}$, it is useful to rewrite the estimate as $\omega^{1/2} \leq 2(\varepsilon\delta)^{1/2}$. Combining these, we have seen that on $\operatorname{supp}a$,

$$-\delta - \varepsilon\delta \leq \eta \leq 2\varepsilon \quad \text{and} \quad \omega^{1/2} \leq 2(\varepsilon\delta)^{1/2}. \tag{6-5}$$

Moreover, on $\operatorname{supp}a \cap \operatorname{supp}\chi_1'$,

$$-\delta - \varepsilon\delta \leq \eta \leq -\delta \quad \text{and} \quad \omega^{1/2} \leq 2(\varepsilon\delta)^{1/2}.$$

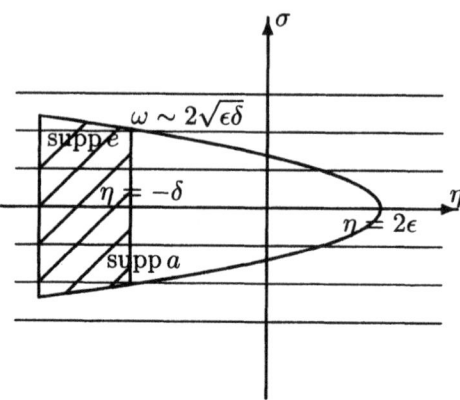

Figure 8. The region $\operatorname{supp} a$ in (η, σ) coordinates. Again, $\operatorname{supp} e$ is the shaded region on the left.

Note that given any neighborhood U of $\bar\xi$, we can thus make a supported in U by choosing ε and δ sufficiently small. Below we illustrate the parabola shaped region given by $\operatorname{supp} a$.

Note that as $\varepsilon \to 0$, but δ fixed, the parabola becomes very sharply localized at $\omega = 0$. In particular, for very small $\varepsilon > 0$ we obtain a picture quite analogous to letting $\operatorname{supp} \chi_2 \to \{0\}$ in Figure 7.

So we have shown that a is supported near $\bar\xi$. We define

$$ e = \chi_0(2 - \frac{\phi}{\varepsilon})H_p\chi_1((\eta + \delta)/(\varepsilon\delta) + 1), $$

so the crucial question is whether $H_p\phi \geq 0$ on $\operatorname{supp} a$. Note that choosing $\delta_0 \in (0,1)$ and $\varepsilon_0 \in (0, \delta_0)$ sufficiently small, one has $H_p\eta \geq c_0 > 0$ where $|\eta| \leq 2\delta_0$, $\omega^{1/2} \leq 2(\varepsilon_0\delta_0)^{1/2}$. So $H_p\phi \geq 0$ on $\operatorname{supp} a$, provided that $|H_p\omega| \leq \frac{c_0}{2}\varepsilon$ there.

But being a sum of squares of functions with nonzero differentials, $H_p\omega$ vanishes at $\omega = 0$ and satisfies $|H_p\omega| \leq C\omega^{1/2}$. Due to (6–5), we deduce that $|H_p\omega| \leq 2C(\varepsilon\delta)^{1/2}$. So $|H_p\omega| \leq \frac{c_0}{2}\varepsilon$ holds if $\frac{c_0}{2}\varepsilon \geq 2C(\varepsilon\delta)^{1/2}$, i.e. if $\varepsilon \geq C'\delta$ for some constant $C' > 0$ independent of ε, δ. Note that this constraint on ε, i.e. that it cannot be too small, gives very rough localization: the width of the parabola at $\eta = -\delta$ is roughly $\omega^{1/2} \sim \delta$, i.e. it is very wide, and in particular insufficient to prove the propagation of singularities along the bicharacteristics. The reason is simple: our localizing function ω has no relation to H_p, so we cannot expect a more precise estimate. Note, however, that the estimate is still nontrivial! Indeed, it shows that singularities propagate in the sense that $\bar\xi$ cannot be an isolated point of $\mathrm{WF}_{\mathrm{sc}}(u)$. (We required $\varepsilon \in (0, \delta]$ beforehand, but in fact we could have dealt with $\varepsilon \leq \mu\delta$, even if $\mu > 1$, if we localized slightly differently.)

We need to adapt ω to H_p to get a better estimate. If we linearize H_p as above, and take $\omega = q_2^2 + \ldots + q_{2n-1}^2$, then $H_p\omega = 0$ and any $\varepsilon > 0$ works. Thus,

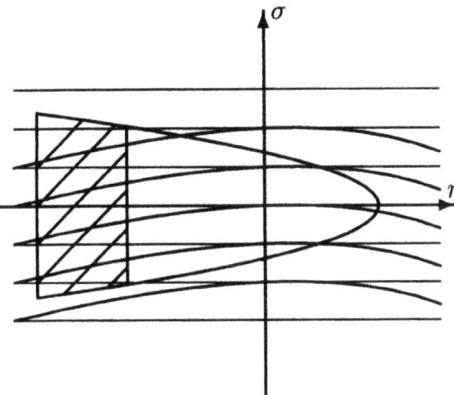

Figure 9. Bicharacteristics and supp a. The labels from Figure 8 have been removed to make the picture less cluttered. The straight horizontal lines are the $\sigma = $ constant lines, while the nearby parabolae are the bicharacteristics.

in this case, we can prove propagation of singularities much like by the previous, simpler, construction.

However, we do not need such a strong relationship to H_p. Suppose instead that we merely get ω 'right' at $\bar{\xi}$, in the sense that $\omega = \sum \sigma_j^2$ and $H_p \sigma_j(\bar{\xi}) = 0$. Then $|H_p \sigma_j| \leq C_0(\omega^{1/2} + |\eta|)$, so $|H_p \omega| \leq C \omega^{1/2}(\omega^{1/2} + |\eta|)$. Using (6–5), we deduce that $|H_p \omega| \leq (c_0/2)\varepsilon$ provided that $(c_0/2)\varepsilon \geq C''(\varepsilon \delta)^{1/2}\delta$, i.e. that $\varepsilon \geq C'\delta^3$ for some constant C' independent of ε, δ. Now the size of the parabola at $\eta = -\delta$ is roughly $\omega^{1/2} \sim \delta^2$, i.e. we have localized along a single direction, namely the direction of H_p at $\bar{\xi}$. By a relatively simple argument, one can piece together such estimates (i.e. where the direction is correct 'to first order') and deduce the propagation of singularities. We emphasize that the lower bound for ε is natural. Indeed, with q_j as above, we may take σ_j e.g. to be $\sigma_j = q_j + q_1^2$, $j \geq 2$. The bicharacteristics are $q_j = $ constant, but we are localizing near $\sigma_j = $ constant, and at $\eta = -\delta$ these differ by δ^2. So any localization better than $\omega^{1/2} \sim \delta^2$ would in fact contradict the propagation of singularities!

The microlocal positive commutator estimates in many-body scattering arise by this method. In particular, one can take $\eta = \frac{z \cdot \zeta}{|z|}$, which is the radial component of the momentum. The function ω needs to be π-invariant, so if $\bar{\xi} \in {}^{sc}T^*\bar{X}_a$, it involves functions on ${}^{sc}T^*\bar{X}_a$ as well as $Z^a = \frac{z^a}{|z|}$ and η. The only additional argument needed is to show that the commutator is indeed positive, which has to be understood in an operator sense. Thus, the key point is that the commutator of $H - \lambda$ with a quantization B of η is positive, modulo compact terms, when localized at λ in the spectrum of H and microlocalized away from the radial set $\mathcal{R}(\lambda)$. Note that, as usual, there is nothing to prove at $\mathcal{R}(\lambda)$, since each point in it is the image of a maximally extended generalized broken bicharacteristic.

This positivity can be proved directly by showing that the indicial operators of the commutator are positive away from $\mathcal{R}(\lambda)$, which follows from an iterative argument. However, it also reduces to the Mourre estimate, involving the generator of dilations $A = \frac{1}{2}(\cdot D_z + D_z \cdot z)$, which has principal symbol at $\zeta \cdot z$. The Mourre estimate states the following. Suppose that $\lambda \notin \Lambda$ and $\varepsilon > 0$. Then there is a $\delta > 0$ such that if $\phi \in C_c^\infty(\mathbb{R})$ is supported in $(\lambda - \delta, \lambda + \delta)$ there exists $K' \in \Psi_{\mathrm{sc}}^{-\infty,1}(\bar{X}, \mathcal{C})$ such that

$$\phi(H)i[A, H]\phi(H)) \geq 2(d(\lambda) - \varepsilon)\phi(H)^2 + K', \tag{6-6}$$

where

$$d(\lambda) = \inf\{\lambda - \lambda' : \lambda' \leq \lambda, \ \lambda' \in \Lambda\} \geq 0$$

is the distance of λ to the next threshold below it if $\lambda \geq \inf \Lambda$, and $d(\lambda)$ an arbitrary positive number if $\lambda < d(\lambda)$. Since $d(\lambda) > 0$ if $\lambda \notin \Lambda$, (6-6) is indeed a positive commutator estimate, which does not even have a 'negative' term E, unlike (6-1). The Mourre estimate, originating in [47], has been well understood since the work of Perry, Sigal and Simon [53] and Froese and Herbst [16]. Here I just outline the argument in the simplest case, namely if no proper subsystem has any L^2-eigenvalues.

In this simplest case, the argument of [16] (see also [8] and [67]) proceeds as follows. In order to prove (6-6), one only needs to show that for all b, the corresponding indicial operators satisfy the corresponding inequality, i.e. that

$$\phi(\hat{H}_b)i\widehat{[A, H]}_b\phi(\hat{H}_b) \geq 2(d(\lambda) - \varepsilon)\phi(\hat{H}_b)^2. \tag{6-7}$$

(This means that the operators on the two sides, which are families of operators on X^b, depending on $(y_b, \zeta_b) \in {}^{\mathrm{sc}}T_{C_b}^*\bar{X}$, satisfy the inequality for all $(y_b, \zeta_b) \in {}^{\mathrm{sc}}T_{C_b}^*\bar{X}$.) It is convenient to assume that ϕ is identically 1 near λ; if (6-7) holds for such ϕ, it holds for any ϕ_0 with slightly smaller support, as follows by multiplication by $\phi_0(\hat{H}_b)$ from the left and right.

Note that for $b = 0$ the estimate certainly holds: it comes from the Poisson bracket formula in the scattering calculus, or from a direct computation yielding $i\widehat{[A, H]}_0 = 2\Delta_{g0}$. Hence, if the the localizing factor

$$\phi(\hat{H}_0) = \phi(|\zeta|^2)$$

is supported in $(\lambda - \delta, \lambda + \delta)$ and $\lambda > 0$, (6-7) holds even with $d(\lambda) - \varepsilon$ replaced with $\lambda - \delta$. Note that $\lambda \geq d(\lambda)$, if $\lambda > 0$, since 0 is a threshold of H. On the other hand, if $\lambda < 0$, both sides of (6-7) vanish for ϕ supported near λ, so the inequality holds trivially.

In general, we may assume inductively that at all clusters c with $C_c \subsetneq C_b$, i.e. $X^b \subsetneq X^c$, (6-7) has been proved with ϕ replaced by a cutoff $\tilde{\phi}$ and ε replaced by ε', i.e. we may assume that for all $\varepsilon' > 0$ there exists $\delta' > 0$ such that for all c with $C_c \subsetneq C_b$, and for all $\tilde{\phi} \in C_c^\infty(\mathbb{R}; [0, 1])$ supported in $(\lambda - \delta', \lambda + \delta')$,

$$\tilde{\phi}(\hat{H}_c)i\widehat{[A, H]}_c\tilde{\phi}(\hat{H}_c) \geq 2(d(\lambda) - \varepsilon')\tilde{\phi}(\hat{H}_c)^2. \tag{6-8}$$

But these are exactly the indicial operators of $\tilde{\phi}(\hat{H}_b)i\widehat{[A, H]}_b\tilde{\phi}(\hat{H}_b)$, so, as discussed in [65, Proposition 8.2], (6–7) implies that

$$\tilde{\phi}(\hat{H}_b)i\widehat{[A, H]}_b\tilde{\phi}(\hat{H}_b) \geq 2(d(\lambda) - \varepsilon')\tilde{\phi}(\hat{H}_b)^2 + K_b, \quad K_b \in \Psi_{\mathrm{sc}}^{-\infty,1}(X^b, \mathcal{C}^b). \quad (6\text{–}9)$$

This implication relies on a square root construction in the many-body calculus, which is particularly simple in this case.

Now, we first multiply (6–9) through by $\phi(H)$ from both the left and the right. Recall that we use coordinates (z_b, z^b) on $X_b \oplus X^b$ and (ζ_b, ζ^b) are the dual coordinates. We remark that $\hat{H}_b = |\zeta_b|^2 + H^b$, so if $\lambda - |\zeta_b|^2$ is not an eigenvalue of H^b, then as $\mathrm{supp}\,\phi \to \{\lambda\}$, $\phi(H^b + |\zeta_b|^2) \to 0$ strongly, so as K_b is compact, $\phi(H^h + |\zeta_b|^2)K_b \to 0$ in norm; in particular it can be made to have norm smaller than $\varepsilon' - \varepsilon > 0$. After multiplication from both sides by $\phi_1(\hat{H}_b)$, with ϕ_1 having even smaller support, (6–7) follows (with ϕ_1 in place of ϕ), with the size of $\mathrm{supp}\,\phi_1$ a priori depending on ζ_b. However, $i\phi_1(\hat{H}_b)\widehat{[A, H]}_b\phi_1(\hat{H}_b)$ is continuous in ζ_b with values in bounded operators on $L^2(X^b)$, so if (6–7) holds at one value of ζ_b, then it holds nearby. Moreover, for large $|\zeta_b|$ both sides vanish as $\hat{H}_b = H^b + |\zeta_b|^2$, with H^b bounded below, so the estimate is in fact uniform if we slightly increase $\varepsilon > 0$.

In general, the proof requires to treat the range of E, the spectral projection of H^b to $\{\lambda\}$, separately. Roughly, the positivity estimate on the range of E comes from the virial theorem, $iE[z^b D_{z^b}, H^b]E = 0$, which is formally clear, and is easy to prove. Thus,

$$iE[A, H_b]E = iE[z^b D_{z^b}, H^b]E + iE[z_b D_{z_b}, \Delta_{X_b}]E = i[z_b D_{z_b}, \Delta_{X_b}]E,$$

and the commutator $i\phi(H_b)[z_b D_{z_b}, \Delta_{X_b}]\phi(H_b)$ is easily computed to be positive. Of course, there are also cross-terms that need to be considered, but they can be estimated by Cauchy-Schwartz estimates, see [16] or [67].

I refer to [66] and [65] for the detailed arguments proving propagation of singularities in the many-body setting, and to [68, Appendix] for weaker estimates with simplified proofs.

7. Asymptotic Completeness

Asymptotic completeness (AC) is an L^2-based statement describing the long-term behavior of solutions of the Schrödinger equation. In the short-range setting it says that for any $\phi \in L^2(X_0)$ in the range of $\mathrm{Id} - E_{\mathrm{pp}}$, E_{pp} being the projection onto the bound states of H (i.e. onto its L^2-eigenfunctions), there exist $\phi_\alpha \in L^2(X_a)$ such that

$$\left\| e^{-iHt}\phi - \sum_\alpha e^{-iH_a t}(\phi_\alpha \otimes \psi_\alpha) \right\| \to 0 \text{ as } t \to +\infty.$$

In the long-range setting, $e^{-iH_a t}$ must be somewhat modified. After the groundbreaking work of Enss [12; 10], AC was first proved by Sigal and Soffer [55] in the short-range setting (see Graf's paper [20] for a different proof), and later by Dereziński [7], and also by Sigal and Soffer [56; 57], in the long-range setting. In the short range case the main ingredient is equivalent to certain estimates of the resolvent at the radial sets in a sense that I now describe. In the long-range setting, as already in two-body scattering, additional constructive steps are needed. The estimates, in a different language, appeared first in the work of Sigal and Soffer [55]. I hope that the following discussion makes it clearer how they relate to the propagation of singularities.

While asymptotic completeness gives a complete long-term L^2-description of solutions of the Schrödinger equation, the question remains whether an analogous description exists on other spaces, such as weighted L^2-spaces. For example, if ϕ is Schwartz, are the ϕ_α Schwartz? Or dually, starting with a tempered distribution ϕ, are there tempered distributions α such that the convergence holds, as $t \to +\infty$, in a suitable sense? A different point of view is the parameterization of generalized eigenfunctions of H using the Poisson operators $P_{\alpha,+}(\lambda)$, and the analogues of these questions can be asked there as well. The answer is affirmative in the two-body setting (even in the geometric setting, see [41; 46]). However, as indicated by the related issue of the mapping properties of the scattering matrices, discussed at the end of this section, it is unlikely that the same holds in the many-body setting. One can then ask weaker question, e.g. whether it holds in weighted spaces L_s^2, s near 0. Or, one may ask whether one can give a precise description of the map $\phi \mapsto \phi_\alpha$ e.g. as some sort of Fourier integral operator.

As a starting point of relating the propagation of singularities to AC, we note that the propagation of singularities is proved by showing its 'relative' versions, i.e. that for any l, $\mathrm{WF}_{\mathrm{sc}}^{*,l}(u)$ is also a union of maximally extended generalized broken bicharacteristics. When considering the resolvent, first recall that for $f \in \dot{C}^\infty(\bar{X})$, $R(\lambda + i0)f \in H^{\infty,l}$ for all $l < -\frac{1}{2}$, so we only need to find

$$\mathrm{WF}_{\mathrm{sc}}^{*,l}(R(\lambda + i0)f)$$

for $l \geq -\frac{1}{2}$. Theorem 3.3 is also valid for $\mathrm{WF}_{\mathrm{sc}}^{*,l}$, i.e. the following holds.

THEOREM 7.1. *If $\lambda \notin \Lambda$, then for $f \in S(\mathbb{R}^n)$, $l \geq -\frac{1}{2}$, $\mathrm{WF}_{\mathrm{sc}}(R(\lambda + i0)f)$ is a subset of the image of $\mathcal{R}_+(\lambda)$ under the forward generalized broken bicharacteristic relation.*

This result allows $u = R(\lambda + i0)f$ *not* to lie in $H^{*,-1/2}$ on the image of $\mathcal{R}_+(\lambda)$ under the forward generalized broken bicharacteristic relation. This is a small set, but it is important to know whether $\mathrm{WF}_{\mathrm{sc}}^{*,l}(u)$ may indeed intersect the forward image of $\mathcal{R}_+(\lambda)$. Of course, we cannot expect an improvement at $\mathcal{R}_+(\lambda)$, as shown already by the example of the free Euclidean Laplacian. The crucial improvement is the following estimate, due to Sigal and Soffer [55].

THEOREM 7.2. *If $\lambda \notin \Lambda$, then for $f \in S(\mathbb{R}^n)$, $\mathrm{WF}_{\mathrm{sc}}^{*,-1/2}(R(\lambda + i0)f) \subset \mathcal{R}_+(\lambda)$.*

REMARK. This theorem also has a time-dependent analogue. If u is a solution of the Schrödinger equation $(D_t + H)u = 0$ with $u|_{t=0} \in \mathcal{S}(X_0)$ then on the one hand $u \in H^{\infty,l}(X_0 \times \mathbb{R})$ for $l < -\frac{1}{2}$, on the other hand $\mathrm{WF}_{\mathrm{sc}}^{*,-1/2}(u) \subset \mathcal{R}$.

In fact, this theorem can be improved along the lines of the distributional statement in Theorem 3.3:

COROLLARY 7.3. *Suppose that* $\lambda \notin \Lambda$, $f \in H^{*,1/2}$ *and* $\mathrm{WF}_{\mathrm{sc}}^{*,1/2+\varepsilon}(f) \cap \mathcal{R}_-(\lambda) = \varnothing$ *for some* $\varepsilon > 0$. *Then* $R(\lambda+i0)f = \lim_{t \to 0} R(\lambda+it)f$ *exists in* $H^{*,-1/2-\varepsilon'}(\bar{X})$, $\varepsilon' > 0$, *and* $\mathrm{WF}_{\mathrm{sc}}^{*,-1/2}(R(\lambda+i0)f) \subset \mathcal{R}_+(\lambda)$.

Theorem 7.2 can be proved rather simply. The main issue is how to obtain a positive commutator at the radial point. Away from the radial sets arbitrary weights can be accommodated by suitable construction, as pointed out in the previous section. At radial points only the weights can give positive commutators. Now, one has to use weights x^{-2l-1} to obtain estimates for $\mathrm{WF}_{\mathrm{sc}}^{*,l}$, and these weights will give a commutator whose sign depends on that of $-2l - 1$, hence on whether $l > -\frac{1}{2}$, $l < -\frac{1}{2}$ or $l = -\frac{1}{2}$. It turns out that the sign is correct for (6–3) to be of use if $l < -\frac{1}{2}$; this yields the limiting absorption principle. The sign is wrong if $l > -\frac{1}{2}$, so no results can be expected then. In the borderline case $l = -\frac{1}{2}$, the weight vanishes. The way to obtain a positive commutator is thus to consider operators A which are microlocally (a multiple of) the identity near $\mathcal{R}_+(\lambda)$. The commutator then vanishes microlocally near $\mathcal{R}_+(\lambda)$, which is reasonable since no estimate on $\mathrm{WF}_{\mathrm{sc}}^{*,-1/2}(u)$ can be expected there.

It is then straightfoward to construct A so that (6–3) can be used to prove Theorem 7.2. Indeed, it suffices to show that on $\mathrm{WF}_{\mathrm{sc}}^{*,-1/2}(u)$, $\eta = \frac{z \cdot \zeta}{|z|}$ must satisfy $\lambda - \eta^2 \in \Lambda$, for then the full statement of the theorem follows by the propagation of singularities for $\mathrm{WF}_{\mathrm{sc}}^{*,-1/2}(u)$. So we proceed to prove this simpler result, namely that if $\lambda - \bar{\eta}^2 \notin \Lambda$ then for any point ξ, $\eta(\xi) = \bar{\eta}$ implies $\xi \notin \mathrm{WF}_{\mathrm{sc}}^{*,-1/2}(u)$.

To do so, we let $a = \chi(\eta)$ where $\chi \in \mathcal{C}_c^\infty(\mathbb{R})$, $\chi \geq 0$, is chosen so that $\chi \equiv 1$ on $[0, \bar{\eta} - \delta]$ for some $\delta > 0$, $\chi' \leq 0$ on $(0, \infty)$, $\chi'(\bar{\eta}) < 0$, and $t \in \operatorname{supp} \chi'$ implies that $\lambda - t^2 \notin \Lambda$. This can be arranged as Λ is closed. We can further make sure that $\sqrt{-\chi'}$ is $\mathcal{C}^\infty$ on $(0, \infty)$. Then the positive commutator methods outlined show the commutator of A, a quantization of a, with $H - \lambda$ is positive, in the region $\eta > 0$, yielding the estimate that proves the theorem. We remark that partial microlocalization, using functions of η, hes been used extensively in many-body scattering, especially by Gérard, Isozaki and Skibsted [18; 19] and Wang [69], to obtain partially microlocal statements such as radiation conditions and uniqueness statements [32; 31], and indeed to prove the smoothness of 2-cluster to 2-cluster scattering matrices [59].

It turns out that there is an even simpler way of proving Theorem 7.2, or indeed a stronger statement, which is due to Yafaev [73]. His estimate states that in a neighborhood of $C_{a,\mathrm{reg}}$, where we write y_a for the coordinates $z_a/|z_a|$ along

$C_{a,\mathrm{reg}}$, $xD_{y_a}R(\lambda + i0)f$ is in $H^{*,-1/2}(\bar{X})$. Since the principal symbol of xD_{y_a} is invertible on $(T^*X_{a,\mathrm{reg}} \cap \dot{\mathrm{Char}}(\lambda)) \setminus \mathcal{R}(\lambda)$, this result implies Theorem 7.2. Yafaev's proof relies on a simple and explicit commutator calculation, which allows one to deal with various error terms that one may, a priori, expect. However, exactly because of its explicit nature, it is presumably hard to generalize to more geometric settings, while the argument we sketched does not face this difficulty.

As discussed by Yafaev [73] in the usual time-dependent version, short-range asymptotic clustering, hence asymptotic completeness, are relatively easy consequences of Corollary 7.3, and we refer to [73] and [8] for more details. However, it is worth pointing out that the reason why Coulomb-type potentials (i.e. those in S^{-1}) are not 'short-range' is that the Hamilton vector field in some subsystem vanishes at radial points. This degeneracy makes even the subprincipal term important in describing the precise behavior of generalized eigenfunctions microlocally near this point.

Before turning to scattering theory on symmetric spaces, we note the implications of Theorem 7.2 for the scattering matrices. Previously, $S_{\alpha\beta}(\lambda)$ was only defined as a map $S_{\alpha\beta}(\lambda) : \mathcal{C}_c^\infty(S_{a,\mathrm{reg}}) \to \mathcal{C}^{-\infty}(S_{b,\mathrm{reg}})$. Indeed, part of the broken bicharacteristic relation connects $\mathcal{R}_+(\lambda)$ with its image, and this can a priori give a singularity in the kernel of $S_{\alpha\beta}(\lambda)$ of the kind that does not even allow one to conclude that $S_{\alpha\beta}(\lambda) : \mathcal{C}_c^\infty(S_{a,\mathrm{reg}}) \to \mathcal{C}^\infty(S_{b,\mathrm{reg}})$. The pairing formula, (4–2), combined with Theorem 7.2, show that in fact

$$S_{\alpha\beta}(\lambda) : L^2(S_a) \to L^2(S_b). \tag{7–1}$$

It is an interesting question whether this can be improved if we restrict $S_{\alpha\beta}(\lambda)$ to $\mathcal{C}_c^\infty(S_{a,\mathrm{reg}})$. Namely, except in special cases such as N-clusters and two-clusters, the best known result is the trivial consequence of (7–1):

$$S_{\alpha\beta}(\lambda) : \mathcal{C}_c^\infty(S_{a,\mathrm{reg}}) \to L^2(S_b).$$

(In the case of N-clusters and 2-clusters, the geometry of generalized broken bicharacteristics gives $S_{\alpha\beta}(\lambda) : \mathcal{C}_c^\infty(S_{a,\mathrm{reg}}) \to \mathcal{C}^\infty(S_{b,\mathrm{reg}})$.) The putative improvement would have to be connected to an improvement of Theorem 7.2, namely to the existence of *some* $l > -\frac{1}{2}$ such that $\mathrm{WF}_{\mathrm{sc}}^{*,l}(R(\lambda+i0)f) \subset \mathcal{R}_+(\lambda)$. It would also be connected to a better understanding of $R(\lambda\pm i0)$ at the thresholds, in which direction Wang's paper [72] is the only one I am aware of.

8. Scattering on Higher Rank Symmetric Spaces

In this section I discuss $\mathrm{SL}(N,\mathbb{R})/\mathrm{SO}(N,\mathbb{R})$ — indeed, mostly I will discuss $\mathrm{SL}(3,\mathbb{R})/\mathrm{SO}(3,\mathbb{R})$. The books [25], [34] and [9] are good general references. $N = 2$ yields the hyperbolic plane $\mathbb{H}^2$, which is a rank one symmetric space on which many aspects of analysis, such as the asymptotic behavior of the resolvent kernel and the analytic continuation of the resolvent are well understood. Indeed, these have been described on asymptotically hyperbolic spaces by Mazzeo and Melrose [37] and Perry [51; 52].

Higher rank symmetric spaces, such as $SL(N)/SO(N)$, $N \geq 3$, are much less understood. For example, using results of Harish-Chandra, and Trombi and Varadarajan (see [25]), Anker and Ji only recently obtained the leading order behavior of the Green's function [2; 3; 4]. Also, while spherical functions, which are most analogous to partial plane-partial spherical waves in the Euclidean setting, have been analyzed by Harish-Chandra, Trombi and Varadarajan, and in particular their analytic continuation is understood, the same cannot be said about the Green's function. The analysis of spherical functions relies on perturbation series expansions, much like in the proof of the Cauchy-Kovalevskaya theorem, and it does not work well at the walls of the Weyl chambers. Here I only illustrate some recent joint results with Rafe Mazzeo [40; 39], that illuminate the connections with many-body scattering, and in particular give rather direct results for the resolvent.

First I describe the space $SL(3)/SO(3)$. The polar decomposition states that any $C \in SL(3)$ can be written uniquely as $C = VR$, $V = (CC^t)^{1/2}$ is positive definite and has determinant 1, $R \in SO(3)$. Thus, $SL(3)/SO(3)$ can be identified with the set M of positive definite matrices of determinant 1; this is a five-dimensional real analytic manifold. The Killing form provides a Riemannian metric g. The associated Laplacian $\Delta = \Delta_g$ gives a self-adjoint unbounded operator on $L^2(M, dg)$, with spectrum $[\lambda_0, +\infty)$, $\lambda_0 = \frac{1}{3}$. Let $R(\lambda) = (\Delta - \lambda)^{-1}$ be the resolvent of Δ_g, $\lambda \notin [\lambda_0, +\infty)$.

Fix a point $o \in M$, which we may as well assume is the image of the identity matrix I in the identification above. The stabilizer subgroup K_o (in the natural $SL(3)$ action on M) is isomorphic to $SO(3)$. The Green function $G_o(\lambda)$ with pole at o and at eigenvalue λ is, by definition $R(\lambda)\delta_o$. It is standard that G_o lies in the space of K_o-invariant distributions on M. It is thus natural to study Δ on K_o-invariant functions.

Perhaps the most interesting property is the analytic continuation of the resolvent, which I state before indicating how it, and other results, relate to many-body scattering.

Fix the branch of the square root function $\sqrt{}$ on $\mathbb{C} \setminus [0, +\infty)$ which has negative imaginary part when $w \in \mathbb{C} \setminus [0, +\infty)$. Let S denote that part of the Riemann surface for $\lambda \mapsto \sqrt{\lambda - \lambda_0}$ where we continue from $\lambda - \lambda_0 \notin [0, +\infty)$ and allow $\arg(\lambda - \lambda_0)$ to change by any amount less than π. In other words, starting in the region $\mathrm{Im}\sqrt{\lambda - \lambda_0} < 0$, we continue across either of the rays where $\mathrm{Im}\sqrt{\lambda - \lambda_0} = 0$ and $\mathrm{Re}\sqrt{\lambda - \lambda_0} > 0$, respectively < 0, allowing the argument of $\sqrt{\lambda - \lambda_0}$ to change by any amount less than $\pi/2$ (so that only the positive imaginary axis is not reached).

THEOREM 8.1. *With all notation as above, the Green function $G_o(\lambda)$ continues meromorphically to S as a distribution. Similarly, as an operator between* *appropriate spaces of K_o-invariant functions, the resolvent $R(\lambda)$ itself has a meromorphic continuation in this region, with all poles of finite rank.*

Having stated the theorem, I indicate how it relates to many-body scattering. To do so, fix the point o – we may as well take it to be the identity matrix I. Now, M is a perfectly nice real analytic manifold and Δ is an elliptic operator on it in the usual sense, so the only question is its behavior at infinity. In order to describe this, we remark that any matrix $A \in M$ can be diagonalized, i.e. written as $A = O\Lambda O^t$, with $O \in SO(3)$ and Λ diagonal, with entries given by the eigenvalues of A. If $\mathfrak{a}$ is the set of diagonal matrices of trace 0, then $\Lambda \in \exp(\mathfrak{a})$. If all eigenvalues of A are distinct, then Λ is determined except for the ordering of the eigenvalues, and there are only finitely many possibilities for O as well. However, at the walls, which are defined to be the places where two eigenvalues coincide, there is much more indeterminacy. For example, if two eigenvalues coincide, only their joint eigenspace is well-defined. Correspondingly, we may replace O by $O'O$ for any $O' \in SO(3)$ preserving the eigenspace decomposition and still obtain the desired diagonalization.

This is closely reflected in the structure of the Laplacian at infinity. In fact, it turns out that on $SO(3)$-invariant functions, Δ is essentially a three-body Hamiltonian on $\mathfrak{a}$ with first order interactions and with collision 'planes' given by the walls (they are lines), see e.g. [25, Chapter II, Proposition 3.9]. So rather than particles, eigenvalues scatter in this case! Consequently, many-body results can be adapted to this setting.

We indicate how this is done. The most succint way of describing the geometry of M at infinity is to compactify it to a manifold $\bar{M}$ with codimension two corners. It has two boundary hypersurfaces, $H_\sharp$ and $H^\sharp$, which are perhaps easiest to describe in terms of a natural system of local coordinates derived from the matrix representation of elements in M. As above, we write $A \in M$ as $A = O\Lambda O^t$, with $O \in SO(3)$ and Λ diagonal. The ordering of the diagonal entries of Λ is undetermined, but in the region where no two of them are equal, we denote them as $0 < \lambda_1 < \lambda_2 < \lambda_3$ (but recall also that $\lambda_1\lambda_2\lambda_3 = 1$). In this region the ratios

$$\mu = \frac{\lambda_1}{\lambda_2}, \qquad \nu = \frac{\lambda_2}{\lambda_3}$$

are independent functions, and near the submanifold $\exp(\mathfrak{a})$ in M we can complete them to a full coordinate system by adding the above-diagonal entries c_{12}, c_{13}, c_{23} in the skew-symmetric matrix $T = \log O$. On M we have $\mu, \nu > 0$, and locally the compactification consists of replacing $(\mu, \nu) \in (0,1) \times (0,1)$ by $(\mu, \nu) \in [0,1) \times [0,1)$. Then $H^\sharp = \{\mu = 0\}$ and $H_\sharp = \{\nu = 0\}$, and this coordinate system gives the C^∞ structure near the corner $H_\sharp \cap H^\sharp$.

On the other hand, in a neighborhood of the interior of $H_\sharp$, for example, we obtain the compactification and its C^∞ structure as follows. Write the eigenvalues of $A \in M$, i.e. the diagonal entries of Λ in the decomposition for A above,

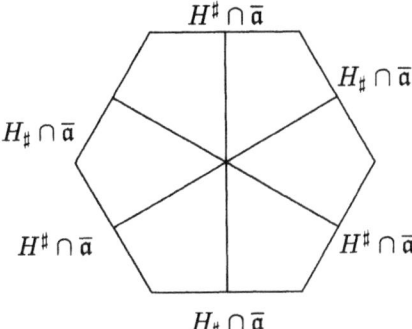

Figure 10. The closure of $\mathfrak{a}$, or rather $\exp(\mathfrak{a})$, in the compactification $\bar{M}$ of M. The lines in the interior are the Weyl chamber walls, playing the role of collision planes in many-body scattering. The side faces $H^\sharp \cap \bar{\mathfrak{a}}$ and $H_\sharp \cap \bar{\mathfrak{a}}$ correspond to the front faces on Figure 5. The main face on Figure 5 would only show up if we did a logarithmic blow-up of all boundary hypersurfaces of $\bar{M}$ and then blew up the corner.

as λ_1, λ_2 and λ_3. Suppose that A lies in a small neighbourhood $\mathcal{U}$ where

$$c < \frac{\lambda_1}{\lambda_2} < \frac{1}{c} < \lambda_3,$$

for some fixed $c \in (0,1)$. Recall also that $\lambda_3 = 1/\lambda_1\lambda_2$. These inequalities imply that $\lambda_1 = (\lambda_1/\lambda_2)^{1/2}\lambda_3^{-1/2} < 1$ and $\lambda_2 = (\lambda_2/\lambda_1)^{1/2}\lambda_3^{-1/2} < 1$, and $\lambda_3 > 1$ in $\mathcal{U}$. Hence there is a well-defined decomposition $\mathbb{R}^3 = E_{12} \oplus E_3$ for any $A \in \mathcal{U}$, where E_{12} is the sum of the first two eigenspaces and E_3 is the eigenspace corresponding to λ_3, regardless of whether or not λ_1 and λ_2 coincide. We could write equivalently $A = OCO^t$, where C is block-diagonal, preserving the splitting $\mathbb{R}^2 \oplus \mathbb{R}$ of $\mathbb{R}^3$. The ambiguity in this factorization is that C can be conjugated by an element of $O(2)$ (acting in the upper left block), and $O(2)$ can be included in the top left corner of $SO(3)$ (the bottom right entry being set equal to ± 1 appropriately). Let C' denote the upper-left block of C; the bottom right entry of C is just λ_3, and so $\lambda_3 \det C' = 1$. In other words, $C' = \lambda_3^{-1/2}C''$ where C'' is positive definite and symmetric with determinant 1, hence represents an element of $SL(2)/SO(2) \equiv \mathbb{H}^2$. Hence for an appropriate neighbourhood $\mathcal{V}$ of I in $SL(2)/SO(2)$, the neighbourhood $\mathcal{U}$ is identified with $(\mathcal{V} \times SO(3))/O(2) \times (0, c^{3/2})$, where the variable on the last factor is $s = \lambda_3^{-3/2}$. The compactification then simply replaces $(0, c^{3/2})_s$ by $[0, c^{3/2})$. Note that although the action of $O(2)$ on $\mathcal{V}$ has a fixed point (namely I), its action on $SO(3)$, and hence on the product, is free. The neighbourhood $\mathcal{V}$ can be chosen larger when λ_3 is larger, and the limiting 'cross-section' $\lambda_3 = \text{const}$ has the form $(\mathbb{H}^2 \times SO(3))/O(2)$. This space is a fibre bundle over $SO(3)/O(2)$ $(= \mathbb{R}P^2)$ with fibre $\mathbb{H}^2$. Notice that the Weyl chamber wall corresponds to the origin (i.e. the point fixed by the $SO(2)$ action) in $\mathbb{H}^2$. I refer to [40] for a more thorough description of $\bar{M}$.

On each boundary hypersurface of M, it is now easy to describe model operators for Δ acting on $SO(3)$-invariant functions. For instance, at $H_\sharp$ this model can be considered as an operator $L_\sharp$ on $\mathbb{R}_s \times \mathbb{H}^2$, acting on $SO(2)$-invariant functions. Explicitly,

$$L_\sharp = \tfrac{1}{4}(sD_s)^2 + i\tfrac{1}{2}(sD_s) + \tfrac{1}{3}\Delta_{\mathbb{H}^2}.$$

This is tensor product type, so its resolvent can be obtained from an integral of the resolvents of $\tfrac{1}{4}(sD_s)^2 + i\tfrac{1}{2}(sD_s)$ and $\tfrac{1}{3}\Delta_{\mathbb{H}^2}$. (Note that I am ignoring the weights of the L^2 spaces on which we are working, hence the appearance of the perhaps strange first order terms.)

This framework allows one to develop the elliptic theory, for example to analyze $(\Delta - \lambda)^{-1}$ for $\lambda \in \mathbb{C} \setminus [\lambda_0, +\infty)$. In particular, one can construct a parametrix for Δ on $\bar{M}$ that has a smoothing error. Since this error has no decay at infinity, it is not compact. However, the error can be improved by pasting together the resolvents of $L_\sharp$ and $L^\sharp$, and applying the result to the error term to remove it modulo a decaying, hence compact, new error term. One of the consequences is then the description of the asymptotic behavior of the Green's function, see [40].

The point of complex scaling is to rotate the essential spectrum of the operator being studied, in this case the Laplacian. To give the reader a rough idea how this works, consider the hyperbolic space $\mathbb{H}^2 = SL(2, \mathbb{R}) / SO(2, \mathbb{R})$, which may be identified with the set of two-by-two positive definite matrices A of determinant 1. In terms of geodesic normal coordinates (r, ω) about $o = I$, the Laplacian is

$$\Delta_{\mathbb{H}^2} = D_r^2 - i \coth r\, D_r + (\sinh r)^{-2} D_\omega^2.$$

Now consider the diffeomorphism $\Phi_\theta : A \mapsto A^w$, $w = e^\theta$, on $\mathbb{H}^2$, $\theta \in \mathbb{R}$. This corresponds to dilation along the geodesics through o, since these have the form $\gamma_A : s \mapsto A^{cs}$, $c > 0$. Thus, in geodesic normal coordinates, $\Phi_\theta : (r, \omega) \mapsto (e^\theta r, \omega)$. Φ_θ defines a group of unitary operators on $L^2(\mathbb{H}^2)$ via

$$(U_\theta f)(A) = (\det D_A \Phi_\theta)^{1/2}(\Phi_\theta^* f)(A), \quad J = \det D_A \Phi_\theta = w \frac{\sinh wr}{\sinh r}, \quad w = e^\theta.$$

Now, for θ real, consider the scaled Laplacian

$$(\Delta_{\mathbb{H}^2})_\theta = U_\theta \Delta_{\mathbb{H}^2} U_\theta^{-1} = J^{1/2} \Phi_\theta^* \Delta_{\mathbb{H}^2} \Phi_{-\theta}^* J^{-1/2}$$
$$= J^{1/2}(w^{-2} D_r^2 - iw^{-1} \coth(wr) D_r + (\sinh(wr))^{-2} D_\omega^2) J^{-1/2}.$$

This is an operator on $\mathbb{H}^2$, with coefficients which extend analytically in the strip $|\operatorname{Im} \theta| < \pi/2$. The square root is continued from the standard branch near $w > 0$. (The singularity of the coefficients at $r = 0$ is only an artifact of the polar coordinate representation.) Note that $(\Delta_{\mathbb{H}^2})_\theta$ and $(\Delta_{\mathbb{H}^2})_{\theta'}$ are unitary equivalent if $\operatorname{Im} \theta = \operatorname{Im} \theta'$ because of the group properties of U_θ. The scaled operator, $(\Delta_{\mathbb{H}^2})_\theta$, is not elliptic on all of $\mathbb{H}^2$ when $0 < |\operatorname{Im} \theta| < \frac{\pi}{2}$ because for

r large enough, $w^2 \sinh(wr)^{-2}$ can lie in $\mathbb{R}^-$. However, it is elliptic in some uniform neighbourhood of o in $\mathbb{H}^2$, and its radial part

$$(\Delta_{\mathbb{H}^2})_{\theta,\mathrm{rad}} = J^{1/2}(w^{-2}D_r^2 - iw^{-1}\coth(wr)D_r)J^{-1/2},$$

which corresponds to its action on $\mathrm{SO}(2)$-invariant functions, is elliptic on the entire half-line $r > 0$. The model operator for $(\Delta_{\mathbb{H}^2})_{\theta,\mathrm{rad}} - \lambda$ at infinity,

$$e^{(w-1)r/2}(w^{-2}D_r^2 - iw^{-1}D_r - \lambda)e^{-(w-1)r/2}$$
$$= e^{(w-1)r/2}((w^{-1}D_r - \tfrac{i}{2})^2 - (\lambda - \tfrac{1}{4}))e^{-(w-1)r/2},$$

is also invertible on the model space at infinity, $L^2(\mathbb{R}; e^r\, dr)$, since this is equivalent to the invertibility of

$$e^{wr/2}((w^{-1}D_r - \tfrac{i}{2})^2 - (\lambda - \tfrac{1}{4}))e^{-wr/2} = w^{-2}D_r^2 - (\lambda - \tfrac{1}{4})$$

on $L^2(\mathbb{R}; dr)$. Thus, a parametrix with compact remainder can be constructed for $(\Delta_{\mathbb{H}^2})_{\theta,\mathrm{rad}}$, and this show that its essential spectrum lies in $\tfrac{1}{4} + e^{-2i\,\mathrm{Im}\,\theta}[0, +\infty)$. Hence $((\Delta_{\mathbb{H}^2})_{\theta,\mathrm{rad}} - \lambda)^{-1}$ is meromorphic outside this set. In fact, it is well known that there are no poles in this entire strip (although there are an infinite number on $\arg\sqrt{\lambda - \lambda_0} = \pi/2$).

Combining this with some more standard technical facts, we are in a position to apply the theory of Aguilar-Balslev-Combes to prove that $((\Delta_{\mathbb{H}^2})_{\mathrm{rad}} - \lambda)^{-1}$, and hence $(\Delta_{\mathbb{H}^2} - \lambda)^{-1}$, has an analytic continuation in λ across $(\tfrac{1}{4}, +\infty)$. This is done by noting that for $\mathrm{SO}(2)$-invariant functions $f, g \in L^2(\mathbb{H}^2)$ and $\theta \in \mathbb{R}$,

$$\langle f, ((\Delta_{\mathbb{H}^2})_{\mathrm{rad}} - \lambda)^{-1}g \rangle = \langle U_{\bar\theta}f, ((\Delta_{\mathbb{H}^2})_{\theta,\mathrm{rad}} - \lambda)^{-1}U_\theta g \rangle,$$

by the unitarity of U_θ. Now if f, g lie in a smaller (dense) class of functions such that $U_\theta f$ and $U_\theta g$ continue analytically from $\theta \in \mathbb{R}$, then the meromorphic continuation in λ of the right hand side is obtained by first making θ complex with imaginary part of the appropriate sign, and then allowing λ to cross the continuous spectrum of $\Delta_{\mathbb{H}^2}$ without encountering the essential spectrum of $(\Delta_{\mathbb{H}^2})_{\theta,\mathrm{rad}}$. Hence the left hand side continues meromorphically also. With some additional care, one can even allow g to be the delta distribution at o, yielding the meromorphic continuation of the Green's function.

The argument on the higher rank symmetric space $M = \mathrm{SL}(3)/\mathrm{SO}(3)$ is similar. We still use the same scaling $\Phi_\theta : A \mapsto A^w$, $w = e^\theta$ with $\theta \in \mathbb{R}$, on M. Again, the first concern is that, allowing θ to become complex, the scaled operator Δ_θ is not elliptic. However, it is elliptic near $o = \mathrm{Id}$, and the scaled models for it near the walls, such as $(L_\sharp)_\theta$, remain elliptic at the walls. After all, for the latter, this is just the ellipticity of $(\Delta_{\mathbb{H}^2})_\theta$ near the origin, which we have already observed. This again allows the elliptic parametrix construction to proceed, supplying the results we needed in order to reach the framework of complex scaling. This in turn finishes the proof of Theorem 8.1.

References

[1] J. Aguilar and J. M. Combes. A class of analytic perturbations for one-body Schrödinger operators. *Comm. Math. Phys.*, 22:269–279, 1971.

[2] J.-P. Anker. La forme exacte de l'estimation fondamentale de Harish-Chandra. *C. R. Acad. Sci. Paris*, Sér. I, 305:371–374, 1987.

[3] J.-P. Anker and L. Ji. Comportement exact du noyau de la chaleur et de la fonction de Green sur les espaces symétriques non-compacts. *C. R. Acad. Sci. Paris*, Sér. I, 326:153–156, 1998.

[4] J.-P. Anker and L. Ji. Heat kernel and Green function estimates on noncompact symmetric spaces. *Geom. Funct. Anal.*, 9(6):1035–1091, 1999.

[5] E. Balslev and J. M. Combes. Spectral properties of many body Schrödinger operators with dilation analytic potentials. *Commun. Math. Phys.*, 22:280–294, 1971.

[6] Alberto-P. Calderón. On an inverse boundary value problem. In *Seminar on Numerical Analysis and its Applications to Continuum Physics (Rio de Janeiro, 1980)*, pages 65–73. Soc. Brasil. Mat., Rio de Janeiro, 1980.

[7] J. Dereziński. Asymptotic completeness of long-range N-body quantum systems. *Ann. Math.*, 138:427–476, 1993.

[8] J. Dereziński and C. Gérard. *Scattering theory of classical and quantum N-particle systems*. Springer, 1997.

[9] P. Eberlein. *Geometry of nonpositively curved manifolds*. University of Chicago Press, 1996.

[10] V. Enss. *Quantum scattering theory two- and three-body systems with potentials of short and long range*, pages 39–176. Lecture notes in mathematics. Springer-Verlag, 1985.

[11] V. Enss and R. Weder. The geometrical approach to multidimensional inverse scattering. *J. Math. Phys.*, 36:3902–3921, 1995.

[12] Volker Enss. Asymptotic completeness for quantum mechanical potential scattering. I. Short range potentials. *Comm. Math. Phys.*, 61(3):285–291, 1978.

[13] Volker Enss and Ricardo Weder. Inverse two-cluster scattering. *Inverse Problems*, 12(4):409–418, 1996.

[14] L. D. Faddeev. The inverse problem in the quantum theory of scattering. II. In *Current problems in mathematics, Vol. 3 (Russian)*, pages 93–180, 259. (loose errata). Akad. Nauk SSSR Vsesojuz. Inst. Naučn. i Tehn. Informacii, Moscow, 1974.

[15] R. G. Froese and I. Herbst. Exponential bounds and absence of positive eigenvalues of N-body Schrödinger operators. *Commun. Math. Phys.*, 87:429–447, 1982.

[16] R. G. Froese and I. Herbst. A new proof of the Mourre estimate. *Duke Math. J.*, 49:1075–1085, 1982.

[17] C. Gérard. Distortion analyticity for N-particle Hamiltonians. *Helv. Phys. Acta*, 66(2):216–225, 1993.

[18] C. Gérard, H. Isozaki, and E. Skibsted. *Commutator algebra and resolvent estimates*, volume 23 of *Advanced studies in pure mathematics*, pages 69–82. 1994.

[19] C. Gérard, H. Isozaki, and E. Skibsted. N-body resolvent estimates. *J. Math. Soc. Japan*, 48:135–160, 1996.

[20] G. M. Graf. Asymptotic completeness for N-body short range systems: a new proof. *Commun. Math. Phys.*, 132:73–101, 1990.

[21] Victor Guillemin and David Schaeffer. On a certain class of Fuchsian partial differential equations. *Duke Math. J.*, 44(1):157–199, 1977.

[22] A. Hassell. Distorted plane waves for the 3 body Schrödinger operator. *Geom. Funct. Anal.*, 10:1–50, 2000.

[23] A. Hassell, R. B. Melrose, and A. Vasy. Spectral and scattering theory for symbolic potentials of order zero. *Advances in Mathematics*, to appear.

[24] B. Helffer and J. Sjöstrand. Équation de Schrödinger avec champ magnétique et équation de Harper. In *Schrödinger operators (Sønderborg, 1988)*, pages 118–197. Springer, Lecture Notes in Physics, No. 345, Berlin, 1989.

[25] S. Helgason. *Groups and geometric analysis.* Academic Press, 1984.

[26] P. D. Hislop and I. M. Sigal. *Introduction to spectral theory.* Springer-Verlag, 1996.

[27] L. Hörmander. *The analysis of linear partial differential operators*, vol. 1-4. Springer-Verlag, 1983.

[28] Lars Hörmander. On the existence and the regularity of solutions of linear pseudodifferential equations. *Enseignement Math. (2)*, 17:99–163, 1971.

[29] M. Ikawa, editor. *Spectral and scattering theory.* Marcel Dekker, 1994.

[30] H. Isozaki. On N-body Schrödinger operators. *Proc. Indian Acad. Sci. Math. Sci.*, 104:667–703, 1993.

[31] H. Isozaki. A generalization of the radiation condition of Sommerfeld for N-body Schrödinger operators. *Duke Math. J.*, 74:557–584, 1994.

[32] H. Isozaki. *A uniqueness theorem for the N-body Schrödinger equation and its applications.* In Ikawa [29], 1994.

[33] V. Ja. Ivriĭ. Wave fronts of solutions of boundary value problems for a class of symmetric hyperbolic systems. *Sibirsk. Mat. Zh.*, 21(4):62–71, 236, 1980.

[34] J. Jost. *Riemannian geometry and geometric analysis.* Springer, 1998.

[35] Peter D. Lax. Asymptotic solutions of oscillatory initial value problems. *Duke Math. J.*, 24:627–646, 1957.

[36] G. Lebeau. Propagation des ondes dans les variétés à coins. *Ann. Scient. Éc. Norm. Sup.*, 30:429–497, 1997.

[37] R. Mazzeo and R. B. Melrose. Meromorphic extension of the resolvent on complete spaces with asymptotically constant negative curvature. *J. Func. Anal.*, 75:260–310, 1987.

[38] R. Mazzeo and R. B. Melrose. Pseudodifferential operators on manifolds with fibred boundaries. *Asian J. Math.*, 2, 1998.

[39] R. Mazzeo and A. Vasy. Analytic continuation of the resolvent of the Laplacian on SL(3)/SO(3). *Amer. J. Math.*, to appear.

[40] R. Mazzeo and A. Vasy. Scattering theory on SL(3)/SO(3): connections with quantum 3-body scattering. *Preprint*, 2002.

[41] R. B. Melrose. *Spectral and scattering theory for the Laplacian on asymptotically Euclidian spaces.* In Ikawa [29], 1994.

[42] R. B. Melrose. *Geometric scattering theory*. Cambridge University Press, 1995.

[43] R. B. Melrose and J. Sjöstrand. Singularities of boundary value problems. I. *Comm. Pure Appl. Math*, 31:593–617, 1978.

[44] R. B. Melrose and J. Sjöstrand. Singularities of boundary value problems. II. *Comm. Pure Appl. Math*, 35:129–168, 1982.

[45] R. B. Melrose and J. Wunsch. Propagation of singularities for the wave equation on conic manifolds. *Preprint*, 2002.

[46] R. B. Melrose and M. Zworski. Scattering metrics and geodesic flow at infinity. *Inventiones Mathematicae*, 124:389–436, 1996.

[47] E. Mourre. Absence of singular continuous spectrum of certain self-adjoint operators. *Commun. Math. Phys.*, 78:391–408, 1981.

[48] R. G. Novikov and G. M. Khenkin. The $\bar{\partial}$-equation in the multidimensional inverse scattering problem. *Uspekhi Mat. Nauk*, 42(3(255)):93–152, 255, 1987.

[49] Roman G. Novikov. On inverse scattering for the N-body Schrödinger equation. *J. Funct. Anal.*, 159(2):492–536, 1998.

[50] Roman G. Novikov. On determination of the fourier transform of a potential from the scattering amplitude. *Inverse Problems*, 17:1243–1251, 2001.

[51] P. Perry. The Laplace operator on a hyperbolic manifold. I. Spectral and scattering theory. *J. Funct. Anal.*, 75:161–187, 1987.

[52] P. Perry. The Laplace operator on a hyperbolic manifold. II. Eisenstein series and the scattering matrix. *J. Reine. Angew. Math.*, 398:67–91, 1989.

[53] P. Perry, I. M. Sigal, and B. Simon. Spectral analysis of N-body Schrödinger operators. *Ann. Math.*, 114:519–567, 1981.

[54] M. Reed and B. Simon. *Methods of modern mathematical physics*. Academic Press, 1979.

[55] I. M. Sigal and A. Soffer. N-particle scattering problem: asymptotic completeness for short range systems. *Ann. Math.*, 125:35–108, 1987.

[56] I. M. Sigal and A. Soffer. Long-range many-body scattering. *Inventiones Math.*, 99:115–143, 1990.

[57] I. M. Sigal and A. Soffer. Asymptotic completeness of N-particle long-range scattering. *J. Amer. Math. Soc.*, 7:307–334, 1994.

[58] Johannes Sjöstrand and Maciej Zworski. Complex scaling and the distribution of scattering poles. *J. Amer. Math. Soc.*, 4(4):729–769, 1991.

[59] E. Skibsted. Smoothness of N-body scattering amplitudes. *Reviews in Math. Phys.*, 4:619–658, 1992.

[60] J. Sylvester and G. Uhlmann. A global uniqueness theorem for an inverse boundary value problem. *Ann. of Math.*, 125:153–169, 1987.

[61] M. Taylor. Grazing rays and reflection of singularities of solutions to wave equations. *Comm. Pure Appl. Math.*, 29:1–38, 1976.

[62] Gunther Uhlmann and András Vasy. Low-energy inverse problems in three-body scattering. *Inverse Problems*, 18(3):719–736, 2002.

[63] A. Vasy. Structure of the resolvent for three-body potentials. *Duke Math. J.*, 90:379–434, 1997.

[64] A. Vasy. Scattering matrices in many-body scattering. *Commun. Math. Phys.*, 200:105–124, 1999.

[65] A. Vasy. Propagation of singularities in many-body scattering. *Ann. Sci. École Norm. Sup. (4)*, 34:313–402, 2001.

[66] A. Vasy. Propagation of singularities in many-body scattering in the presence of bound states. *J. Func. Anal.*, 184:177–272, 2001.

[67] A. Vasy. Exponential decay of eigenfunctions in many-body type scattering with second order perturbations. *J. Func. Anal.*, to appear.

[68] A. Vasy and X. P. Wang. Smoothness and high energy asymptotics of the spectral shift function in many-body scattering. *Commun. in PDEs*, 27:2139–2186, 2002.

[69] X. P. Wang. Microlocal estimates for N-body Schrödinger operators. *J. Fac. Sci. Univ. Tokyo Sect. IA, Math.*, 40:337–385, 1993.

[70] X. P. Wang. High energy asymptotics for N-body scattering matrices with arbitrary channels. *Ann. Inst. H. Poincaré Phys. Théor.*, 65(1):81–108, 1996.

[71] X. P. Wang. An inverse problem related to channel scattering operators. *Asymptot. Anal.*, 18(1-2):147–164, 1998.

[72] X. P. Wang. Spectral analysis of N-body Schrödinger operators near a threshold. *Preprint*, 2001.

[73] D. Yafaev. Radiation conditions and scattering theory for N-particle Hamiltonians. *Commun. Math. Phys.*, 154:523–554, 1993.

ANDRÁS VASY
DEPARTMENT OF MATHEMATICS
MASSACHUSETTS INSTITUTE OF TECHNOLOGY
CAMBRIDGE MA 02139
UNITED STATES
andras@math.mit.edu

A Mathematical and Deterministic Analysis of the Time-Reversal Mirror

CLAUDE BARDOS

ABSTRACT. We give a mathematical analysis of the "time-reversal mirror", in what concerns phenomena described by the genuine acoustic equation with Dirichlet or impedance boundary conditions. An ideal situation is first considered, followed by the boundary-data, impedance and internal time-reversal methods. We explore the relationship between local decay of energy and accuracy of the method, and explain the positive effect of ergodicity.

1. Introduction: Principle of the Method

In all time reversal experiments, a finite time $0 < T < \infty$ is chosen. At time $t = 0$ waves are emitted from a localized source, recorded in time (for $0 < t < T$) by an array of receivers-transducers, time-reversed and retransmitted in the media during the time $(T < t < 2T)$; for instance the first signal to arrive is reemitted last and the last to arrive is remitted first. In this second step $(t > T)$ one can introduce amplification. The process is possibly repeated several times, leading in some cases to an automatic focusing on the most reflective target in a multiple target media. This has several applications in nondestructive testing, medical techniques such as lithotripsy and hyperthermia, underwater acoustics, etc. See [13].

The intuitive reasons why such a process may work are:

(1) The wave equation is invariant with respect to the symmetry $t \in (0, T) \mapsto 2T - t \in (T, 2T)$.
(2) At high frequencies waves propagate as rays.
(3) Inhomogeneities, randomness and ergodicity contribute to much better refocusing.

Point (1) is really at the origin of the method; the same type of technique in diffusive media (phenomena governed by the diffusion equation) seems more difficult to analyse.

Point (2) gives a good intuitive description of the phenomena. Also it can be used to understand point (3) through multipathing, observing that the number of rays which connect the source with the transducers is greatly enhanced by the complexity of the media. For the case of random or ergodic media, it is necessary to study and derive some high frequency asymptotics, and this study is the backbone of the present contribution. Using classical analysis as our general tool, we start from the basic properties of the wave equation and continue with a mathematical version of the high frequency asymptotics called microlocal analysis. We depart from other points of view where randomness is of crucial importance (see for instance [4] and [1]). We do not assume any randomness but we consider the effect of ergodicity which is the deterministic counterpart of randomness.

We will consider three basic examples of time-reversal methods: the boundary-data time-reversal method, or BDTRM, the impedance time reversal mirror, or IMTRM, and the internal time-reversal method, or INTRM. In all three cases the phenomena are described by solutions to the acoustic equation in a homogeneous medium of dimension d:

$$\partial_t^2 u - \Delta u = 0. \qquad (1\text{--}1)$$

In the BDTRM the solution is defined in the complement $\Omega \subset \mathbb{R}^d$ of a bounded obstacle $K \subset \mathbb{R}^d$ that forms a cavity $\mathcal{C} \subset \Omega$ with an aperture Γ. The boundary of $\mathcal{C}$ is therefore the union of Γ and $\Gamma_c = \bar{\mathcal{C}} \cap \partial\Omega$ (Figure 1). Given an arbitrary but possibly large time T, the solution of (1–1) is assumed to solve a homogeneous boundary condition on $\partial\Omega$ — say, for simplicity, the Dirichlet boundary condition

$$u_i(x, t) = 0 \quad \text{on } \partial\Omega \times (0, T).$$

In the mean time the value of $u_i(x, t)$ is observed on Γ. For $t > T$ one considers the reversed solution u_r defined for $T < t < 2T$ by the equations

$$\partial_t^2 u_r - \Delta u_r = 0,$$

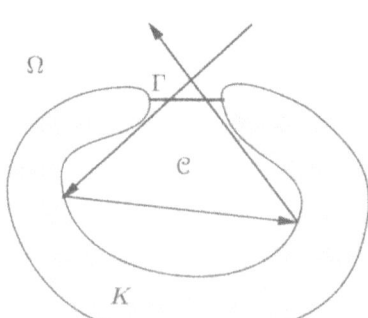

Figure 1. Cavity $\mathcal{C}$ with an aperture Γ contained in the complement Ω of a bounded obstacle K. A broken ray is shown.

with initial conditions

$$u_r(x,T) = u_i(x,T), \qquad \partial_t u_r(x,T) = \partial_t u_i(x,T)$$

and boundary conditions

$$u_r(x,t) = \begin{cases} 0 & \text{on } \Gamma_c = \bar{\mathcal{C}} \cap \partial\Omega, \\ u(x, 2T{-}t) & \text{on } \Gamma. \end{cases}$$

In the impedance time-reversal problem, IMTRM (Figure 2), one considers the same wave equation

$$\partial_t^2 u_i - \Delta u_i = 0$$

in a (closed) bounded cavity $\mathcal{C}$ whose boundary has a region Γ thought of as being covered with transducers (sensors). Away from Γ a homogeneous Dirichlet boundary condition holds, while on Γ an *impedance boundary condition* holds:

$$\begin{aligned} \partial_t u_i + Z(x)\partial_n u_i(x,t) = 0 &\quad \text{on } \Gamma \times (0,T), \\ u_i(x,t) = 0 &\quad \text{on } \partial\mathcal{C} \setminus \Gamma \times (0,T), \end{aligned} \tag{1-2}$$

where $Z(x)$ is a strictly positive function representing the impedance of the transducers that cover the region Γ. Here and below ∂_n denotes the outward normal to the boundary. For $0 \le t \le T$, the value of $\partial_t u_i$ on $\Gamma \times (0,T)$ is recorded and $\partial_n u_i$ is computed using (1–2). Then for $T < t < 2T$ one considers the reversed solution u_r defined by the equations

$$\partial_t^2 u_r - \Delta u_r = 0,$$

with initial conditions

$$u_r(x,T) = u_i(x,T),$$
$$\partial_t u_r(x,T) = \partial_t u_i(x,T).$$

and the Neumann–Dirichlet boundary conditions

$$\begin{aligned} \partial_n u_r(x,t) = \partial_n u_r(x, 2t - T) &\quad \text{on } \Gamma \times (0,T), \\ u_r(x,t) = 0 &\quad \text{on } \partial\mathcal{C} \setminus \Gamma \times (0,T). \end{aligned}$$

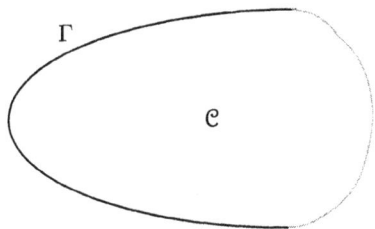

Figure 2. Cavity $\mathcal{C}$ with an impedance time-reversal mirror on a subset Γ of the boundary (in black).

In the internal time-reversal problem, INTRM, one considers for $0 < t < T$ the solution of a homogeneous boundary value problem (for instance with the Dirichlet boundary condition) in a bounded set $\mathcal{C}$,

$$\partial_t^2 u_i - \Delta u_i = 0 \quad \text{in } \mathcal{C},$$
$$u_i(x,t) = 0 \quad \text{on } \partial\mathcal{C} \times (0,T),$$

with initial conditions

$$u(\,\cdot\,,0) \equiv 0, \quad \partial_t u(x,0) = \phi(x),$$

and one introduces a bounded function $\Xi(x)$ with support in a subset σ of $\mathcal{C}$. The support of $\Xi(x)$ represents the domain of action of the transducer. This is where the signal is recorded and reemitted. For $0 < t < T$, record the value of $\partial_t u_i(x,t)$ on σ and for $T < t < 2T$ consider the solution of the problem

$$\partial_t^2 u_r - \Delta u_r = \Xi(x)\partial_t u(x,2T{-}t) \quad \text{in } \Omega,$$
$$u(x,t) = 0 \qquad\qquad\qquad \text{on } \partial\mathcal{C} \times (0,T), \tag{1-3}$$

with initial conditions as above:

$$u_r(x,T) = u_i(x,T),$$
$$\partial_t u_r(x,T) = \partial_t u_i(x,T). \tag{1-4}$$

It is mainly the IMTRM and the INTRM that correspond to real physical experiments. In the laboratory the impedance time-reversal mirror is usually performed by using a setup that measures and records the field on Γ, and after a time T transmits the time-reversed field in $\mathcal{C}$. Such a time-reversal mirror setup is made of an array of reversible piezoelectric transducers on Γ, which can be used now as microphones to record the field, now as loudspeakers to retransmit the time-reversed field ([11], [12] and [13]). When the transducers are used as microphones, due to their elastic properties, the boundary condition is usually an absorbing condition relating the normal derivative of the field to its time derivative through a local impedance condition of type

$$\partial_t u_i + Z(x)\partial_n u_i = 0.$$

In the first step, for $t < T$, the microphones measure the incident acoustic pressure field which is directly proportional to the time derivative of the acoustic potential $\partial_t u$.

In the second step, for $T < t < 2T$, the loudspeakers impose on Γ the normal velocity field which results from the time reversal of the component measured in the first step according to the formula:

$$\partial_n u_r(x,t) = \partial_n u_i(x,2T{-}t) = -\frac{1}{Z(x)}\partial_t u_i(x,2T{-}t).$$

The INTRM has been the object of several practical and numerical experiments trying to evaluate how the ergodicity of the cavity would contribute to the refocusing of the wave. An experiment due to C. Draeger is shown in Figure 5; other experimental or numerical results can be found in [9] and [10].

Even if it is not so close to applications, the BDTRM is studied because in very special cases an exact reversal is obtained. This elucidates how the difference between real and ideal time reversal method is related to the question of local decay of energy.

Therefore this chapter is organized as follows.

(i) Section 2 gives an ideal example of exact time reversal, based only on the strong form of Huygens' principle.
(ii) Section 3 analyzes the BDTRM, mainly in relation with the problem of the local decay, well known in the mathematical community.
(iii) Section 4 is devoted to the IMTRM, which appears closely related to the question of stabilization.
(iv) Section 5 concerns the refocusing by the INTRM in an ergodic cavity. It is shown how such phenomena can be explained in term of recent theorems about quantum mixing.

The present chapter follows with some improvements the ideas of an earlier article [2], which included a comparison with the classical theory of control. The experimental and numerical results were performed by Casten Draeger, who pionnered the study of the ergodic cavity.

2. Example of an Exact Time-Reversal Method

One can fully understand why the method works and what its limitations are by starting with an "academic case" as described below. Consider in $\mathbb{R}^3$ the acoustic equation

$$\partial_t^2 u_i - \Delta u_i = 0, \qquad (2\text{–}1)$$

with prescribed initial conditions $u_i(x, 0)$ and $\partial_t u_i(x, 0)$ having their support in a ball $B_{\rho_1} = \{x : |x| < \rho_1\}$ of radius $0 < \rho_1 < \infty$.

Assume that the observation region Γ is the boundary of a bounded open set $\mathcal{C}$ containing the ball B_{ρ_1} and contained in a bigger ball $B_{\rho_2} = \{x : |x| < \rho_2\}$:

$$\text{supp } u_i(x, 0) \cup \text{supp } \partial_t u_i(x, 0) \subset B_{\rho_1} \subset \mathcal{C} \subset B_{\rho_2}$$

(see Figure 3). Observe this solution (defined in $\mathbb{R}^3 \times (0, T)$) on $(\partial \mathcal{C} = \Gamma) \times (0, T)$. For $0 < T < t \leq 2T$ introduce the solution of the reversed problem:

$$\partial_t^2 u_r - \Delta u_r = 0 \quad \text{in } \mathcal{C} \times \{T < t < 2T\} \qquad (2\text{–}2)$$

with initial conditions

$$u_r(x, T) = u_i(x, T),$$
$$\partial_t u_r(x, T) = \partial_t u_i(x, T) \qquad (2\text{–}3)$$

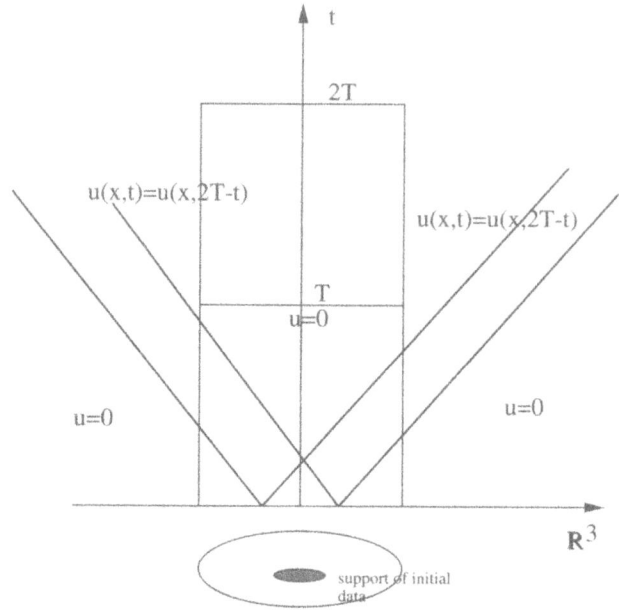

Figure 3. Finite speed of propagation and Huygens' principle in three space variables.

and boundary conditions

$$u_r(x,t) = u_i(x, 2T-t) \quad \text{on } \Gamma \times (T, 2T), \tag{2-4}$$

where as usual $\Gamma = \partial \mathcal{C}$. Then the following easy theorem, a direct consequence of Huygens' principle, precisely indicates the validity of the method:

THEOREM 2.1. *Consider the solution u_r defined in $\mathcal{C} \times (T, 2T) \subset \mathbb{R}^3 \times (T, 2T)$ by equations (2–1), (2–2), (2–3) and (2–4). Then, under the hypothesis*

$$T > \rho_1 + \rho_2$$

one has in $\mathcal{C}$

$$u_r(x, 2T) = u_i(x, 0), \qquad \partial_t u_r(x, 2T) = -\partial_t u_i(x, 0).$$

PROOF. Consider $U(x,t)$ defined in $\mathcal{C} \times (0, 2T)$ by the formulas

$$U(x,t) = \begin{cases} u_i(x,t) & \text{for } 0 < t < T, \\ u_r(x,t) & \text{for } T < t < 2T. \end{cases}$$

Such a function is a solution of a mixed time-dependent boundary value problem.

As a consequence of the strong form of Huygens' principle ([17, Theorem 1.3, p. 96 and figure 3]) the initial solution $u_i(x,t)$ is zero in the cone

$$\{(x,t) : |x| \le t - \rho_1\}$$

and one has, for $t = T$:

$$U(x,T) \equiv \partial_t U(x,T) \equiv 0 \text{ in } \Omega.$$

Furthermore for the boundary condition one has, for $T < t < 2T$, by construction:

$$U(x,t) = U(x, 2T-t) \quad \text{on } \partial\mathcal{C} \times (T, 2T). \tag{2-5}$$

The function obtained by time symmetry around T is a solution of the same problem. Both the data and the equation are therefore invariant with respect to the time symmetry around $t = T$. the uniqueness of the mixed boundary-value problem for the wave equation [6] on $\mathcal{C} \times (0, 2T)$ implies the relation

$$U(x,t) = U(x, 2T-t) \quad \text{on } \Omega \times (T, 2T), \tag{2-6}$$

and the result follows. $\qquad\qquad\qquad\qquad\qquad\qquad\qquad\qquad\qquad\qquad$ $\square$

3. The Boundary-Data Time-Reversal Method (BDTRM)

The preceding example, together with recent results on the decay of the solution of the exterior problem, leads to an understanding of the possibilities and limitations of the method in a cavity. Once again for simplicity the problem is considered in $\mathbb{R}^3$ or $\mathbb{R}^d$ with d odd. (The case d even introduces some algebraic decay of the local energy.) As mentioned in the introduction for the exterior problem, the solution is defined in the complement Ω of a bounded obstacle $K \subset \mathbb{R}^d$ which forms a cavity $\mathcal{C} \subset \Omega$ with an aperture Γ and the boundary of $\mathcal{C}$ is therefore the union of Γ and $\Gamma_c = \bar{\mathcal{C}} \cap \partial\Omega$ (Figure 2).

Given an arbitrary but possibly large time T, the solution $u_i(x,t)$ of (1–1) is assumed to evolve with homogeneous Dirichlet boundary condition on $\partial\Omega$. In the mean time the value of $u_i(x,t)$ is observed on Γ and for $T < t < 2T$ one considers the reversed solution u_r defined for $T < t < 2T$ by the equations

$$\partial_t^2 u_r - \Delta u_r = 0, \tag{3-1}$$

with initial conditions

$$u_r(x,T) = u_i(x,T),$$
$$\partial_t u_r(x,T) = \partial_t u_i(x,T), \tag{3-2}$$

and boundary conditions

$$u_r(x,t) = \begin{cases} 0 & \text{on } \Gamma_c = \bar{\mathcal{C}} \cap \partial\Omega, \\ u_i(x, 2T-t) & \text{on } \Gamma. \end{cases} \tag{3-3}$$

Observe that $u_r(x,t)$ decomposes in $\mathcal{C} \times (T, 2T)$ into the sum of two functions

$$u_r(x,t) = u_D(x,t) + u_R(x,t),$$

which are solutions of

$$\partial_t^2 u_D - \Delta u_D = 0 \quad \text{in } \mathcal{C} \times (0, 2T),$$

$$u_D(x, t) = 0 \quad \text{on } \left(\partial \mathcal{C} = (\bar{\mathcal{C}} \cap \partial \Omega) \cup \Gamma\right) \times (0, 2T),$$

$$u_D(x, T) = 0, \quad \partial_t u_D(x, T) = \partial_t u(x, T)$$

and

$$\partial_t^2 u_R - \Delta u_R = 0 \quad \text{in } \mathcal{C} \times (0, 2T),$$

$$u_R(x, t) = 0 \quad \text{on } (\bar{\mathcal{C}} \cap \partial \Omega) \times (T, 2T),$$

$$u_R(x, t) = u_i(x, t) \quad \text{on } \Gamma \times (0, T),$$

$$u_R(x, t) = u_i(x, 2T{-}t) \quad \text{on } \Gamma \times (T, 2T),$$

$$u_R(x, T) = u_i(x, T), \quad \partial_t u_R(x, T) = 0 \quad \text{on } \mathcal{C}.$$

We have:

(i) $u_R(x, t)$ is time symmetric with respect to T,

(ii) $u_D(x, t) + u_R(x, t)$ coincide with $u_i(x, t)$ for $0 < t < T$ and with $u_r(x, t)$ for $T < t < 2T$.

Thus the difference between $\left(u_i(x, 0), \partial_t u_i(x, 0)\right)$ and $\left(u_r(x, 2T), -\partial_t u_r(x, 2T)\right)$ is bounded in the energy norm

$$E_{\mathcal{C}}(u) = \frac{1}{2} \int_{\mathcal{C}} \left(|\nabla u|^2 + |\partial_t u|^2\right) dx$$

by twice the energy norm of $u_D(x, t)$, which is time invariant and equal for $t = T$ to

$$\int_{\mathcal{C}} |\partial_t u_i(x, T)|^2 dx.$$

Using the standard notation concerning the energy norm and the Sobolev space $H_0^1(\mathcal{C})$ one obtains:

PROPOSITION 3.1. *Assume that u_r is constructed with the algorithm* (3–1), (3–2) *and* (3–3). *Then*

$$\left\| (u_r(x, 2T), -\partial_t u_r(x, T)) - (u(x, 0), \partial_t u(x, 0)) \right\|_{H_0^1(\mathcal{C}) \times L^2(\mathcal{C})}^2 \leq \int_{\mathcal{C}} |\partial_t u(x, T)|^2 dx.$$

A consequence of this proposition is that the validity of the time-reversal method can be estimated in terms of the local energy decay of the solution of the wave equation in an exterior problem. Such problems have been studied in detail; some historical information can be found in the revised version of Lax and Phillips [17]. Since the first edition of this book it has become known that with initial data of compact support and finite energy the local energy decays to zero as $t \to \infty$. However it is also known that this decay depends on the geometry of the classical Hamiltonian flow, which describes the evolution of rays of geometrical optics in $\bar{\Omega} \times \mathbb{R}_t$. In the present case, where the coefficients are constant, and with the canonical identification between tangent and cotangent space, these rays are defined as continuous maps $t \mapsto \gamma(t) = (x(t), \xi(t))$ from $\mathbb{R}_t$ with values in $\bar{\Omega} \times S^{d-1}$, according to the following prescription:

In $\Omega \times S^{d-1}$ rays propagate with constant velocity

$$\dot{x}(t) = \xi, \quad \dot{\xi}(t) = 0.$$

Then the interaction with the boundary is described as follow.

First consider only the rays that, coming from the interior, intersect transversally $\partial\Omega$ at a point x_b and at a time t_b. Extend them for further time by specular reflection according to the formula

$$\dot{x}_b^+ = \xi_b^+ = \xi_b^- - 2(\xi_b^-, n_b)n_b. \tag{3-4}$$

With several reflections one obtains broken rays which are continuous maps from $\mathbb{R}_t$ with value in $\bar{\Omega} \times S^{d-1}$. The compressed broken hamiltonian flow $t \mapsto (x(t), \xi(t))$ is defined as the closure of these broken rays in $\bar{\Omega} \times S^{d-1}$, for the C^0 topology (Figure 1). Under very general hypothesis (but not always; see [14, vol. 3, p. 438]) the curves of the compressed broken hamiltonian flow (which are called bicharacteristics) realize a "foliation" of $\bar{\Omega} \times S^{d-1}$ and the singularities of the solutions propagate along these bicharacteristics. As a consequence, the following definition and theorem are now classic in microlocal analysis:

DEFINITION 3.1. A bounded obstacle $K \subset \mathbb{R}^n$ is *nontrapping* if there exists a ball B_ρ with $\Omega = \mathbb{R}^n \setminus K$, $K \subset B_\rho$, and a time $T > 0$ such that for any compressed broken ray $t \mapsto (x(t), \xi(t))$ with initial data satisfying

$$x(0) \in \Omega \cap B_\rho,$$

one has

$$x(t) \notin B_\rho \quad \text{for } t > T.$$

When this is not the case, the obstacle is *trapping*.

THEOREM 3.1. [17] *Consider the exterior problem with homogeneous boundary conditions (say Dirichlet or Neumann boundary conditions) and initial data with compact support and finite energy the local energy decays always to zero. When the obstacle is nontrapping this decay is uniform (and exponential when the dimension of the space is odd). When the obstacle is trapping this decay may be arbitrarily slow. More precisely, in odd dimensions, the solutions of*

$$\partial_t^2 u - \Delta u = 0 \quad in \ \Omega,$$
$$u(x, t) = 0 \quad on \ \partial\Omega \tag{3-5}$$

with initial conditions $(u(x, 0), \partial_t u(x, 0))$ of finite energy and compact support satisfy the following assertions:

(i) *If the obstacle $K = \mathbb{R}^n \setminus \Omega$ is not trapping, there exists a constant β such that*

$$E_\rho(u)(t) = \frac{1}{2} \int_{\Omega \cap B_\rho} \left(|\nabla u(x,t)|^2 + |\partial_t u(x,t)|^2 \right) dx$$

$$\leq \frac{1}{2} e^{-\beta t} \int_\Omega \left(|\nabla u(x,0)|^2 + |\partial_t u(x,0)|^2 \right) dx.$$

(ii) *If the obstacle is trapping, for any pair ε, T there exists a solution u_ε of (3–5) such that*

$$E_\rho(u_\varepsilon)(t) \geq \frac{1}{2}(1 - \varepsilon) \int_\Omega \left(|\nabla u_\varepsilon(x,0)|^2 + |\partial_t u_\varepsilon(x,0)|^2 \right) dx. \qquad (3\text{–}6)$$

for all $t \in (0, T)$.

The proof of (3–6) is constructed with the concentration of high frequency solutions along a trapped ray for which higher norms would blow up with ε.

On the other hand, if the solution is uniformly bounded for all time in a subspace of higher regularity, the Rellich and Banach Steinhaus theorems imply the existence of a uniform rate of decay.

For a precise statement it is convenient to write the wave equation as a group of transformations in the energy space $\mathcal{E}$ introduced by Lax and Phillips [17]. This space is the closure for the norm

$$\|(u,v)\|^2 = \frac{1}{2} \int_\Omega \left(|\nabla u|^2 + |v|^2 \right) dx \qquad (3\text{–}7)$$

of the set of smooth functions (u, v) with compact support in Ω. The generator $\mathcal{A}$ of this wave group and its domain $D(\mathcal{A})$ are defined by the formulas

$$\mathcal{A} = \begin{pmatrix} 0 & I \\ \Delta & 0 \end{pmatrix}$$

and

$$D(\mathcal{A}) = \{ U = (u, v) \in \mathcal{E} : \mathcal{A}U \in \mathcal{E} \}. \qquad (3\text{–}8)$$

The quantity $\|\mathcal{A}^s U\|_\mathcal{E} + \|U\|_\mathcal{E} = \|U\|_{D(\mathcal{A}^s)}$ is invariant under the action of the wave group and the conjunction of the Rellich and Banach–Steinhaus Theorems implies, for $s > 0$, the existence of a positive continuous function $\phi(t, s)$ going to zero with $t \to \infty$ such that, for any solution with initial data having support in B_ρ,

$$E_\rho(u)(t) = \frac{1}{2} \int_{\Omega \cap B_\rho} \left(|\nabla u(x,t)|^2 + |\partial_t u(x,t)|^2 \right) dx$$

$$\leq \phi(t, s) \|(u(x,0), \partial_t u(x,0))\|^2_{D(\mathcal{A}^s)}.$$

The optimal result (involving no hypotheses on the geometry) on the decay of $\phi(t, s)$ has been obtained by Burq, using Carleman estimates:

PROPOSITION 3.2. [5] *For any solution of the wave equation with Dirichlet boundary data* (3–5) *and initial data supported in* $B_\rho \cap \Omega$ *one has:*

$$E_\rho(u)(t) = \frac{1}{2} \int_{\Omega \cap B_\rho} \left(|\nabla u(x,t)|^2 + |\partial_t u(x,t)|^2 \right) dx$$

$$\leq \frac{C}{\log(2+t)^{2s}} \|(u(x,0), \partial_t u(x,0))\|^2_{D(A^s)},$$

where the constant C depends only on the domain Ω and the number ρ.

Proposition 3.1, Theorem 3.1 and Proposition 3.2 together have the following consequence for the analysis of the boundary time-reversal method in a cavity:

THEOREM 3.2. *Assume that u_r is constructed with the algorithm* (3–1), (3–2), (3–3), *with Γ, where the time symmetry is done, being the aperture of the cavity. Then:*

(i) *For a nontrapping obstacle in odd dimensions, there is a constant β for which*

$$\left\| (u_r(x,2T), -\partial_t u_r(x,T)) - (u(x,0), \partial_t u(x,0)) \right\|_{H_0^1(\mathcal{C}) \times L^2(\mathcal{C})}$$

$$\leq C e^{-\beta T} \int_\Omega \left(|\nabla u(x,0)|^2 + |\partial_t u(x,0)|^2 \right) dx$$

(ii) *For either trapping or nontrapping obstacles, the following estimate holds if the initial data are smooth:*

$$\left\| (u_r(x,2T), -\partial_t u_r(x,T)) - (u(x,0), \partial_t u(x,0)) \right\|_{H_0^1(\mathcal{C}) \times L^2(\mathcal{C})}$$

$$\leq C \frac{C}{\log(2+T)^{2s}} \|(u(x,0), \partial_t u(x,0))\|^2_{D(A^s)},$$

and this estimate is optimal. It is saturated when a stable periodic orbit is contained in $\mathcal{C}$ and does not meet Γ [5].

REMARK 3.1. This theorem gives qualitative results on intuitive facts. It shows that the BDTRM always works at least for smooth solutions and large time T. The larger T and the bigger Γ, the better the reconstruction. The reconstruction is obtained with an accuracy $e^{-\beta T}$ when the dimension is odd and the aperture is large enough to capture all the rays of geometric optic. In the worst case when an essential part of the initial signal propagates near a closed stable geodesic which does not meet the aperture the accuracy of the reconstruction is in $O((\log T)^k)$ with k depending on the smoothness of the initial data. In some situations where the set of geodesics which do not meet the aperture in a finite time is "unstable and small", the reconstruction of a smooth signal is obtained with an error of the order of T^{-k}; see [15].

4. The Impedance Time-Reversal Mirror (IMTRM)

In the IMTRM the initial wave u_i evolves for $0 \leq t \leq T$ in a bounded cavity $\mathcal{C}$:

$$\partial_t^2 u_i - \Delta u_i = 0 \quad \text{in } \mathcal{C}, \tag{4-1}$$

with boundary conditions

$$
\begin{aligned}
u_i(x,t) &= 0 \quad \text{on } (\partial \mathcal{C} \setminus \Gamma), \\
\partial_t u_i(x,t) + Z(x) \partial_n u_i(x,t) &= 0 \quad \text{on } \Gamma.
\end{aligned}
\tag{4-2}
$$

Then, for $T \leq t \leq 2T$, one considers the solution u_r of the reversed problem:

$$\partial_t^2 u_r - \Delta u_r = 0 \quad \text{in } \mathcal{C}, \tag{4-3}$$

with boundary conditions

$$
\begin{aligned}
\partial_n u_R(x,t) &= \partial_n u_R(x, 2t - T) \quad \text{on } \Gamma \times (0,T), \\
u_R(x,t) &= 0 \quad \text{on } \partial \mathcal{C} \setminus \Gamma \times (0,T).
\end{aligned}
\tag{4-4}
$$

and initial data

$$u_r(x,T) = u_i(x,T), \quad \partial_t u_r(x,T) = \partial_t u_i(x,T). \tag{4-5}$$

PROPOSITION 4.1. *Assume that u_r is constructed with the algorithm (4–3), (4–4) and (4–5). Then*

$$\left\| (u_r(x,2T), -\partial_t u_r(x,T)) - (u(x,0), \partial_t u(x,0)) \right\|_{H^1(\mathcal{C}) \times L^2(\mathcal{C})}^2 \leq \int_{\mathcal{C}} |\partial_t u(x,T)|^2 dx.$$

PROOF. One introduces the T-symmetric solution u_R of of the Neumann–Dirichlet boundary-value problem:

$$\partial_t^2 u_R - \Delta u_R = 0 \quad \text{in } \mathcal{C} \times (0, 2T)$$

with boundary conditions

$$
\begin{aligned}
u_R(x,t) &= 0 \quad \text{on } (\partial \mathcal{C} \setminus \Gamma) \times (0, 2T), \\
\partial_n u_R(x,t) &= \partial_n u_i(x,t) \quad \text{on } \Gamma \times (0,T), \\
\partial_n u_R(x,t) &= \partial_n u_i(x, 2T - t) \quad \text{on } \Gamma \times (T, 2T)
\end{aligned}
$$

and initial data (at time $t = T$):

$$u_R(x,T) = u_i(x,T), \quad \partial_t u_R(x,T) = 0. \tag{4-6}$$

With this symmetric function the proof is completed along the lines of the Proposition 3.1. $\qquad\square$

Proposition 4.1 implies that the accuracy of the method relies on the decay of the energy norm

$$\frac{1}{2} \int_{\mathcal{C}} \left(|\nabla u_i(x,t)|^2 + |\partial_t u_i(x,t)|^2 \right) dx.$$

This decay has been extensively studied in connection with the problem of the stabilization by boundary feedback ([3], [5], [18]) and it appears that the properties are exactly of the same nature as for the local decay of the exterior problem. It is convenient to introduce a function space $\mathcal{E}$ and an unbounded operator $\mathcal{A}$ adapted as above to the introduction of the variable $v = \partial_t u$ and $U = (u, v)$, according to the formulas

$$\mathcal{E} = \{ U = (u, v) \in H^1(\mathcal{C}) \times L^2(\mathcal{C}) : u = 0 \text{ on } \partial\mathcal{C} \setminus \Gamma \},$$

$$\|U\|_{\mathcal{E}}^2 = \|(u, v)\|_{\mathcal{E}}^2 = \frac{1}{2} \int_{\mathcal{C}} \left\{ |\nabla u(x,t)|^2 + |v(x,t)|^2 \right\} dx,$$

$$D(\mathcal{A}) = \{ U = (u, v) \in \mathcal{E} : \Delta u \in L^2(\mathcal{C}), \ v \in H^1(\mathcal{C}), \ v + \partial_n u = 0 \text{ on } \Gamma \}.$$

Multiplication of the equation

$$\partial_t^2 u - \Delta u = 0$$

by $\partial_t u$ and integration over $\mathcal{C}$ gives, with the boundary condition (4–2), the energy identity

$$\frac{d}{dt} \left(\frac{1}{2} \int_{\mathcal{C}} \left(|\nabla u_i(x,t)|^2 + |\partial_t u_i(x,t)|^2 \right) dx \right) + \int_{\Gamma} Z(x) |\partial_n u|^2 \, d\sigma_x = 0,$$

which leads through classical functional analysis to the following statement.

The operator $\mathcal{A}$ is, in $\mathcal{E}$, the generator of a strongly continuous contraction semigroup, and

$$\lim_{t \to \infty} e^{t\mathcal{A}} U_0 = 0$$

for any initial data $U_0 = (u(x,0), \partial_t u(x,0) = v(x,0))$.

Once again the rate of decay depends on the geometry. Following [3] one says that Γ *geometrically stabilizes* the cavity $\mathcal{C}$, if there exists a time T such that any generalized ray $t \in [0, T] \mapsto x(t) \in \bar{\mathcal{C}}$ intersects Γ at least once in a nondiffractive point. The following results are now well known; see [3], [18], [5].

THEOREM 4.1. (i) *If Γ geometrically stabilizes $\mathcal{C}$, there exists a constant $\beta > 0$ such that*

$$\frac{1}{2} \int_{\mathcal{C}} \left(|\nabla u(x,t)|^2 + |\partial_t u(x,t)|^2 \right) dx = \|U(t)\|_{\mathcal{E}}^2 \leq e^{-\beta t} \|U(0)\|_{\mathcal{E}}^2$$

$$= \frac{1}{2} \int_{\mathcal{C}} \left(|\nabla u(x,0)|^2 + |\partial_t u(x,0)|^2 \right) dx.$$

(ii) *If Γ does not geometrically stabilize $\mathcal{C}$, the decay may be arbitrary slow in the sense of Theorem 3.1(ii).*

(iii) *However, for any sufficiently smooth initial data the following estimate is always valid (and optimal in the absence of hypotheses on the geometry)*

$$\frac{1}{2}\int_{\Omega\cap B_\rho}\left(|\nabla u(x,t)|^2+|\partial_t u(x,t)|^2\right)dx\le\frac{C}{\log(2+t)^{2s}}\left\|(u(x,0),\partial_t u(x,0))\right\|^2_{D(A^s)}.$$

With Proposition 4.1 and Theorem 4.1 one concludes as in the previous section how the accuracy of the method depends on the size of Γ and on the time of observation.

5. Internal Time-Reversal Method in an Ergodic Cavity (INTRM)

Intuition suggests that the domain where the time-reversal process occurs can be much smaller if the compressed hamiltonian flow is ergodic and if the time of "action" is large enough. This has been corroborated by numerical simulation and ultrasonic experiments made on a silicium wafer by C. Draeger and M. Fink ([9], [10]). What is observed with a time-reversal experiment conducted on one single point is a very good refocusing of a localized initial signal. The mathematical explanation, as described below, relies on (1) an asymptotic formula which in [9] is called the "cavity formula", and (2) the notion of quantum ergodicity, which is closely related to classical ergodicity ([21], [8], [23], [24]).

As in the previous section, $\mathcal{C}$ denotes a bounded open set and Δ is the Laplace operator with Dirichlet boundary condition on $\partial\mathcal{C}$. It is convenient to introduce the operators

$$\exp\left(it(-\Delta)^{1/2}\right),\quad\sin\left(t(-\Delta)^{1/2}\right),\quad\cos\left(t(-\Delta)^{1/2}\right).$$

The solution of the initial value problem

$$\begin{aligned}\partial_t^2 u-\Delta u=0\quad&\text{in }\mathcal{C},\\u(x,t)\equiv 0\quad&\text{on }\partial\mathcal{C}\end{aligned}\tag{5-1}$$

with the initial condition

$$u(x,0)=0\qquad\partial_t u(x,0)=\psi(x)\tag{5-2}$$

is given by

$$u(x,t)=(-\Delta)^{-1/2}\sin(t(-\Delta)^{1/2})\psi.\tag{5-3}$$

The solution of the problem

$$\begin{aligned}\partial_t^2 u-\Delta u=f(x,t)\quad&\text{in }\mathcal{C},\\u(x,t)\equiv 0\quad&\text{on }\partial\mathcal{C}\end{aligned}$$

with initial conditions

$$u(x,0)=0,\qquad\partial_t u(x,0)=0,$$

is given by

$$u(x,t) = \int_0^t (-\Delta)^{-1/2} \sin\big((t-s)(-\Delta)^{1/2}\big) f(s)\, ds. \qquad (5\text{--}4)$$

Observe that (5–3) and (5–4) are well defined (this can be done by duality) not only for L^2 functions but also for distributions in $\mathcal{D}'(\mathcal{C})$ and that, with the introduction of the eigenvalues and eigenvectors of $-\Delta$, namely

$$-\Delta\phi_k = \omega_k^2\phi_k, \quad \phi_k(x) = 0 \quad \text{on } \partial\mathcal{C}, \quad 1 \le k \le \infty$$

the kernel of the operator

$$(-\Delta)^{-1/2}\sin(t(-\Delta)^{1/2})$$

is the distribution

$$k(x,y,t) = \sum_{1 \le k \le \infty} \frac{\sin t\omega_k}{\omega_k} \phi_k(x) \otimes \phi_k(y),$$

which turns out to be the (fundamental) solution of the problem

$$\partial_t^2 k(x,y,t) - \Delta_x k(x,y,t) = \delta_t \otimes \delta_y.$$

For the INTRM one observes the solution u_i of (5–1) and (5–2) on a subset $\sigma \subset \mathcal{C}$ (which may be arbitrary small), introduces an L^∞ function $\Xi(x)$ with support contained in σ and eventually introduces for $T < t < 2T$ the solution of the problem

$$\begin{aligned} \partial_t^2 u_r - \Delta u_i &= K\Xi(x)\partial_t u_i(x, 2T{-}t) \quad \text{in } \mathcal{C}, \\ u_i(x,t) &= 0 \quad \text{on } \partial\mathcal{C}, \end{aligned} \qquad (5\text{--}5)$$

with initial conditions

$$u_r(x,T) = u_r(x,T), \qquad \partial_t u_r(x,T) = \partial_t u_r k(x,T).$$

In (5–5) K represents an amplification factor which may be large. Therefore

$$\partial_t u_r(x,2T) = \cos\big(2T(-\Delta)^{1/2}\big)\psi$$
$$+ K \int_T^{2T} \cos\big((2T{-}t)(-\Delta)^{1/2}\big)\Xi \cos\big((2T{-}t)(-\Delta)^{1/2}\big)\psi\, dt.$$

To use the ergodicity property T will be taken large enough. This also reinforces the influence of the reemitted signal which is also amplified by the factor Ampl. Accordingly one writes

$$\partial_t u_r(x,2T) = T\Big(\frac{1}{T}\cos\big(2T(-\Delta)^{1/2}\big)\psi$$
$$+ \frac{\text{Ampl}}{T} \int_T^{2T} \cos\big((2T{-}t)(-\Delta)^{1/2}\big)\Xi \cos\big((2T{-}t)(-\Delta)^{1/2}\big)\psi dt\Big).$$

In any convenient sense, and in particular for the energy norm (with initial data of finite energy), one has

$$\lim_{T \to \infty} \frac{1}{T} \cos(2T(-\Delta)^{1/2})\psi = 0.$$

Therefore

$$u_r(x, 2T) \simeq \frac{\text{Ampl } T}{T} \int_T^{2T} \cos((2T-t)(-\Delta)^{1/2}))\Xi \cos((2T-t)(-\Delta)^{1/2})\psi \, dt$$

whenever this limit exists.

This result follows from the cavity equation and the quantum chaos principle. One has

$$\frac{1}{T} \int_T^{2T} \cos((2T-t)(-\Delta)^{1/2})\Xi \cos((2T-t)(-\Delta)^{1/2})\psi \, dt$$

$$= \frac{1}{4T} \int_0^T (e^{it(-\Delta)^{1/2}} + e^{-it(-\Delta)^{1/2}})\Xi(e^{it(-\Delta)^{1/2}} + e^{it(-\Delta)^{1/2}}) dt\psi.$$

which is written as the sum of two terms:

$$M(T)\psi = \frac{1}{4T} \int_{-T}^T e^{it(-\Delta)^{1/2}}\Xi e^{it(-\Delta)^{1/2}} \psi \, dt,$$

$$N(T)\psi = \frac{1}{4T} \int_{-T}^T e^{it(-\Delta)^{1/2}}\Xi e^{-it(-\Delta)^{1/2}} \psi \, dt.$$

For $M(T)$ we have:

PROPOSITION 5.1 (CAVITY FORMULA). *The family of operators* $T \mapsto M(T)$ *is uniformly equibounded in* $L^2(\Omega)$ *and for* $T \to \infty$ *it converges weakly to* 0.

PROOF. Observe that one has

$$\|M(T)\| \leq \tfrac{1}{2}|\Xi\|_{L^\infty(\mathcal{C})}. \tag{5-6}$$

Then for any pair of eigenvectors $(\phi_k(x), \phi_l(x))$,

$$\lim_{T \to 0} (M(T)\phi_k, \phi_l) = \lim_{T \to 0} \frac{1}{4T} \int_{-T}^T (\Xi e^{it(-\Delta)^{1/2}} \phi_k, e^{-it(-\Delta)^{1/2}} \phi_l) \, dt$$

$$= \lim_{T \to 0} \frac{\sin(T(\omega_k + \omega_l))}{2T(\omega_k + \omega_l)}(\Xi\phi_k, \phi_l) = 0,$$

and the result follows by density.   □

REMARK 5.1. By Rellich's theorem, it follows from the above proposition that

$$\lim \|M(T)\psi\|_{L^2(\mathcal{C})} = 0.$$

for any $\psi \in H^s(\mathcal{C})$ with $s > 0$.

From the cavity formula one deduces the relation

$$u_R(x, 2T) \simeq TN(T) \simeq \tfrac{1}{2}\mathrm{Ampl}\left(\frac{K}{2T} \int_{-T}^{T} e^{it(-\Delta)^{1/2}} \Xi(e^{-it(-\Delta)^{1/2}})dt \right)\psi.$$

Start from the Hamiltonian system

$$\dot{x}(t) = \xi, \qquad \dot{\xi}(t) = 0.$$

For any function f write, whenever that makes sense (i.e., when the trajectory $x(s)$ for $s \in [0, t]$ remains in $\mathcal{C}$)

$$V(t)f = f\big(x(t), \xi(t)\big).$$

With the introduction of the broken hamiltonian flow, extend this operator on the functions defined on $S^*(\bar{\mathcal{C}}) = \{(x, \xi) : x \in \mathcal{C}, |\xi| = 1\}$ and denote by

$$|S^*(\bar{\mathcal{C}})| = \iint_{\mathcal{C}\times\{|\xi|=1\}} dx\, d\xi$$

the volume of this cosphere bundle. Further, denote by σ_P the principal symbol of any zero order pseudodifferential operator P.

DEFINITION 5.1. (i) The flow is *classically ergodic* if

$$\lim_{t\to\infty} V(t)f = \bar{f} = \frac{1}{|S^*(\bar{\mathcal{C}})|} \int_{\mathcal{C}\times\{|\xi|=1\}} f(x, \xi)\, dx\, d\xi.$$

in the weak L^* topology, for any continuous function $f \in C^0(S^*(\bar{\mathcal{C}}))$.

(ii) Let $\Pi_l = \sum_{1\le k\le l} \phi_k \otimes \phi_k^*$ be the projection onto the space spanned by the l first eigenvectors of $-\Delta$. An operator $K \in \mathcal{L}(L^2(\mathcal{C}))$ is *spectrally regularizing* if it satisfies the bound

$$\|\Pi_l K \Pi_l\|_{HS}^2 = o(l).$$

(iii) The flow is *quantum ergodic* if, for any zero-order pseudodifferential operator P and in the weak operator limit, one has

$$\lim_{T\to\infty} \frac{1}{2T} \int_{-T}^{T} e^{it(-\Delta)^{1/2}} P e^{-it(-\Delta)^{1/2}}\, dt = \langle P\rangle I + K$$

with K spectrally regularizing and

$$\langle P \rangle = \frac{1}{|S^*(\bar{\mathcal{C}})|} \int_{\mathcal{C}\times\{|\xi|=1\}} \sigma_P(x, \xi)\, dx\, d\xi = \lim_{l\to\infty} \sum_{1\le k\le l} \frac{1}{l}(P\phi_k, \phi_k).$$

It has been proved that classical ergodicity implies quantum ergodicity. See [8], [21], [23], [24].

Therefore it follows from Proposition 5.1 that:

THEOREM 5.1. *The INTRM solution constructed (see (5–5) and (1–3)) satisfies, as $T \to \infty$:*

$$u_R(x, 2T) \simeq \tfrac{1}{2} T \operatorname{Ampl}\left(\langle \Xi \rangle \psi + K \psi\right)$$

with K spectrally regularizing and

$$\langle \Xi \rangle = \lim_{l \to \infty} \sum_{1 \leq k \leq l} \frac{1}{l}(\Xi \phi_k, \phi_k) = \sum_{1 \leq k \leq l} \frac{1}{l} \int_{\mathfrak{C}} \left(\Xi(x)\phi_k(x), \phi_k(x)\right) dx.$$

REMARK 5.2. The notion of spectrally regularizing is not very explicit in its present form. However, for any pseudodifferential operator P of zero order,

$$\lim_{l \to \infty} \frac{1}{l}\|\Pi_l P \Pi_l\|_{HS}^2 = \langle P \rangle = \frac{1}{|S^*(\bar{\mathfrak{C}})|} \int_{\mathfrak{C} \times \{|\xi|=1\}} |\sigma_P(x, \xi)|^2 \, dx \, d\xi;$$

see [25, Prop. 1.1(ii)]. Therefore any spectrally regularizing pseudodifferential operator P has its principal symbol equal to zero and has a regularizing effect. Similarly one shows [23, p. 921] that in general (at least when the spectra of Δ has bounded multiplicity) K is compact. This is why the preceding theorem carries pertinent information when the initial data ψ is a distribution with a single singularity located at one point, say A, then the reversed solution is a sum of a more regular term $\tfrac{1}{2} T K \psi$ and of a leading term proportional to

$$\tfrac{1}{2} T \operatorname{Ampl}\langle \Xi \rangle \psi,$$

which also has a singularity at the point A. In this sense for large time the refocusing is perfect and this is in agreement with the experiment of C. Draeger (see Figure 5) and the numerical simulations of [9] and [10].

Acknowledgment. The author thanks Gunther Ulhmann and the series editor, Silvio Levy, for their careful reading of and improvements to this chapter.

References

[1] G. Bal and L. Ryzhik: Time reversal for classical waves in random media. *C. R. Acad. Sci. Paris Sér. I Math.* **333** (2001), no. 11, 1041–1046.

[2] C. Bardos and M. Fink: Mathematical foundations of the time reversal mirror, *Asymptot. Anal.* **29** (2002), no. 2, 157–182.

[3] C. Bardos, G. Lebeau, and J. Rauch: Sharp sufficient conditions for the observation, Control and stabilization of waves from the boundary , *SIAM Journal on Control Theory and Applications* **30** (1992), 1024–1065.

[4] P. Blomgren, G. Papanicolaou and H. Zhao: Super-resolution in time-reversal acoustics. Submitted to *J. Acoust. Soc. Amer.*

[5] N. Burq: Décroissance de l'énergie locale de l'équation des ondes pour le problème extérieur et absence de résonance au voisinage du réel. *Acta Math.* **180**, (1988) 1–29.

[6] J. Chazarain and A. Piriou: *Introduction à la théorie des équations aux dérivées partielles linéaires*, Gauthier-Villars, Paris, 1981.

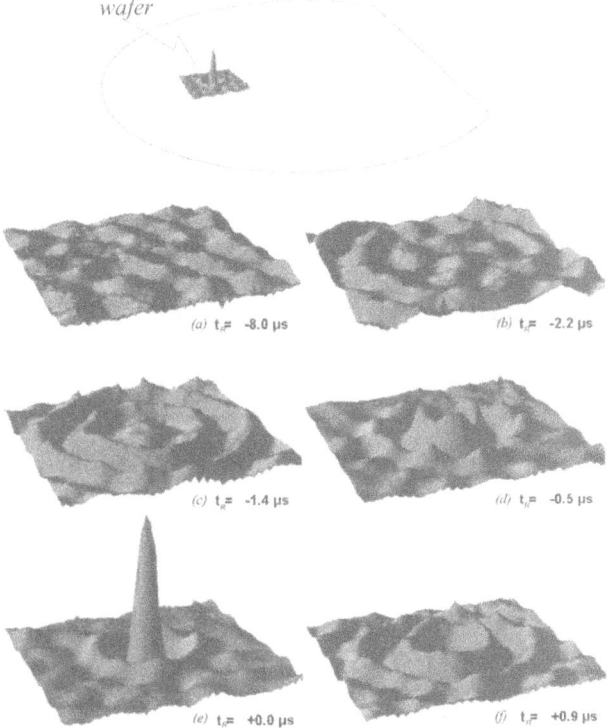

Figure 4. Linear representation of the acoustic field around point A during the refocusing after a time reversal of $T = 2\,\text{ms}$. Measurements have been made ([9], [10]) on a square of side 15 mm, with a spatial step of 0.25 mm.

[7] D. Cassereau, M. Fink: Time reversal of ultrasonic fields: theory of the closed time-reversal cavity, *IEEE Trans. Ultrasonics, Ferroelectric and Frequency Control* **39**, 5, (1992) 579–592.

[8] Y. Colin de Verdière: Ergodicité et fonctions propres du Laplacien, *Comm. Math. Phys* **102**, (1985) 497–502.

[9] C. Draeger, M. Fink: One-channel time-reversal in chaotic cavities: theoretical limits, *Journal of Acoustical Society of America* **105**:2, (1999), 611–617.

[10] C. Draeger and M. Fink: One-Channel time reversal of elastic waves in a chaotic 2D-silicon cavity, *Phys. Rev. Letters* **79**:3 (1997), 407–410.

[11] M. Fink: Time reversal of ultrasonic fields: basic principles, *IEEE Trans. Ultrasonics, Ferroelectric and Frequency Control* **39**:5 (1992), 555–566.

[12] M. Fink: Time reversed acoustics, *Physics Today*, **20** (1997), 34–40.

[13] M. Fink, D. Cassereau, A. Derode, C. Prada, P. Roux, M. Tanter, J. L. Thomas and F. Wu: Time-reversed acoustics, *Rep. Prog. Phys.* **63** (2000), 1933–1995.

[14] L. Hörmander: *The analysis of linear partial differential operators* (4 volumes), Grundlehren der math. Wissenschaften **256**, **257**, **274**, **275**, Springer, Berlin, 1983–1985.

[15] Ikawa: Mitsuru Asymptotics of scattering poles for two strictly convex obstacles, pp. 171–187 in *Long time behaviour of classical and quantum systems* (Bologna, 1999), Ser. Concr. Appl. Math. **1**, World Scientific, River Edge, NJ, 2001.

[16] F. John, On linear partial differential equation with analytic coefficients, *Commun. Pure Appl. Math.* **2** (1949), 209–253.

[17] P. Lax and R. Phillips: *Scattering theory*, revised edition, Academic Press, 1989.

[18] G. Lebeau: Equations des ondes amorties, pp. 73–109 in *Algebraic and Geometric Methods in Mathematical Physics*, edited by A. Boutet de Monvel et V. Marchenko, Kluwer (1996).

[19] G. Lebeau, Contrôle analytique 1: estimations à priori, *Duke Math. J.* **68** (1992), 1–30.

[20] J.-L. Lions: *Contrôlabilité exacte, perturbation et stabilisation des systèmes distribués*, Masson, Paris, 1988.

[21] A.I Schnirelman: Ergodic properties of eigenfunctions, *Usp. Math. Nauk* **29** (1994), 181–182.

[22] M.Tanter, J-L. Thomas, M. Fink: Influence of boundary conditions on time reversal focusing through heterogeneous media, *Applied Physics Letters* **72**:20 (1998), 2511–2513.

[23] S. Zelditch: Uniform distribution of eigenfunctions on compact hyperbolic surfaces, *Duke Math. J.* **55** (1987), 919–941.

[24] S. Zelditch: Quantum transition amplitudes for ergodic and for completely integrable systems, *J. Funct. Anal.* **94** (1990), 415–436.

[25] S. Zelditch: Quantum mixing, *J. Funct. Anal.* **140** (1996), 68–86.

CLAUDE BARDOS

UNIVERSITÉ DENIS DIDEROT AND

LABORATOIRE JACQUES LOUIS LIONS
UNIVERSITÉ PIERRE ET MARIE CURIE
BOÎTE COURRIER 187
75252 PARIS CEDEX 05
FRANCE
bardos@math.jussieu.fr

For EU product safety concerns, contact us at Calle de José Abascal, 56–1°,
28003 Madrid, Spain or eugpsr@cambridge.org.

www.ingramcontent.com/pod-product-compliance
Ingram Content Group UK Ltd.
Pitfield, Milton Keynes, MK11 3LW, UK
UKHW040951090126
466816UK00019B/358